Handbook of
Teratology

Edited by

JAMES G. WILSON
*The Children's Hospital Research Foundation
and Department of Pediatrics, University of Cincinnati
Cincinnati, Ohio*

and

F. CLARKE FRASER
*Departments of Biology and Paediatrics
McGill University and The Montreal Children's Hospital
Montreal, Canada*

2   Mechanisms
   and Pathogenesis

PLENUM PRESS · NEW YORK AND LONDON

Library of Congress Cataloging in Publication Data

Main entry under title:

Handbook of teratology.

Includes bibliographies and index.
CONTENTS: v. 1. General principles and etiology.—v. 2. Mechanisms and patho-
genesis.
1. Abnormalities. 2. Teratogenic agents. I. Wilson, James Graves, 1915-   II.
Fraser, F. Clarke, 1920-   [DNLM: 1. Abnormalities. QS675 H236]
QM691.H26   616′.043   76-41787
ISBN 0-306-36242-2 (v. 2)

© 1977 Plenum Press, New York
A Division of Plenum Publishing Corporation
227 West 17th Street, New York, N.Y. 10011

Printed in the United States of America

Contributors

G. J. BINGLE, Laboratory of Developmental Biology and Anomalies, National Institute of Dental Research, National Institutes of Health, Bethesda, Maryland

ROBERT PAUL BOLANDE, Director of the Department of Pathology, The Montreal Children's Hospital, and Professor of Pathology and Paediatrics, McGill University, Montreal, Quebec, Canada

ALFRED J. COULOMBRE, Laboratory of Vision Research, National Eye Institute, National Institutes of Health, Bethesda, Maryland

JANE L. COULOMBRE, Late of the Laboratory of Vision Research, National Eye Institute, National Institutes of Health, Bethesda, Maryland

F. CLARKE FRASER, Departments of Biology and Paediatrics, McGill University, and Director, Department of Medical Genetics, The Montreal Children's Hospital, Montreal, Quebec, Canada

DAVID FROMSON, Department of Biology, McGill University, Montreal, Quebec, Canada

SALOME GLUECKSOHN-WAELSCH, Department of Genetics, Albert Einstein College of Medicine, Bronx, New York

ALLEN S. GOLDMAN, The Teratology Center, Children's Hospital of Philadelphia, Philadelphia, Pennsylvania

CASIMER T. GRABOWSKI, Department of Biology, University of Miami, Coral Gables, Florida

KURT HIRSCHHORN, Department of Pediatrics, Division of Medical Genetics, Mount Sinai School of Medicine of the City University of New York, New York

LILLIAN Y. F. HSU, Department of Pediatrics, Division of Medical Genetics, Mount Sinai School of Medicine of the City University of New York, New York

OSCAR C. JAFFEE, Department of Biology, University of Dayton, Dayton, Ohio

M. C. JOHNSTON, Laboratory of Developmental Biology and Anomalies, National Institute of Dental Research, National Institutes of Health, Bethesda, Maryland. Present address: Dental Research Center, University of North Carolina, Chapel Hill, North Carolina

D. M. KOCHHAR, Department of Anatomy, School of Medicine, University of Virginia, Charlottesville, Virginia. Present address: Department of Anatomy, Jefferson Medical College, Philadelphia, Pennsylvania

RALF KROWKE, Institut für Toxikologie und Embryonal-Pharmakologie der Freien Universität Berlin, Berlin, Germany

D. C. KUSHNER, Laboratory of Developmental Biology and Anomalies, National Institute of Dental Research, National Institutes of Health, Bethesda, Maryland

K. SUNE LARSSON, Laboratory of Teratology, Karolinska Institutet, Stockholm, Sweden

IAN W. MONIE, Department of Anatomy, University of California, San Francisco, San Francisco, California

G. M. MORRISS, Laboratory of Developmental Biology and Anomalies, National Institute of Dental Research, National Institutes of Health, Bethesda, Maryland

DIETHER NEUBERT, Institut für Toxikologie und Embryonal-Pharmakologie der Freien Universität Berlin, Berlin, Germany

EDMOND J. RITTER, Children's Hospital Research Foundation and Department of Pediatrics, University of Cincinnati, Cincinnati, Ohio

LAURI SAXÉN, Third Department of Pathology, University of Helsinki, Helsinki, Finland

WILLIAM J. SCOTT, JR., Children's Hospital Research Foundation and Department of Pediatrics, University of Cincinnati, Cincinnati, Ohio

DAPHNE G. TRASLER, Department of Biology, McGill University, Montreal, Quebec, Canada

KENNETH MANAO YAMADA, Laboratory of Molecular Biology, National Cancer Institute, National Institutes of Health, Bethesda, Maryland

Contents

Initiating Mechanisms and Early Pathogenesis

Gene Expression and Its Regulation

1

DAVID FROMSON

I. INTRODUCTION

The major purpose of this review is to outline and indicate the major steps in the regulation of gene activity in eukaryotes at the molecular level. This review will not include an analysis of the inhibition of genetic expression by various drugs and teratogens. Rather, it will concentrate on describing the sites of regulation of genetic activity; genetic activity is defined as the production and utilization of messenger RNA (mRNA).

There are several regulatory sites for the production and utilization of mRNA in eukaryotic cells. These are (1) amplification of specific genes thus increasing the amount of DNA coding for a specific RNA sequence, (2) differential gene expression—the specific selection of gene(s) for transcription, (3) specificity of the RNA polymerase involved in the transcriptive events, (4) selection and processing of large, heterogeneous nuclear RNA (HnRNA) transcripts, some of which will be exported from the nucleus to the cytoplasm, (5) selection of specific mRNA molecules from a cytoplasmic pool for translation, and (6) regulation of the interaction between ribosomes and mRNA to control the initiation of protein synthesis. Each of these major regulatory sites will be explored in greater detail utilizing examples to illustrate each mode of regulation.

DAVID FROMSON • Department of Biology, McGill University, Montreal, Quebec, Canada, H3A 1B1.

II. SITES OF GENETIC REGULATION

A. Gene Amplification

Gene amplification is differential replication of specific genes without concomitant replication of the entire genome. A very direct way to affect the amount of a specific protein produced in a cell is to increase differentially the number of genes coding for that protein. This differential increase in these genes is called gene amplification, and is illustrated by the following examples.

The first example involved the dramatic increase in ribosomal RNA (rRNA) that occurs in the developing amphibian oocyte (Davidson, 1968), in apparent anticipation of the need for ribosomes on which to synthesize polypeptides in early development. The DNA coding for rRNA is localized in the nucleolus and is referred to as rDNA. Cytological observation of amphibian oocytes during oogenesis has revealed the striking increase in the number of nucleoli. Amphibian somatic cell nuclei usually contain one nucleolus per haploid genome; however, during oogenesis the number of nucleoli per nucleus increases to 600 in *Triturus* and 1000 in *Xenopus* oocytes (Brown and Dawid, 1968). The new nucleoli are functionally and structurally the same as somatic cell nucleoli (Brown and Dawid, 1968).

Molecular hybridization of rRNA to somatic cell has shown that 0.057% of the somatic cell DNA is complementary to rRNA, suggesting that this proportion of the somatic cell DNA codes for rRNA. Brown and Dawid (1968) have measured the amount of nucleolar rDNA in somatic cells and germinal vesicles and have shown that there are 0.014 picograms and 5.3 picograms, respectively. These figures represent a 340-fold increase in the amount of rDNA in oocytes, and this figure is probably too low due to the contamination by mitochondrial DNA. There was no appreciable increase in the amount of total DNA in oocytes, indicating that there was selective replication of rDNA. This point was further corroborated by CsCl gradient analysis of the DNA isolated from amphibian germinal vesicles and somatic cells. The data showed clearly that there was selective synthesis of a high-density (1.729 g/cc) rDNA with no increased synthesis of the nuclear DNA. Molecular hybridization of labeled rRNA to the high-density DNA confirmed its identity as the DNA coding for rRNA. Additional hybridization experiments confirmed amplification of the rRNA genes. However, there was no increased synthesis of 4 S RNA or 5 S RNA genes during amphibian oogenesis; these genes are already present in large enough numbers to account for the production of these species of RNA.

Amplification of the rRNA genes was shown to occur during the pachytene stage of oogenesis by Gall (1968), using ovaries from newly metamorphosed *Xenopus laevis* cultured in medium containing radioactive thymidine and analyzing the newly synthesized DNA by CsCl gradient centrifugation. The data showed that there is differential synthesis of rDNA.

Cytological evidence suggests that selective replication of the nucleoli and extensive transcription of the rDNA genes occur at this time as well. These phenomena are restricted to the pachytene stage, and gene amplification does not continue throughout development. This illustrates a very efficient means to control the production of the primary gene product rRNA. This situation is unique, for rRNA represents a terminal gene product and is not itself translated to yield a protein. It is very important to know whether the mechanism of selective gene replication can account for the increased production of specific protein(s) in a population of highly specialized cells; i.e., does it play a role in differentiation?

Many differentiated cells produce specific proteins in large quantities; hemoglobin in red blood cells, actin and myosin in muscle cells, and fibroin in silk glands of the silkworm, for example. A possible regulatory mechanism to account for specific protein production is amplification of the genes coding for a specific protein or set of proteins in differentiated cells. Several studies have been carried out to test whether this type of gene amplification occurs. Bishop *et al.* (1972) isolated the hemoglobin mRNA from immature duck red blood cells that had been incubated with [^{3}H]uridine to radiolabel the RNA. The radioactive hemoglobin mRNA was hybridized to duck DNA. The results of these experiments and subsequent calculations indicated that the absolute maximum reiteration frequency of the hemoglobin genes is 10, and the authors present cogent arguments suggesting that this value is probably too high. After considering all the technical limitations, Bishop *et al.* (1972) estimated that there are 1–2 genes coding for hemoglobin mRNA per haploid genome. These experiments show that specific gene amplification does not account for the extensive hemoglobin synthesis in duck red blood cells.

Suzuki and Brown (1972) and Suzuki *et al.* (1972) have thoroughly studied the fibroin genes and mRNA in the posterior silk gland of the silkworm *Bombyx mori*. Fibroin, the silk protein, constitutes most of the protein synthesized in these cells. It is quite well characterized; the amino acid sequence is known, and thus the nucleotide sequence and size of the fibroin mRNA was correctly predicted. This enabled isolation of relatively pure samples of fibroin mRNA using size criteria. Fibroin mRNA has a high molecular weight, a G + C content of 59%, and a very long half-life. It comprises as much as 1.5% of the total RNA in posterior silk gland cells. Once pure fibroin mRNA was isolated, the number of genes coding for fibroin in the silk gland cells could be measured and the role of gene amplification in differentiation could be assessed.

The DNA coding for fibroin was isolated from bulk DNA and rDNA by $Ag^+/CsSO_4$ centrifugation which enabled RNA–DNA hybridization of fibroin mRNA to DNA isolated from three different tissues: the posterior silk gland, a middle portion of the silk gland, and silkworm carcass. Radioactive fibroin mRNA was annealed to these three DNA preparations, and the data showed that the labeled fibroin mRNA hybridized to the three DNA preparations to the same extent. These data support the conclusion that there is no amplification of the fibroin genes in the posterior silk gland of the silkworm *Bombyx mori*. Furthermore, Suzuki *et al.* (1972) estimated that there are no more than 3 fibroin genes per haploid complement of DNA. These data, like those previously discussed, indicate that gene amplification does not account for the synthesis of large amounts of specific proteins in differentiating cells.

Another possible regulatory mechanism to account for the production of large amounts of cell-specific protein could be reiteration of the DNA sequences (genes) in the genome coding for the specific protein, as is the case with the histone genes in sea urchin embryos which are repeated 500–700 times per haploid genome (Kedes and Birnstiel, 1971), but this also does not seem to be a mechanism utilized by differentiated cells.

B. Differential Gene Expression

Probably the most widely accepted explanation for the profound differences in various cell and tissue types is that the genes or sets of genes, specific for the proteins characterizing those cells as differentiated, are active (transcribed) only in those cells and tissues. Precise mechanisms regulating gene expression in eukaryotic cells are presently unknown. However, there are several models that have been proposed to explain gene regulation. Two of these are presented.

Britten and Davidson (1969,1971) proposed a scheme to account for gene regulation in higher cells that accounts for many of the observed properties of these systems. The DNA content of higher cells is 1000 times that of bacteria, and analysis of the DNA has shown that a significant portion of the DNA nucleotide sequences are repeated many times, although a portion of the nucleotide sequences are unique. This model also attempts to explain the stimulation of genetic activity in target cells by effectors such as hormones, the coordinated appearance of a specific set of enzymes in developing tissues such as the pancreas, and the synthesis of heterogeneous nuclear RNA (HnRNA) in higher cells. Approximately 80–90% of the HnRNA is completely confined to the nucleus and never functions as mRNA. The scheme proposed by Britten and Davidson (1969) accounts for these features of higher cells. It is a complicated scheme. The components of the model are: (1) producer gene—a nucleotide sequence in the DNA that is transcribed to yield mRNA which after translation yields protein (a structural gene); (2) receptor gene—a DNA sequence that is linked to the producer gene and specifically interacts with activator RNA to activate the producer gene; (3) activator RNA—the RNA that forms complexes with the receptor gene, the complex then activating the producer gene; (4) integrator gene—the DNA sequence that codes for activator RNA, which could interact with several receptor genes at different sites and result in the coordinated activation of a large number of genes; and (5) sensor gene—a DNA sequence which binds and/or interacts with incoming signals such as hormones or specific regulatory proteins.

Simplification of the model allows the following series of events to be postulated for gene regulation. An effector molecule enters the cell nucleus and interacts with the sensor gene which in turn activates the integrator gene to produce activator RNA. This activator RNA leaves its site of synthesis and interacts with receptor gene(s) which in turn interact with producer genes, and the subsequent production of mRNA results. An important feature of the Britten–Davidson model is the redundancy of receptor genes and/or integrator genes which may account for some of the reiterated DNA sequences in eukaryotes and which could account for some of the observed properties of gene activation in higher cells (see Britten and Davidson, 1969). Note that this model postulates the existence of genes other than structural genes which have not yet been positively identified. Although it has many attractive features, it is still a theory lacking final confirmation.

Another model to explain transcriptional control takes into account the structure of the chromosomal material (chromatin) and attempts to explain genetic regulation at the level of the chromosome. Chromatin prepared by extensively washing isolated nuclei of water-soluble and saline-soluble materials contains DNA, histones, nonhistone proteins, and a very small amount of RNA. In the early 1960s, J. Bonner and his coworkers (see Bonner, 1966) used this chromatin as a template for added DNA-dependent RNA polymerase and observed less RNA synthesis with chromatin templates than with purified DNA templates. This observation led Bonner to suggest that DNA gene sites were masked by histones and that there was restricted access to the DNA by the RNA polymerase as a result of this histone–DNA interaction. More recent results have shown that these earlier hypotheses were probably incorrect and that it is unlikely that histones alone regulate gene activity. In more recent experiments, Paul and Gilmour (1968) used RNA–DNA hybridization to examine the RNA synthesized *in vitro* using chromatin templates from different mammalian tissues and exogenous RNA polymerase. Not only was less RNA transcribed from chromatin templates than from DNA templates, but there seemed to be organ specificity, for the *in vitro* synthesized RNA was comparable to the *in vivo* synthesized RNA as judged by competition hybridization experiments (Paul and Gilmour, 1968). This observation suggested that the masking of chromatin changes during cell and tissue specialization.

Paul and Gilmour (1968) and Gilmour and Paul (1969) have been able to fractionate chromatin into DNA, histones, and nonhistone proteins, and these components can be recombined and reconstituted under controlled conditions to yield chromatin. Partial reconstitution was also achieved. These reconstituted and partially reconstituted chromatin preparations were used for *in vitro* RNA transcription studies in which the amount of RNA synthesis was measured, and Paul (1971) reached the following conclusions: (1) The template activity of DNA that had been recombined with histone and nonhistone proteins was essentially the same as native chromatin. (2) DNA combined with histones was a very poor template for *in vitro* RNA synthesis. (3) The template properties of chromatin resulting from the addition of histones to previously dehistoned chromatin were similar to those of native chromatin. (4) When DNA was combined with only nonhistone proteins, the template activity of the resulting material was intermediate between DNA and native chromatin.

These results suggested that the histones interact with the DNA in a nonspecific way to reduce transcription, but the nonhistone proteins interact with the DNA in a specific fashion both to repress the template activity of the DNA and to introduce some specificity into the histone–DNA interaction. This suggests that the specificity of transcription in reconstituted chromatin, and perhaps in native chromatin as well, is determined by the nonhistone protein. The precise mechanisms involved in the interactions between DNA, histone, and nonhistone proteins are not well understood. However, it is clear from Paul's work, even with certain limitations, that the nonhistone proteins are very important in regulation of gene activity. This idea is supported by the work of Stein et al. (1974). Nonhistone proteins have been isolated from chromatin and characterized, and many different kinds of protein have been identified. The nonhistone proteins in different tissues are not all the same. Also, cells undergoing differentiation display different profiles of nonhistone proteins during different phases of their differentiation, as do cells in various phases of the cell cycle (Stein et al., 1974). Therefore, it is possible that these nonhistone proteins specifically regulate gene activity. There is extensive in vivo phosphorylation and dephosphorylation of the nonhistone chromosomal proteins, and this has led to the idea that these processes are involved in the regulation of gene activity by the nonhistone proteins. One possibility is that phosphorylation of the nonhistone proteins increases the repulsive forces between the DNA and protein components of chromatin and thereby forms a site for RNA polymerase to initiate transcription (Stein et al., 1974).

C. Multiple Forms of RNA Polymerase

RNA synthesis could be regulated if there was a high degree of specificity between the species of RNA synthesized and a particular RNA polymerase. There is a large body of data showing that there are multiple RNA polymerase molecules in eukaryotic cells and each of these RNA polymerases transcribes an individual class of RNA. Roeder and Rutter (1969) isolated and partially characterized three chromatographically separate RNA polymerase molecules from sea urchin embryo nuclei and two chromatographically distinct RNA polymerases from rat liver nuclei. The chromatographic peaks were designated I, II, and III. These three RNA polymerases show maximal activity at Mn^{2+} concentrations of 1–2 mM; the Mg^{2+} concentration optima cover a broader range. The three forms of the enzyme have different ammonium sulfate concentration optima. In addition, they have absolute DNA template requirements and can utilize either native or denatured DNA. The three forms of RNA polymerase are not interconvertible. Roeder and Rutter (1969, 1970) suggested that RNA polymerase I is localized in the nucleolus and catalyzes the synthesis of ribosomal RNA (rRNA) and that RNA polymerase II is localized in the nucleoplasm and synthesizes DNA-like-RNA (HnRNA).

Tocchini-Valentini and Crippa (1970) described two forms of RNA polymerase activity in *Xenopus laevis* oocytes. These investigators found one species of RNA polymerase that was confined to the nucleolus, and probably responsible for rRNA synthesis, and another enzyme RNA polymerase II, which is localized in the nucleoplasm. Only RNA polymerase II is inhibited by the drug α-amanitin, and this selective inhibition offers a means for distinguishing these polymerases (Roeder *et al.*, 1970). Tocchini-Valentini and Crippa (1970) also attempted to show template specificity of these enzymes by showing that RNA polymerase I had a greater affinity for ribosomal DNA than RNA polymerase II did, but affinity measurements do not necessarily mean template specificity.

Roeder *et al.* (1970), pursuing the question of whether polymerase level could play a role in gene regulation, measured the levels of RNA polymerases in developing *Xenopus laevis* embryos. RNA polymerase activity increases on a per embryo basis from 4.8 activity units in 2–6-cell-stage embryos to 5.2 units in 40,000-cell gastrulae to 21 units in 400,000-cell swimming tadpoles. Approximately one third of this enzyme activity is α-amanitin-sensitive, showing that at all stages examined two-thirds of the enzyme activity is due to the presence of RNA polymerase I. One of the most interesting observations was that both normal tadpoles and anucleolate tadpoles, which do not synthesize any rRNA, have the same levels of RNA polymerase activity, and furthermore, approximately one third of the enzyme activity is due to the presence of RNA polymerase I even though there is no rRNA synthesis in these embryos. Therefore, although there is specificity between the class of RNA synthesized and the specific RNA polymerase involved, there does not seem to be a strict correlation between the levels of specific RNA polymerase activity and the amounts of RNA synthesized at any specific developmental stage. These observations suggested that there is no regulation of gene expression by limiting the quantities of the various RNA polymerase present at a specific developmental stage. It is still possible that polymerase activity is independently regulated in another way. The enzyme's accessibility to the DNA could be regulated as well. These studies and many more (Sugden and Sambrook, 1970; Keller and Goor, 1970; Chambon *et al.*, 1970) show that there are multiple RNA polymerases and that each type of enzyme seems to synthesize a specific class of RNA. There does not seem to be a correlation between the level of enzyme activity in a cell and the amount of each class of RNA synthesized in that cell. Thus, the regulation of gene expression through modulation of the levels of RNA polymerase does not appear to be operating in eukaryotic cells.

D. Heterogeneous Nuclear RNA

Gene expression does not appear to be completely regulated by the levels of RNA polymerase nor is it regulated solely by differential gene activity. Another mechanism is suggested by the fact that there is a large amount of DNA-like RNA synthesized in eukaryotic cells which is not translated. There have been numerous (see Darnell, 1968) descriptions of this heterogeneous, high-molecular-weight, rapidly synthesized, rapidly turning-over nuclear RNA in eukaryotic cells. This RNA, designated HnRNA, is thought to be a precursor to cytoplasmic RNA (see Greenberg, 1975; Lewin, 1975). Approximately 80–90% of the HnRNA is degraded in the nucleus, and only about 10% is transported from the nucleus to the cytoplasm where it may be translated. The HnRNA, like the high-molecular-weight rRNA precursor molecules, is probably specifically processed so that only a portion of the molecule is transported to the cytoplasm. This scheme, then, imposes additional steps in the regulation of gene activity. The processes of selection of which HnRNA molecules will be completely degraded and which will be further processed for export to the cytoplasm may be important regulatory steps. This mechanism, of course, assumes that HnRNA is precursor to mRNA. (The relationship between HnRNA and mRNA is reviewed by Brandhorst, 1976).

1. Poly(A)

An important advance was the discovery that a portion of HnRNA and mRNA molecules in cultured mammalian cells contained a homopolynucleotide tract of adenylic acid residues (Darnell *et al.*, 1971; Edmonds *et al.*, 1971; Lee *et al.*, 1971). These tracts were estimated to be approximately 150–250 nucleotides in length. In subsequent experiments (Molloy *et al.*, 1972; Nakazato *et al.*, 1973) it was shown that these polyadenylic [poly(A)] tracts are located at the 3' end of mRNA and HnRNA molecules. The function of these tracts is presently unknown, but there has been an enormous amount of work done to try to elucidate the function and significance of poly(A). It was soon clearly demonstrated that there was no poly(A) detected in histone mRNA from HeLa and L cells (Adesnik and Darnell, 1972; Greenberg and Perry, 1972). Furthermore, these authors suggested that virtually all the nonhistone mRNA molecules were polyadenylated. A large number of mRNA molecules coding for tissue-specific proteins are polyadenylated, including hemoglobin mRNA (Burr and Lingrel, 1971), calf lens mRNA (Lavers *et al.*, 1974), leg hemoglobin (Verma *et al.*, 1974), silkworm fibroin mRNA (Lizardi *et al.*, 1975), silkmoth chorion mRNA (Vournakis *et al.*, 1974), and ovalbumin mRNA (Palmiter, 1975).

However, a number of recent reports have suggested that a significant proportion of eukaryotic cell mRNA is not polyadenylated. Milcarek *et al.* (1974) have estimated that approximately 30% of HeLa cell mRNA is not polyadenylated. Fromson and Duchastel (1974, 1975) determined the proportion of newly synthesized polyribosomal RNA that is polyadenylated in sea urchin embryos and found that a large proportion of early embryo RNA lacks poly(A). Approximately 70% of cleaving embryo mRNA, 60% of early blastula mRNA, and 50% of mesenchyme blastula mRNA is not polyadenylated. Nemer *et al.* (1974) also found that about half of sea urchin blastula mRNA lacks poly(A). Therefore, it is unlikely that the original ideas stating that all nonhistone mRNA molecules are polyadenylated is correct. The functional significance of the poly(A) tract on the mRNA is not understood at the present time. Poly(A) is not required for transport of mRNA from the nucleus to the cytoplasm nor is it required for translation of the mRNA, for histone mRNA is both transported and translated (Schochetman and Perry, 1972). In addition, mRNA molecules from which the poly(A) was enzymatically removed were effectively translated *in vitro*. Bard *et al.* (1974) depolyadenylated L cell mRNA and translated it in the wheat embryo cell-free protein-synthesizing system; Williamson *et al.* (1974) removed the poly(A) from mouse globin mRNA and translated it in the Krebs ascites cell-free system; Huez *et al.* (1974) injected intact and depolyadenylated rabbit globin mRNA into *Xenopus* oocytes and measured the rate and duration of the resulting *in vivo* translation. Both types of mRNA were translated with similar initial rates. However, the stability of the unaltered, polyadenylated mRNA was far greater, suggesting that the poly(A) tract may be implicated in the regulation of mRNA stability. This conclusion is based on data obtained from heterologous systems—rabbit globin mRNA translated in *Xenopus* oocytes—and therefore may not be entirely valid. Perry and Kelley (1973) measured the stability of histone mRNA and poly(A)$^+$ mRNA and found that the two populations of RNA molecules have similar lifetimes, although the regulation of stability of these two kinds of molecules may be different. Thus, there are various interpretations of the available data as to whether the poly(A) is involved in regulating mRNA stability.

Another possibility is that poly(A)$^+$ and poly(A)$^-$ mRNA code for different populations of proteins. Fromson and Verma (1976) have translated sea urchin blastula poly(A)$^+$ and poly(A)$^-$ mRNA in the wheat embryo cell-free protein-synthesizing system. Both types of RNA supported the incorporation of [^{3}H]leucine and [^{3}H]tryptophan into protein. Since histones lack tryptophan, the fact that poly(A)$^-$ mRNA stimulated the incorporation of tryptophan suggests that poly(A)$^-$ mRNA codes for cellular proteins other than histones. A comparison of the H_2SO_4-soluble/H_2SO_4-insoluble leucine and tryptophan incorporation ratios of the peptides synthesized *in vitro* in response to added poly(A)$^+$ and poly(A)$^-$ mRNA suggests that these two mRNA populations may code for different populations of proteins. Although there is a tremendous volume of literature concerned with poly(A), its precise function is still not elucidated, and it is possible that poly(A) performs important regulatory functions at both the transcriptional and translational levels (see Darnell *et al.*, 1973; Molloy *et al.*, 1974).

E. Translational Control

Protein synthesis can be controlled not only by regulating the amount or type of mRNA synthesized and transported from the nucleus to the cytoplasm, but also by regulating the entry of cytoplasmic mRNA into polyribosomes where it is translated. For the latter mechanism to operate, there must be a pool of mRNA available for translation but which is somehow held away from the translational machinery. There are many examples of this type of regulation occurring in eukaryotic cells (see Hogan, 1975).

Upon fertilization of sea urchin eggs, the rate of protein synthesis and the amount of newly synthesized protein that accumulates increases markedly (see Giudice, 1973, p. 304). Epel (1967) has shown that the rate of protein synthesis is 15 times greater in embryos than in eggs. The elevated protein synthesis was explained by suggesting that more mRNA was synthesized following fertilization (Wilt, 1963). However, *de novo* mRNA synthesis is not required for this increased protein synthesis. Gross and Cousineau (1964) treated eggs with actinomycin D (115 μg/ml), under conditions where virtually all the RNA synthesis was inhibited, and observed no effect on protein synthesis. Both control and actinomycin-D-treated embryos incorporated [^{14}C]valine into TCA-precipitable material (protein) to the same extent for the first 12 hr after fertilization. A more refined series of experiments was carried out by Gross *et al.* (1964). In these experiments embryos were continually exposed to 20 μg/ml actinomycin D and were incubated with [^{14}C]leucine for 15 min. The incorporation into protein was the same for control and drug-treated embryos for the first 5 hr postfertilization, showing that the rate of protein synthesis is unaffected by the drug.

The fact that drug-induced suppression of RNA synthesis had no appreciable effect on protein synthesis indicated that new mRNA synthesis is not required for the increased protein synthesis following fertilization. This finding suggests that mRNA templates are present in the egg prior to fertilization and are translated at a higher rate following fertilization. Similar conclusions were reached by Denny and Tyler (1964), who separated unfertilized sea urchin eggs into nucleated and nonnucleated fragments by ultracentrifugation, measured *in vivo* protein synthesis in untreated and parthenogenically activated nonnucleate egg fragments, and observed greater protein synthesis in the activated egg fragments. These experiments clearly indicated that mRNA synthesis was not required for the increased amount and rate of protein synthesis following fertilization. This led to the formulation of the idea that there is sufficient mRNA synthesized during oogenesis to support the postfertilization protein synthesis and, furthermore, that this untranslated mRNA is present in the cells and enters polyribosomes where it is translated in a precisely regulated manner. Unfortunately, this mechanism is not presently understood.

The idea that mRNA–protein complexes (mRNP) exist in eukaryotic cells is well documented, and these mRNA particles have been extensively characterized in a number of systems (see Spirin, 1969). It has been suggested that the mRNA emerges from the nucleus after being processed from HnRNA (Molloy *et al.*, 1974) as messenger ribonucleoprotein (mRNP) particles. These particles then enter the cytoplasm where they form a compartment of available mRNA molecules, some of which eventually are translated. Scherrer *et al.* (1970) has suggested that a portion of this cytoplasmic messenger-like RNA is degraded and not translated. Although these mRNA particles have been extensively studied and well characterized, it is not yet known whether or not they comprise a precursor pool of molecules that are actually translated.

Another possible mechanism for translational control is the specific interaction between mRNA and ribosomes. Heywood (1969) demonstrated that salt-washed ribosomes from muscle and reticulocytes supported *in vitro* poly-U stimulated phenylalanine incorporation to the same extent. However, when putative myosin mRNA was added to an *in vitro* protein-synthesizing system utilizing reticulocyte ribosomes, peptide synthesis occurred only when the ribosomes were washed in high salt to remove the reticulocyte-specific ribosome factor and muscle ribosome factor was added to the reaction mixtures (ribosome factors are contained in a high-salt wash). Furthermore, the translation of myosin mRNA was dependent on muscle ribosome factor, and reticulocyte ribosome factor could not be used as a substitute. Myosin mRNA was translated *in vitro* using salt-washed reticulocyte ribosomes and muscle ribosome factor. These results suggested that there is a specific interaction between certain mRNA molecules and specific ribosome factors that is necessary for translation to occur. Further characterization of the ribosome factors has suggested that there may be cell-specific initiation factors required for protein synthesis. Heywood (1970) further characterized the ribosome factor required for myosin synthesis and reported that this factor was sensitive to digestion by proteolytic enzymes. In addition, Heywood (1970) fractionated the muscle ribosome factor by column chromatography and found that one specific subfraction was necessary to bind myosin mRNA to ribosomes.

In fact when either muscle or reticulocyte salt-washed ribosomes were used, myosin mRNA bound to the ribosomes only when this specific subfraction of muscle ribosome factor was also present. In addition, globin mRNA was preferentially bound to salt-washed ribosomes only in the presence of a specific reticulocyte ribosome factor. These data, therefore, suggest that cell-specific or RNA-specific factors may be required for the initiation of protein synthesis.

Recently Heywood *et al.* (1974) isolated red muscle initiation factor 3 and further fractionated this material by phosphocellulose chromatography and obtained four fractions. The effects of these four fractions were tested in a cell-free protein-synthesizing system with both myosin and myoglobin mRNA. Only one of these four fractions stimulated myoglobin synthesis, and another fraction specifically stimulated only myosin synthesis. These data again suggest a very specific interaction between a specific mRNA and a messenger-specific factor which was originally isolated from ribosomes.

These workers also have isolated an RNA fraction from muscle initiation factor 3 which they termed translational control RNA (tcRNA). These workers maintained that this RNA prevented the translation of heterologous RNA when it was added to an *in vitro* protein-synthesizing system. Rabbit reticulocyte tcRNA prevented the synthesis of myoglobin and myosin but had no affect on globin synthesis, and similarly tcRNA isolated from muscle initiation factor 3 depresses the *in vitro* synthesis of globin but has no effect on myosin and myoglobin synthesis. The molecular weight of the tcRNA is 6500, and its biological activity is lost when it was treated with RNase. The mechanism of action of the tcRNA has not been elucidated, and the mechanism of action proposed by Heywood *et al.* (1974) does not explain the observed results. The existence of a biological role for the tcRNA awaits confirmation.

These experiments do, however, indicate that genetic expression is controlled at the translational level.

III. SUMMARY

In this review I have attempted to list and briefly describe the possible sites of control of genetic activity. It is important to be aware of these genetic control points, for when one interferes with normal genetic expression, the mechanisms of action of these interfering agents should be understood. It is unlikely that this is presently possible. However, it is important to attempt to elucidate the mode of action of various drugs, metabolic inhibitors, and teratogens.

Genetic regulation in higher organisms depends on a series of enormously complicated processes that are not fully understood at the present time. It is probable that several of the listed control mechanisms are operating simultaneously in a single cell or population of cells. Thus, the precise genetic regulation that is seen in eukaryotic cells is the result of the interaction between several individual regulatory events. Further research will hopefully elucidate other modes of genetic regulation as well as the nature of the interactions between the various regulatory elements.

REFERENCES

Adesnik, M., and Darnell, J. E., 1972, Biogenesis and characterization of histone messenger RNA in HeLa cells, *J. Mol. Biol* **67**:397.

Bard, E., Efron, D., Marcus, A., and Perry, R. P., 1974, Translational capacity of deadenylated messenger RNA, *Cell* **1**:101.

Bishop, J. O., Pemberton, R., and Baglioni, C., 1972, Reiteration frequency of haemoglobin genes in the duck, *Nature (London), New Biol.* **235**:231.

Bonner, J., 1966, The template activity of chromatin, *J. Cell Comp. Physiol.* **66**(Suppl. 1):77.

Brandhorst, B. P., 1976, Heterogeneous nuclear RNA of animal cells and its relationship to messenger RNA, *in: Protein Synthesis* (E. H. McConkey, ed.), Vol. 2, p. 1, Marcel Dekker, New York.

Britten, R. J., and Davidson, E. H., 1969, Gene regulation for higher cells: A theory, *Science* **165:**349.

Brown, D. D., and Dawid, I., 1968, Specific gene amplification in oocytes, *Science* **160:**272.

Burr, H., and Lingrel, J. B., 1971, Poly(A) sequences at the 3′ termini of rabbit globin mRNA's, *Nature (London), New Biol.* **233:**41.

Chambon, P., Gissinger, F., Mandel, J. L., Kedinger, C., Gmazdowski, M., and Meihlac, M., 1970, Purification and properties of calf thymus DNA-dependent RNA polymerase A and B, *Cold Spring Harbor Symp. Quant. Biol.* **35:**693.

Darnell, J. E., 1968, Ribonucleic acids from animal cells, *Bacteriol. Rev.* **32:**262.

Darnell, J. E., Wall, R., and Tushinski, R. J., 1971, An adenylic acid-rich sequence in messenger RNA of HeLa cells and its possible relationship to reiterated sites in DNA, *Proc. Natl. Acad. Sci. U.S.A.* **68:**1321.

Darnell, J. E., Jelinek, W. R., and Molloy, G. R., 1973, Biogenesis of mRNA: Genetic regulation in mammalian cells, *Science* **181:**1215.

Davidson, E. H., 1968, *Gene Activity in Early Development,* p. 202, Academic Press, New York.

Davidson, E. H., and Britten, R. J., 1971, Note on the control of gene expression during development, *J. Theor. Biol.* **32:**123.

Denny, P. C., and Tyler, A., 1964, Activation of protein biosynthesis in nonnucleate fragments of sea urchin eggs, *Biochem. Biophys. Res. Commun.* **14:**245.

Edmonds, M., Vaughan, M. H., Jr., and Nakazato, H., 1971, Polyadenylic acid sequences in the heterogeneous nuclear RNA and rapidly-labelled polyribosomal RNA of HeLa cell: Possible evidence for a precursor–product relationship, *Proc. Natl. Acad. Sci. U.S.A.* **68:**1336.

Epel, D., 1967, Protein synthesis in sea urchin eggs: A "late" response to fertilization, *Proc. Natl. Acad. Sci. U.S.A.* **57:**899.

Fromson, D., and Duchastel, A., 1974, Poly(A)-containing polyribosomal RNA in sea urchin embryos, *J. Cell Biol.* **63:**105a.

Fromson, D., and Duchastel, A., 1975, Poly(A)-containing polyribosomal RNA in sea urchin development: Changes in proportion during development, *Biochim. Biophys. Acta* **378:**394.

Fromson, D., and Verma, D. P. S., 1976, Translation of nonpolyadenylated messenger RNA of sea urchin embryos, *Proc. Natl. Acad. Sci. U.S.A.* **73:**148.

Gall, J., 1968, Differential synthesis of the genes for ribosomal RNA during amphibian oogenesis, *Proc. Natl. Acad. Sci. U.S.A.* **60:**553.

Gilmour, R. S., and Paul, J., 1969, RNA transcribed from reconstituted nucleoprotein is similar to natural RNA, *J. Mol. Biol.* **40:**137.

Giudice, G., 1973, *Developmental Biology of the Sea Urchin Embryo,* Academic Press, New York.

Greenberg, J. R., 1975, Messenger RNA metabolism of animal cells' possible involvement of untranslated sequences and mRNA associated proteins, *J. Cell Biol.* **64:**269.

Greenberg, J. R., and Perry, R. P., 1972, Relative occurrence of polyadenylic acid sequences in messenger RNA and heterogeneous nuclear RNA of L cells as determined by hydroxylapatite chromatography, *J. Mol. Biol.* **72:**91.

Gross, P. R., and Cousineau, G. H., 1964, Macromolecule synthesis and the influence of actinomycin on early development, *Exp. Cell. Res.* **33:**368.

Gross, P. R., Malkin, L. I., and Mayer, W. A., 1964, Templates for the first proteins of embryonic development, *Proc. Natl. Acad. Sci. U.S.A.* **51:**407.

Heywood, S. M., 1970, Specificity of mRNA binding factor in eukaryotes, *Proc. Natl. Acad. Sci. U.S.A.* **67:**1782.

Heywood, S. M., Kennedy, D. S., and Bester, A. J., 1974, Separation of specific initiation factors involved in the translation of myosin and myoglobin messenger RNAs and isolation of a new RNA involved in translation, *Proc. Natl. Acad. Sci. U.S.A.* **71:**2428.

Hogan, B. L. M., 1975, Post-transcriptional control of protein synthesis, *in: The Biochemistry of Animal Development* (R. Weber, ed.), Vol. III, p. 183, Academic Press, New York.

Huez, G., Marbaix, G., Hubert, E., Leclereq, M., Nudel, U., Soreq, H., Salomon, R., Lebleu, B., Revel, M., and Littauer, V. Z., 1974, Role of polyadenylate segment in the translation of globin messenger RNA in *Xenopus* oocytes, *Proc. Natl. Acad. Sci. U.S.A.* **71:**3143.

Kedes, L. H., and Birnstiel, M. L., 1971, Reiteration and clustering of DNA sequences complementary to histone messenger RNA, *Nature (London), New Biol.* **230**:165.

Keller, W., and Goor, R., 1970, Mammalian RNA polymerase: Structural and functional properties, *Cold Spring Harbor Symp. Quant. Biol.* **35**:671.

Lavers, G. C., Chen, J. H., and Spector, A., 1974, The presence of polyadenylic acid sequences in calf lens messenger RNA, *J. Mol. Biol.* **82**:15.

Lee, S. Y., Mendecki, J., and Brawerman, G., 1971, A polynucleotide segment rich in adenylic acid in the rapidly-labelled polyribosomal RNA component of mouse sarcoma 180 ascites cells, *Proc. Natl. Acad. Sci. U.S.A.* **68**:1331.

Lewin, B., 1975, Units of transcription and translation: The relationship between heterogeneous nuclear RNA and messenger RNA, *Cell* **4**:11.

Lizardi, P. M., Williamson, R., and Brown, D. D., 1975, The size of fibroin messenger RNA and its polyadenylic acid content, *Cell* **4**:199.

Milcarek, C., Price, R., and Penman, S., 1974, The metabolism of a poly(A) minus mRNA fraction in HeLa cells, *Cell* **3**:1.

Molloy, G. R., Sporn, M. B., Kelly, D. E., and Perry, R. P., 1972, Localization of polyadenylic acid sequences in messenger ribonucleic acid of mammalian cells, *Biochemistry* **11**:3256.

Molloy, G. R., Jelinek, W., Salditt, M., and Darnell, J. E., 1974, Arrangement of specific oligonucleotides within poly(A) terminated HnRNA molecules, *Cell* **1**:43.

Nakazato, H., Kopp, D. W., and Edmonds, M., 1973, Localization of the polyadenylate sequences in messenger RNA and in heterogeneous nuclear RNA of HeLa cells, *J. Biol. Chem.* **248**:1472.

Nemer, M., Graham, M., and Dubroff, L. M., 1974, Co-existence of non-histone messenger RNA species lacking and containing polyadenylic acid in sea urchin embryos, *J. Mol. Biol.* **89**:435.

Palmiter, R. D., 1975, Quantitation of parameters that determine the rate of ovalbumin synthesis, *Cell* **4**:189.

Paul, J., 1971, Transcriptional regulation in mammalian chromosomes, *in: Control Mechanisms of Growth and Differentiation*, p. 117, Symposium of the Society for Experimental Biology, Academic Press, London.

Paul, J., and Gilmour, R. S., 1968, Organic-specific restriction of transcription in mammalian chromatin, *J. Mol. Biol.* **34**:305.

Perry, R. P., and Kelley, D. E., 1973, Messenger RNA turnover in mouse L-cells, *J. Mol. Biol.* **79**:681.

Roeder, R. G., and Rutter, W. J., 1969, Multiple forms of RNA polymerase in eukaryotic organisms, *Nature* **224**:234.

Roeder, R. G., and Rutter, W. J., 1970, Multiple ribonucleic acid polymerases and ribonucleic acid synthesis during sea urchin development, *Biochemistry* **9**:2543.

Roeder, R. G., Reeder, R. H., and Brown, D. D., 1970, Multiple forms of RNA polymerase in *Xenopus laevis:* Their relationship to RNA synthesis *in vivo* and their fidelity of transcription *in vitro*, *Cold Spring Harbor Symp. Quant. Biol.* **35**:727.

Scherrer, K., Spohr, G., Granboulan, N., Morel, C., Grosclaude, J., and Chezzi, C., 1970, Nuclear and cytoplasmic messenger-like RNA and their relation to the active messenger RNA in polyribosomes of HeLa cells, *Cold Spring Harbor Symp. Quant. Biol.* **35**:539.

Schochetman, G., and Perry, R. P., 1972, Early appearance of histone messenger RNA in polyribosomes of cultured L cells, *J. Mol. Biol.* **63**:591.

Spirin, A. S., 1969, Informosomes, *Eur. J. Biochem.* **10**:20.

Stein, G. S., Spelsberg, T. C., and Kleinsmith, L. J., 1974, Nonhistone chromosomal proteins and gene regulation, *Science* **183**:817.

Sugden, B., and Sambrook, J., 1970, RNA polymerase from HeLa cells, *Cold Spring Harbor Symp. Quant. Biol.* **35**:663.

Suzuki, Y., and Brown, D. D., 1972, Isolation and identification of the messenger RNA from silk fibroin from *Bombyx mori*, *J. Mol. Biol.* **63**:409.

Suzuki, Y., Gage, L. P., and Brown, D. D., 1972, The genes for silk fibroin in *Bombyx mori*, *J. Mol. Biol.* **70**:637.

Tocchini-Valentini, G. P., and Crippa, M., 1970, Ribosomal RNA synthesis and RNA polymerase, *Nature* **228**:993.

Verma, D. P. S., Nash, D. T., and Shulman, H. M., 1974, Isolation and *in vitro* translation of soybean legaemoglobin mRNA, *Nature* **251**:74.

Vournakis, J. N., Gelinas, R. E., and Kafatos, F. C., 1974, Short polyadenylic acid sequences in insect chorion messenger RNA, *Cell* **3**:265.

Williamson, R., Crossley, J., and Humphries, S., 1974, Translation of mouse globin messenger ribonucleic acid from which the poly(adenylic acid) sequence has been removed, *Biochemistry* **13**:703.

Wilt, F. H., 1963, The synthesis of RNA in sea urchin embryos (*Strongylocentrotus purpuratus*), *Biochem. Biophys. Res. Commun.* **11**:447.

Developmental Genetics 2

SALOME GLUECKSOHN-WAELSCH

I. INTRODUCTION

The concept of "developmental genetics" has undergone significant changes in recent years, particularly since molecular biology and genetics began to focus attention on mechanisms of gene transcription and translation within the context of problems of cell differentiation. As a consequence, the field has broadened and at present covers a wide area ranging from the molecular level of transcription to that of organogenesis and its genetic control. The present chapter will not attempt to cover the entire field but will be restricted to aspects of developmental genetics closest to problems of teratology, in particular those dealing with developmental mechanisms and their genetic control in the normal as well as the abnormal organism.

The historical affinity between experimental studies of development and teratology is the object of an essay which discusses the evidence for the close relationship between these two sciences and stresses the important role played by developmental genetics in the approach to problems of teratology (Oppenheimer, 1968). The relevance of developmental genetics to teratology is the determining factor in the choice of material to be included here, and it also guides the theoretical considerations in this chapter.

Studies of developmental genetics serve the purpose of identification and subsequent analysis of (1) genes concerned with processes of development and differentiation, (2) mechanisms of gene action, and (3) mechanisms of development. One kind of approach to these problems makes use of deviations from normal development caused by gene alterations which serve as experimental material for the identification of normal processes of develop-

SALOME GLUECKSOHN-WAELSCH • Department of Genetics, Albert Einstein College of Medicine, Bronx, New York 10461.

ment. This aspect of developmental genetics comes closest to the area of teratology concerned with experimental teratogenesis and its contributions to the analysis of developmental mechanisms.

II. THE ROLE OF THE NUCLEUS IN DEVELOPMENT

The correlation between the total genome, as represented in the chromosomes, and processes of development was emphasized first by Boveri (1902) in his classical experiments with polyspermy in sea urchins. The developmental abnormalities observed by Boveri in sea urchin embryos, originating from eggs fertilized by more than one sperm, led him to conclude that there was a correlation between the developmental disturbances and the abnormal distributions of chromosomes at the time of cleavage. These in turn were the result of formation of a tetrapolar spindle in the zygote containing one haploid maternal nucleus and two haploid paternal nuclei. Boveri's study provided the first demonstration of the causative role of chromosomal aberrations in abnormal development. Since that time numerical and structural chromosome abnormalities have, of course, become the central issue in investigations of the cytogenetics of congenital defects, particularly in humans.

Both nuclear and cytoplasmic factors are involved in developmental mechanisms and their normal as well as abnormal functioning. Boveri defined the respective roles of nucleus and cytoplasm in development and differentiation in the same paper (1902) as follows: "cytoplasmic differentiation serves to start the machinery whose essential and probably most complicated mechanism is located in the nuclei." In recent years, similar concepts have guided the experimental work of Gurdon (1970) in his search for mechanisms of control of gene activity during development. Individual nuclei obtained from differentiated cells, e.g., frog skin, are able to support perfectly normal development when injected into previously enucleated frog oocytes. Cytoplasmic components of the host serve to promote DNA synthesis in the donor nucleus; the sequential synthesis of nuclear, transfer, and ribosomal RNA follows the normal host pattern. Whereas the effects of cytoplasmic factors in promoting RNA synthesis appear to be species specific, those promoting DNA synthesis are neither species nor class specific.

III. DEVELOPMENTAL CONCEPTS AND GENETIC CONTROL

A. Determination

The concept of determination plays a prominent role in theories of embryonic differentiation. In vertebrates, an undifferentiated embryonic cell has multiple potentials and may eventually differentiate into a variety of different cell types. Gradually, these multiple potencies become restricted until finally demonstrated by transplanting test pieces of the cultured imaginal disks periodically into host larvae where they undergo metamorphosis together with their host, and differentiate. The products of differentiation in the transplanted disk usually correspond to the normal potential fate of the original imaginal disk. Cell heredity is responsible for the continuous replication of the initial state of determination in the individual cells of such disk cultures. Of particular interest are exceptions to the rule of unchanging cell heredity which occur occasionally; the phenomenon involved in such a change is referred to as "transdetermination." In these latter cases, test implants from cultured disks determined to form head structures (autotypic) may differentiate into totally different (= allotypic) structures, e.g., legs.

Certain mutations in *Drosophila* produce developmental changes similar to those found as the result of transdetermination in the Hadorn system. Examples of such "homeotic" mutations are: "nasobemia" and "aristapedia" where single gene mutations transform head structures into those of a leg (Gehring, 1969). Transdetermination has obviously been caused in the cells of the head imaginal disk by the mutant genes. The mechanisms involved in the effects of such mutations would be expected to be similar to those instrumental in the experimental alteration of determination referred to as transdetermination. The relevance of their elucidation for problems of teratology and teratogenesis is obvious.

Processes operating in normal determination and normal differentiation, as well as those of abnormal development and malformation, are open to further analysis with the help of this model system and the conceptual and technical methodology used by Hadorn and his school. (For further discussion see Hadorn, 1974.)

B. Inductive Interaction

Among the most characteristic mechanisms of vertebrate development is that of inductive interaction between cells and tissues which is an essential process in the course of the normal succession of states of determination in different tissues.

The classical studies of Spemann and of Harrison and their schools (Hadorn, 1974) laid the experimental foundation for the concept of inductive interaction and its role in determination and differentiation mechanisms of the vertebrate embryo. The interplay between gene-controlled developmental processes in the course of inductive interaction is exemplified strikingly by the results of the xenoplastic transplantations carried out by Spemann and Schotté (1932). In these experiments, the induction of mouth parts in donor skin by the underlying host tissue was studied by reciprocal exchanges between developing larvae of two amphibian orders, urodeles and anurans, of undifferentiated belly skin transplated into the region of the mouth.

These experiments proved that the inducing effect was nonspecific: The underlying urodele tissue interacted with anuran skin by communicating its one and only one potential remains. The cell then is said to have become determined. Determination is a gradual process, not a sudden event. Its molecular basis is not understood, even though recent definitions of the concept of determination make reference to transcriptional or translational mechanisms. At best, such a definition provides an updated description of determination as the state in which a particular cell type has developed a mechanism to select the nucleotide sequence of a particular DNA for purposes of transcription while all other DNA remains repressed. Alternatively, this selection might occur at the translational level, for example, with differential distribution or properties of ribosomes (see Chapter 1, this volume).

The genetic basis of determination, and the possible role of individual genes in controlling it, would be open to study if appropriate mutations of such genes were available (cf. Hadorn, 1966). However, as yet no such mutations have been identified unequivocally. Brief mention might be made here of a mutation in the mouse with the potential of perhaps providing suitable experimental material for molecular studies of determination. This is one of the recessive alleles at the T locus in linkage group IX (chromosome 17). Homozygous embryos (t^9/t^9) are characterized by a striking excess of neural tissue in contrast to an equally striking deficiency of mesodermal tissue. One possible cause for this may be that in such homozygous embryos an abnormality of the determination mechanism stimulates an abnormally high number of as yet noncommitted embryonic cells to activate their neural genome, whereas the mesodermal genome becomes activated in a correspondingly smaller than normal number of cells. Further studies of the mechanism of such mutational effects might yield information that could reveal molecular aspects of the determination mechanism (Moser and Gluecksohn-Waelsch, 1967).

Among the characteristic properties of determination is its stability. Once a cell or a tissue has been determined, the particular determined state continues to be expressed and is perpetuated by cell heredity. The actual expression of differentiated traits may follow immediately after determination or may occur at a later time. A system with a considerable lag between attainment of the determined state and expression of differentiated cellular traits would provide a type of "slow-motion" picture for the identification of various parameters of determination over an extended period of time. Such a system exists in the imaginal disks of *Drosophila* which Hadorn utilized for an elegant experimental approach to problems of determination and differentiation. The future fate of individual cells of imaginal disks is determined during an early phase of embryogenesis. Normally, these cells remain in the determined but undifferentiated state for about 8–9 days before they undergo differentiation. But it is possible to prolong this period almost indefinitely by culturing imaginal disks in the abdomen of adult flies where lack of specific molting hormones keeps the larval material from undergoing metamorphosis, thus maintaining it in an undifferentiated state. Toward the end of the host's life span, the implanted imaginal disk is transferred to another young adult host, and this is repeated many times. The absence of possible adverse effects of such serial transfers on the cells' capacity for normal differentiation is inducing signal to form mouth parts and vice versa. However, the genome of the reacting system, while changing its program from skin to mouth parts, remained true to its species and carried out the new program according to the demands of the species, i.e., anuran-type mouth parts in anuran skin, and urodele-type balancers in the reverse combination. The genome involved in the output of the inducing signal serves to elicit the proper response in a foreign species, but the specific nature of the response is determined by the genome of the reacting system.

Inductive interaction in mammalian kidney development was studied by Grobstein (1955) using tissue culture techniques; he demonstrated the interdependence of the two components of renal primordia, i.e., ureteric bud and metanephrogenic mesenchyme, in normal kidney differentiation in the mouse. Inductive interaction between these primordia leads to branching of the ureteric bud and the differentiation of excretory tubules, while secretory tubule formation occurs in the kidney mesenchyme. Neither the mechanisms of this inductive interaction nor the possible role of genes in its control are known.

A mutation in the mouse causing abnormalities of kidney differentiation seemed to offer material for a study of these problems. The mutation is an incomplete dominant affecting heterozygotes (*Sd/* +) less severely than homozygotes (*Sd/Sd*) which lack kidneys altogether. Kidney development in such *Sd* homozygotes is characterized by abnormalities of growth and differentiation of the ureteric buds. The metanephrogenic mesenchyme either does not differentiate at all or is retarded in its differentiation. The mutant effect on the ureter seemed to precede that on the metanephrogenic mesenchyme. Organ-culture studies showed that both mutant ureter and metanephrogenic blastema differentiated normally in all possible combinations of normal and mutant rudiments (Rota and Glueksohn-Waelsch, 1963). The mutation, therefore, does not affect either the inducing or the reacting potential of the tissues concerned. The mechanism by which mutant kidney differentiation is suppressed *in vivo* may reside in differential retardation of growth and differentiation of both rudiments, interfering with the spatial and temporal pattern of their inductive interaction and the essential synchronization of inductive and reacting processes. *In vitro* conditions appear to speed up growth and differentiation of both mutant rudiments differentially so that the inducing capacity of one and the competence to react of the other once more coincide in time. Recently the concept of genetic control of temporal aspects of differentiation has received much impetus from the work of Kenneth Paigen (1971) and his concept of temporal genes. Even though the *Sd* mutation has not been identified at this time as a "temporal" gene in the sense of controlling the time of activation of another structural gene, temporal aspects of differentiation in general are no doubt under genetic control and can be the target of mutations, as seems to be the case in the *Sd* mutation in the mouse. Its analysis calls attention to the importance of as yet unknown normal developmental mechanisms of synchronization and temporal integration of the multitude of interacting processes of differentiation and their genetic control. It is likely also that disturbances of temporal aspects of development play a significant role in the etiology of congenital abnormalities. (See also Chapter 2, Vol. 1.)

IV. DEVELOPMENTAL MUTATIONS

Developmental processes are subject to the effects of mutations, causing deviations from normal development, which are able to serve the identification of the corresponding normal events and their genetic control. Consequently, developmental studies of mutations in the mouse have helped to illuminate the origin of abnormalities of various tissue and organ systems, and in turn have contributed to knowledge of normal development. Examples that illustrate this point are numerous. Obviously, a complete review of all such work is not possible here. Particular cases have been chosen in order to demonstrate various aspects and concepts of developmental genetics. Among mutations studied developmentally, the *T* locus in the mouse has received particular attention.

A. The *T* Locus in the Mouse

The *T* locus is a complex locus with a large number of mutations which are alleles or closely linked to each other. Many of these mutations have been shown to interfere with mechanisms of early embryonic differentiation, and have served as useful tools for the analysis of normal differentiation and its genetic control in the mammalian embryo. Studies of the *T* locus and its developmental implications are the subject of a detailed review (Gluecksohn-Waelsch and Erickson, 1970) to which reference should be made for additional information.

The original Brachyury mutation, *T*, was shown to affect the differentiation of the embryonic axial system, including notochord, neural tube, and spinal column, resulting in heterozygotes in shortening of the tail. Embryos homozygous for *T* have severe and extensive abnormalities of notochord, neural tube, and somites; the posterior body region and posterior limb buds are missing altogether (Chesley, 1935). These abnormalities become apparent at about 8 days of gestational age; death of the homozgyote which lacks any circulatory connection with the mother occurs at about 10 days when maternal circulation becomes essential for the embryo's nutrition (Gluecksohn-Schoenheimer, 1944).

The complex *T* locus in the mouse is part of linkage group IX, i.e., chromosome 17 (Miller and Miller, 1972), which also carries the complex *H-2* (histocompatibility) locus. In the same linkage group is a gene serving a useful marker function *tufted* (*tf*). The dominant mutations *T*, *Fu*, and *Fu^{Ki}* are almost certainly point mutations; a large number of recessive alleles at and near the *T* locus probably result from quantitative and/or qualitative changes of the entire chromosome region between T and Fu. From the point of view of their developmental effects the following attributes of these recessive "alleles" are significant:

1. The majority of t alleles interact with the dominant mutation T enhancing its effect on the tail and axial skeleton and causing taillessness.
2. Many t alleles are embryonic lethals when homozygous.
3. Different lethal alleles show partial, but not complete, complementation with each other.
4. Many t alleles suppress crossing over in an area of T extending for about eight crossover units.
5. Their mutation frequency is high, approximately 2×10^{-3} as compared to the average mutation frequency of approximately 1×10^{-5}.
6. The occurrence of rare, exceptional, and unequal crossing over within the T region in the presence of a particular t allele gives rise to new t "alleles" with properties different from those of the t allele from which they originated.

All observations of the genetic behavior and of the effects of the recessive mutations at and near the T locus are compatible with the assumption that a considerable segment of chromosome 17 has undergone a change, the exact nature of which remains to be identified but which could be quantitative or qualitative.

The analysis of the developmental effects of T, Fu, Fu^{Ki}, as well as of the lethal t alleles, has shown the entire complex T locus to be concerned with the control of early processes of morphogenesis, in particular those of cell and tissue interactions.

Of the t alleles investigated in some detail, t^{12} produces the earliest developmental defect and t^{12}/t^{12} homozygous embryos stop development as 30-cell morulae at about 80 hr after fertilization. Cytoplasmic, primarily ribosomal RNA appears reduced in mutants when compared with normal littermates, and abnormalities of size and shape of mutant nucleoli are the most characteristic distinguishing criteria (Smith, 1956; Calarco and Brown, 1968). As the result of in $vitro$ studies of t^{12}/t^{12} morulae, on the other hand, the occurrence of nuclear or cytoplasmic abnormalities prior to developmental arrest has been disclaimed (Hillman et $al.$, 1970), and nucleolar abnormalities of homozygotes are considered to be secondary to postdegenerative cellular changes. As yet, the problem of the primary effect of t^{12} is not solved.

Embryos homozygous for t^0 can be distinguished from their normal littermates at approximately 5 days after fertilization, in the early postimplantation egg cylinder stage (Gluecksohn-Schoenheimer, 1940). While elongation and mesoderm formation characterize the normal embryo, the homozygous mutant shows no signs of differentiation, the inner cell mass remains unorganized and surrounded by a heavy entoderm layer. This stage is maintained for about 36 hr until signs of cell degeneration and necrosis appear before the entire embryo is eventually resorbed.

Chronologically, the mutations t^4 and t^9 follow t^0 in their effects on embryonic development. These t alleles arose independently, but cannot be distinguished from each other genetically nor in their developmental effects (Moser and Gluecksohn-Waelsch, 1967). Homozygous embryos show a disproportion of ectodermal and mesodermal components: It appears that the balance is disturbed between ectodermal (neural) derivatives and those of the mesoderm. These mutations were discussed in another context earlier (p. 21).

Striking effects on development and differentiation of the early mouse embryo have been reported also in studies of the mutation Fu^{ki} (kink) located at a distance of about 7 crossover units from T. Embryos homozygous for Fu^{ki} are characterized by more or less complete duplications of the entire embryonic axis, individual organs or parts of them, at about 7–8 days after fertilization. The existence of such malformations in the mutants indicates that normally embryonic regulation is a typical property of mammalian embryos in analogy to that revealed experimentally in amphibian embryos (Gluecksohn-Schoenheimer, 1949). Duplications of axial organs have been described also in embryos homozygous for t^{w18}, a recessive lethal allele at the T locus (Bennett and Dunn, 1960). Finally, an allele of Fu^{ki}, i.e., Fu, produces duplications of the neural tube in Fu homozygous embryos (Theiler and Gluecksohn-Waelsch, 1956).

The unique genetic and developmental properties of the T locus and its neighboring region have as yet not found any explanation, and the mechanisms by which they operate are largely unknown. It has been speculated that the abnormalities of embryonic differentiation caused by the alleles at the T locus may involve defects in processes including those of cell-to-cell recognition and morphogenetic movements. Since cellular interactions and cell associations are functions of cell-surface properties, T-locus products may be located at the cell surface and T-locus mutations may change the nature of such cell-surface substances (Gluecksohn-Waelsch and Erickson, 1970).

An additional characteristic of the recessive t alleles is their striking effect on male reproduction. Analysis of these effects has made unique contributions to spermatozoan genetics and has provided strong support for the existence of postsegregational gene action in spermatogenesis (Beatty and Gluecksohn-Waelsch, 1972). The unusual properties of the t alleles in respect to male reproduction may be summarized as follows: (1) *Segregation distortion:* The transmission ratio of the offspring of males heterozygous for t and either $+$ or T, departs from the expected Mendelian frequency of 0.5. (2) In the case of t alleles which segregate in excess in heterozygotes, males doubly heterozygous for two such different t alleles are *sterile*.

Among attempts to explain these unusual observations, a model has been proposed which assumes interaction between haploid spermatozoa. The abnormal gene product of t alleles is assumed to be located on the sperm cell surface and to interact with the non-t, i.e., normal, gene product on the genetically normal sperm to render the normal spermatozoon dysfunctional. In double heterozygotes, spermatozoa, each carrying a different t allele, interact rendering each other dysfunctional, thus accounting for the sterility of such males. This model assumes the site of T-locus products to be on the cell surface, in accord with conclusions of the developmental studies referred to above. (For further discussion, see Braden *et al.*, 1972; Gluecksohn-Waelsch and Erickson, 1971.

B. Developmental Genetics of Erythropoiesis

Mechanisms of erythropoiesis have been subjected to analysis with the help of two particular mouse mutations, the dominant *W* (causing macrocytic anemia in homozygotes) and the recessive *flexed, f* (causing siderocytic anemia), (Russell and McFarland, 1966). In homozygotes for either of these mutations, the first stage of erythropoiesis, i.e., that with large nucleated erythrocytes, is affected. The severe macrocytic anemia of *W/W* fetuses is apparent at 12 days and persists throughout the short life of the homozygote which dies soon after birth. The mutation interferes with the proliferation of hemopoietic cells, but hemoglobin is deposited normally in those red cells which are present. This indicates that during hemopoiesis in the mutant fetuses, enzyme activity in the pathway of heme synthesis keeps pace with cell proliferation (Russell *et al.*, 1968).

Flexed homozygotes (*f/f*) are also anemic at 12 days of gestational age, and the number of red cells of the primitive nucleated type is reduced. The rate of cell proliferation in *f* homozygotes is normal, and the relative deficiency of red blood cells remains stationary until 16 days. But, in contrast to *W* homozygotes, much less hemoglobin than normal is deposited in mutant *f/f* red cells of 12–16 days gestational age. The enzymes in the pathway of heme synthesis appear to be unable to keep up an activity commensurate with the normal and rapid rate of cell proliferation in the mutants. As yet, the primary defect underlying both the original reduction in red cell number and the later inhibition of hemoglobin synthesis in nonnucleated red cells of 13–16-day *f/f* fetuses has not been identified. Fetuses homozygous for both *W* and *f* die at 16–17 days of prenatal life: Proliferation of nucleated and nonnucleated red cells is decreased by *W* in double dose, and hemoglobin deposition in nonnucleated red cells is inhibited by *f/f*.

These mutations and others with a variety of effects on erythropoiesis provide tools with a high power of resolution for the identification of morphogenetic and biochemical processes during red blood cell differentiation and their genetic control (cf. also p. 29).

C. Developmental Genetics of Myogenesis

Great interest attaches to a mutation in the mouse which specifically affects the differentiation of striated skeletal muscle. The mutation, known as *mdg* (*muscular dysgenesis*), is responsible for an extensive syndrome of abnormalities in newborn homozygous mice, including a severe generalized deficiency of skeletal, voluntary musculature. In contrast to this, cardiac and smooth muscle appear normal. The skeletal muscle deficiency is due to an inherent defect of myoblast differentiation and is not the result of a degenerative process. Perinatal death of homozygous *mdg* mice results from their total inability to breathe because of the muscular deficiency of thorax and diaphragm. Deficiencies of skeletal structures normally serving as sites of muscle attachment and failure of development of joint cavities between limb and girdle primordia are secondary results of the abnormality of fetal myogenesis (Pai, 1964a,b).

Abnormal skeletal muscle differentiation in *mdg* homozygotes is independent of time and place of development, whereas differentiation of cardiac and smooth musculature is normal throughout. Skeletal myogenesis becomes noticeably abnormal at the stage of fusion of mononucleated myoblasts into multinucleated myotubes; such fusion is only rarely observed in *mdg* homozygotes, as is the formation of multinucleated myotubes. Deficient differentiation of striated muscle is expressed also in the failure of acetylcholinesterase activity to appear at the motor end plates which may be the result of a deficiency of the membrane-derived subneural folds or of the enzyme itself. Failure of differentiation of the mutant muscle cell thus includes cellular products other than merely the contractile apparatus.

A developmental study of the ultrastructure of mutant muscle-cell differentiation revealed abnormalities of the sarcoplasmic reticulum as the earliest and most significant mutant symptom. This ultrastructural deficiency correlates with the functional incompetence of the contractile system of mutant muscle (Platzer and Gluecksohn-Waelsch, 1972).

In the developmental study of the *mdg* mutation, genetic control of membrane structure and of the differentiation of one particular muscle histotype, i.e., the skeletal muscle, is revealed which is independent of the genetic control of cardiac and smooth-muscle differentiation. Current interest in mechanisms of cell fusion, and the interference of a particular mutation with fusion in that cell type which normally undergoes spontaneous and physiological fusion, lend particular significance to the *mdg* mutation with its intriguing experimental possibilities.

D. Developmental Genetics of Pleiotropy

The analysis of multiple, i.e., pleiotropic, effects of single genes has played an important role in studies of developmental genetics throughout its history. But the concept of pleiotropy has undergone significant changes since methods and approaches of molecular biology have made possible more direct analysis of fine genetic structure and primary gene action. As long as the gene remained an abstraction defined primarily in operational terms, i.e., by its effects, the preferred method of approach was that of retrograde analysis: Multiple effects were traced back throughout development in the hope of finding the responsible basic gene-caused deficiency. Although this approach has served a useful purpose, and continues to do so in being able to establish, in some instances, a "pedigree of causes" (Grüneberg, 1943), its limitations are obvious. The level of phenotypic analysis characteristic of such studies is so far removed from the level of primary gene action that significant advances in the solution of fundamental problems of gene action cannot be expected to emerge from this approach. Even on the level of polypeptide synthesis the relation of genotype to phenotype is not necessarily simple. The control of a single polypeptide by two genes in the case of β-galactosidase as the result of splicing of the protein has been reported by Apte and Zipser (1973). Cases in which the primary gene product is known to be responsible for pleiotropic effects include mutations of the hemoglobin genes, e.g., sickle hemoglobin, as well as mutations of enzyme subunits resulting in abnormalities of more than one enzyme.

The multiplicity of single-gene effects in higher organisms should be looked at as a function of their particular developmental pattern which is epigenetic and includes sequences of cell and tissue interactions, well defined in space and time, and causally related to specific types of differentiation. Consequently, a specific gene product originating in a particular biosynthetic event, and in a particular cell, may disseminate its effect to a variety of cells and tissues, all of them participating in epigenetic interactions. Retrograde studies of pleiotropic gene effects, therefore, serve more frequently to reveal unknown developmental mechanisms and interactions than to identify the single error on the level of primary gene action.

Several examples may serve to illustrate this statement: A dominant mutation in the mouse, *Splotch*, affects the pattern of pigmentation in heterozygotes, while homozygotes show a deficiency of closure of the neural tube, growth abnormalities of neural tissues, and failure to differentiate of spinal ganglia. These homozygotes die at about 14 days gestational age, i.e., before the stage of pigment differentiation in the skin, so that the effect on pigmentation remained unknown. Gene expression beyond the lethal phase of an embryo may be studied by transplanting parts of the lethal embryo to an environment favorable to its survival and differentiation. Transplantation of *Splotch* homozygous mutant skin into the eye cavity of albino host mice showed its actual inability to form pigment. The association of deficient pigment differentiation, absent dorsal root ganglia, and abnormalities of neural tube closure pointed to the neural crest as the target of the abnormal gene effects. The identification of this typical neural-crest syndrome confirmed the developmental role of the neural crest in the mammalian embryo in analogy to that in amphibians (Auerbach, 1954).

It is interesting, and certainly relevant to teratological phenomena in man, that in the mouse genetically caused pigment abnormalities are typically associated with simultaneous pleiotropic effects on various other developmental systems. Such association of pigment deficiencies and those of the erythropoietic system is sufficiently frequent and caused by a variety of mutations to be considered more than mere coincidence. Among these are the W alleles (cf. p. 27 and Russell and McFarland, 1966) with their effects on pigmentation, red blood cells, and germ cells. Developmental studies point at cell migration as the feature common to the three relevant cell types, but the mechanism of effect remains unknown. The close developmental relationship between the three systems is underscored also by the existence of a totally different mutation (*Steel*) with identical targets of effect although through a different mechanism (Bennett, 1956). Transplantation experiments have shown that the *Steel* mutant gene affects fetal hepatic erythropoiesis but that the site of action is the microenvironment of the hematopoietic tissue, not the erythroid cell lineage itself (Chui and Russell, 1974).

A final example of association of effects on pigmentation with those on other organ systems is that combining spotting and abnormalities of the inner ear (Deol, 1970). A developmental analysis of this pleiotropy is intriguing since it may reveal as yet unknown mechanisms of interaction during differentiation and also have a bearing on the genetic and teratogenic origin of deafness in man.

One of the striking examples of pleiotropic gene effects is a mutation in the mouse, *dominant hemimelia* (*Dh*) which causes abnormalities of the hind limbs, spleen, digestive tract, and urogenital system, all of which seem to trace back to the failure of normal transient thickening of the epithelium of the splanchnic mesoderm in 11-day embryos to which, as the result of developmental studies of a mutation, an important role in normal morphogenetic mechanisms of the affected structures may now be assigned. This is a good illustration of contributions which the analysis of a mutation may make to the elucidation of inductive interactions in development (Green, 1967).

The emergence of the characteristic pattern of differentiation of cells, tissues, and organs in the course of development of higher organisms is a striking and puzzling phenomenon. Teratogenic abnormalities frequently present a severe disturbance of this normal pattern. In normal development the typical spatial arrangement of embryonic primordia depends on an integrated series of morphogenetic movements of cells. Although the mechanisms governing cell migrations and aggregations are largely unknown, various mutations indicate their genetic control. One such mutation in the mouse is *phocomelia*, a recessive causing a syndrome of widespread skeletal abnormalities. Abnormalities of the aggregation pattern of mesenchymal cells in the limb buds precede the skeletal symptoms (Sisken and Glueksohn-Waelsch, 1959). The genetic control of cell-surface properties governing cell interactions during differentiation is emphasized by this mutation. A gene-caused defect of cell adhesion would be expected to be widespread through different cells and tissues, thus accounting for pleiotropic gene effects. It is tempting to speculate that the teratogenic drug thalidomide, with effects similar to those of the mutation *phocomelia*, might cause an analogous interference with cell-surface properties and cell adhesion. Complexities of cell and tissue interactions throughout the development of higher organisms no doubt account for similarities of gene- and teratogen-caused abnormalities in a large number of instances.

V. CHIMERAS, AN EXPERIMENTAL SYSTEM FOR STUDIES OF DEVELOPMENTAL GENETICS

In recent years a powerful new method has been developed for the study of gene effects, gene control, and gene interactions during development. This is the successful fusion of late cleavage stages or of morulae of mouse embryos of different genotypes. The fusion products are cultured for a day and subsequently transferred to the uterus of a female made pseudopregnant by mating with a vasectomized male. Surviving embryos are tetraparental chimeras. The original experimental approach and methodology were suggested and put to test by Tarkowski (1961) and subsequently developed extensively by Mintz (1974). In the course of years various developmental processes and their genetic control have been studied with this method. They include those of differentiation of somites, muscle, skeleton, hair, pigmentation, sensory organs, also hematopoiesis, sex development, and the ontogeny of the immune system. In the context of teratology, the methodology of producing genetic mosaics and the evaluation and interpretation have provided approaches to a variety of problems of which only some are listed here: localization of the focus of a mutant defect, the question of cell autonomy of a specific genetic defect as against one acting through cell and tissue interaction, and the identification of the developmental stage of gene activation. Abnormalities of hematopoiesis, hereditary retinal degeneration, and hermaphroditism are some genetically controlled developmental defects studied in tetraparental mice.

In the case of lethal mutations, experimental "rescue" of mutant genotypes appears possible when the mutant effect is restricted to certain cell and tissue types only and the gene product does not diffuse into all cells. Thus, an attempt was made to rescue mice of the W/W genotype which suffer from a lethal macrocytic anemia (cf. Mintz, 1974), by producing chimeras between normal and mutant embryos. Survivors were found to have normal red blood cells (cf. Mintz, 1974) indicating rescue of the lethal genotype. This method is preferable to that of transplanting normal blood cells into mutants at various postnatal stages because of the avoidance of any, even early embryonic, stages at which the mutant genotype is expressed and also because of the absence of immunogenetic incompatibility.

Chimeras produced by *in vitro* fusion of cleavage-stage mouse embryos were used by McLaren (1972) for a study of interactions of genotype and cellular environment within the gonad during gametogenesis. In particular, the development of XX germ cells in an XY environment was shown to proceed as far as the beginning of meiosis in the fetal testis with subsequent degeneration of all such cells in the pachytene stage.

There exists a certain degree of confusion in the literature concerning terminology. The product of embryo fusion was called "chimaera" by Tarkowski (1961), which seems appropriate for various reasons. The term dates back to Greek mythology where it specifically referred to monsters that were part lion, part goat, and part serpent. In studies of experimental embryology at the time of Spemann, the term was applied to transplantation chimeras made up of components of different species or even genera, e.g., chimeric gills and extremities consisting of primordia from two different amphibian species (Rotmann 1931). Race and Sanger (1968), discussing blood cell chimeras in twins, refer to Medawar's definition of "chimeras" as being organisms whose cells derive from two or more distinct zygote lineages, adding the term "genetical chimera." This also applies to the results of experimental transplantation of hematopoietic tissue between different genotypes in studies of developmental genetics. Mintz (1974) continues to prefer the term "allophenic" mice to denote the existence of different genotypic cell populations within one organism. This has caused some confusion since the term "allophenic" gained wide usage after having been introduced by Hadorn (1945) to refer to the nonautonomous behavior of transplanted tissues and cells. Specifically, it applies to cases where, as the result of the effects of other cell systems in the environment, the transplant expresses a phenotype different (= allo) from that for which the cells were programed in their original environment. In agreement with McLaren (1972) it seems appropriate at this time to accept normal priority rules of scientific usage and to refer to genetically composite mice as "chimeras" and not "allophenic" mice.

An elegant method of obtaining mouse chimeras was developed by Gardner (1968), who injected single cells from various donors into the blastocoel cavity of mouse blastocysts. Such injected cells became incorporated into host embryos many of which showed extensive mosaicism of skin and coat color. This method offers extremely interesting opportunities for studies of differentiation both in normal and mutant organisms.

VI. X-CHROMOSOME INACTIVATION AND DEVELOPMENT

The inactivation of one of the two X chromosomes of the homogametic mammalian female has become an established fact since it was first proposed by Mary Lyon (1961). In the context of this chapter, greatest interest focuses on the identification of the particular developmental stage at which inactivation of one of the X chromosomes takes place. Several approaches have been used to determine this stage, and two of these will be described.

Gardner and Lyon (1971) produced chimeric mice by injecting cells from a donor embryo homozygous for the albino (c) allele in chromosome 7 and heterozygous for the translocation-carrying X chromosome [with the wild type (C) color allele] into recipient embryos homozygous for pink (p) in chromosome 7. If at the blastocyst stage when donor cells were obtained, X inactivation had not yet occurred, three types of pigmentation could be expected: (1) wild type, derived from donor cells in which the normal X became inactivated after transplantation, (2) albino, derived from donor cells in which the translocation carrying X had become inactivated, (3) pink, derived from host cells.

If, on the other hand, X inactivation had already taken place at the time the donor cells were removed from the blastocyst, only two types of pigmentation would be expected in any particular chimera: either albino and pink, or wild type and pink. The actual results included a mixture of 3-colored, and 2-colored chimeras when donor cells were obtained from 3½- to 4½-day-old blastocyst donors. The conclusion was that X inactivation occurs around the 3½- to 4½-day stage.

Another developmental question relates to the problem of whether X inactivation may be correlated with the state of differentiation and thus may occur at different times in different tissues. Deol and Whitten (1972a,b), in a study of retinal melanocytes and migratory melanocytes, compared the pigmentation pattern in X-translocation-carrying mice with that in chimeras resulting from the fusion of +/+ and c/c (albino) morulae. In studies of migratory melanocytes they found a thorough intermingling of pigmented and unpigmented areas in various parts of the eyes and inner ears of mice heterozygous for Cattanach's translocation, and homozygous for the albino allele (X^T/X; c/c), whereas in the chimeras of +/+ and c/c morulae large areas tended to be colonized by one type of cell only. Since the migratory melanocytes originate in the neural crest only at about 9 days of gestational age, and undergo many cell divisions after reaching their destination, the intermingling of differently pigmented cells in various parts of the eye in the translocation carriers indicates relatively late X inactivation; it is estimated to occur after the 11th day. Based on the results of a similar experimental system of comparing translocation carriers and chimeras, X inactivation in the retinal nonmigratory melanocytes is estimated to occur shortly before the formation of the neural plate, i.e., the start of the 7th day or possibly somewhat later. The differences in time of X inactivation between the two types of melanocytes may indicate a correlation between time of cell determination and that of X inactivation.

In oogenesis the activity of both X chromosomes has been inferred to be the case in oocytes as reported by Epstein (1969). Interestingly, only one X chromosome appears to be active in primordial germ cells, as Ohno (1963) reports one condensed chromosome in yolk-sac primordial germ cells. This was further established by Gartler *et al.* (1975) using G6PD alleles as markers.

In all heterogametic males the single X chromosome appears to pass from an active to an inactive state at a critical stage of spermatogenesis. Such inactivation is an essential control in the male, not a step toward dosage compensation as in the female. Whenever X inactivation fails to occur, sterility of the male results (Lifschytz, 1972). The assumption of obligatory X inactivation in spermatogenesis rests primarily on genetic evidence. Included in the latter are examples of sex-chromosome aneuploidy with resulting sterility. Also, male sterility is observed in cases where translocations of parts of the X chromosome to autosomes occur with resulting failure of inactivation.

Mechanisms of X inactivation are as yet in the speculative stage. Obviously, their identification will be most relevant to the elucidation of mechanisms of gene activation and inactivation during development and differentiation. It is in this connection that the significance of X inactivation to problems of teratology and abnormal development may be found.

VII. BIOCHEMICAL GENETICS OF DEVELOPMENT

The preceding sections of this chapter have dealt with genetically controlled abnormalities of development expressed morphologically. Developmental defects, caused genetically or otherwise, include those of metabolism and of biochemical traits. Studies of the developmental genetics of biochemical defects have revealed specific mechanisms of abnormal differentiation of enzymes and other biochemical entities; furthermore, they have added information concerning normal biochemical differentiation, and they are beginning to provide some understanding of the correlation between structural or ultrastructural morphogenesis and biochemical differentiation.

Two approaches to these problems have been selected for discussion as examples of different model systems serving the analysis of differentiation on various levels.

In an attempt to reduce the study of differentiation and its genetic control to experimentally accessible dimensions, Paigen and his collaborators have chosen for their model system "enzyme realization" as the set of processes instrumental in producing the final phenotype of a protein. The molecular mechanisms of enzyme realization can be studied with the help of mutations affecting such protein differentiation (Paigen, 1971). Different classes of mutations are distinguished: structural mutations which change an enzyme's amino acid sequence and those which affect "the remaining cell machinery," that regulates and controls protein synthesis. The latter were divided into three types: (1) regulatory mutations which alter enzyme activity, synthesis, and breakdown; (2) architectural mutations affecting the site of an enzyme within a cell; and (3) temporal mutations which change the temporal aspects of the enzyme's developmental pattern, e.g., the time of activation of the corresponding structural gene. Obviously, the effects of these latter three classes of mutations may be mimicked by specific structural gene mutations. Normal development and differentiation must be expected to depend on normal spatial as well as temporal integration of enzyme differentiation into the entire pattern of differentiation of an organism.

Several attempts to identify these types of controlling genes will be discussed here. In specific strains of mice the enzymes β-glucuronidase and β-galactosidase appear to share a common developmental program. A single structural gene is responsible for the synthesis of each enzyme, and the two structural genes are not linked. A hereditary alteration of the β-glucuronidase developmental pattern was ascribed to a gene linked to the structural gene for this enzyme in chromosome 5. Furthermore, a hereditary alteration of the developmental pattern of β-galactosidase was shown to be caused by a gene linked to the structural gene of β-galactosidase on chromosome 9. It is interesting that the developmental pattern shared by the two unlinked structural genes is controlled by two independent temporal genes (Felton *et al.*, 1974).

The apparently close linkage of a controlling gene to the respective structural gene has been reported recently also in *Drosophila*. This gene influences the temporal expression of the enzyme aldehyde oxidase; it is closely linked to the structural gene, and variants of the gene have been described (Dickinson, 1975). A disturbance of the normal temporal pattern of development would be expected to act as a teratogenic factor and cause abnormalities of development.

A model system which makes possible the study of genes controlling the spatial integration of structural gene products into the cell is that of several radiation-induced albino alleles in the mouse. Six such alleles were shown to be perinatal or prenatal lethals when homozygous. Cytological examinations revealed a demonstrable chromosomal deletion in at least one of these alleles (Miller *et al.*, 1974). A number of enzymes were found to be deficient in these homozygotes, including glucose-6-phosphatase, tyrosine aminotransferase, serine dehydratase, glutaminesynthetase, and UDP-glucuronyltransferase. Furthermore, serum protein levels were reduced by about 20% in the mutants. In contrast to the general rule for structural gene mutations which predicts a gene dosage effect in heterozygotes with enzyme activities intermediate between normal and mutant homozygotes, no such dosage effect was found in heterozygotes for these alleles, and activities of all five enzymes were equal to those of normal homozygotes. The mutations, therefore, were not likely to have occurred in the respective structural genes but were representing a genome affecting the cellular machinery that controls protein synthesis and concentration. In fact, subsequent electron-microscopic studies of mutant homozygotes with enzyme deficiencies revealed severe ultrastructural abnormalities of the rough endoplasmic reticulum, the Golgi apparatus, and the nuclear membrane. These abnormalities were restricted to liver and kidney cells and were parallel, in location as well as pattern of fetal differentiation, to the sites and pattern of expression of the enzyme glucose-6-phosphatase. The state of intercellular, as well as intracellular, differentiation in late fetal stages is revealed by these mutational effects: Liver and kidney cells are differentiated from other cells by the characteristics of their ultrastructural organelles which are susceptible to the mutational effects, in contrast to the same organelles in other cell types. Intracellularly, the specific membranes affected must of necessity differ in their molecular structure from those subcellular organelles which are resistant to the mutational effects and remain normal. The parallelism of biochemical and ultrastructural differentiation in the mutants, and the identification of a genome concerned with the control of that part of the cellular machinery that regulates specific protein synthesis and concentration, lends particular interest to the syndrome of abnormalities caused by the radiation-induced deletions at the *albino* locus in the mouse. Two of the radiation-induced *albino* alleles cause death of homozygotes in early embryonic stages, one in the 4–8 cell cleavage stage, the other soon after implantation in the uterus. The possible correlation of biochemical defects with those of differentiation opens up intriguing questions of the biochemistry of normal embryonic development, which may be studied with the help of these mutants (Gluecksohn-Waelsch and Cori, 1970; Thorndike *et al.,* 1973; Trigg and Gluecksohn-Waelsch, 1973; Gluecksohn-Waelsch *et al.,* 1974).

The etiology of inborn errors of metabolism in man, with their frequent pleiotropic effects, is usually thought to lie in mutations of structural genes for the respective enzymes. The model system of the radiation-induced albino alleles reveals mechanisms of causation of biochemical errors that may well concern morphogenetic and ultrastructural attributes. Thus, morphogenesis may not only be affected secondarily by genetically caused metabolic errors but may in turn be a causative factor in the origin of errors of metabolism. The mutual interdependence of enzyme differentiation and morphogenesis is a concept that must be kept in mind in the consideration of problems of teratology, i.e., the science of congenital defects.

VIII. CONCLUSIONS

In this chapter an attempt was made to call attention to studies of developmental genetics that are relevant to problems of teratology. Throughout its development the higher organism, in particular the vertebrate, has at its disposal only a limited number of choices of reactions when confronted with unexpected and abnormal events caused by mutations or teratogens. To quote L. C. Dunn (1940): "the recurrence of certain types of variation, regardless of the immediate stimuli which call them forth, testifies to the unity and the diversity which are preserved by heredity." Studies of developmental genetics on a variety of levels and with a variety of systems, therefore, serve to reveal and at times elucidate mechanisms involved in deviations from normal differentiation. These in turn have helped to uncover the corresponding normal mechanisms and thus promoted progress of the analysis of development. It is to be hoped that studies in teratology will continue to find such knowledge useful as a basis for further analysis of normal and abnormal development.

NOTE ADDED IN PROOF

Since the submission of this chapter for publication, significant progress has been reported in several areas of developmental genetics with the use of new analytical systems and approaches. At this time, it is possible to provide references only to the most relevant publications, and a more detailed discussion of specific experiments and interpretations cannot be presented. The areas are as follows.

(1) *Antigen expression during development and mutational effects*. Further studies of the mutations at the *T* locus in the mouse (cf. pp. 24–26) give indications of the expression of antigens on early embryos and effects on them by individual mutations. Experimentally, the actual presence as well as the nature of particular embryonic antigens (F^9 and t antigens) on the surface of mutant early embryonic cells are still a matter of controversy (Kemler *et al.*, 1976; Dewey *et al.*, 1977). Nevertheless, the causal relation between specific cell-surface properties and differentiation as well as the control of cell and tissue interactions by the *T* locus postulated earlier (cf. p. 25) (Glucksohn-Waelsch and Erickson, 1970), remain strong working hypotheses (Bennett, 1975).

(2) *Teratocarcinomas* have come into the center of interest of developmental geneticists as *a system for the study of mammalian embryogenesis* and its genetic control. Embryonal carcinoma cells from teratocarcinomas resemble cells of the early mammalian embryo in several respects. Like cells of the early embryo they are pluripotent, have similar ultrastructural and biochemical properties, and share at least one cell-surface antigen. They thus represent a valuable model system for studies of normal embryonic differentiation and its genetic control (Martin, 1975; Sherman and Solter, 1975; Stern *et al.*, 1975).

The injection of single teratocarcinoma cells into normal mouse blastocysts shown to result in the colonization of the host by derivatives of embryonal carcinoma cells, and the production of chimeras (cf. p. 32), was used in recent studies of their differentiation potential. Consequently, teratocarcinoma cells were shown to be totipotent and able to give rise to a variety of somatic as well as gonadal tissues (Illmensee and Mintz, 1976; Papaioannou *et al.*, 1975).

A further potential use of this system is the production of teratocarcinomas from early mouse embryos homozygous for particular lethal mutations thus making possible the "rescue" of lethals and providing a system for studies of mutants and the genetic control of differentiation (Bennett *et al.*, 1977).

ACKNOWLEDGMENTS

The investigations reported from the laboratory of the author were supported in part by grants from the National Institutes of Health (HD 00193, GM 00110, GM 19100) and the American Cancer Society (VC 64).

Sincere thanks are extended to Dr. Susan Lewis for helpful comments on the manuscript.

REFERENCES

Apte, B. N., and Zipser, D., 1973, *In vivo* splicing of protein: One continuous polypeptide from two independently functioning operons (β-galactosidase/lactose operon), *Proc. Natl. Acad. Sci. U.S.A.* **70:**2969.

Auerbach, R., 1954, Analysis of the developmental effects of a lethal mutation in the house mouse, *J. Exp. Zool.* **127**:305.

Beatty, R. A., and Gluecksohn-Waelsch, S., eds., 1972, *Edinburgh Symposium on the Genetics of the Spermatozoon* (R. A. Beatty and S. Gluecksohn-Waelsch, eds. and publishers), Edinburgh, U.K.

Bennett, D., 1956, Developmental analysis of a mutation with pleiotropic effects in the mouse, *J. Morphol.* **98**:199.

Bennett, D., 1975, The *T*-locus of the mouse, *Cell* **6**:441.

Bennett, D., and Dunn, L. C., 1960, A lethal mutant (t^{w18}) in the house mouse showing partial duplications, *J. Exp. Zool.* **143**:203.

Bennett, D., Artzt, K., Spiegelman, M., and Magnusen, T., 1977, Experimental teratomas as a way of analyzing the development of the *T/t* locus mutations, Proc. VI Sigrid Suselius Symposium, *Cell Interactions in Differentiation,* Academic Press (in press).

Boveri, T., 1902, On multipolar mitosis as a means of analysis of the cell nucleus, *in: Foundations of Experimental Embryology* (B. H. Willier and J. M. Oppenheimer, eds.), Prentice-Hall, Englewood Cliffs, N.J.

Braden, A. W. H., Erickson, R. P., Gluecksohn-Waelsch, S., Hartl, D. L., Peacock, W. J., and Sandler, L., 1972, Comparison of effects and properties of segregation distorting alleles in the mouse (*t*) and in Drosophila (*SD*), *in: Edinburgh Symposium on the Genetics of the Spermatozoon* (R. A. Beatty and S. Gluecksohn-Waelsch, eds. and publishers), Edinburgh, U.K.

Calarco, P G., and Brown, E. H., 1968, Cytological and ultrastructural comparisons of t^{12}/t^{12} and normal mouse morulae, *J. Exp. Zool.* **168**:169.

Chui, D., and Russell, E. S., 1974, Fetal erythropoiesis in Steel mutant mice: I. A. Morphological study of erythroid cell development in fetal livers, *Dev. Biol.* **40**:256.

Deol, M. S., 1970, The relationship between abnormalities of pigmentation and of the inner ear, *Proc. R. Soc. London* **175**:201.

Deol, M. S., and Whitten, W. K., 1972a, X-chromosome inactivation: Does it occur at the same time in all cells of the embryo? *Nature (London) New Biol.* **240**:277.

Deol, M. S., and Whitten, W. K., 1972b, Time of X-chromosome inactivation in retinal melanocytes of the mouse. *Nature (London), New Biol.* **238**:159.

Dewey, M. J., Gearhart, J. D., and Mintz, B., 1977, Cell surface antigens of totipotent mouse teratoma cells grown *in vivo:* Their relation to embryo, adult and tumor antigens, *Dev. Biol.* (Feb., 1977; in press).

Dickinson, W. J., 1975, A genetic locus affecting the developmental expression of an enzyme in *Drosophila melanogaster, Dev. Biol.* **42**:131.

Dunn, L. C., 1940, Heredity and development of early abnormalities in vertebrates, *Harvey Lect.* **35**:135.

Epstein, C. J., 1969, Mammalian oocytes: X-chromosome activity, *Science* **163**:1078.

Felton, J., Meisler, M, and Paigen, K., 1974, A locus determining β-galactosidase activity in the mouse, *J. Biol. Chem.* **249**:3267.

Gardner, R. L., 1968, Mouse chimaeras obtained by the injection of cells into the blastocyst, *Nature* **220**:596.

Gardner, R. L., and Lyon, M. F., 1971, X chromosome inactivation studied by injection of a single cell into the mouse blastocyst, *Nature (London)* **231**:385.

Gartler, S. M., Andina, R., and Gant, N., 1975, Ontogeny of X-chromosome inactivation in the female germ line, *Exp. Cell. Res.* **91**:454.

Gehring, W. J., 1969, Problems of cell determination and differentiation in Drosophila, *in: Problems in Biology: RNA in Development* (E. W. Hanly, ed.), Univ. of Utah Press, Salt Lake City, Utah.

Gluecksohn-Schoenheimer, S., 1940, The effect of an early lethal (t^0) in the house mouse, *Genetics* **25**:391.

Gluecksohn-Schoenheimer, S., 1949, The effects of a lethal mutation responsible for duplications and twinning in mouse embryos, *J. Exp. Zool.* **110**:47.

Gluecksohn-Waelsch, S., and Cori, C. F., 1970, Glucose-6-phosphate deficiency: Mechanisms of genetic control and biochemistry, *Biochem. Genet.* **4:**195.

Gluecksohn-Waelsch, S., and Erickson, R. P., 1970, The *T*-locus of the mouse: Implications for mechanisms of development, *Curr. Top. Dev. Biol.* **5:**281.

Gluecksohn-Waelsch, S., and Erickson, R. P., 1971, Cellular membranes, a possible link between *H-2* and *T*-locus effects, *in: Proc. Symp. Immunogenetics of the H-2 System* (A. Lengerova and M. Vojtiskova, eds.), Liblice-Prague, Karger, Basel.

Gluecksohn-Waelsch, S., Schiffman, M. B., Thorndike, J., and Cori, C. F., 1974, Complementation studies of lethal alleles in the mouse causing deficiencies of glucose-6-phosphatase, tyrosine aminotransferase, and serine dehydratase, *Proc. Natl. Acad. Sci. U.S.A.* **71:**825.

Green, M. C., 1967, A defect of the splanchnic mesoderm caused by the mutant gene dominant hemimelia in the mouse, *Dev. Biol.* **15:**62.

Grobstein, C., 1955, Inductive interaction in the development of the mouse metanephros, *J. Exp. Zool.* **130:**319.

Grüneberg, H., 1943, Congenital hydrocephalus in the mouse, a case of spurious pleiotropism, *J. Genet.* **45:**1.

Gurdon, J. B., 1970, Nuclear transplantation and the control of gene activity in animal development, *Proc. R. Soc. London, Ser. B* **176:**303.

Hadorn, E., 1945, Zur Pleiotropie der Genwirkung, *Arch. Julius Klaus-Stift., Suppl.* **20:**82.

Hadorn, E., 1966, Dynamics of determination, *in: Major Problems in Developmental Biology* (M. Locke, ed.), Academic Press, New York.

Hadorn, E., 1974, *Experimental Studies of Amphibian Development,* Springer-Verlag, Berlin.

Hillman, N., Hillman, R., and Wileman, G., 1970, Ultrastructural studies of cleavage stage t^{12}/t^{12} mouse embryos, *Am. J. Anat.* **128:**311.

Illmensee, K., and Mintz, B., 1976, Totipotency and normal differentiation of single teratocarcinoma cells cloned by injection into blastocysts, *Proc. Natl. Acad. Sci.* **73:**549.

Jacob, F., 1977, Mouse teratocarcinoma and embryonic antigens, *Transpl. Rev.* (in press).

Kemler, R., Babinet, C., Condamine, H., Gachelin, G., Guenet, J. L., and Jacob, F., 1976, Embryonal carcinoma antigens and the *T/t* locus of the mouse, *Proc. Natl. Acad. Sci.* **73:**4080.

Lifschytz, E., 1972, X-chromosome inactivation. An essential feature of normal spermiogenesis in male heterogametic organisms, *in: Genetics of the Spermatozoon* (R. A. Beatty and S. Gluecksohn-Waelsch, eds. and publishers), Edinburgh, U.K.

Lyon, M., 1961, Gene action in X-chromosome of the mouse (*Mus musculus* L.), *Nature (London)* **190:**373.

Martin, G. R., 1975, Teratocarcinomas as a model system for the study of embryogenesis and neoplasia. *Cell* **5:**229.

McLaren, A., 1972, Germ cell differentiation in artificial chimaeras of mice, *in: Genetics of the Spermatozoon* (R. A. Beatty and S. Gluecksohn-Waelsch, eds. and publishers), Edinburgh, U.K.

Miller, D. A., and Miller O. J., 1972, Chromosome mapping in the mouse, *Science* **178:**949.

Miller, D. A., Dev. V. G., Tantravahi, R., Miller, O. J., Schiffman, M. B., Yates, R. A., and Gluecksohn-Waelsch, S., 1974, Cytological detection of the c^{25H} deletion involving the albino (*c*) locus on chromosome 7 in the mouse, *Genetics* **78:**905.

Mintz, B., 1974, Gene control of mammalian differentiation, *Annu. Rev. Genet.* **8:**411.

Moser, G. C., and Gluecksohn-Waelsch, S., 1967, Developmental genetics of a recessive allele at the complex *T*-locus in the mouse, *Dev. Biol.* **16:**564.

Ohno, S., 1963, Life history of female germ cells in mammals, *in: Congenital Malformations* (M. Fishbein, ed.), International Medical Congress Ltd., New York.

Oppenheimer, J. M., 1968, Some historical relationships between teratology and experimental embryology, *Bull. Hist. Med.* **42:**145.

Pai, A. C., 1964a, Developmental genetics of a lethal mutation, muscular dysgenesis (mdg), in the mouse. I. Genetic analysis and gross morphology, *Dev. Biol.* **11:**82.

Pai, A. C., 1964b, Developmental genetics of a lethal mutation, muscular dysgenesis (mdg), in the mouse. II. Developmental analysis, *Dev. Biol.* **11:**93.

Paigen, K., 1971, The genetics of enzyme realization, *in: Enzyme Synthesis and Degradation in Mammalian Systems* (M. Rechcigl., Jr., ed.), Karger, Basel.

Papaioannou, V., McBurney, M., Gardner, R., and Evans, M., 1975, Fate of teratocarcinoma cells injected into early mouse embryos, *Nature* 258:70.

Platzer, A. C., and Gluecksohn-Waelsch, S., 1972, Fine structure of mutant (muscular dysgenesis) embryonic mouse muscle, *Dev. Biol.* 28:242.

Race, R. R., and Sanger, R., 1968, *Blood Groups in Man,* F. A. Davis, Philadelphia, Pa.

Rota, T. R., and Gluecksohn-Waelsch, S., 1963, Development in organ tissue culture of kidney rudiments from mutant mouse embryos, *Dev. Biol.* 7:432.

Rotmann, E., 1931, Die Rolle des Ektoderms und Mesoderms bei der Formbildung der Kiemen und Extremitäten von Triton, *Arch. Entwicklungsmeck Org.* 124:747.

Russell, E. S., and McFarland, E. C., 1966, Analysis of pleiotropic effects of *W* and *F* genic substitutions in the mouse, *Genetics* 53:949.

Russell, E. S., Thompson, M. W., and McFarland, E. C., 1968, Analysis of effects of *W* and *f* genetic substitutions on fetal mouse hematology, *Genetics* 58:259.

Sherman, M. I., and Solter, D. (eds.), 1975, *Teratomas and Differentiation,* Academic Press, New York.

Sisken, B. F., and Gluecksohn-Waelsch, S., 1959, A developmental study of the mutation "phocomelia" in the mouse, *J. Exp. Zool.* 142:623.

Smith, L. J., 1956, A morphological and histochemical investigation of a preimplantation lethal (t^{12}) in the house mouse, *J. Exp. Zool.* 132:51.

Spemann, H., and Schotté, O., 1932, Über xenoplastische Transplantation als Mittel zur Analyse der embryonalen Induktion, *Naturwissenschaften* 20:463.

Stern, P. L., Martin, G. R., and Evans, M. J., 1975, Cell surface antigens of clonal teratocarcinoma cells at various stages of differentiation, *Cell* 6:455.

Tarkowski, A. P., 1961, Mouse chimaeras developed from fused eggs, *Nature (London)* 190:857.

Theiler, K., and Gluecksohn-Waelsch, S., 1956, The morphological effects and the development of the fused mutation in the mouse, *Anat. Rec.* 125:83.

Thorndike, J., Trigg, M. J., Stockert, R., Gluecksohn-Waelsch, S., and Cori, C. F., 1973, Multiple biochemical effects of a series of X-ray induced mutations at the albino locus in the mouse, *Biochem. Genet.* 9:25.

Trigg, M. J., and Gluecksohn-Waelsch, S., 1973, Ultrastructural basis of biochemical effects in a series of lethal alleles in the mouse, *J. Cell Biol.* 58:549.

Numerical and Structural Chromosome Abnormalities

3

LILLIAN Y. F. HSU and
KURT HIRSCHHORN

I. INTRODUCTION

Within the past 15 years, the rapid progress in human cytogenetics has resulted not only in the delineation of a wide variety of syndromes, but also in the demonstration of a causal relationship between various chromosomal abnormalities and phenotypic manifestations. Multiple congenital abnormalities and mental retardation are an almost consistent finding in individuals with autosomal unbalanced aberrations and to a lesser frequency and degree in individuals with sex chromosomal abnormalities.

Among 43,558 consecutive liveborns studied in six large surveys (Sergovitch *et al.*, 1969; Gerald and Walzer, 1970; Lubs and Ruddle, 1970; Hamerton *et al.*, 1972; Friedrich and Nielsen, 1973; Jacobs *et al.*, 1974a), 0.56% had a chromosome aberration detectable with conventional staining techniques. The frequencies of various chromosomal abnormalities in consecutive liveborns are listed in Table 1.

Up to 1975, more than 3000 spontaneous abortuses have been cytogenetically studied. Approximately 42% have been found to have chromosomal abnormalities (Table 2). Of these 49% were trisomies, 18% were 45,XO, and

LILLIAN Y. F. HSU and KURT HIRSCHHORN • Department of Pediatrics, Division of Medical Genetics, Mount Sinai School of Medicine of the City University of New York, New York, New York 10029.

Table 1. Frequencies of Various Chromosomal Abnormalities in Consecutive Liveborns

Type of chromosomal aberration	Frequency			References (series)	Overall frequency of 6 series[a]		
	%	Ratio	Number affected / Number studied		%	Ratio	Number affected / Number studied
47,XYY	0.11	1/910 males	3/2,615 males	Friedrich and Nielsen, 1973 (Denmark)	0.09	1/1100 males	26/28,582 males
	0.13	1/770 males	10/7,849 males	Jacobs et al., 1974a (U.K.)			
	0.14	1/714 males	3/2,176 males	Lubs and Ruddle, 1970 (U.S.A.)			
47,XXY	0.11	1/910 males	9/7,849 males	Jacobs et al., 1974a (U.K.)	0.10	1/1,000 males	30/28,582 males
	0.15	1/666 males	4/2,615 males	Friedrich and Nielsen, 1973 (Denmark)			
	0.18	1/555 males	4/2,176 males	Lubs and Ruddle, 1970 (U.S.A.)			
47,XXX	0.12	1/833 females	3/2,434 females	Friedrich and Nielsen, 1973 (Denmark)	0.09	1/1,000 females	13/14,976 females
	0.13	1/770 females	5/3,831 females	Jacobs et al., 1974a (U.K.)			
	0.14	1/714 females	3/2,177 females	Lubs and Ruddle, 1970 (U.S.A.)			
45,X	0.04	1/2500 females	1/2,434 females	Friedrich and Nielsen, 1973 (Denmark)	0.013	1/7,500	2/14,976 females
	0.05	1/2000 females	1/2,177 females	Lubs and Ruddle (U.S.A.)			
47,+21	0.15	1/666	17/11,680	Jacobs et al., 1974a (U.K.)	0.10	1/1,000	45/43,558
	0.11	1/910	12/11,347	Hamerton et al., 1972 (Canada)			
47,+18	0.017	1/5880	2/11,680	Jacobs et al., 1974 (U.K.)	0.01	1/10,000	4/43,558
	0.02	1/5000	1/4,353	Lubs and Ruddle, 1970 (U.S.A.)			
47,+13	0.02	1/5000	1/5,049	Friedrich and Nielsen, 1973 (Denmark)	0.01	1/10,000	3/43,558
	0.01	1/10,000	1/11,347	Hamerton et al., 1972 (Canada)			
45,−D,−D, +t(DqDq)	0.05	1/2000	6/11,680	Jacobs et al., 1974a (U.K.)	0.07	1/1,430	31/43,558
45,−D,−G, +t(DqGq)	0.09	1/1100	10/11,347	Hamerton et al., 1972 (Canada)	0.02	1/5,000	8/43,558
	0.03	1/3333	4/11,680	Jacobs et al., 1974a (U.K.)			
	0.02	1/5000	2/9,048	Gerald and Walzer, 1970 (U.S.A.)			

[a] Jacobs et al., 1974a, Edinburgh, U.K. (Total: 11,680: males, 7,849; females, 3,831). Friedrich and Nielsen, 1973, Arhus, Denmark. (Total: 5,049; males, 2,615; females, 2,434). Sergovich et al., 1969, Ontario, Canada (Total: 2,081; males, 1,066; females, 1,015). Hamerton et al., 1972, Winnipeg, Canada (Total: 11,347; males, 5,828; females, 5,519). Gerald and Walzer, 1970, Boston, U.S.A. (Total: all males, 9,048). Lubs and Ruddle, 1970, New Haven, U.S.A. (Total: 4,353; males, 2,176; females, 2,177).

Table 2. Frequencies of Chromosome Abnormalities in Unselected Spontaneous Abortions

Number of abortuses studied	Number of abortuses with chromosome aberration	%	Different types of chromosome aberrations (%)		Polyploids			Reference
			Trisomy	45,X	Triploid	Tetraploid	Other	
1026	222	21.7%	45.5	19	18.5	6	11	Carr, 1971 (a review of 20 unselected studies)
650	198	30.5%	39.4	30.8	11.6	4.6	13.6	Combined data from Dhadial et al., 1970; Arakaki and Waxman, 1970; Lauritsen et al., 1972
1498	921	61.4%	52.0	15.3	19.8	6.2	6.7	Boué et al., 1975
Total 3194	1341	42.25%	49.0	18.2	18.4	5.9	8.4	Overall

24% were polyploidies (especially triploid). It has also been shown that there is an increased incidence of chromosome aberrations in couples with a history of fetal wastage. In 749 parents studied, 18 were found to have chromosome abnormalities (reviewed by Lucas *et al.*, 1972). Of these 14 were translocations. This represents a frequency of 1.9% (1 in 53 parents or 1 in 26 couples) which is significantly higher than the frequency of translocation (0.4% or 4/1020) found in the general adult population (Court Brown, 1967).

With the recent development of new banding techniques for accurate identification of each chromosome and its parts, the estimated frequencies of chromosome aberrations in liveborns, spontaneous abortuses, couples with a history of repeated abortions, and various other selected subpopulations are rapidly being revised upward (Kim *et al.*, 1975).

II. NEW BANDING TECHNIQUES FOR CHROMOSOME IDENTIFICATION

Prior to the development of the new banding techniques for chromosome identification, the 46 human chromosomes were distinguishable mainly as seven major groups (A, B, C, D, E, F, and G). Only numbers 1, 2, and 3 in the A group, number 16 in the E group, and the Y in the G group are readily recognizable by conventional staining (Chicago Conference, 1966). With autoradiographic techniques, individual chromosomes in the B, D, and E groups and the sex chromosomes became identifiable by their different DNA replicating patterns (Chicago Conference, 1966). In 1968, Caspersson and his coworkers discovered that specific chromosome regions become fluorescent after staining with quinacrine mustard. Subsequently several other banding techniques for chromosome identification were developed. These new advances have opened a new era in the field of cytogenetics. The five major banding methods will be briefly reviewed here (Paris Conference, 1972).

A. Quinacrine Fluorescent (Q Banding)

By 1970, Caspersson and his associates were able to show that each human chromosome developed specific patterns of fluorescent bands which could be analyzed either photographically or by photoelectric measurements along each metaphase chromosome (Caspersson *et al.*, 1970a,b, 1971a). The same fluorescent patterns (Q banding) have been produced by quinacrine dihydrochloride (Lin *et al.*, 1971). A Q-banding karyotype is shown in Figure 1. In addition to the characteristic banding patterns of each chromosome pair, there is an intense flourescence of the distal two-thirds of the long arm of the Y chromosome. However, this bright fluorescent area of the Y chromosome, as well as other regions designated as "v.b." (variable band) (Figure 1), are

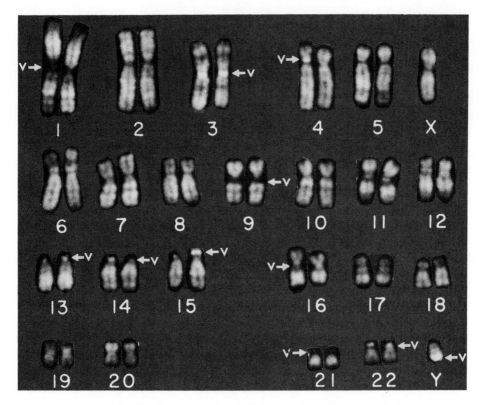

Fig. 1. A Q-banding karyotype of 46,XY. Arrows (V) indicate variable bands.

polymorphic; that is, they vary between individuals in terms of intensity and size of fluorescence. These variations are inherited from parent to child (Paris Conference, 1971).

Because the brightly fluorescent portion of the Y can be detected in interphase nuclei, it has been used as a screening technique for surveying groups of individuals for the presence and the number of Y chromosomes (Caspersson *et al.*, 1970c; Pearson *et al.*, 1970). The X chromatin can also be stained with quinacrine derivatives, although it is not as brightly fluorescent as that of the Y body (Wyandt and Hecht, 1971). Thus quinacrine staining of the interphase nuclei has provided a simple technique for discovering sex chromosome aberrations.

B. Constitutive or Centric Heterochromatin Staining (C Banding)

Arrighi and Hsu (1971) described a method which resulted in intense staining with Giemsa of the centromeric regions of chromosomes as well as of the distal two thirds of the long arm of the Y chromosome (Figure 2). This

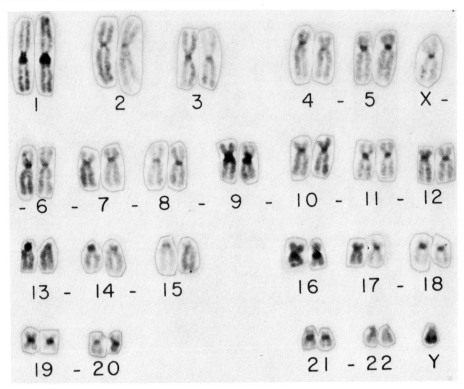

Fig. 2. A C-banding karyotype of 46,XY.

method was based on the observation of Pardue and Gall (1970) that the fraction of DNA consisting of highly repetitive copies of short segments renatures more rapidly and stains more intensely than other types of DNA after alkaline denaturation. With this technique, the intensely staining portions include the distal two thirds of the long arm of the Y chromosome and the secondary constriction areas adjacent to the centromeres on the long arm of chromosomes 1, 9, and 16. The centromeric regions of other chromosomes are also distinctly stained but less intensely.

C. Giemsa Banding (G Banding)

Giemsa banding of human chromosomes was first reported by Sumner *et al.* (1971). They were able to produce bands of metaphase chromosomes by treating acetic alcohol-fixed preparations with 2 × SSC (i.e., 0.3 M NaCl and 0.03 M sodium citrate) at 60°C for 1 hr and then staining with Giemsa at pH 6.8. Several other methods were developed soon after, including the denaturation–renaturation methods of Schnedl (1971) and Drets and Shaw (1971), the Giemsa 9 method by Patil *et al.* (1971), and methods using proteolytic enzymes such as trypsin by Seabright (1972), pronase by Dutrillaux *et al.* (1971), or chymotrypsin by Finaz and de Grouchy (1971). A trypsin G-banding human karyotype is shown in Figure 3. Comparison of G banding with Q banding has shown identity for all bands except those associated with secondary constrictions, other polymorphic areas, and the distal two thirds of the long arm of the Y chromosome. In our experience the easiest and most reproducible G-banding method is the trypsin method with a modification (Hirschhorn *et al.*, 1973).

D. Reverse Banding (R Banding)

Dutrillaux and Lejeune (1971) were able to produce reverse Giemsa banding patterns of metaphase chromosomes by treating the preparation with a hot (87°C) 0.02 M phosphate buffer (pH 6.5) for 10–12 min with subsequent cooling to 70° and staining with Giemsa. The banding patterns are the reverse of the patterns obtained with G banding or Q banding, i.e., those regions which stain densely with Giemsa or fluoresce brightly with quinacrine do not stain densely with this R-banding technique and vice versa. The Y chromo-

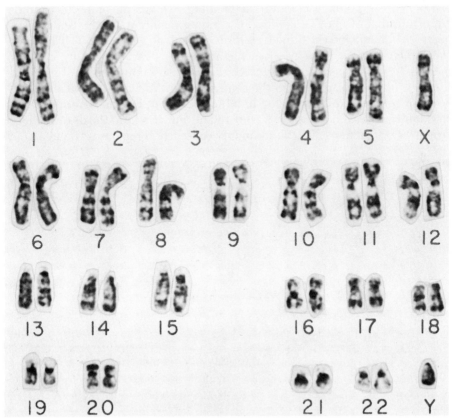

Fig. 3. A G-banding karyotype of 46,XY.

some was uniformly lightly stained, and the secondary constriction regions of chromosomes 1, 9, and 16 were indistinct.

E. Terminal Banding (T Banding)

Methods to produce terminal banding patterns have been developed by Dutrillaux (1973) and Bobrow and Madan (1973). There are two basic techniques: one using a treatment of hot phosphate buffer (pH 6.7 at 87°C) for 2 min and then staining with Giemsa, the other using a treatment of hot 2 × SSC buffer (85°C) for 5–10min and then staining with 0.01% acridine orange for fluorescence-microscope examination. A T-banding karyotype is shown in Figure 4.

F. Summary and Mechanisms of Banding

Comparison of these five major banding patterns has shown that Q and G banding are essentially identical; R banding exhibits an opposite banding pattern to that of Q or G banding. C banding mainly stains the centromeric heterochromatin and the distal two thirds of the long arm of the Y. T banding stains the telomeric portions of each chromosome. Differences are noted in the staining characteristics of the secondary constriction regions of chromosomes 1, 9, and 16 adjacent to the centromere on the long arm (Table 3). C banding produces dense staining in all four regions, i.e., the secondary constriction regions of numbers 1, 9, and 16 and the distal two thirds of the long arm of Y. G banding stains the secondary constriction regions of 1 and 16 but

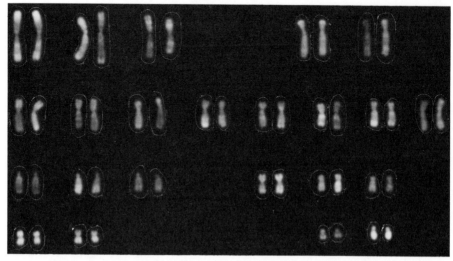

Fig. 4. A fluorescent karyotype of 46,XX showing both T and R banding (kindly provided by Dr. Hausjakob Müller).

**Table 3. Differences in the Staining Properties
of the Four Major Banding Techniques**

Chromosome region	C band	G band	Q band	R band
1qh	+	+	−	−
9qh	+	−	−	−
16qh	+	+	−	−
Distal Yq	+	Variable	Brilliant	Variable

not of 9. The Y is not as densely stained as that of C banding. Q banding does not stain any of these secondary constriction regions but stains the distal two thirds of the long arm of Y. R banding does not stain any of these secondary constriction regions. The finding that the secondary constrictions in chromosomes 1, 9, and 16 stain with the C-banding technique, do not stain with the Q-banding technique, and stain differentially with the G-banding technique (1 and 16 positive, 9 negative) indicates that there must be different species of DNA containing short-chain repetitive units. This has already been confirmed by the exclusive staining of the secondary constriction of chromosome 9 by Giemsa at pH 11 (Bobrow *et al.*, 1972) and by the finding that one fraction of human satellite DNA separated by differential centrifugation is primarily in the secondary constriction of chromosomes 1 and 16 (Jones and Corneo, 1971), while another class is chiefly in the secondary constriction of chromosome 9 (Saunders *et al.*, 1972).

As to the mechanisms underlying the various banding patterns, studies indicate that quinacrine binds preferentially to chromosome regions high in content of adenine and thymine, whereas DNA rich in guanine quenches the fluorescence (Weisblum and de Haseth, 1972; Ellison and Barr, 1972). The indication that Q bands are associated with A–T-rich DNA is further supported by the fact that Q bands can also be produced by using fluorescein-tagged, anti-adenosine antibody (Dev *et al.*, 1972). Recently, Schreck *et al.* (1973), using anti-cytosine-specific antibody, were able to produce fluorescent R-banding patterns of metaphase chromosomes. These results indicate that the fluorescent banding produced by the nucleoside-specific antibodies are a reflection of the major composition of chromosomes. As to the mechanism of G banding, it appears that both protein and DNA play important roles. The fact that proteolytic treatment results in G-banding patterns for all chromosomes implies that either some sections of DNA are covered by a different class of protein or that the DNA–protein complex in some areas is more susceptible to proteolytic activity. Sumner and Evans (1973) presented evidence suggesting that chromosome banding is a consequence of the loosening of chromosome structure in certain places, permitting DNA chains to move further apart. As a result, these regions are weakly stained. A number of treatments that may reduce the binding of Giemsa to DNA because of loosened chromosome structure are those which break hydrogen bonds, such as hot saline, or treatment which affects peptide linkages in the protein such as proteolytic enzymes. Sumner (1973) further suggested that G banding appears to be a consequence of a varying concentration of protein disulfides and sulfhydryls along the chromosomes, as would be indicated by urea, which also produces G banding. Dark bands seem to be relatively rich in disulfides, whereas pale bands are relatively rich in sulfhydryls. According to the studies of Comings *et al.* (1973), it appears that interaction between DNA and nonhistone protein plays an important role in the production of C bands. The complete C-banding technique removes an average of 60% of the DNA. However, C bands can be produced by prolonged trypsin treatment (Merrick *et al.*, 1973) which removes very little DNA.

These banding techniques have provided new tools to identify various chromosomal aberrations which could not be diagnosed with the previous conventional methods, such as reciprocal translocation with two exchanged pieces of equal size, paracentric inversion, and various duplication–deficiency types of abnormalities, as well as various types of autosomal trisomies. We have recently identified a reciprocal translocation involving two exchanged pieces of equal size (Kim *et al.*, 1975). Caspersson *et al.* (1972) have identified 4 patients with trisomy 8 mosaicism. By combining Q-, G-, and C-banding techniques, we were able to distinguish the true centric fusion type of translocation from reciprocal translocation of two number 13 chromosomes (Hsu *et al.*, 1973). Dutrillaux and his associates (1973) demonstrated an unbalanced translocation which was not detected with G banding. Apparently, certain reciprocal translocations involving the regions which only stain lightly or are not stained by G- or Q-banding methods require R banding for accurate diagnosis.

None of these current banding methods can identify the late-replicating X chromosome. Autoradiography remains a good method for this purpose. However, recently Latt (1973), using a special bisbenzimidazole dye, 33528 Hoechst, was able to distinguish late-replicating chromosomes by differential fluorescence, since the fluorescence intensity of this dye is less when bound to BUdR (Bromodeoxyuridine) than when bound to thymidine. In cells exposed to BUdR during early S (DNA synthetic period) and thymidine during the late S period, the binding of this dye results in brighter fluorescence in the thymidine-rich region than the chromosome regions incorporating BUdR. This approach may replace the time-consuming autoradiographic technique.

III. MECHANISMS FOR CHROMOSOME ABERRATIONS

A. Numerical Aberrations

1. Nondisjunction

Nondisjunction means failure of homologous chromosomes or sister chromatids to separate in a dividing cell. This leads to an extra chromosome in one daughter cell and a chromosome missing in the other. It can occur either in meiosis or in mitosis.

a. Meiotic Nondisjunction. In the developing gamete, nondisjunction of homologous chromosomes results in one daughter cell with two homologs and one with neither. If the former cell is fertilized by a normal gamete, a trisomic condition is produced. If the latter cell is fertilized, a monosomic zygote is formed. Except for the occasional viability of 45,X embryos, zygotes with autosomal monosomies are not viable in man.

It is now generally accepted that most trisomes are caused by meiotic nondisjunction, which can occur either at the first or the second meiotic division. In trisomy 21, Down syndrome, Q-banding studies have shown that nondisjunction of chromosome 21 most commonly occurs at the first meiotic division (Robinson, 1973). Nondisjunction is termed primary if the original germ cell is normal or secondary if the germ cell is trisomic. The latter condition has been seen in the offspring of females with trisomy 21 Down syndrome or parents mosaic for normal and trisomic cells.

Nondisjunction during the first meiotic division theoretically can result from either absence of synapsis or failure of disjunction after synapsis. Distinction between these two causes is theoretically possible by distinguishing whether the nondisjunctional event has taken place before or after the process of crossing over from studying whether there is recombination in the end products.

The pathophysiology of meiotic nondisjunction is unknown. There is no doubt that advanced maternal age is a major factor. Human oocytes are particularly susceptible to divisional errors because they remain in the dictyotene state of first meiotic prophase for many years prior to ovulation and are thus more likely to have been affected by metabolic or environmental factors. In a study reported by Mikamo (1968), aging of eggs in *Xenopus* was found to cause degeneration of the meiotic spindle fibers, leading to meiotic nondisjunction. Conceivably, chronologic aging of human oocytes might progressively impair the meiotic spindle fibers. It is also possible that the adverse effect of advanced maternal age may reside within the cytoplasm of the oocyte. Chronologic aging might also lead to terminalization of chiasma between bivalents and consequent reduction of synapsis and disorientation of the subsequent separation of homologous chromosomes. Although this phenomenon has not been confirmed in man, it has been observed in mouse oocytes by Henderson and Edwards (1968).

German (1968) has suggested that the advanced parental age found in trisomic syndromes may actually reflect less frequent sexual relationship between older couples which would increase the probability of delayed fertilization of ova. However, even if this factor does play a role, it cannot lead to the exponential increase with age of Down syndrome and therefore cannot be the only cause.

A possible role of autoimmune disease in the etiology of Down syndrome in the younger maternal age group was suggested by Fialkow (1966). He found that the frequency of thyroid autoantibodies was significantly higher among mothers of patients with Down syndrome (28%) than in controls (14%). This difference became more striking when only younger women were considered (29% vs. 9%).

Environmental factors such as irradiation and infection have also been incriminated in the pathogenesis of trisomy 21. Uchida and Curtis (1961) reported a possible association between maternal exposure to radiation and Down syndrome. However, Schull and Neel (1962) were not able to find an increased incidence of Down syndrome among children of Japanese women exposed to an atomic bomb explosion. A study of Alberman *et al.* (1972) showed that the mothers of Down syndrome children had more total X-rays both in number and dose before conception, especially more than 10 years prior to conception, than the mothers of controls.

A significant correlation between infectious hepatitis epidemics and the increased frequency of births 9 months later of children with Down syndrome was found by Stoller and Collmann (1965) and Kucera (1970), but not by Stark and Fraumeni (1966) and Leck (1966). Evans (1967) suggested that DNA viruses, by increasing the persistence of the nucleolus through meiosis may predispose to nondisjunction.

Parental structural chromosomal abnormalities such as translocation (Turpin and Lejeune, 1961) or inversion (Sparkes *et al.*, 1970) have also been implicated as a predisposing factor for meiotic nondisjunction. We have recently found a reciprocal balanced translocation between 4 and 11 in a young mother who has one child with Down syndrome, one early spontaneous abortion, and one child with multiple abnormalities (Kim *et al.*, 1975).

Genetic control of chromosome nondisjunction is another possibility. This is well known in organisms other than man, such as *Drosophila* (White, 1954). In reviewing the occurrence of double trisomy, Hamerton *et al.* (1965) suggested a genetic or environmental predisposition to nondisjunction in such a double trisomy as 48,XXY,+21, since the incidence of 48,XXY,+21 is too high to be explained as a chance event. A significantly increased frequency of acrocentric association in the parents of Down syndrome was observed by Curtis (1974). This may well be one of the factors for predisposition of nondisjunction of acrocentric chromosomes.

b. Mitotic Nondisjunction. Mitotic nondisjunction is a postzygotic event. Cytologically it is identical to nondisjunction in the second meiotic division, but it is completely different from nondisjunction in the first meiotic division.

Mitotic nondisjunction during early cleavage divisions of a zygote results in chromosomal mosaicism. If mitotic nondisjunction occurs at the very first cleavage division of a normal zygote, mosaicism of 45 and 47 chromosomal constitutions will result. Cells with 45,X may be viable, but cells with autosomal monosomy will be eliminated leaving only cells with 47 chromosomes, i.e., trisomic cells. Mitotic nondisjunction after the first cleavage division results in mosaicism of 45/46/47 chromosomal constitutions. Again, except for 45,X cells, monosomic cells are not viable and mosaicisms of 46/47 with trisomy will persist.

The cause of mitotic nondisjunction is not known, although we know that advanced maternal age is not involved (Penrose and Smith, 1966). Genetic and environmental factors again must be considered. In familial chromosomal mosaicism, a genetically determined factor, either inherited as an autosomal recessive factor (Hsu *et al.*, 1970) or in a dominant fashion (Zellweger and Abbo, 1965), has been suggested, although the support for the dominant hypothesis is tenuous, to say the least. So far there have been no good studies regarding possible environmental factors in mitotic nondisjunction. A preliminary report by Fabricant and Schneider (1975) suggested that maternal age may play a role in causing chromosomal mosaicism in mice.

Environmental and genetic factors may also influence the selective advantage or disadvantage of cells with different chromosome constitutions. The proportion of cell lines with different chromosome constitutions in a mosaic individual may vary from tissue to tissue, or with time. Neu *et al.* (1969) found a disappearance of a trisomy C cell line in the leukocytes from a child with trisomy C mosaicism demonstrated early in life; this was probably due to a selective advantage of the normal cell line. According to Penrose and Smith (1966), trisomic cells appear to occur about twice as frequently in fibroblasts as in leukocytes of mosaic individuals. Taysi *et al.* (1970) noted that prolonged skin fibroblast culture of trisomy 21 mosaic tissue enhances the proportion of trisomic cells. However, stability of trisomic 18 cells in long-term mosaic skin fibroblast culture was demonstrated by our group (Beratis *et al.*, 1972).

2. Anaphase Lag

Anaphase lag is another genetic error occurring during mitotic cell division. One of the paired chromatids lags behind during anaphase and fails to be included in either daughter cell. This produces one daughter cell with a normal complement and one that is monosomic. Anaphase lag is the simplest explanation for 45,X/46,XY or 45,X/46,XX mosaicism. The cause is not really known. It has been speculated that a structural abnormality of any chromosome may predispose to anaphase lag. In a report by Starkman and Jaffe (1967), a structurally abnormal Y chromosome apparently caused 45,X/46,XY mosaicism in both father and son. Of 35 cases with an isochromosome (Xqi) collected by Ferguson-Smith (1965) 23 were mosaic for 45,X/46,XXqi. Anaphase lag, delayed separation of chromatids, and delayed inclusion of chromosome arms were noted in a variety of structurally abnormal chromosomes during anaphase, telophase, and early interphase (Allderdice and Miller, 1967). Recently we found that a nonfluorescent, nonheterochromatic Y chromosome may be predisposed to anaphase lag or mitotic nondisjunction (Hsu *et al.*, 1974a). In a total of eight cases of XO/XY mosaicism studied with Q banding, three had a nonfluorescent Y (reviewed by Hsu *et al.*, 1974a) and in three cases of XO/XY/XYY studied by Caspersson *et al.* (1971b), all three Y chromosomes were nonfluorescent.

3. Formation of Triploidy

Triploidy (69 chromosomes, three times the basic haploid number) can arise either by diandry through fertilization of one egg by two sperm or by digyny through fertilization of both the ovum and an unseparated polar body by one sperm. Aging of the ovum due to delayed fertilization could lead to polyspermy; increased frequency of triploidy has been found in the mouse (Vickers, 1969) and other animals after delayed fertilization (reviewed by Lanman, 1968). There is the suggestion that triploidy is increased among abortuses from women who became pregnant within six months of discontinuing oral contraception (Carr, 1970).

Triploidy has been found mostly in early first-trimester spontaneous abortuses (Carr, 1971). Thirteen fetuses with triploidy without mosaicism are known to have survived to term or near term (Walker *et al.*, 1973). At least seven fetuses with triploid–diploid mosaicism have been reported (Schindler and Mikamo, 1970). Triploid–diploid mosaicism may arise from double fertilization of one ovum and one polar body, one of which is fertilized by two sperm. The other possible mechanism is a triploid zygote with subsequent disturbance of the tripolar spindle during a subsequent division, resulting in cells with triploid, diploid, and haploid constitutions. Since haploid cells are not viable, the result is triploid–diploid mosaicism.

B. Structural Aberrations

Chromosomal structural aberrations result from chromosome breakage and rearrangement. The mechanism of chromosome breakage is not yet completely understood. Various environmental agents such as ionizing radiation, chemicals, and viruses are capable of producing chromosomal breakage (reviewed by Bloom, 1972). Increased frequency of chromosomal breakage and rearrangement have been found in at least four syndromes, namely, Fanconi's anemia, Bloom's syndrome, ataxia-telangiectasia, and xeroderma pigmentosum (reviewed by German, 1972). Homozygous xeroderma pigmentosum fibroblasts cannot repair ultraviolet-induced damage to DNA (Cleaver, 1969), apparently because they are defective in endonuclease-mediated chain breakage, which is most likely the initial step in dimer excision following ultraviolet irradiation and subsequent DNA repair (Setlow *et al.*, 1969). Defective DNA repair has been demonstrated in fibroblasts of a patient with Fanconi's anemia due to faulty excision of DNA lesions, most likely as a result of a specific exonuclease deficiency (Poon *et al.*, 1974). Bloom's syndrome is associated with retarded DNA chain growth (Hand and German, 1975). Further studies on the effects of these mutant genes on the enzymes involved in DNA synthesis, and their relation to malignancy, should be illuminating.

Chromosome breakage appears to occur more often in the heterochromatic regions of the chromosomes, e.g., the centromeres and secondary constrictions, than in other parts of the chromosome (Hirschhorn and Cohen,

1968). With the current banding techniques, break points of structural rearrangements have been demonstrated (Jacobs *et al.*, 1974b; Holmberg and Jonasson, 1973). There appears to be a nonrandom distribution of breakpoints. In reciprocal translocations, there was an excess of breaks in the terminal regions, an excess of terminal/centromeric translocation where ascertainment was through a balanced carrier, and a possible excess of terminal/median translocations where ascertainment was through an unbalanced carrier; the frequency of break points in chromosome arms, with the possible exception of the long arms of chromosome 11, was correlated with their relative lengths.

1. Translocation

There are three types of translocation: reciprocal, centric fusion (Robertsonian), and insertion. Reciprocal translocation is a result of chromosome breaks of two nonhomologous chromosomes or of two homologous chromosomes at different points, with exchange of chromosome segments through attachment of the nonhomologous broken fragments. The two new chromosomes, each containing a centromere, will function normally during mitosis. When two chromosomes break in the centromeric region, a centric fusion (Robertsonian) type of translocation may take place. This type of translocation usually involves two acrocentric chromosomes and retains almost all the genetic material of the long arms of the acrocentric chromosomes. Insertion-type translocation is a result of insertion of one fragment of a chromosome into another chromosome, preceded by two breaks in the donor chromosome and one break in the recipient chromosome.

When a cell with a reciprocal translocation undergoes meiosis, the two translocation chromosomes and their homologs will form a quadrivalent configuration. As a result, 36 different types of segregation products could be formed. The six major forms from three different types of disjunction (adjacent 1, adjacent 2, and alternate) are shown in Figure 5. Four of these products contain a duplication deficiency, only one arrangement contains a normal chromosomal constitution, and one type again results in a balanced reciprocal translocation. There is, however, no evidence that each result has an equal probability; in fact, there is evidence to the contrary. Spontaneous abortions occur frequently during gestation of translocation heterozygotes (Lucas *et al.*, 1972), since most of the meiotic segregation products in these individuals result in abnormal gametes. When the abnormal gamete is fertilized, the genetic imbalance may be so severe that it is incompatible with life and leads to spontaneous abortion or even lack of implantation.

2. Deletion

A deletion can be either interstitial, resulting from two breaks in one chromosome arm with loss of the internal fragment (Figure 6), or caused by a

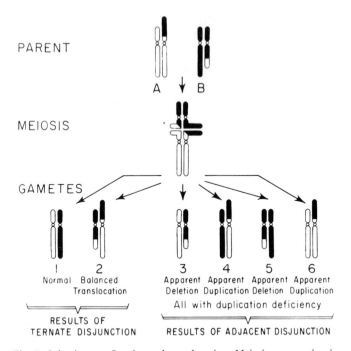

Fig. 5. Inheritance of reciprocal translocation. Meiotic segregation in a balanced A/B translocation carrier illustrating how the six major types of gametes may be produced. (30 other types of gametes which could result from meiotic nondisjunction and crossing-over, followed by meiotic segregation, are not shown here.)

translocation, either inherited from a reciprocal balanced translocation carrier parent (Figure 5) or as a result of a rearrangement during meiosis. In the latter case the exact extent of the deletion may be difficult to define, since the tip of the partially deleted chromosome actually derives from another chromosome. However, the banding techniques are helpful in defining the extent of deletion or duplication of chromosomal material. There are at least five well-known deletion syndromes, namely 4p−, 5p−, 13q−, 18q−, and 22q− (reviewed by Hsu and Hirschhorn, 1972), but there will be many more deletion syndromes identified with the banding techniques now available (see Section IV.-A.8).

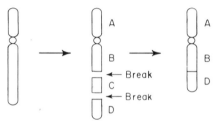

Fig. 6. Mechanism of formation of an interstitial deletion.

Ring chromosomes are formed by the union of two broken ends of a chromosome and thus represent a deletion of the end of both the short and long arms (Figure 7). Ring chromosomes are known to be unstable during mitosis but in some cases can be self-perpetuating; the ring chromosome may be lost during cell division or form dicentric rings or two rings in one cell (Neu and Kajii, 1969). Rings have been reported in chromosomes of various groups (Neu and Kajii, 1969; Polani, 1969).

3. Inversion

An inversion results when there has been a double break in a chromosome, reversal of a segment, and repair, in the inverted sequence. If the inversion includes the centromere, a "pericentric" inversion results. If it is confined to a single arm of the chromosome, it is called "paracentric." Pericentric inversion can usually be detected in mitotic chromosomes if the arm ratio of the chromosome is noticeably altered. But if the break points are at equal distance from the centromere, such pericentric inversions are not recognizable by conventional staining. Only with the current banding techniques are such pericentric inversions, as well as paracentric inversions, diagnosable and accurate identification of the break points possible. Jacobs *et al.* (1974b) found that three of nine inversions analyzed had identical break points in chromosome 8.

4. Isochromosomes

An isochromosome is a chromosome in which the two arms are of equal length and genetically identical. It may result from a misdivision of the centromere (Darlington, 1940) or from a small pericentric inversion followed by a crossover event within the inversion; this leads to a chromosome with essentially identical short and long arms (Nusbacher and Hirschhorn, 1968). Another possibility is a centric fusion translocation between homologs. Autosomal isochromosomes are rare in comparison to the frequently reported X isochromosomes. Since an isochromosome represents duplication of one arm and absence of the other arm, it results in partial trisomy for one arm and partial monosomy for the other arm. It is conceivable that the rare occurrence of autosomal isochromosomes is due to their lethal effect. Isochromosome for the long arm of chromosome 21 is well known and results in 100% of offspring with Down syndrome (Hamerton *et al.*, 1961). An isochromosome of the long arm of chromosome 18 has been described (Neu and Kajii, 1969).

5. Aneusomy by Recombination

In a number of families carrying reciprocal translocations, probands were found to have a translocation morphologically identical to those found in their

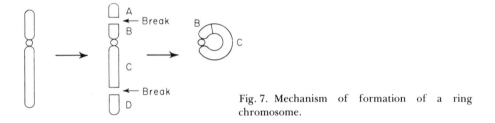

Fig. 7. Mechanism of formation of a ring chromosome.

phenotypically normal parents, yet the probands had multiple congenital abnormalities. This may be explained by ascertainment bias. However, it was suggested by Lejeune and Berger (1965) that a mechanism called "aneusomy by recombination" might explain this phenomenon. Such recombinations are supposedly the result of meiotic crossover within the rearrangement segments (Figure 8). Such a process can produce a morphologically identical yet imbalanced chromosomal complement in the offspring. In fact, one such case derived from a crossover within an inversion was diagnosed by Hirschhorn *et al.* (1973) with the G-banding technique.

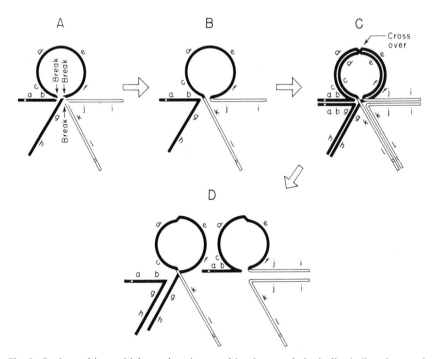

Fig. 8. Reciprocal interstitial translocation resulting in morphologically similar abnormal chromosomes in parent and child, but balanced in parent (normal phenotype) and unbalanced in child (abnormal phenotype). A: Three breaks in two chromosomes; B: balanced interstitial translocation resulting from A; C: crossover in meiosis within translocated segment; D: unbalanced products in gametes. (Redrawn from Lejeune and Berger, 1965.)

IV. CONGENITAL MALFORMATIONS

A. Autosomal Aberrations

Individuals with unbalanced autosomal aberrations generally have multiple congenital abnormalities. Mental retardation and small stature are two rather constant features. Low birth weight and failure to thrive are frequent. Congenital malformations have involved all systems. For each autosomal trisomy, as well as other structural aberrations, there are definite patterns of congenital malformations, yet there is a significant clinical variability within groups of patients with the same cytogenetic abnormality.

1. Trisomy 21 (Down Syndrome, Mongolism)

The characteristics associated with trisomy 21 are flat facial profile, hypotonia, hyperflexibility of joints, upward slanting of palpebral fissures, small anomalous auricles, excess skin in the back of the neck, dysplasia of the pelvis, simian crease, and distinctive dermatoglyphics such as distally located axial triradius ulnar loops on the index fingers, radial loops on the fourth and/ or fifth finger, and arch tibial on the hallucal areas (Hall, 1966; Preus and Fraser, 1972). Mental deficiency is a constant feature; the degree of mental retardation does not correlate with the severity of the physical stigmata (Baumeister and Williams, 1967). Congenital heart disease occurs in about 20% of patients with Down syndrome (Warkany, 1971). Ventricular and atrial septal defects are most common. Atrioventricularis communis and ostium primum occur more often in patients with Down syndrome than in nonmongoloid patients with congenital heart disease (Warkany, 1971). Annular pancreas also appears to occur more often in children with Down syndrome (de Wolff, 1964).

2. Trisomy 18

The abnormalities which appear to be most specific for trisomy 18 include spasticity or hypertonia, micrognathia, small mouth, prominent occiput, short sternum, horseshoe kidney, ectopic pancreatic tissue, small pelvis, flexed fingers with overriding of index fingers on third fingers (Smith, 1970), and increased arch patterns on fingertips with 7 or more arches in more than 80% of patients (Preus and Fraser, 1972). Some 30% die within the first month and 50% by the second month, while 10% may survive to 1 year. Severe mental retardation is especially noticeable in the survivors over 1 year of age. Congenital heart disease occurs in more than 50% of patients. Ventricular septal defect and patent ductus arteriosus are the leading cardiovascular malformations (Smith, 1970). The gastrointestinal system is involved in about 75% of cases, particularly Meckel's diverticulum and malrotations (Warkany, 1971).

3. Trisomy 13

The anomalies most specific for trisomy 13 include absence of the olfactory nerve or bulb, sloping forehead, severe eye defect (especially microphthalmia and colobomata), deafness, cleft lip and/or palate, capillary hemangiomas, postaxial polydactyly and narrow convex fingernails, seizures, and failure to thrive (Smith, 1970). The characteristic dermatoglyphic findings of trisomy 13 include a tibial or fibular loop, or fibular arch, or fibular S pattern in the hallucal area and a distally located axial triradius (t''') in the palm (Preus and Fraser, 1972). The vast majority of infants with trisomy 13 die before 6 months of age. Severe mental deficiency has been noted in the few survivors past 1 year of age.

4. Trisomy 22

With the current banding techniques, a trisomy 22 syndrome has been well established. Of 25 patients, all had growth and mental retardation; 70% or more of patients had microcephaly, micrognathia, typical facies, preauricular skin tags and sinuses, lowset and/or malformed ears, cleft palate, and congenital heart disease including atrial septal defect, patent ductus arteriosus, and ventricular septal defect; 50% had a long philtrum, a long and/or beaked nose, a finger-like or malopposed thumb and deformed lower extremities (Uchida *et al.*, 1968; Hsu *et al.*, 1971; Punnett *et al.*, 1973; Bass *et al.*, 1973; Gustavson *et al.*, 1972; reviewed by Hsu and Hirschhorn, 1977). The natural history of this syndrome is not yet known. Their life-span is apparently longer than that of patients with trisomy 18 or 13.

5. Trisomy 8

At least 20 cases of trisomy 8 have been identified by banding techniques; 11 of these were mosaic (46/47,+8) (Caspersson *et al.*, 1972; Tuncbilek *et al.*, 1974; Walraven *et al.*, 1974; Crandall *et al.*, 1974; Jacobsen *et al.*, 1974, Aller *et al.*, 1975, Fineman *et al.*, 1975; Cassidy *et al.*, 1975). Mental retardation was present in all cases but one, who was a phenotypically normal, but mosaic, adult female. The most common anomalies of trisomy 8 included strabismus, abnormal ears, jaw, and mouth, congenital heart disease, skeletal anomalies, clinodactyly, and abnormal deep skin furrows especially on the soles.

6. Trisomy 9

Three cases of trisomy 9, including two with mosaicism, have been identified with banding techniques (Feingold and Atkins, 1973; Haslam *et al.*, 1973; Bowen *et al.*, 1974). The common features were small palpebral fissures (2/3), sagging or low-set ears (3/3), micrognathia (3/3), undescended testes

(3/3), hypotonia (2/3), anatomic defects in the brain (2/3), and multiple joint dislocations (3/3).

7. Other Trisomies or Partial Trisomies

Partial trisomy of various autosomes has been identified since the development of the banding techniques. Almost all of these were derived from reciprocal translocation carrier parents and were the results of adjacent segregation during parental gametogenesis. A partial list of partial trisomies and their associated congenital malformations is shown in Table 4. It must be emphasized here that these cases with partial trisomy also were partially monosomic for other autosomes as a result of the translocation. Thus, the phenotypic anomalies were caused by chromosomal deficiency as well.

8. Deletion Syndromes

The development of the modern banding techniques has led to a sharp increase in the number of reports of multiple defects associated with deletion of a specific segment of a particular chromosome. As one would expect, the phenotypes described are quite variable, depending presumably on precisely which region is missing; nevertheless a number of recognizable deletion syndromes are beginning to emerge (Lewandowski and Yunis, 1975).

a. The 5p− Deletion Syndrome (Cat-Cry Syndrome). The cat-cry syndrome was first described by Lejeune *et al.* (1963) in a group of patients with a mewlike cry and other anomalies including low birth weight, failure to thrive, hypotonia, psychomotor retardation, microcephaly, hypertelorism, epicanthal folds, antimongoloid slant of eyes, low-set and/or malformed ears, and simian crease (Smith, 1970; Breg, 1969). Among the less-frequent findings are congenital heart disease, cleft lip, cleft palate, preauricular skin tag, and short neck. Dermatoglyphically, there is an increased frequency of bilateral simian creases and of digital whorls (Preus and Fraser, 1972).

b. The 4p− Deletion Syndrome. The common abnormalities found in more than 50% of patients with 4p− were low birth weight, growth and mental retardation, microcephaly, seizure disorder, prominent glabella or midline scalp defect, hypertelorism, broad misshapen nose, low-set ears with little cartilage, cleft lip or palate, strabismus, iris defects, micrognathia, cranial asymmetry, and hypospadias in males (Wolf *et al.*, 1965; Hirschhorn *et al.*, 1965; Wilson *et al.*, 1970; Arias *et al.*, 1970). Dermatoglyphically, there are frequent simian creases, dermal ridge dysplasia, and an excess of arches. Survival has varied from 24 hr to late childhood.

c. The 18q− Deletion Syndrome. The phenotypic manifestations of the 18q− syndrome are mainly midfacial retraction or hypoplasia, prominent antihelix, high incidence of fingertip whorl pattern, and severe mental deficiency (de Grouchy *et al.*, 1964; Lejune *et al.*, 1966). The other frequent

Table 4. Partial Trisomy and Its Associated Abnormalities

Type of partial trisomy	Number of patients	Congenital malformations and clinical features	References (*=review)
1q	2	1 had small eyes and jaw, malformed thorax, persistent arteriosus, extensive atrial septal defect;	Van den Berghe et al., 1973
		1 had downward slanting of eyes, cleft lip and palate, receding chin, absence of gall bladder, microgyria, thick corpus callosum.	Norwood and Hoehn, 1974
3p	4 (3 were sibs)	4/4 had square-shaped facies with hypertelorism and macrostomia; 2/4 had congenital heart disease, 1/4 had esophageal atresia, cleft palate, mesenterium commune.	Rethore et al., 1972; Ballesta and Vehi, 1974
4p	8	Mental retardation, short stature, microcephaly, low hair line, folded helix, stubby nose, elongated chin, receding jaw, short neck, heart defect, skeletal abnormalities	Schwanitz and Grosse, 1973*; Owen et al., 1974; Furbetta et al., 1974*
4q	3	Mental retardation, renal abnormalities, ambiguous external genitalia or undescended testes, low birth weight, hypotonia	Schrott et al., 1974*
7q	6	Asymmetrical skull, small palpebral fissures, malformed and low-set ears, cleft palate, large tongue, micrognathia, short neck, hyperconvex nails	Vogel et al., 1973*; Alfi et al., 1974
8q	4	Mental retardation, craniofacial dysmorphy	Lejeune et al., 1972; Fryns et al., 1974a
9p	13	Mental retardation, microcephaly, enophthalamos, antimongoloid slanting of eyes, protruding ears, globular nose, downward slanting of mouth, hypoplasia of phalanges, simian lines and partial webbing of toes and fingers	Rethore et al., 1973*; Podruch and Weisskopf, 1974; Baccichetti and Tenconi, 1974
10p	3	Severe mental and psychomotor retardation, dysmorphic skull, peculiarly shaped face, epicanthal folds, broad nasal bridge, hypertelorism, and multiple malformations of the extremities	Schleiermacher et al., 1974; Hustinx et al., 1974

10q	7	Mental retardation, abnormal muscle tone, microcephaly, hypertelorism, microphthalmia, blepharophimosis, ptosis of eyelids, low-set and/or malformed ears, small or depressed nasal bridge, micrognathia, flat and round face, webbing of fingers and toes	de Grouchy et al., 1972; Yunis and Sanchez, 1974
11p	2	Mental retardation, prominent frontal bossing, nystagmus, antimongoloid slant, strabismus, arched or cleft palate, broad fingers and toes	Sanchez et al., 1974*
11q	5	Cranial deformities, agenesis of corpus callosum, neural tube defects, micrognathia, congenital heart disease, biliary system defects, agenesis of kidney	Tusques et al., 1972; Wright et al., 1974*
12p	2	Mental retardation, hypotonia, epicanthus, Brushfield spots, flat nasal bridge, nystagmus, flat facies, flat occiput, microcephaly, short neck with abundant skin, simian lines, gap between first and second toe	Uchida and Lin, 1973*; Fryns et al., 1974b
12q	1	Mental and growth retardation, square shaped head, bridged nose, low-set and poorly lobulated ears, simian lines	Hobolth et al., 1974
14q−	3	Mental retardation, microcephaly, hypertelorism, microphthalmia, congenital glaucoma, low-set ears, deafness, congenital heart disease	Pfeiffer et al., 1973; Muldal et al., 1973; Short et al., 1973
15q−	2	Mental retardation and other nonspecific minimal phenotypic changes	Crandall et al., 1973
17p	1	Low birth weight, small placenta, failure to thrive, hypertonia, microcephaly, ptosis, low-set and malformed ears, micrognathia, extra skin at nape, congenital heart disease, contractures of big joints, flexion anomaly of fingers, short dorsiflexed hallux	Latta and Hoo, 1974
20p	1	Mild mental retardation, epicanthal folds, gothic palate, hypoplasia of the maxillary region	Subrt and Brychnac, 1974

features are small stature, microcephaly, hypotonia, poor coordination, nystagmus, atretic ear canals, ocular funduscopic anomalies, carp-shaped mouth, vertical tali, long tapering fingers, abnormal toe placement, skin dimples over the acromion and/or knuckles, and congenital heart disease.

d. The 13q− Deletion Syndrome. The 13q− deletion syndrome was suggested by Allderdice *et al.* (1969). Characteristic features are facial dysmorphy, microcephaly, trigonocephaly, broad prominent nasal bridge, hypertelorism, microphthalmos, epicanthal fold, large prominent low-set ears, facial asymmetry, imperforate anus or perineal fistula, hypoplastic or absent thumbs, and psychomotor retardation.

e. The Gq− Deletion Syndromes. In the review of the cases of Gq deletion, either ring G or Gq−, Warren and Rimoin (1970) proposed that these syndromes be divided into two groups: "G deletion syndrome I or antimongolism" and "G deletion syndrome II." The constant features confined to G deletion type I were down-slanting (antimongoloid) of palpebral fissures, prominent nasal bridge, large external ears, a somewhat elongated skull, and hypertonia. The features confined to G deletion type II include hypotonia, epicanthal folds, syndactyly, ptosis, bifid uvula, and clinodactyly. Both syndromes shared a few common features: mental retardation, microcephaly, high-arched palate, and large or low-set ears. With the use of banding techniques, it is now clear that G deletion syndrome I is 21q− and G deletion syndrome II is 22q− (Richmond *et al.*, 1973; Stoll *et al.*, 1973).

f. Other Deletion Syndromes. With the banding techniques, a ring chromosome 6 has been identified in two cases (Moore *et al.*, 1973; Fried *et al.*, 1975). Both patients showed microcephaly, depressed nasal bridge, and mental retardation; one had microphthalmia, micrognathia, and microstomia. 46,9p− was diagnosed in two infants with trigonocephaly, wide flat nasal bridge, anteverted nostrils, long upper lip, short neck, and muscle hypertonia (Alfi *et al.*, 1973). 46,10p− has been detected in two children (Francke *et al.*, 1975; Shokeir *et al.*, 1975). One, a 5-year-old girl, showed mental and growth retardation, abnormally shaped head, antimongoloid slant of eyes, hypotelorism, epicanthal folds, ptosis, strabismus, dysplastic nose, microdontia, small low-set posteriorly rotated ears, and asymmetric thorax (Francke *et al.*, 1975); the second, a newborn boy, had cleft lip and palate, preauricular pits, low-set malpositioned ears, antimongoloid slant of eyes, microcephaly, micrognathia, congenital heart disease, and cryptorchidism (Shokier *et al.*, 1975). 46,12p− was found in a severely retarded man with short stature, microcephaly, antimongoloid slant of eyes, big ears, short and webbed neck, short arms, hands, and fingers, and broad thumbs (Magnelli and Therman, 1975).

B. Sex Chromosome Aberrations

The relatively high incidence of sex chromosome abnormalities in comparison to autosomal abnormalities (Table 1) can be explained by the less lethal effect of sex chromosome aberrations than that of autosomes. Most zygotes with unbalanced autosomal aberrations die *in utero* or during infancy, while most individuals with multiple X or Y chromosomes survive to adulthood, often without much physical disability. This difference has been explained by the Lyon hypothesis of genetic inactivation of any X chromosome material in excess of one, and the relatively less important genetic role of the Y chromosome than that of the X chromosome, except in male determination.

1. Turner Syndrome

This syndrome is a disorder of phenotypic females characterized by three cardinal features, short stature, streak gonads, sexual infantilism, and associated congenital abnormalities, so-called Turner stigmata: shield chest, webbed neck, lymphedema, cubitus valgus, short 4th metacarpals, hypoplastic nails, pigmented nevi, and congenital heart disease, usually coarctation of the aorta. Mental retardation is mild, if present at all. Approximately 60% of the patients with Turner syndrome are found to have 45 chromosomes with only one sex chromosome, an X (45,X) while almost 40% of the patients are 45,X mosaics with various different sex chromosome constitutions in the second or even a third cell line (Ferguson-Smith, 1965). Diversity of clinical features has been noted in the mosaic individuals. Presumably the expression of clinical manifestations is modified by the presence of the second cell line and, perhaps, the proportion of the different cell lines in the gonads. The most common mosaic pattern is 45,X/46,XX. Individuals with this type of mosaicism are generally short, but 20% of them may have spontaneous menstruation.

In addition to the absence of the second sex chromosome, structural abnormalities of the second X chromosome with short arm deletion (Xp−) or isochromosome formation of the long arm (Xqi) have also been found in individuals with typical Turner syndrome (Ferguson-Smith, 1965).

2. Klinefelter Syndrome

The clinical abnormalities associated with this syndrome are aspermia, gynecomastia, small testes, and elevated urinary gonadotropins. In addition, the patients may have eunuchoid habitus and decreased facial, sexual, and body hair. Their testicular biopsies show hyalinized and hypoplastic tubules with clumped Leydig cells (Court Brown, 1967). The majority of patients have a 47,XXY chromosome constitution. Various additional sex chromosome aberrations with multiple X's and Y's with and without mosaicism have also been identified (Court Brown, 1967). In general, the intelligence quotient of an XXY individual is low or average, with a greater-than-expected frequency of borderline and defective intelligence (Money, 1964). When the number of X chromosomes increases, such as in XXXY and XXXXY, the mental retardation becomes more severe. XXY individuals also appear to be more vulnerable to mental illness such as schizophrenia (Court Brown, 1967).

3. XXXXY Syndrome

Individuals with a 49,XXXXY chromosomal constitution have a number of distinctive features distinguishing them from those with the 47,XXY Klinefelter syndrome, and they are usually severely retarded (Smith, 1970; Zaleski et al., 1966). In addition to hypogonadism, they frequently have hypogenitalism, with small penis, hypoplastic scrotum, and cryptorchidism. A typical finding is radioulnar synostosis. Other frequently associated features are mongoloid slanting of the eyes, epicanthal folds, hypertelorism, strabismus, wide flat nose, abnormal ears, mandibular prognathism, short neck, and short stature. Dermatoglyphically there is a high frequency of low arch patterns, the finger ridge counts are low (mean of 50) (Penrose, 1967), and there is a high frequency of clinodactyly (Smith, 1970).

4. Females with Multiple X's

From a review of 155 females with XXX (Barr et al., 1969), it appears that the majority of these individuals are physically normal. The diversity of physical defects found in a minority speaks against an XXX syndrome in a clinical sense. The presence of three X chromosomes may predispose the individual to mental retardation or mental illness. However, the magnitude of this risk is not known because of the possible bias of ascertainment. Apparently, XXX is compatible with a normal reproductive system; 70% of postpubertal women with XXX had a normal menstrual history and 28 women bore 67 children, all chromosomally normal.

Individuals with XXXX or XXXXX are generally mentally and physically more retarded than XXX individuals. In addition, many skeletal malformations which have been found in XXXXY males, such as radioulnar synostosis, clinodactyly, coxa valga, have been found in XXXXX females (Sergovich et al., 1971). Poly-X females also show lower finger ridge counts as the number of X chromosomes increases (Penrose, 1967).

5. XYY and XXYY Males

Sandberg et al. (1961) first described the 47,XYY sex chromosome complement in an apparently normal male. It was not until the work of Jacobs et al. in 1965 that the most important features were recognized, i.e., tall stature and possible antisocial behavior (Hook, 1973). It is now clear that the Y chromosome affects linear growth, probably to a greater extent than the X.

On the average, XYY males are taller than XY males who are taller than XX females. However, it is not completely clear whether XYY males are indeed more aggressive or have greater criminal tendencies than XY males, although the incidence of XYY appears to be increased in several studies of certain male criminal groups (The National Institute of Mental Health, Report on the XYY chromosomal abnormality, 1970; Hook and Kim, 1970). The intelligence levels in patients with XYY may vary from low to above average (Leff and Scott, 1968). A recent study by Witkin *et al.* (1976) indicates that the elevated crime rate of XYY males is not related to aggression but is probably due to poor judgment associated with low intelligence. XXYY males have similar physical features to XYY males (Court Brown, 1967), but they also have Klinefelter syndrome.

6. Sex Chromosome Mosaicism with Y Chromosome and a 45,X Cell Line

Mosaicism of 46,XY and 45,X is one of the most common sex chromosome abnormalities. In a total of 88 cases of XY/XO mosaicism reviewed by Hsu *et al.* (1974a), the phenotypic manifestations vary from phenotypic females with clinical Turner syndrome (25%), to "mixed gonadal dysgenesis" (ambiguous external genitalia and asymmetrical abdominal gonads (60%), to phenotypic males with hypospadias and abnormal testes (13%). It is important to point out that the incidence of gonadal tumor in XY/XO mosaicism is rather high. In 88 cases reviewed, intra-abdominal gonadal tumors were reported in 7. It is essential to remove the intra-abdominal dysgenic gonad containing testicular elements that may have a malignant potential.

In 10 cases of XYY/XY/XO mosaicism reviewed by Hsu and Hirschhorn (1971), the phenotypic manifestations vary from phenotypic females with primary amenorrhea, absence of secondary sex characteristics, and increased stature (in 2), to intersex with ambiguous genitalia (in 5), to phenotypic males with abnormal testes, hypospadias, and small penis (in 3).

7. True Hermaphroditism

By definition, true hermaphrodites have gonadal tissue of both sexes. The ductal and external genital differentiation of these patients varies from almost normal female to male with some degree of hypospadias (Butler *et al.*, 1969). Of 70 cases reviewed by Hsu and Hirschhorn (1971), 38 cases were found to have a normal female karyotype, 46,XX; 10 had XX/XY mosaicism which most likely represents chimerism due to double fertilization; 9 had a normal male karyotype 46,XY; and the others consisted of various different types of sex chromosome mosaicism. Since the gonadal tissues of most true hermaphrodites were not studied cytogenetically, it is impossible to correlate the phenotypic manifestations and the cytogenetic findings.

C. Syndromes Related to Chromosome Breakage

Increased incidence of chromosome breakage and rearrangement has been described in four, recessively inherited, syndromes: Fanconi's anemia, Bloom's syndrome, ataxia–telangiectasia, and xeroderma pigmentosum (see Section III.B). The syndrome of Fanconi's anemia consists of hypoplasia to aplasia of the thumb, pancytopenia, short stature, small cranium, hyperpigmentation, renal defects, hypogonadism, and a variety of other malformations (Smith, 1970). Bloom's syndrome consists of telangiectatic erythema appearing in infancy and affecting mainly the face, sensitivity to sunlight, short stature, and malar hypoplasia (Smith, 1970). Ataxia–telangiectasia consists of progressive cerebellar ataxia, oculocutaneous telangiectasia beginning on the bulbar conjunction at age 3–6 years, evidence of immunological disorder, and frequent otic and sinorespiratory tract infections (Smith, 1970). The immunological abnormalities include a decrease in peripheral lymphoid tissue, lymphopenia, absence or decreased amounts of IgA, deficiency of delayed hypersensitivity, impaired rejection of skin grafts, and sluggish circulating antibody response to some antigens (Eisen *et al.*, 1965). Fanconi's anemia, Bloom's syndrome, and ataxia–telangiectasia are associated with high incidence of leukemia or other neoplasms. The syndrome of xeroderma pigmentosum consists of early photosensitivity, photophobia to the ultraviolet range of light, development of premalignant and malignant skin lesions in areas exposed to sunlight, development of keratitis and scarring of the cornea with the malignant tumors of the conjunctiva and eyelids, hyper- and hypopigmentation of skin (Haddida *et al.*, 1963).

D. Chromosome-Breaking Agents and Teratogenesis

It has been well established that there are three major classes of exogenous agents capable of producing chromosomal damage: ionizing radiation, certain chemicals, and certain viruses (reviewed by Bloom, 1972). When a human being is given a large enough dose of ionizing radiation, an increased frequency of chromosomal aberrations will be detectable in the peripheral lymphocyte culture immediately after the exposure. The chromosome aberrations consist of two different types: unstable aberrations (including acentric fragments, ring chromosomes, and dicentric chromosomes) and stable aberrations (including reciprocal translocations, inversions, and partial deletions). The cells with unstable chromosomal aberrations quickly decrease and may finally disappear within 3–5 years, but the stable aberrations will usually persist for over 20 years (Buckton *et al.*, 1972). The chemicals which have been found to cause chromosome damage include antibiotics such as streptonigrin, actinomycin, and mitomycin; alkylating agents such as nitrogen mustard, myleran, and cytoxan; base analogues; and other miscellaneous groups (Hirschhorn and Cohen, 1968; Bloom, 1972). These chemicals can be generally subdivided into those affecting DNA and RNA synthesis and those affecting DNA precursors and related substances. The viruses which have been found to induce chromosome damage are oncogenic viruses such as SV40 and Rous sarcoma virus, adenoviruses, and other common viruses such as rubella, measles, varicella, mumps, herpes simplex type 1 and 2, etc. (Nichols, 1966; Nusbacher *et al.*, 1967, Bloom, 1972).

It has long been recognized that ionizing radiation can produce all manner of developmental defects in the human embryo if it is exposed to high enough doses during the crucial period of organogenesis. This subject is considered in detail in Chapter 5, Vol. 1.

The congenital rubella syndrome is another example of the teratogenic effect of an agent that also causes chromosome damage. The major malformations seen in rubella syndrome are cataract, cardiac defect, deafness, and psychomotor retardation. Plotkin *et al.* (1965) demonstrated that the rubella virus can induce chromosome breaks *in vitro.* Subsequently, increased chromosome breaks were also found in children with congenital rubella syndrome as compared to normal controls (Nusbacher *et al.*, 1967). The teratogenic effect of this virus could thus be explained by its action upon genetic material at a critical stage in embryonic development. It is possible that chromosomal damage is responsible for the congenital anomalies because of cell loss during rapid organ development in fetal life, although a causal relationship between rubella-induced teratogenicity and chromosome breakage has yet to be demonstrated.

V. PRENATAL DIAGNOSIS

The usefulness of amniocentesis in prenatal diagnosis was first demonstrated by Serr *et al.* (1955), who diagnosed the sex of unborn fetuses by sex chromatin determination of amniotic fluid cells. Since 1966, the technique of culturing amniotic fluid cells has been developed and improved, and the cultured cells have been used for both cytogenetic and biochemical studies (Nadler, 1972; Milunsky, 1973; Hsu and Hirschhorn, 1974). Amniotic fluid cells have been shown to be derived from the amnion and the fetus, including fetal urine, skin, and buccal and vaginal mucosa; these cells are of two types: polygonal epithelial cells and smaller round cells. Within several days to two weeks, the epithelial cells gradually disappear and the cells resembling fibroblasts predominate. It is the fibroblast-like cells which are generally used for cytogenetic and biochemical studies.

The technique of transabdominal amniocentesis has gained widespread acceptance since 1965. The chances of a successful tap and cell culture are best after 13 weeks of gestation. The risk to the mother or baby appears to be small.

It would seem ideal to do prenatal cytogenetic diagnosis on all pregnancies so as to eliminate all chromosomally abnormal fetuses. However, limitation in manpower and economic resources, as well as lack of data as to just how low the risk to mother and fetus may be, has restricted such study to pregnancies with an increased risk of having a child with a chromosomal or detectable metabolic abnormality. These include parents carrying translocations, parents with chromosomal mosaicism, mothers of advanced maternal age (over 35–40 years), and mothers with a history of having a previous child with Down syndrome or other chromosome abnormalities. In addition, prenatal diagnosis of the sex of the fetus is indicated in a mother who is likely to be a carrier of an X-linked disorder not diagnosable prenatally since there is a 50% chance that a male fetus will be affected. Parents who have been exposed to radiation or radiomimetic drugs prior to or (in the case of the mother) during early pregnancy present a difficult problem since, if an increase in chromosome breaks were demonstrated, one would not be able to say what this meant in terms of risk to the fetus.

A total of 1368 cases for prenatal cytogenetic diagnosis was collated by Milunsky (1973) from centers in the U.S.A. and Canada. In 485 cases studied because of a previous child with Down syndrome, there was a recurrence rate of about 1%. In the group of advanced maternal age, 347 mothers over 40 years of age produced 9 fetuses with abnormal chromosomes (i.e., affected frequency of 2.89%); from 255 mothers between 35 and 40 years, 4 fetuses were found to have abnormal chromosomes, i.e., 1.56%. About 20% unbalanced karyotypes were found in pregnancies from translocation-carrier parents (93 cases). Another 188 cases were studied for other reasons and only 1 was found to have abnormal chromosomes.

From 1969 to 1974, we performed 550 successful prenatal cytogenetic diagnoses. The indications, the number of cases, and the results (therapeutic abortion is indicated by an asterisk) are: (1) advanced maternal age: 305 cases (40 years and over: 169 cases, 7* affected; 4–47,+21; 2–47,+18; 1–47,XXY; 35–39 years: 136 cases, 8* affected; 4–47,+21; 1–47,+18; 1–48,XYY,+21; 1–45,X/46,XX; 1–45,X/46,XY); (2) previous child with Down syndrome: 127 cases, 2 recurrences of 47,+21*; 1 case of 46/47,+20*; (3) previous child with multiple anomalies: 20 cases (all normal); (4) translocation carrier: 7 cases (6 normal, 1 carrier); (5) previous child with trisomy 18 or 13: 8 cases (all normal); (6) parental mosaicism: 4 cases (all normal); (7) X-linked disorders: 11 cases (8 cases of 46,XY, 3 cases of 46,XX), 7 cases XY*; and (8) others: 68 cases (67– normal; 1–46,13p−) (Hsu et al., in press). Thus our results also show about 1% recurrence rate in the group with a previous child with Down syndrome and 3–4% affected in the advanced maternal age group (40 years and over).

From our experience, we recommend separate harvests from different initial tissue culture flasks be used for accurate diagnosis of true mosaicism. In cases with enlarged satellites, cells in late prophase or early metaphase must be used for analysis in order to eliminate confusion with translocation. The enlarged satellites can be further identified by the quinacrine fluorescent staining technique. As to prenatal sex determination, karyotype analysis of the cultured amniotic fluid cells is the only accurate means, since a small Y may fail to fluoresce and bright fluorescent spots may be produced by certain areas of several autosomes (the centromeric region of chromosome 3, and the short arms or satellites of chromosomes 13, 21, or 22). Furthermore, most of the uncultured amniotic fluid cells are pyknotic and not suitable for X-chromatin determination. One must keep proper control of the culture conditions, including pH, and test for mycoplasma contamination, since karyotypes of cultured amniotic fluid cells may be altered by such factors.

VI. CONCLUDING REMARKS

The study of human cytogenetics has provided a great deal of information as to the relationship between chromosomal aberrations and congenital malformations. Yet we still do not understand the effects of abnormal chromosomal constitutions on cellular metabolism. Nor do we know the nature or mechanism of the teratological effects of the various different chromosome abnormalities during embryonic differentiation and fetal development.

The two recent major advances in human genetics, i.e., the development of prenatal diagnosis and discovery of the banding techniques for accurate chromosome identification, have brought new dimensions to medical genetics and genetic counseling. Prenatal diagnosis can prevent the birth of children with abnormal chromosomal constitutions, and the banding techniques enable us to diagnose various previously undetectable chromosomal abnormalities.

While the field of human cytogenetics has been highly productive for the past two decades, new advances in this field should lead to further exciting achievements and, hopefully, understanding of the relationship of chromosomal constitution to teratologic abnormality.

ACKNOWLEDGMENTS

Supported by Genetics Center grant GM-19443 and U.S. Public Health Service grant HD 02552. The authors are Career Scientists of the Health Research Council of the City of New York (LYFH: I-761; KH: I-513).

We wish to thank Ms. Minnie Woodson for her help in preparation of the manuscript, and Mrs. Felice Yahr, Drs. Judy Willner, and Karen David for proofreading.

REFERENCES

Alberman, E., Polani, P. E., Fraser-Roberts, J. A., Spicer, C. C., Elliott, M., and Armstrong, E., 1972, Parental exposure to X-irradiation and Down's syndrome, *Ann. Hum. Genet.* **36:**195.

Alfi, O., Donnell, G. N., Crandall, B. F., Derencsenyi, A., and Menon, R., 1973, Deletion of the short arm of chromosome #9 (46,9p−): A new deletion syndrome *Ann. Genet.* **16:**17.

Alfi, O. S., Donnell, G. N., and Kramer, S. L., 1974, Partial trisomy of the long arm of chromosome No. 7, *J. Med. Genet.* **10:**187.

Allderdice, P. W., and Miller, O. J., 1967, Chromosome behavior in human leukocyte cultures not exposed to colchicine or hypotonic solution, 6th Conference on Mammalian Cytology and Somatic Cell Genetics, Abstract, November 5, 1967, Asilomar, California.

Allderdice, P. W., Davis, J. G., Miller, O. J., Klinger, H. P., Warburton, D., Miller, D. A., Allen, F. H., Jr., Abrams, C. A. L., and McGilvray, E., 1969, The 13q− deletion syndrome, *Am. J. Hum. Genet.* **21:**499.

Aller, V., Abrisqueta, J. A., Perez, A., Martin, M. A., Goday, C., and Del Mazo, J., 1975, A case of trisomy 8 mosaicism 47,XX,+8/46,XX, *Clin. Genet.* **7:**232.

Arias, D., Passarge, E., Engle, M. A., and German, J., 1970, Human chromosomal deletion: Two patients with the 4p− syndrome, *J. Pediatr.* **76:**82.

Arakaki, D. T., and Waxman, S. H., 1970, Chromosome abnormalities in early spontaneous abortions, *J. Med. Genet.* **7:**118.

Arrighi, F. E., and Hsu, T. C., 1971, Localization of heterochromatin in human chromosomes, *Cytogenetics* **10:**81.

Baccichetti, C., and Tenconi, R., 1974, A new case of trisomy for the short arm of #9 chromosome, *J. Med. Genet.* **10:**296.

Ballesta, F., and Vehi, L., 1974, Partial trisomy of distal end of 3p, *Ann. Genet.* **17:**290.

Barr, M. L., Sergovich, F. R., Carr, D. H., and Shaver, E. L., 1969, The triplo-X female: An appraisal based on a study of 12 cases and a review of the literature, *Can. Med. Assoc. J.,* **101:**247.

Bass, H. N., Crandall, B. F., and Sparkes, R. S., 1973, Probable trisomy 22 identified by fluorescent and trypsin-Giemsa banding, *Ann. Genet.* **16:**189.

Baumeister, A. A., and Williams, J., 1967. Relationship of physical stigmata to intellectual functioning in mongolism, *Am. J. Ment. Defic.* **71:**586.

Beratis, N. G., Hsu, L. Y. F., Kutinsky, E., and Hirschhorn, K., 1972, Stability of trisomic (47,18+) cells in long-term mosaic skin fibroblast culture, *Can. J. Genet. Cytol.* **14:**869.

van den Berghe, H., van Eygen, M., Fryns, J. P., Tanghe, W., and Verresen, H., 1973, Partial trisomy 1, karotype 46,XY, 12−, t(1q,12p)+, *Humangenetik* **18:**225.

Bloom, A. D., 1972, Induced chromosomal aberrations in man, *Adv. Hum. Genet.* **2:**99.

Bobrow, M., and Madan, K., 1973, The effect of various banding procedures on human chromosomes, studied with acridine orange, *Cytogenet. Cell Genet.* **12:**145.

Bobrow, M., Madan, K., and Pearson, P. L., 1972, Staining of some specific regions of human chromosomes, particularly the secondary constriction of No. 9, *Nature (London), New Biol.* **238:**122.

Boué, J., Boué, A., and Lazar, P., 1975, Retrospective and prospective epidemiological studies of 1500 karyotyped spontaneous human abortions, *Teratology* **12:**11.

Bowen, P., Ying, K. L., and Chung, G. S. H., 1974, Trisomy 9 mosaicism in a newborn infant with multiple malformations, *J. Pediatr.* **85:**95.

Breg, W. R., 1969, Cri du chat syndrome, *in: Endocrine and Genetic Diseases of childhood* (L. I. Gardner, ed.), pp. 632–638, W. B. Saunders, Philadelphia, Pa.

Buckton, K. E., Jacobs, P. A., Court-Brown, W. M., and Doll, R., 1972, A study of the chromosome damage persisting after X-ray therapy for ankylosing spondylitis, *Lancet* **2:**676.

Butler, L. J., Snodgrass, G. J. A. I., France, N. E., Russell, A., and Swain, V. A. J., 1969, True hermaphroditism or gonadal intersexuality, *Arch. Dis. Child.* **44:**666.

Carr, D. H., 1970, Chromosome studies in selected spontaneous abortions: 1. Conception after oral contraceptives, *Can. Med. Assoc. J.* **103:**343.

Carr, D. H., 1971, Cytogenetics and malformation in abortions, *Fed. Proc.* **30:**102.

Caspersson, T., Farber, S., Foley, G. E., Kudynowski, J., Modest, E. J., Simonsson, E., Wahg, U., and Zech, L., 1968, Chemical differentiation along metaphase chromosomes, *Exp. Cell Res.* **49:**219.

Caspersson, T., Zech, L., and Johansson, C., 1970a, Analysis of human metaphase chromosome set by aid of DNA-binding fluorescent agents, *Exp. Cell Res.* **62:**490.

Caspersson, T., Zech, L., Johansson, C., and Modest, E. J., 1970b, Identification of human chromosomes by DNA-binding fluorescent agents, *Chromosoma* **30:**215.

Caspersson, T., Zech, L., Johansson, C., Lindsten, J., and Hulten, M., 1970c, Fluorescent staining of heteropycrotic chromosome regions in human interphase nuclei, *Exp. Cell Res.* **61:**472.

Caspersson, T., Lomakka, G., and Zech, L., 1971a, The 24 fluorescence patterns of the human metaphase chromosome distinguishing characters and variability, *Hereditas* **67:**89.

Caspersson, T., Hulten, M., Johasson, J., Lindsten, J., Thekelsen, A., and Zech, L., 1971b, Translocations causing non-fluorescent Y chromosomes in human XO/XY mosaic, *Heriditas* **68:**317.

Caspersson, T., Lindsten, J., Zech, L., Buckton, K. E., Price, W. H., 1972, Four patients with trisomy 8 identified by the fluorescence and Giemsa banding techniques, *J. Med. Genet.* **9:**1.

Cassidy, S. B., McGee, B. J., van Eys, J., Nance, W. E., and Engel, E., 1975, Trisomy 8 syndrome, *Ped.* **56:**826.

Chicago Conference, 1966, Standardization in human cytogenetics, Birth defects, Original Article Series, Vol. II, No. 2. The National Foundation-March of Dimes.

Cleaver, J. E., 1969, Xeroderma pigmentosum: A human disease in which an initial stage of DNA repair is defective, *Proc. Natl. Acad. Sci. U.S.A.* **63:**428.

Cohen, M. M., and Hirschhorn, R., 1971, Lysosomal and non-lysosomal factors in chemically induced chromosome breakage, *Exp. Cell Res.* **64:**209.

Comings, D. E., Avelino, E., Okada, T. A., and Wyandt, H. E., 1973, The mechanism of C and G-banding chromosomes, *Exp. Cell Res.* **77:**469.

Court Brown, W. M., 1967, Human population cytogenetics, North-Holland Research Monographs, *Frontiers of Biology,* Vol. 5 (A. Neuberger and E. L. Tatum, eds.), North-Holland Publishing Company, Amsterdam.

Crandall, B. F., Muller, H. M., and Bass, H. N., 1973, Partial trisomy of chromosome number 15 identified by trypsin-Giemsa banding, *Am. J. Ment. Defic.***77:**571.

Crandall, B. F., Bass, H. N., Marcy, S. M., Glovsky, M., and Fish, C. H., 1974, The trisomy 8 syndrome: Two additional mosaic cases, *J. Med. Genet.* **11:**393.

Curtis, D. J., 1974, Acrocentric associations in mongol populations, *Humangenetik* **22:**17.

Darlington, C. D., 1940, The origin of iso-chromosomes, *J. Genet.* **39:**351.

Dev, V. G., Warburton, D., Miller, O. J., Miller, D. A., and Erlanger, B. F., 1972, Consistent pattern of binding of anti-adenosine antibodies to human metaphase chromosomes, *Exp. Cell Res.* **74:**288.

Dhadial, R. K., Machin, A. M., and Tait, S. M., 1970, Chromosomal anomalies in spontaneously aborted human fetuses, *Lancet* **2:**20.

Drets, M. E., and Shaw, M. W., 1971, Specific banding patterns of human chromosomes, *Proc. Natl. Acad. Sci. U.S.A.* **68:**2073.

Dutrillaux, B., 1973, Nouveau système de marquage chromosomique: Les bandes T, *Chromosoma* **41:**395.

Dutrillaux, B., and Lejeune, J., 1971, Sur une nouvelle technique d'analyse du caryotype humain, *C.R. Acad. Sci.* **272:**2638.

Dutrillaux, B., Grouchy, J. de, Finaz, C., and Lejeune, J., 1971, Mise en évidence de la structure fine des chromosomes humains par digestion enzymatique (pronase en particulier), *C.R. Acad. Sci.* **273:**587.

Dutrillaux, B., Jonasson, J., Lauren, K., Lejeune, J., Lindsten, J., Petersen, G. B., and Saldana-Garcia, P., 1973, An unbalanced *4q/21q* translocation identified by the R but not the G and Q chromosome banding techniques, *Ann. Genet.* **16:**11.

Eisen, A. H., Karpati, G., Laszlo, T., Andermann, F., Robb, J. P., and Bacal, H. L., 1965, Immunologic deficiency in ataxia telangiectasia, *N. Engl. J. Med.* **272:**18.

Ellison, J. R., and Barr, H. J., 1972, Quinacrine fluorescence of specific chromosome regions. Late replication and high A:T content in *Samoaia leonensis*, *Chromosoma* **36:**375.

Evans, H. J., 1967, The nucleolus, virus infection and trisomy in man, *Nature* **214:**361.

Fabricant, J. D., and Schneider, E. L., 1975, Maternal age effect: A mouse model, Proceeding of 10th International Congress of Gerontology, June, 1975, Jerusalem.

Feingold, M., and Atkins, L., 1973, A case of trisomy 9, *J. Med. Genet.* **10**:184.

Ferguson-Smith, M. A., 1965, Karyotype-phenotype correlations in gonadal dysgenesis and their bearing on the pathogenesis of malformations, *J. Med. Genet.* **2**:142.

Fialkow, P. J., 1966, Autoimmunity and chromosomal aberrations, *Am. J. Hum. Genet.* **18**:93.

Finaz, C., and de Grouchy, J., 1971, Le caryotype humain après traitement par l'α-chymotrypsine, *Ann Genet.* **14**:309.

Fineman, R. M., Ablow, R. C., Howard, R. O., Albright, J., and Breg, W. R., 1975, Trisomy 8 mosaicism syndrome, *Ped.* **56**:762.

Francke, U., Kernahan, C., and Bradshaw, C., 1975, Del(10)p autosomal deletion syndrome: Clinical, cytogenetic and gene marker studies, *Humangenetik* **26**:343.

Fried, K., Rosenblatt, M., Mundel, G., and Krikler, R., 1975, Mental retardation and congenital malformations associated with a ring chromosome 6, *Clin. Genet.* **7**:192.

Friedrich, U., and Nielsen, J., 1973, Chromosome studies in 5,049 consecutive newborn children, *Clin. Genet.* **4**:333.

Fryns, J. P., Verresen, H., Kerckvoorde, J. V., and Cassiman, J. J., 1974a, Partial Trisomy 8: Trisomy of the distal part of the long arm of chromosome number 8+(8q2) in a severely retarded and malformed girl, *Humangenetik* **24**:241.

Fryns, J. P., Berghe, H. V., Van Herck, G., and Cassiman, J. J., 1974b, Trisomy 12p due to familial 1(12p−,6q+) translocation, *Humangenetik* **24**:252.

Furbetta, M., Rosi, G., Cossu, P., and Cao, A., 1975, A case of trisomy of the short arms of chromosome no. 4 with translocation t(4p21p;4q21q) in the mother, *Humangenetik* **26**:87.

Geneva Conference, 1966, Standardization of procedures for chromosome studies in abortion, *Cytogenetics* **5**:361.

Gerald, P. S., and Walzer, S., 1970, Chromosome studies of normal newborn infants, *in: Human Population Cytogenetics*, Pfizer Medical Monographs No. 5 (P. A. Jacobs, W. H. Price, and P. Law, eds.), p. 143, Edinburgh University Press, Edinburgh, U.K.

German, J., 1968, Mongolism, delayed fertilization and human sexual behavior, *Nature* **217**:516.

German, J., 1972, Genes which increase chromosomal instability in somatic cells and predispose to cancer, *in: Progress in Medical Genetics*, Vol. VIII, (A. G. Steinberg and A. G. Bearn, eds.), p. 61, Grune and Stratton, New York.

de Grouchy, J., Royer, P., Salmon, C., and Lamy, M., 1964, Deletion partielle du bras long du chromosome 18, *Pathol. Biol.* **12**:579.

de Grouchy, J., Finaz, C., Roubin, M., and Roy, J., 1972, Deux translocations familiales survenues ensemble chez chacune de deux soeurs, l'une équilibrée, l'autre trisomique partielle 10q, *Ann Genet.* **15**:85.

Gustavson, K. H., Hitrec, V., and Santesson, B., 1972, Three non-mongoloid patients of similar phenotype with an extra G-like chromosome, *Clin. Genet.* **3**:135.

Haddida, E., Marill, F. G., and Sayog, J., 1963, Xeroderma pigmentosum (a propos de 48 observations personnels), *Ann. Dermatol. Syphiligr.* **90**:467.

Hall, B., 1966, Mongolism in newborn infants. An examination of the criteria for recognition and some speculations on the pathogenic activity of the chromosomal abnormality, *Clin. Pediatr.* **5**:4.

Hamerton, J. L., Briggs, J. M., Giannelli, F., and Carter, C. O., 1961, Chromosome studies in detection of parents with high risk of second child with Down's syndrome, *Lancet* **2**:788.

Hamerton, J. L., Giannelli, F., and Polani, P. E., 1965, Cytogenetics of Down's syndrome (mongolism). I. Data on a consecutive series of patients referred for genetic counselling and diagnosis, *Cytogenetics* **4**:171.

Hamerton, J. L., Ray, M., Abbott, J., Williamson, C., and Ducasse, G. C., 1972, Chromosome studies in a neonatal population, *Can. Med. Assoc. J.* **106**:776.

Hand, R., and German, J., 1975, Retarded DNA chain growth in Bloom's syndrome, *Proc. Natl. Acad. Sci. U.S.A.,* **72**:758.

Haslam, R. H. A., Broske, S. P., Moore, C. M., Thomas, G. H., and Neil, C. A., 1973, Trisomy 9 mosaicism with mutiple congenital anomalies, *J. Med. Genet.* **10**:180.

Henderson, S. A., and Edwards, R. G., 1968, Chiasma frequency and maternal age in mammals, *Nature* **218:**22.

Hirschhorn, K., and Cohen, M. M., 1968, Drug induced chromosomal aberrations, *Ann. N.Y. Acad. Sci.* **151:**977.

Hirschhorn, K., Cooper, H. L., and Firschein, L., 1965, Deletion of short arms of chromosome 4–5 in a child with defects of midline fusion, *Humangenetik* **1:**479.

Hirschhorn, K., Lucas, M., and Wallace, I., 1973, Precise identification of various chromosomal abnormalities, *Ann. Hum. Genet.* **36:**375.

Hobolth, N., Jacobsen, P., and Mikkelsen, M., 1974, Partial trisomy 12 in a mentally retarded boy and translocation (12;21) in his mother. *J. Med. Genet.* **10:**299.

Holmberg, M., and Jonasson, J., 1973, Preferential location of X-ray induced chromosome breakage in the R-bands of human chromosomes, *Heriditas* **74:**57.

Hook, E. B., 1973, Behavioural implications of the human XYY genotype, *Science,* **179:**150.

Hook, E. B., and Kim, D. S., 1970, Prevalence of XYY and XXY karyotypes in 337 non-retarded young offenders, *N. Engl. J. Med.* **283:**410.

Hsu, L. Y. F., and Hirschhorn, K., 1971, Sex chromosome aberrations, *Proc. XIII. Int. Congr. Pediatr.* **5:**75.

Hsu, L. Y. F., and Hirschhorn, K., 1972, Cytogenetic aspects of brain dysfunction, *in: Biology of Brain Dysfunction* (G. E. Gaull, ed.), Chapter 3, pp. 89–142, Plenum Press, New York.

Hsu, L. Y. F., and Hirschhorn, K., 1974, Prenatal diagnosis of genetic disorders (A minireview), *Life Sci.* **14:**2311.

Hsu, L. Y. F., and Hirschhorn, K., 1977, The trisomy 22 syndrome and cat eye syndrome, *in: New Chromosomal Syndromes* (J. Yunis, ed.), Academic Press, New York (in press).

Hsu, L. Y. F., Hirschhorn, K., Goldstein, A., and Barcinski, M., 1970, Familial chromosomal mosaicism, genetic aspect, *Ann. Hum. Genet.* **33:**343.

Hsu, L. Y. F., Shapiro, L. R., Gertner, M., Lieber, E., and Hirschhorn, K., 1971, Trisomy 22: A clinical entity, *J. Pediatr.* **79:**12.

Hsu, L. Y. F., Kim, H. J., Sujansky, E., Kousseff, B. G., and Hirschhorn, K., 1973, Reciprocal translocation versus centric fusion between two no. 13 chromosomes, *Cytogenet. Cell Genet.* **12:**235.

Hsu, L. Y. F., Kim, H. J., Paciuc, S., Steinfeld, L., and Hirschhorn, K., 1974, Non-fluorescent and non-heterochromatic Y chromosome in 45,X/46,XY mosaicism, *Ann. Genet.* **17:**5.

Hsu, L. Y. F., Yahr, F., Kim, H. J., Kerenyi, T., and Hirschhorn, K., 1977, 550 cases of prenatal cytogenetic diagnosis: Experience and prospects, *Mt. Sinai J. Med.* (Festschrift for Dr. Horace Hodes) (in press).

Hustinx, Th. W. J., ter Haar, B. G. A., Scheres, J. M. J. C., and Rutten, F. J., 1974, Trisomy for the short arm of chromosome no. 10, *Clin. Genet.* **6:**408.

Jacobs, P. A., and Strong, J. A., 1959, A case of human intersexuality having a possible XXY sex-determining mechanism, *Nature* **183:**302.

Jacobs, P. A., Melville, M., Ratcliffe, S., Keay, A. J., and Syme, J., 1974a, A cytogenetic survey of 11,680 newborn infants, *Ann. Hum. Genet.* **37:**359.

Jacobs, P., Buckton, K., Cunningham, C., and Newton, M., 1974b, An analysis of the break points of structural rearrangements in man, *J. Med. Genet.* **11:**50.

Jacobsen, P., Mikkelsen, M., and Rosleff, F., 1974, The trisomy 8 syndrome: report of two further cases, *Ann. Genet.* **17:**87.

Jones, K. W., and Corneo, G., 1971, Location of satellite and homogeneous DNA sequences on human chromosomes, *Nature (London), New Biol.* **233:**268.

Kim, H. J., Hsu, L. Y. F., Paciuc, S., Cristian, S., Quintana, A., and Hirschhorn, K., 1975, Cytogenetics of fetal wastage, *New Eng. J. Med.,* **293:**844.

Krmpotic, E., Rosenthal, I. M., Szego, K., and Bocian, M., 1971, Trisomy F (?20). Report of a 14q/F (?20) familial translocation, *Ann. Genet.* **14:**291.

Kucera, J., 1970, Down's syndrome and infectious hepatitis, *Lancet* **1:**569.

Lanman, J. T., 1968, Delays during reproduction and their effects on the embryo and fetus. 2. Aging of eggs, *N. Engl. J. Med.* **278:**1047.

Latt, S. A., 1973, Microfluorometric detection of deoxyribonucleic acid replication in human metaphase chromosomes. *Proc. Natl. Acad. Sci. U.S.A.* **70:**3395.

Latta, E., and Hoo, J. J., 1974, Trisomy of the short arm of chromosome 17, *Humangenetik* **23:**213.

Lauritsen, J. G., Jonasson, J., Therkelsen, A. J., Lass, F., Lindsten, J., and Patersen, G. B., 1972, Studies on spontaneous abortions. Fluorescence analysis of abnormal karyotypes, *Heriditas* **71:**160.

Leck, I., 1966, Incidence and epidemicity of Down's syndrome, *Lancet* **2:**457.

Leff, J. P., and Scott, P. D., 1968, XYY and intelligence, *Lancet* **1:**645.

Lejeune, J., and Berger, R., 1965, Sur deux observations familiales de translocations complexes, *Ann. Genet.* **8:**21.

Lejeune, J., LaFourcade, T., Berger, R., Vialatte, J., Boeswillwold, M., Seringe, P., and Turpin, R., 1963, Trois cas de délétion partielle des bras courts d'un chromosome 5, *C.R. Acad. Sci.* **257:**3098.

Lejeune, J., Berger, R., LaFourcade, J., and Rethore, M. O., 1966, La délétion partielle du bras long du chromosome 18. Individualization d'un nouvel état morbide, *Ann. Genet.* **9:**32.

Lejeune, J., Rethore, M. O., Dutrillaux, B., and Martin, G., 1972, Translocation 8–22 sans changement de longueur et trisomic partielle 8q., *Exp. Cell Res.* **74:**293.

Lewandowski, R. C., and Yunis, J. J., 1975. New chromosomal syndrome, *Am. J. Dis. Child.* **129:**515.

Lin, C. C., Uchida, I. A., and Byrnes, E., 1971, A suggestion for the nomenclature of the fluorescent banding patterns in human metaphase chromosomes, *Can. J. Genet. Cytol.* **13:**361. **13:**361.

Lubs, H. A., and Ruddle, F. H., 1970, Chromosomal abnormalities in the human population: Estimation of rates based on New Haven newborn study, *Science* **169:**495.

Lucas, M., Wallace, I., and Hirschhorn, K., 1972, Recurrent abortions and chromosome abnormalities, *J. Obstet. Gynaecol. Br. Commonw.* **79:**1119.

Magnelli, N. C. and Therman, E., 1975, Partial 12p deletion: A cause for a mental retardation, multiple congenital abnormality syndrome, *J. Med. Genet.* **12:**105.

Merrick, S., Ledley, R. S., and Lubs, H. A., 1973, Production of G and C banding with progressive trypsin treatment, *Pediatr. Res.* **7:**39.

Mikamo, K., 1968, Intrafollicular overripeness and teratological development, *Cytogenetics* **7:**212.

Milunsky, A., 1973, *The Prenatal Diagnosis of Hereditary Disorders,* Charles C Thomas, Springfield, Ill.

Money, J., 1964, Two cytogenetic syndromes: Psychologic comparisons. I. Intelligence and specific-factor quotients, *J. Psychiatr. Res.* **2:**223.

Moore, C. M., Heller, R. H., and Thomas, G. H., 1973, Developmental abnormalities associated with a ring chromosome 6, *J. Med. Genet.* **10:**299.

Muldal, S., Enoch, B. A., Ahmed, A., and Harris, R., 1973, Partial trisomy 14q− and pseudoxanthoma elasticum, *Clin. Genet.* **4:**480.

Nadler, H. L., 1972, Prenatel detection of genetic disorders, *Adv. Hum. Genet.* **3:**1.

National Institute of Mental Health, Report on the XYY Chromosomal Abnormality, 1970, Public Health Service Publication No. 2103, October.

Neu, R. L., and Kajii, T., 1969, Other autosomal abnormalities, *in: Endocrine and Genetic Diseases of Childhood* (L. I. Gardner, ed.), pp. 652–667, W. B. Saunders, Philadelphia, Pa.

Neu, R. L., Bargman, G. J., and Gardner, L. I., 1969, Disappearance of a 47,XX,+C leukocyte cell line in an infant who had previously exhibited 46,XX/47,XX,C+ mosaicism, *Pediatrics* **43:**624.

Nichols, W. W., 1966, The role of viruses in the etiology of chromosomal abnormalities, *Am. J. Hum. Genet.* **18:**81.

Norwood, T. H., and Hoehn, H., 1974, Trisomy of the long arm of human chromosome 1, *Humangenetik* **25:**79.

Nusbacher, J., and Hirschhorn, K., 1968, Autosomal anomalies in man, *Adv. Teratol.* **3:**1.

Nusbacher, J., Hirschhorn, K., and Cooper, L. Z., 1967, Chromosomal abnormalities in congenital rubella, *N. Engl. J. Med.* **276:**1409.

Owen, L., Martin, B., Blank, C. E., and Harris, F., 1974, Multiple congenital defects associated with trisomy for the short arm of chromosome 4, *J. Med. Genet.* **10:**291.

Pardue, M. L., and Gall, J. G., 1970, Chromosomal localization of mouse satellite DNA, *Science* **168:**1356.

Paris Conference, 1972, Standardization in human cytogenetics, birth defects. Original Article Series, Vol. VIII, No. 7, The National Foundation–March of Dimes.

Patil, S. R., Merrick, S., and Lubs, H. A., 1971, Identification of each human chromosome with modified Giemsa stain, *Science* **173:**821.

Pearson, P. L., Bobrow, M., and Vosa, C. G., 1970, Technique for identifying Y chromosomes in human interphase nuclei, *Nature* **226:**78.

Penrose, L. S., 1967, Finger-print pattern and the sex chromosomes, *Lancet* **1:**298.

Penrose, L. S., and Smith, G. F., 1966, *Down's Anomaly*, Little, Brown, Boston.

Pfeiffer, R. A., Buttinghaus, K., and Struck, H., 1973, Partial trisomy 14, *Humangenetik* **20:**187.

Plotkin, S. A., Boué, A., and Boué, J. G., 1965, *In vitro* growth of rubella virus in human embryo cells, *Am. J. Epidemiol.* **81:**71.

Podruch, P. E., and Weisskopf, B., 1974, Trisomy for the short arms of chromosome 9 in two generations, with balanced translocations t(15p+;9q−) in three generations, *J. Pediatr.* **85:**92.

Polani, P. E., 1969, Autosomal imbalance and its syndromes, excluding Down's *Br. Med. Bull.* **25:**81.

Poon, P. K., O'Brien, R. L., and Parker, J. W., 1974, Defective DNA repair in Fanconi's anemia, *Nature* **250:**223.

Preus, M., and Fraser, F. C., 1972, Dermatoglyphics and syndromes, *Am. J. Dis. Child.* **124:**1972.

Punnett, H. H., Kistenmacher, M. L., Toro-Solo, M. A. and Kohn, G., 1973, Quinacrine fluorescence and Giemsa banding in trisomy 22, *Theor. Appl. Genet.* **43:**134.

Rethore, M. O., Lejeune, J., Carpentier, S., Prieur, M., Dutrillaux, B., Seringe, Ph., Rossier, A., and Job, J. C., 1972, Trisomie pour la partie distale du bras court du chromosome 3 chex trois germains. Premier exemple d'insertion chromosomique: ins (7;3) (q31;p21p26), *Ann. Genet.* **15:**159.

Rethore, M. O., Hoehn, H., Rott, H. D., Couturier, J., Dutrillaux, B., and Lejeune, J., 1973, Analyse de la trisomie 9p par denaturation menagee, *Humangenetik* **18:**129.

Richmond, H. G., MacArthur, P., and Hunter, D., 1973, A "G" deletion syndrome antimongolism, *Acta Paediatr. Scand.* **62:**216.

Riis, P., and Fuchs, F., 1966, Sex chromatin and antenatal sex diagnosis, *in: The Sex Chromatin* (K. L. Moore, ed.), Chapter 13, pp. 220–228, W. B. Saunders, Philadelphia, Pa.

Robinson, J. A., 1973, Origin of extra chromosome in trisomy 21, *Lancet* **1:**131.

Sanchez, O., Yunis, J. J., and Escobar, J. I., 1974, Partial trisomy 11 in a child resulting from a complex maternal rearrangement of chromosomes 11, 12 and 13, *Humangenetik* **22:**59.

Saunders, G. F., Hsu, T. C., Getz, M. J., Simes, E. L., and Arrighi, F. E., 1972, Locations of a human satellite DNA in human chromosomes, *Nature (London), New Biol.* **236:**244.

Schindler, A. N., and Mikamo, K., 1970, Triploidy in man: Report of a case and discussion on etiology, *Cytogenetics* **9:**116.

Schleiermacher, E., Schliebitz, U., and Steffens, C., 1974, Brother and sister with trisomy 10p: A new syndrome, *Humangenetik* **23:**163.

Schnedl, W., 1971, Analysis of the human karyotype using a reassociation technique, *Chromosoma* **34:**448.

Schreck, R. R., Warburton, D., Miller, O. J., Beiser, S. M., and Erlanger, B. F., 1973, Chromosome structure as revealed by a combined chemical and immunochemical procedure, *Proc. Natl. Acad. Sci. U.S.A.* **70:**804.

Schrott, H. G., Sakaguchi, S., Francke, U., Luzzatti, L., and Fialkow, P. J., 1974, Translocation, t(4q−;13q+), in three generations resulting in partial trisomy of the long arm of chromosome 4 in the fourth generation, *J. Med. Genet.* **11:**201.

Schull, W. J., and Neel, J. V., 1962, Maternal radiation and mongolism, *Lancet* **1:**537.

Schwanitz, G., and Grosse, K. P., 1973, Partial trisomy 4p with translocation 4p−, 22p+ in the father, *Ann. Genet.* **16:**263.

Seabright, M., 1972, The use of proteolytic enzymes for the mapping structural rearrangements in the chromosomes of man, *Chromosoma* **36:**204.

Sergovich, F., Valentine, G. H., Chen, A. T. L., Kinch, R. A. H., and Smout, M. S., 1969, Chromosome aberrations in 2159 consecutive newborn babies, *N. Engl. J. Med.* **280:**851.

Sergovich, F., Uilenberg, C., and Pozsonyi, J., 1971, The 49,XXXXX chromosome constitution: Similarities to the 49,XXXXY condition, *J. Pediatr.* **78:**285.

Serr, D. M., Sachs, L., and Danon, M., 1955, Diagnosis of sex before birth using cells from the amniotic fluid, *Bull. Res. Council Israel* **58:**137.

Setlow, R. B., Regan, J. D., German, J., and Carrier, W. L., 1969, Evidence that xeroderma pigmentosum cells do not perform the first step in the repair of ultraviolet damage to their DNA, *Proc. Natl. Acad. Sci. U.S.A.* **64:**1035.

Shokeir, M. H. K., Ray, M., Hamerton, J. L., Bauder, F. and O'Brien, H. O., 1975, Deletion of the short arm of chromosome no. 10, *J. Med. Genet.* **12:**99.

Short, E. M., Solitare, G. B., and Berg, W. R., 1973, A case of partial 14 trisomy 47,XY,(14q–)+ and translocation t(9p+;14q–) in mother and brother, *J. Med. Genet.* **9:**367.

Smith, D. W., 1970, Recognizable patterns of human malformation; genetic, embryologic and clinical aspects, *Major Problems in Clinical Pediatrics* (D. W. Smith, ed.), Vol. VII, pp. 38–53, W. B. Saunders, Philadelphia, Pa.

Sparkes, R. S., Muller, H. M., and Veomett, I. C., 1970, Inherited pericentric inversion of a human Y chromosome in trisomic Down's syndrome, *J. Med. Genet.* **7:**59.

Stark, C. R., and Fraumeni, J. F., Jr., 1966, Viral hepatitis and Down's syndrome, *Lancet* **1:**1036.

Starkman, M. N., and Jaffe, R. B., 1967, Chromosome aberrations in XO/XY mosaic individuals and their fathers, *Am. J. Obstet. Gynecol.* **99:**1056.

Stewart, A. L., Keay, A. J., Jacobs, P. A., and Melville, M. M., 1969, A chromosome survey of unselected live-born children with congenital abnormalities, *J. Pediatr.* **74:**449.

Stoll, C., Rohmer, A., and Sauvage, P., 1973, Chromosome 22 en anneau R (22): Identification per denaturation thermique menagee, *Ann. Genet.* **16:**193.

Stoller, A., and Collmann, R. D., 1965, Incidence of infective hepatitis followed by Down's syndrome nine months later, *Lancet* **2:**1221.

Subrt, I., and Brychnac, V., 1974, Trisomy for short arm of chromosome 20, *Humangenetik* **23:**219.

Sumner, A. T., 1973, Involvement of protein disulphides and sulphydryls in chromosome banding, *Exp. Cell Res.* **83:**438.

Sumner, A. T., and Evans, H. J., 1973, Mechanisms involved in the banding of chromosomes with quinacrine and Giemsa II. The interaction of the dyes with the chromosomal components, *Exp. Cell Res.* **81:**223.

Sumner, A. T., Evans, H. J., and Buckland, R. A., 1971, A new technique for distinguishing between human chromosomes, *Nature (London), New Biol.* **232:**31.

Sumner, A. T., Evans, H. J., and Buckland, R. A., 1973, Mechanisms involved in the banding of chromosomes with quinacrine Giemsa, I. The effects of fixation in methanolacetic acid, *Exp. Cell Res.* **81:**214.

Surana, R. B., and Conen, P. E., 1972, Partial trisomy 4 resulting from a 4/18 reciprocal translocation, *Ann. Genet.* **15:**191.

Taysi, K., Kohn, G., and Mellman, W. J., 1970, Mosaic mongolism II. Cytogenetic studies, *J. Pediatr.* **76:**880.

Tuncbilek, E., Haliciouglu, C., and Say, B., 1974, Trisomy 8 syndrome, *Humangenetik* **23:**23.

Turpin, R., and Lejeune, J., 1961, Chromosome translocations in man, *Lancet* **1:**616.

Tusques, J., Grislain, J. R., Andre, M. J., Mainard, R., Rival, J. M., Cadudal, J. L., Dutrillaux, B., and Lejeune, J., 1972, Trisomie partielle 11q identifiée grace à l'étude en "dénaturation ménagée" par la chaleur, de la translocation équilibrée paternelle, *Ann. Genet.* **15:**167.

Uchida, I. A., and Curtis, E. J., 1961, A possible association between maternal radiation and mongolism, *Lancet* **2:**848.

Uchida, I. A., and Lin, C. C., 1973, Identification of partial 12 trisomy by quinacrine fluorescence, *J. Pediatr.* **82:**269.

Uchida, I. A., Ray, M., McRae, K. N., and Besant, D. F., 1968, Familial occurrence of trisomy 22, *Am. J. Hum. Genet.* **20**:107.

Vickers, A. D., 1969, Delayed fertilization and chromosomal anomalies in mouse embryos, *J. Reprod. Fertil.* **20**:69.

Vogel, W., Siebers, J. W., and Reinwein, H., 1973, Partial trisomy 7q, *Ann. Genet.* **16**:277.

Vosa, C. G., 1970, Heterochromatin recognition with fluorochromes, *Chromosoma* **30**:366.

Walker, S., Andrews, J., Gregson, N. M., and Gault, W., 1973, Three further cases of triploidy in man surviving to birth, *J. Med. Genet.* **10**:135.

Walraven, P., Greensher, A., Sparks, J. W., and Wesenberg, R. L., 1974, Trisomy 8 mosaicism, *Am. J. Dis. Child.* **128**:564.

Warkany, J., 1971, Part VIII, Cardiovascular malformations, *in: Congenital Malformations,* pp. 576–578, Year Book Medical Publishers, Chicago.

Warren, R. J., and Rimoin, D. L., 1970, The G deletion syndromes, *J. Pediatr.* **77**:658.

Weisblum, B., and de Haseth, P., 1972, Quinacrine a chromosome stain specific for deoxyadenylate-deoxythymidylate-rich regions in DNA, *Proc. Natl. Acad. Sci. U.S.A.* **69**:629.

White, M. J. D., 1954, *Animal Cytology and Evolution,* 2nd ed. Cambridge Univ. Press, London and New York.

Wilson, M. G., Towner, J. W., and Negus, L. D., 1970, Wolf-Hirschhorn syndrome associated with an unusual abnormality of chromosome No. 4, *J. Med. Genet.* **7**:164.

Witkin, H. A., Mednick, S. A., Schulsinger, F., Bakkestrøm, E., Christiansen, K. O., Goodenough, D. R., Hirschhorn, K., Lundsteen, C., Owen, D. R., Philip, J., Rubin, D. B., and Stocking, M., 1976, Criminality in XYY and XXY men, *Science* **193**:547.

Wolf, U., Reinwein, H., Porsch, R., Schroter, R., and Baitsch, H., 1965, Defizienz an den kurzen Armen eines Chromosome No. 4, *Humangenetik* **1**:397.

de Wolff, E., 1964, Étude clinique de 134 mongolien, *Ann. Paediatr.* **202**(Suppl. 1):1.

Wright, Y. M., Clark, W. E., and Breg, W. R., 1974, Craniorachischisis in partially trisomic 11 fetus in a family with reproductive failure and a reciprocal translocation, t(6p+;11q−), *J. Med. Genet.* **11**:69.

Wyandt, H. E., and Hecht, F., 1971, Detection of the X-chromatin body in human fibroblasts by quinacrine fluoromicrocopy, *Lancet* **2**:1379.

Yunis, J. J., and Sanchez, O., 1974, A new syncrome resulting from partial trisomy for the distal third of the long arm of chromosome 10, *J. Pediatr.* **84**:567.

Zaleski, W. A., Houston, C. S., Pozsonyi, J., and Ying, K. L., 1966, The XXXXY chromosome anomaly: Report of three new cases and review of 30 cases from the literature, *Can. Med. Assoc. J.* **94**:1143.

Zellweger, H., and Abbo, G., 1965, Familial mosaicism attributable to a new gene, *Lancet* **1**:455.

Cell Death and Reduced Proliferative Rate

4

WILLIAM J. SCOTT, JR.

I. INTRODUCTION

Numerous teratological investigations have provided evidence that chemical or physical insult to the developing embryo often produces within a few hours or days obvious signs of cell necrosis in tissues destined to be malformed. This observation has been made so often that one hardly need hesitate in ascribing the malformation to the earlier cytotoxicity. However, further observations cloud the issue somewhat: (1) Agents which at high doses produce cell death and teratogenesis, at low doses continue to kill cells (albeit at a lower rate) but may not produce malformations, and (2) many cytotoxic, teratogenic agents produce cell death in tissues which appear normal at birth. This means that cell death above and beyond physiological levels does not inevitably lead to malformation. Thus, it becomes necessary to know the secondary consequences of cell death in a particular organ system so that the temporal and quantitative aspects of necrosis needed to produce malformation become explainable.

Teratologists are presently unable to follow in detail the complete pathogenesis of a single abnormality, whatever its cause or primary mechanism. This chapter will be partially devoted to reviewing some of the studies which have attempted to explain how cell death leads to malformation.

Another tantalizing question is why, in response to a cytotoxic agent, some embryo cells die while others remain unaffected. Beyond its ample description this phenomenon has received little attention; thus discussion of it

WILLIAM J. SCOTT, JR. • Children's Hospital Research Foundation and Department of Pediatrics, University of Cincinnati, Cincinnati, Ohio 45229.

in this review, while hopefully being informative, is meant mainly to stimulate further investigation.

Many agents which are cytotoxic also delay the progression of cells through the proliferative cycle without killing them. It seems plausible that this effect in itself might lead to abnormal development. Thus, a discussion of proliferation kinetics in embryos is included. Some aspects of cell death will be given little or no mention in this review. One of these, cell death in normal embryogenesis, has been very competently reviewed by others (Glücksmann, 1951, 1965; Menkes *et al.*, 1970; Pexieder, 1975; Saunders, 1966; Schweichel, 1970). Likewise the morphology and cytochemistry of cell death will not be discussed; excellent discussions of these subjects were provided by Crawford *et al.* (1972), Kauffman and Herman (1968), Menkes *et al.* (1970), Mottet and Hammar (1972), Sauer and Duncan (1962), and Schweichel and Merker (1973). Discussions of the mechanisms by which embryo cells die are available by Forsberg and Källén (1968), Lockshin and Beaulaton (1974), and Webster and Gross (1970).

II. SURVEY OF AGENTS

Table 1 is a list of environmental agents which have been shown to produce cell death in organ primordia followed by malformation of that organ as seen in the near-term fetus or newborn individual. The list would undoubtedly be much longer if thorough study had been made shortly after teratogen administration when cytotoxicity is likely to be most pronounced. Conditions in which genetic factors lead to cell death and ensuing malformation are not included; otherwise the list would be cumbersome indeed. Many of these conditions were reviewed by Grüneberg (1963).

The fact that such a wide variety of teratogenic agents produce a cytotoxic response in the embryo has raised the question whether all teratogens act in this manner. Although such a hypothesis is probably not tenable, one is hard pressed to find well-documented cases of teratogenesis without some cytotoxicity.

III. PROLIFERATION KINETICS

As stated earlier, some cytotoxic agents at lower doses inhibit proliferative activity, and there is reason to believe that reduced proliferative activity in itself may contribute to teratogenesis. A great deal of effort in recent years has been concentrated on studying morphological and especially biochemical events which occur in proliferating cells (for reviews see Baserga, 1968; Cleaver, 1967; Harrisson, 1971). Current dogma divides the interval from one cell division to the next into four major periods: (1) G_1, a period of preparation for DNA synthesis, (2) S, at which time the chromosomal DNA is

Table 1. Agents Producing Cytotoxicity in an Organ Anlagen Followed by Malformation of that Organ

Agent	Species	Reference
Drugs		
Cyclophosphamide	Rat	Köhler and Merker, 1973
Hydroxyurea	Rat	Scott *et al.*, 1971
Cytosine arabinoside	Rat	Ritter *et al.*, 1973
Cytosine arabinoside palmitate	Rat	Ritter *et al.*, 1973
Aminothiadiazole	Rat	Scott *et al.*, 1973a
6-Mercaptopurine	Rat	Merker *et al.*, 1975
5-Fluorodeoxyuridine	Mouse	Andreoli *et al.*, 1973
	Rat	Maruyama *et al.*, 1968
5-Fluorouracil	Chick	Kury and Craig, 1966
5-Fluoroortic acid	Chick	Kury and Craig, 1966
Mitomycin-C	Chick	Kury and Craig, 1967
Dinitrophenol	Chick	Bowman, 1967
5-Fluoro-2-deoxycytidine	Mouse	Fränz, 1971
Nitrogen mustard	Mouse	Jurand, 1961
	Chick	Jurand, 1961
Hadacidin	Rat	Lejour-Jeanty, 1966
Ethyl alcohol	Chick	Sander, 1968
5-Fluorouracil	Mouse	Skalko and Sax, 1974
6-Aminonicotinamide	Rat	Chamberlin, 1970
Actinomycin D	Chick	Wolkowski, 1970
Busulfan	Rat	Forsberg and Olivecrona, 1966
Aminopterin	Chick	O'Dell and McKenzie, 1963
Thalidomide	Armadillo	Marin-Padilla and Benirschke, 1965
Urethane	Mouse	Sinclair, 1950
5-Azacytidine	Mouse	Seifertová *et al.*, 1973
5-Azauridine	Mouse	Seifertová *et al.*, 1973
Colchicine	Hamster	Ferm, 1964
Vincristine	Hamster	Ferm, 1964
Vinblastine	Hamster	Ferm, 1964
Methadone	Chick	Jurand, 1973
Ethylnitrosourea	Rat	Wechsler, 1973
Methylnitrosourea	Rat	Wechsler, 1973
Methylbenzanthracene	Rat	Crawford *et al.*, 1972
Vitamin Excess or Deficiency		
Riboflavin deficiency	Rat	Shepard *et al.*, 1968
Vitamin A deficiency	Pig	Palludan, 1966
Pantothenate deficiency	Rat	Giroud *et al.*, 1955
Vitamin A excess	Hamster	Marin-Padilla and Ferm, 1965
Selenium excess	Chick	Grunewald, 1958
Folate deficiency	Rat	Johnson, 1964
Hormones		
Adrenalin	Rabbit	Jost *et al.*, 1969

(cont'd)

Table 1. (cont'd)

Agent	Species	Reference
Vasopressin	Rat	Jost, 1951
ACTH	Rat	Jost, 1951
Pituitary extract	Rat	Jost, 1950
Hydrocortisone	Mouse	Jurand, 1968
Insulin	Chick	Zwilling, 1959
Infectious Agents		
Rubella	Human	Töndury and Smith, 1968
Reovirus	Hamster	Kilham and Margolis, 1974
Cytomegalovirus	Human	Naeye, 1967
Coxiella burnetti	Rat	Giroud *et al.*, 1968
Panleucopenia virus	Cat	Kilham *et al.*, 1967
Pasturella pseudo TB	Mouse	Flamm *et al.*, 1964
H-1 Virus	Hamster	Ferm and Kilham, 1965
Newcastle disease virus	Chick	Williamson *et al.*, 1965
Influenza A	Chick	Williamson *et al.*, 1965
Vaccinia virus	Mouse	Thalhammer, 1957
Mumps	Chick	Williamson *et al.*, 1965
Physical Trauma		
Amniocentesis	Rat	Singh and Singh, 1973
Hypoxia	Chick	Ilies, 1970
	Rat	Ilies, 1970
Electrocoagulation	Chick	Rickenbacher, 1968
Hyperthermia	Rat	Skreb and Frank, 1963
Electric light	Chick	Källén and Rüdeberg, 1966
Ultraviolet irradiation	Chick	Sandor and Elias, 1968
X-irradiation	Rat	Hicks and D'Amato, 1966
Natural Products		
Apivene (bee poison)	Chick	Ruch *et al.*, 1962
Viper venom	Mouse	Clavert and Gabriel-Robez, 1974

duplicated, (3) G_2, that period following the completion of DNA synthesis when preparation is made for division, and (4) M, mitosis, when condensation and separation of the chromosomes occurs on the mitotic spindle.

It has been known for many years that proliferative rate in embryonic tissue may be relatively rapid. For example, Köhler *et al.* (1972) calculated that between days 8 and 10 of pregnancy the rat embryo would have to proceed through a minimum of 10 mitotic divisions to account for the demonstrated increase in cell number. They reasoned that this allowed, on the average, 5 hours for one cell cycle but cautioned that not all cells are proliferating at the same rate and, therefore, assumed that "the cell cycle of some cell types must be even lower than 5 hours." No normal adult tissue, even a continuously proliferating population such as the intestinal crypt epithelium, displays such rapid mitotic activity.

A. Methods Used to Measure the Rate of Cell Proliferation

1. Mitotic Index

Mitosis is easily visible in the light microscope, a fact which allows estimation of proliferative rates within the embryo. Mitotic index, which is the percentage of cells in the population that is in some stage of mitosis at a given time, indicates whether a teratogenic influence has altered proliferative activity within a particular tissue. Using this method Johnson (1964) was able to demonstrate a reduced proliferative rate in various areas of the neural tube and adjacent mesenchyme in the 10-day rat embryo subjected to folate deficiency.

Some of the technical difficulties and limitations of this method were summarized by Cameron (1971). One of the major limitations relates to the short time which the cell remains in mitosis. The data of Kauffman (1968) on studies of the generation cycle in the neural tube of the mouse embryo illustrate this point vividly. It was demonstrated that the total generation time was 10.5 hr, made up of 2.74 hr in G_1, 5.38 hr in S, 1.15 hr in G_2, and 1.23 hr in M. Assuming that the entire population is proliferating at a uniform rate, only about 12% (1.23/10.5) of the cells will be in mitosis. Since such a small percentage of the population is examined, it is impossible to detect small perturbations in proliferative activity by this method. To overcome this limitation the technique of accumulating cells in mitosis was developed.

2. Accumulated Mitoses

A group of chemicals termed spindle poisons, represented by colchicine and the vinca alkaloids, arrest dividing cells at metaphase by interfering with the formation of the mitotic spindle which is responsible for the separation of chromosomes to daughter cells. At proper dosage these agents do not seem to interfere with the progression of cells through the other phases of the proliferative cycle. This causes cells to accumulate in metaphase, an easily visible stage, and thus a larger percentage of the population is represented in the calculation, thereby providing more reliable information on proliferative activity than the mitotic index. Cameron (1971) and Lehmiller (1971) review the conditions which must be fulfilled for this procedure to be valid. Corliss (1953) used this technique to examine proliferative activity in the postimplantation mouse embryo. Contrary to earlier studies he found proliferative activity uniformly distributed throughout the embryo. Specifically, the higher rate of ectodermal activity compared to mesoderm reported by Preto (1938) and Pasteels (1943) was not evident. Besides using colchicine to accumulate mitoses, Corliss reported his findings as mitoses/unit volume rather than the standard mitoses × 100/total cells, and it is likely that this is responsible for the differing results.

More recently Jelinek and Dostal (1974) used this technique to examine mitotic activity in the developing palatal shelves of mouse embryos. In a subsequent paper (Jelinek and Dostal, 1975), these authors described a reduced proliferative rate in the palatal shelves prior to horizontalization, following administration of corticoids, and postulate that the reduced shelf size was the basis for cleft palate.

This method has not gained wide popularity, probably due to the recent preference for tritiated thymidine autoradiography. However, there are certain instances when thymidine is an unsuitable precursor (antifolates and fluorinated pyrimidines) and the accumulated mitosis technique should be remembered as an alternative.

3. Incorporation of Labeled Thymidine

A wide variety of radioactive isotope precursors are commercially available and have been used experimentally to label DNA in proliferating cells. None can match the popularity of thymidine for this purpose, the major reason being its unique specificity in labeling. Other labeled precursors are incorporated into a few or many biochemical products, but thymidine is incorporated only into DNA. Thus, when using other precursors, the radioactivity elsewhere must be separated from that in DNA, either by isolating the DNA chemically or by removing the other molecules containing radioactivity (e.g., RNase to remove RNA). This is not to say that the use of thymidine to label DNA is without pitfalls. In fact a number of assumptions must be made which have not been critically examined in mammalian embryos. These assumptions are stated succinctly by Cameron (1968) and discussed at length by Cleaver (1967) and Feinendegen (1967). The newcomer to the field is strongly urged to consult these sources before embarking on an experimental course which may be fruitless.

An assumption that is probably questionable in the mammalian embryo is the presumed short time over which an injection of thymidine is available for incorporation. It has been repeatedly shown in adult animals (reviewed in Cleaver, 1967) that injected thymidine is available for only a short time (30–60 min) due to breakdown of the thymidine by hepatic enzymes. Studies in this author's laboratory (Ritter et al., 1971; Scott et al., 1971), in which thymidine incorporation was determined in rat embryos 2 hr after maternal intraperitoneal injection, revealed a large amount of unincorporated radioactivity in the tissues. Whether this radioactivity is still in a form utilizable for DNA synthesis has not been determined, but a study in mouse embryos (Kauffman, 1969) reported the percentage of cells incorporating thymidine in control embryos increased from 50.6% at 30 min to 58.2% at 60 min to 65.1% at 90 min, suggesting that thymidine once sequestered in the embryo may be relatively free from catabolic degradation. If such is the case, the validity of those kinetic measurements (e.g., percent labeled mitoses) which assume that thymidine is available only for a brief pulse will need re-examination.

Also worthy of mention is the difficulty encountered in labeling the cells of the rodent embryo shortly after implantation (Jollie, 1968; Solter *et al.* 1971), despite the fact that many can be seen histologically to be undergoing mitosis. The reasons for this apparent paradox are not entirely clear, but a recent study in mice (Miller and Runner, 1975) suggests that the visceral yolk sac epithelium acts as a barrier to the passage of thymidine into the embryo. In an attempt to circumvent this problem, Miller (1974) has devised an *in vitro* method for labeling early rodent embryos.

The studies which have demonstrated a reduction in the incorporation of labeled precursor into DNA in embryos destined to be malformed are too numerous to cite here. Caution must be exercised, however, in interpreting any decrease in DNA synthesis rate as a direct cause of malformation. It has been shown (Ritter *et al.*, 1971; Scott *et al.*, 1971) that at least some agents which cause malformations at high doses may produce an appreciable depression in embryonic DNA synthesis at nonteratogenic doses.

IV. HOW DOES CELL DEATH OR REDUCED PROLIFERATIVE RATE LEAD TO MALFORMATION?

It is not difficult to imagine that an agent which destroys cells or reduces their rate of division could produce a teratogenic effect. As Kalter warns (1968, p. 138), however, such a theory says very little "but sounds so logical that it engenders complacence." In fact a number of investigators have suggested that a direct correlation does not always exist between the incidence of embryotoxicity and the antiproliferative or cytotoxic effect of a teratogen (Ferm, 1964; Gibson and Becker, 1968; Joneja and LeLiever, 1973; Răska *et al.*, 1966; Škreb and Frank, 1963; Wilson, 1954). In addition it has been shown that agents which reduce cell number are capable of producing tissue-excess deformities such as polydactyly (Scott *et al.*, 1975). Thus modifying factors such as embryonic age, regenerative capacity, and tissue interactions, to mention a few, must be considered when attempting to explain the mechanism and pathogenesis of a defect induced by an antiproliferative or cytotoxic agent.

A number of investigators believe, however, that the reduced cell number resulting from either of these effects plays a central role in many malformations. Jelinek and Dostal (1975) suggest that the antiproliferative effects of corticoids on developing mouse palatal shelves causes a growth retardation which then induces a chain of sequential events, e.g., inhibition of RNA synthesis and reduction of mucopolysaccharide, hydroxyproline (Shapira and Shoshan, 1972), and phospholipid contents in the shelves (Stepanovich and Gianelly, 1971). According to Jelinek and Dostal some or all of these effects then converge to produce the classic corticoid-induced cleft palate by one or more of three mechanisms: (1) reduced contact of normally elevated shelves as a result of the decreased amount of cellular and/or extracellular materials, (2) prolonged course of shelf movement, or (3) delayed commencement of horizontalization. Very similar conclusions were reached by Mott *et al.* (1969) using tritiated thymidine autoradiography to monitor proliferative rate following cortisone administration to pregnant mice.

A similar pathogenesis for vitamin A-induced cleft palate in rat embryos has been put forward by Kochhar (1968). Histological sections through the palatal regions revealed that in treated embryos the mass of cells forming the palatal process was much smaller than in controls. Tritiated thymidine autoradiographs made 24 hr after the last vitamin A treatment showed a reduced number of labeled cells, in agreement with the smaller size.

In the author's laboratory the pathways by which reduced cell number resulting from treatment with cytotoxic DNA synthesis inhibitors can lead to malformations has been studied (Ritter *et al.*, 1971, 1973; Scott *et al.*, 1971, 1973a). The agents used in these studies, hydroxyurea (HU), cytosine arabinoside (ara-C), aminothiadiazole (ATD), and cytosine arabinoside palmitate (ara-CP), produced a profound depression in the rate of whole embryo DNA synthesis. Also evident was a severe cytotoxic response which was thought to be more influential in determining the rate and type of malformation. Restricting attention to the limb, a close correlation was found between the time of appearance of maximal cell death and the pattern of forelimb ectrodactyly (Table 2). In sharp contrast, the duration and degree of DNA synthesis inhibition showed little relationship to the number of missing digits. Figure 1 compares the pattern of DNA synthesis inhibition following teratogenically equivalent doses of aminothiadiazole, hydroxyurea, and cytosine arabinoside. It can be seen that HU and ara-C produce a rapid and profound

Table 2. Correlation between Time of Maximal Cell Death Seen after Day-12 Treatment and Degree of Forelimb Ectrodactyly Seen in 20-Day Fetus[a]

Drug	Interval between drug administration and appearance of maximal cell death (hr)	Forelimbs missing one digit (%)	Forelimbs missing more than one digit (%)
Hydroxyurea	3	44	0
Aminothiadiazole	23	50	13
Cytosine arabinoside	23–29	28	37

[a]From Scott *et al.*, 1973a.

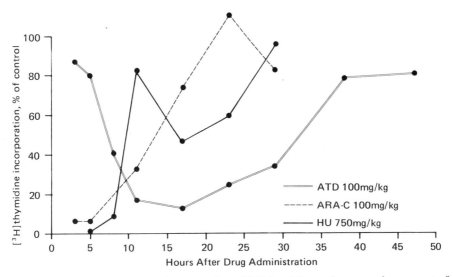

Fig. 1. DNA synthesis reflected by incorporation of [³H]thymidine and expressed as percent of control. Each point represents the mean of scintillation counts on the embryos in one uterine horn from each of at least 3 pregnant rats 2 hr after thymidine. (Reprinted from Scott *et al.*, 1973a.)

depression of DNA synthesis, followed by a gradual return to control rates within 23–29 hr. ATD, on the other hand, produced a gradual depression in DNA synthesis which did not become maximal until 17 hr. Recovery was also gradual with return to near-normal rates occurring at 38 hr. Thus, HU given at 9 AM of day 12 produces a rapid antiproliferative and cytotoxic effect, resulting in loss of one digit per limb. Ara-C produced a rapid antiprolifera- tive effect but a delayed cytotoxic effect, resulting in absence of more than one digit from several limbs. ATD produced a delayed antiproliferative and cytotoxic effect and results in many limbs with only one missing digit, but some limbs with more than one missing digit.

Further evidence in support of this concept was obtained by administer- ing HU on day 13 (Ritter *et al.*, 1973), so that the maximal expression of cytotoxicity from this fast-acting drug would coincide temporally with that produced by ara-C or ATD administered on day 12. In this study the pattern of missing digits on the forelimbs, following HU administration, resembled that produced by day 12 treatment with the other drugs, i.e., absence of more than one digit per limb. To reconcile these results with the idea that reduced cell number leads to specific limb malformations, it was necessary to postulate the existence of an unalterable critical time or event in digit morphogenesis after which the addition of newly regenerated cells would not alter the pattern of digit formation. The difference in the number of digits missing following HU treatment on days 12 and 13 can thus be explained by assuming that treatment on day 12 produced serious damage in the limb but that the long regenerative period allowed restoration of much of this damage so that only a single digit was missing. Treatment on day 13 presumably produces a similar amount of damage but, because there was less time for repair, more than one digit was usually missing. The importance of repair or regeneration has been given little attention in this review, and generally teratologists have not been adequately concerned with the consequences of repair upon the final expression of malformations. Further insight into the tremendous capacity of embryonic tissue for repair following injury can be gained from the work of Rickenbacher (1968) and Andreoli *et al.* (1973).

The above views concerning reduced cell number as a mediator of teratogenesis had to be questioned when it was found that one of these agents, ara-C, could produce extra digits in the limb when given at certain times (Scott *et al.*, 1973b). Following an examination of proliferation kinetics in the limbs destined to be polydactylous after ara-C, a number of possible explanations were considered. These included altered ectodermal–mesodermal relations, abolition or alteration of normal ectodermal cell death patterns, and synchronization of cell cycles within the limb (Scott *et al.*, 1975). These ideas remain speculative and warrant no further discussion at this time.

Limb malformations in chicks after administration of the cytotoxic agent nitrogen mustard have been studied by Wolff (1966) and colleagues. It was found that this agent produces a high percentage of phocomelic limbs when applied at the proper developmental stage. Histological examination of the limbs revealed a severe cytotoxic response 18 hr following treatment, with cytotoxicity restricted mainly to the mesodermal tissue, the ectoderm appearing relatively normal. The studies of Hampé (1966) from the same laboratory purported to show that the two tissues must be of similar age or developmental stage in order for limb morphogenesis to proceed normally. Combining these observations on cytotoxicity and on coincident development of germ layers suggested that the ectoderm is emitting signals which specify proximal limb development during the period when the mesoderm is damaged by nitrogen mustard. By the time the mesoderm recovers from the injury the ectoderm has progressed developmentally to the point where it is emitting signals for distal development. Thus, proximal structures are not formed and phocomelia results. A recent study by Rubin and Saunders (1972) questions the validity of this hypothesis, on the basis of their observation that the inductive signal emitted by the apical ectodermal ridge is qualitatively constant throughout its active phase. The latter authors conclude that "information for proper sequencing of level-specific patterns in the limb must therefore be programmed intrinsically within the mesoderm." What effects a severe cytotoxic response might have on this program were not speculated upon.

These examples indicate that there is a paucity of information concerning the secondary effects of cell death in an organ or region. The importance of secondary changes was further illustrated by Grunewald (1958). Following the administration of selenium to chick embryos, he found necrosis in many tissues, most of which were later malformed. However, defects also appeared in tissues not known to have been affected by necrosis. Speculation centered upon necrosis occurring in a tissue which was developmentally related to the defective structure. This same idea was offered by Lash and Saxén (1972) to explain thalidomide-induced limb defects. These authors found that in cultures of human embryos thalidomide inhibited an interaction between mesonephric mesenchyme and limb tissues which is necessary for limb chondrogenesis. Chondrogenesis in isolated limb tissue was not affected. Further evidence strengthening this concept was gained from studies of drug distribution in cultured embryos. Following exposure to ^{14}C-labeled thalidomide, very little radioactivity was demonstrable in limb tissue but high levels were found in the adjacent mesonephric mesenchyme.

V. SELECTIVITY OF CYTOTOXIC EFFECT

The remaining question to be discussed concerns the basis for the occurrence of cytotoxicity in certain cells while others in the same or adjacent tissues appear unharmed. This phenomenon, exemplified by the findings of Wolff (1966) discussed in the previous section, needs no further reference citations except to note that most of the references to specific agents listed in Table 1 relate to cell death in one or more specific tissues, often including observations on normal cells in the same organism. The sparing of some cells is probably necessary for the production of malformations since, if all tissues were severely affected by necrosis, the embryo would probably die. On the other hand, if cell death were uniform but not severe, the embryo would probably develop normally or be retarded in growth but not structurally malformed.

A definitive basis for specificity of cell death cannot be given, but the subject was reviewed as regards ionizing radiation in the mammalian embryo by Hicks and D'Amato (1966). These authors have carefully catalogued the response of many cell populations, mainly within the central nervous system. In several instances they found a positive correlation between mitotic rate and radiosensitivity, but this was by no means absolute. For example, circulating blood cells which were proliferating as evidenced by a mitotic index of 2–3% appeared resistant to 200 R, a dose which caused severe damage in many other populations; nonproliferating, primitive differentiating cells were highly sensitive to 200 R, strongly supporting the belief that mitotic rate alone does not determine radiosensitivity. These authors also found, using tritiated thymidine autoradiography, that in proliferating cells G_2 was the most radiosensitive phase of the cell cycle. Despite these facts, however, the reason why one cell dies and its apparently similar neighbor survives remains unknown.

Speculation on this subject has most often been centered on the category of drugs in Table 1. These are nearly all antitumor drugs which inhibit proliferation, and accordingly, the often-used explanation has been proferred that tissues with high proliferative activity are more likely to show cell death than tissues with low proliferative activity (Maruyama *et al.*, 1968). Although this may form at least a partial basis for sensitivity differences to such agents, there are cases which cannot be explained solely on these terms. It has been shown in this author's laboratory, in response to the cytotoxic agents hydroxyurea, cytosine arabinoside, and aminothiadiazole, that mesodermal cytotoxicity in the limb bud is severe while the limb ectoderm appears unaffected. In separate experiments the proliferative rate of these two tissues in 12-day rat embryos was determined using tritiated thymidine autoradiography. Following a 1-hr pulse, approximately 60% of the mesodermal cells were labeled, in contrast to about 30% of the ectodermal cells. With an agent such as hydroxyurea, which is thought to kill only those cells synthesizing DNA (S phase), one would thus expect twice as much cell death in the mesoderm as in ectoderm, if mitotic or proliferative rate was the only factor determining sensitivity.

Köhler and Merker (1973) have examined the effects of cyclophosphamide on RNA polymerases in various parts of the rat embryo during day 13 of development. They found a differential response between tissues, as well as a positive correlation between the severity of biochemical effect and the ability of cyclophosphamide to affect most severely those cells that are still proliferating but have begun to differentiate, as is the case in the forelimb of the rat embryo on day 13. In contrast, the hindlimb, which is rapidly proliferating at this time but has not begun to differentiate, is less severely affected. These authors suggest that the disturbance of mRNA metabolism, which varies according to the state of differentiation of the cell, is the basis for the selectivity of cyclophosphamide.

Another possible mechanism for the difference in cytotoxic response concerns the amount of cytotoxic agent reaching the cell. Factors such as differential drug distribution, permeability of cells to the agent, or the amount of intracellular binding might vary between sensitive and resistant cells. Pertinent studies are few in number and fail to implicate any of these factors. Billett *et al.* (1971) showed that actinomycin D affects neural structures and somites, sparing the heart and blood islands, but autoradiographic studies with radioactive drug did not reveal local accumulation in any of these tissues. Wolkowski (1970) also examined the distribution of actinomycin D in chick embryos and found regionally differing biochemical and morphological responses, despite the presence of drug in all regions examined. A greater demand for cells in the region of morphological deficit was offered as an explanation for this phenomenon.

An interesting study reported by Tahara and Kosin (1967) provides evidence that intrinsic cell differences, rather than drug distribution, is the major determinant of whether or not a cell will die. These authors discovered a condition (possibly viral) in turkey embryos in which massive cell death led to malformation. Cytotoxicity was most visible in the nervous system and mesenchymal cells. When tissues from these abnormal embryos were transplanted to normally developing turkey embryos, cell death was again seen in the nervous system and mesenchymal cells even when the site of transplant was at some distance from the affected tissues.

Another case of tissue sensitivity in which the quantity of agent reaching the cell does not appear to be decisive was reported by Williamson *et al.* (1965). These authors examined chick embryos following inoculation with Newcastle disease virus for the presence of antigen and tissue necrosis. They found that the cells of some tissues, such as the general body ectoderm and extraembryonic structures, seemed to be filled with virus without extensive cytotoxicity. The cells of other organs, in which viral antigen sometimes appears more slowly, were rapidly killed. The authors pointed out that organs which were severely affected were the same organs that reveal a degree of cell death during normal development. The significance of this association is not understood.

Another theory purporting to explain sensitivity differences among cells was put forward by Neubert *et al.* (1971) using cyclophosphamide as the cytotoxic agent. These authors suggest that the sensitivity of cells within a given tissue is based on the nutritional state of each cell. Since nutrition is carried out by diffusion during early embryonic stages, these authors postulated that cells furthest from the source of nutrition are most likely to die when extrinsic insult such as a cytotoxic agent is added. To strengthen this postulate they administered cyclophosphamide to pregnant rats and 2–4 hr later removed the embryos and cultured them for 24 hr in medium without cyclophosphamide. A more severe cytotoxic response was evident in these embryos than in embryos remaining *in vivo* for a corresponding length of time, leading to the speculation that unfavorable conditions *in vitro* had a potentiating effect by causing a poor nutritive status throughout the embryo as compared with normal conditions *in vivo*. An additional instance of specificity of morphological effect was reported by these authors, this relating to varying sensitivities of different germ layers. Using the nicotinamide analog, 6-aminonicotinamide, a severe vacuolization was found in ectodermal cells, while mesodermal and endodermal derivatives seemed to be completely free of such changes, but they could only speculate as to the reasons for these differences.

VI. SUMMARY

Many agents are undoubtedly teratogenic due to their ability to kill cells, especially proliferating cells. Such information, however, has not illuminated the full range of pathogenesis of a single congenital malformation.

This review has attempted to gather and discuss the limited information available on the secondary effects of cell death and reduced proliferative rate. The lowered cell number which must result from these effects certainly appears to play an important role in the genesis of many, but probably not all, malformations. Of greatest importance in this author's opinion is the imbalance in interactions between different tissues, as suggested by Wolff and colleagues. Inability to understand the intricacies of these interactions at present prevents a full understanding of the processes of abnormal development.

Also discussed was the basis for variations in sensitivity among cells of an embryo to various cytotoxic agents. Although little is known on this subject, it seems safe to say that agents which only kill proliferating cells (e.g., hydroxyurea) do so in proportion to the percentage of proliferating cells in the embryo or later in susceptible tissues. Even this generalization is not always borne out, indicating that there remain unknown factors which operate to determine sensitivity to cell death.

REFERENCES

Andreoli, J., Rodier, P., and Langman, J., 1973. The influence of a prenatal trauma on formation of Purkinje cells, *Am. J. Anat.* **137**:87–102.

Baserga, R., 1968, Biochemistry of the cell cycle: A review, *Cell Tissue Kinet.* **1**:167–191.

Billett, F. S., Bowman, P., and Pugh, D., 1971, The effects of actinomycin D on the early development of quail and chick embryos, *J. Embryol. Exp. Morphol.* **25**:385–403.

Bowman, P., 1967, The effect of 2,4-dinitrophenol on the development of early chick embryos, *J. Embryol. Exp. Morphol.* **17**:425–431.

Cameron, I. L., 1968, A method for the study of cell proliferation and renewal in the tissues of mammals, *in: Methods in Cell Physiology III* (D. M. Prescott, ed.), pp. 261–276, Academic Press, New York and London.

Cameron, I. L., 1971, Cell proliferation and renewal in the mammalian body, *in: Cellular and Molecular Renewal in the Mammalian Body* (I. L. Cameron and J. D. Thrasher, eds.), pp. 45–85, Academic Press, New York and London.

Chamberlain, J. G., 1970, Early neurovascular abnormalities underlying 6-aminonicotinamide (6-AN)-induced congenital hydrocephalus in rats, *Teratology* **3**:377–387.

Clavert, J., and Gabriel-Robez, O., 1974, The effects on mouse gestation and embryo development of an injection of viper venom (*Vipera aspis*), *Acta Anat.* **88**:11–21.

Cleaver, J. E., 1967, *Thymidine Metabolism and Cell Kinetics*, North Holland Publishing, Amsterdam.

Corliss, C. E., 1953, A study of mitotic activity in the early rat embryo, *J. Exp. Zool.* **122**:193–227.

Crawford, A. M., Kerr, J. F. R., and Currie, A. R., 1972, The relationship of acute mesodermal cell death to the teratogenic effects of 7-OHM-12-MBA in the foetal rat, *Br. J. Cancer* **26**:498–503.

Feinendegen, L. E., 1967, *Tritium Labeled Molecules in Biology and Medicine*, Academic Press, New York.

Ferm, V. H., 1964, Effect of transplacental mitotic inhibitors on the fetal hamster eye, *Anat. Rec.* **148**:129–133.

Ferm, V. H., and Kilham, L., 1965, Histopathologic basis of the teratogenic effects of H-1 virus on hamster embryos, *J. Embryol. Exp. Morphol.* **13**:151–158.

Flamm, H., Friedrich, F., Kovac, W., and Wiedermann, G., 1964, Experimentelle Infektion des Mäusefetus mit *Pasteurelle pseudotuberculosis*, *Biol. Neonate* **6**:52–75.

Forsberg, J.-G., and Källén, B., 1968, Cell death during embryogenesis, *Rev. Roum. Embryol. Cytol. Ser. Embryol.* **5**:91–102.

Forsberg, J.-G., and Olivecrona, A., 1966, The effect of prenatally administered busulphan on rat gonads, *Biol. Neonate* **10**:180–192.

Fränz, J., 1971, Cytological aspects of embryotoxicity and teratogenicity, *in: Malformations Congénitales des Mammifères* (H. Tuchmann-Duplessis, ed.), pp. 151–158, Masson & Cie., Paris.

Gibson, J. E., and Becker, B. A., 1968, Modification of cyclophosphamide teratogenicity by phenobarbital and SKF 525-A in mice, *Teratology* **1:**214.

Giroud, A., LeFebvres, J., Prost, H., and Dupuis, R., 1955, Malformation des membres dues à des lésions vasculaires chez le foetus de rat déficient en acide pantothenique, *J. Embryol. Exp. Morphol.* **3:**1–12.

Giroud, A., Giroud, P., Martinet, M., and Deluchat, Ch., 1968, Inapparent maternal infection by *Coxiella burnetti* and fetal repurcussions, *Teratology* **1:**257–262.

Glücksmann, A., 1951, Cell death in normal vertebrate ontogeny, *Biol. Rev.* **26:**59–86.

Glücksmann, A., 1965, Cell death in normal development, *Arch. Biol. Liége* **76:**419–437.

Grüneberg, H., 1963, *The Pathology of Development,* John Wiley, New York.

Grunewald, P., 1958, Malformations caused by necrosis in the embryo, *Am. J. Pathol.* **34:**77–95.

Hampé, A., 1966, Sur l'induction et la compétence dans les relations entre l'épiblaste et le mesenchyme de la patte de poulet, *J. Embryol. Exp. Morphol.* **8:**246–250.

Harrisson, C. M. H., 1971, The arrangement of chromatin in the interphase nucleus with reference to cell differentiation and repression in higher organisms, *Tissue Cell* **3:**523–550.

Hicks, S. P., and D'Amato, C. J., 1966, Effects of ionizing radiation on mammalian development, *in: Advances in Teratology* (D. H. M. Woollam, ed.), pp. 195–250, Academic Press, New York and London.

Ilies, A., 1970, Nécrose sélective au niveau du système nerveux embryonnaire et foetal, effet de l'hypoxie, *Rev. Roum. Embryol. Cytol. Ser. Embryol.* **7:**101–109.

Jelinek, R., and Dostal, M., 1974, Morphogenesis of cleft palate induced by exogenous factors. VII. Mitotic activity during formation of the mouse secondary palate, *Folia Morphol (Praha)* **22:**94–101.

Jelinek, R., and Dostal, M., 1975, Inhibitory effect of corticoids upon the proliferation pattern within the mouse palatal processes, *Teratology* **11:**193–198.

Johnson, E. M., 1964, Effects of maternal folic acid deficiency on cytologic phenomena in the rat embryo, *Anat. Rec.* **149:**49–56.

Jollie, W. P., 1968, Radioautographic evidence of materno-embryonic transport of thymidine into implanting rat embryos, *Acta Anat.* **70:**434–446.

Joneja, M. G., and LeLiever, W. C., 1973, In vivo effects of vinblastine and podophyllin on dividing cells of DBA mouse fetuses, *Can. J. Genet. Cytol.* **15:**491–495.

Jost, A., 1950, Dégénérescence des extrémitiés du foetus de rat sous l'action de certaines préparations hypophysaires, *C.R. Soc. Biol.* **144:**1324–1327.

Jost, A., 1951, Sur le rôle de la vasopressine et de la corticostimuline (ACTH) dans la production expérimentale de lésions des extrémitiés foetales (hemorrhagies, nécroses, amputations, congénitales), *C.R. Soc. Biol.* **145:**1805–1809.

Jost, A., Roffi, J., and Courtat, M., 1969, Congenital amputations determined by the br. gene and those induced by adrenalin injection in the rabbit fetus, *in: Limb Development and Deformity* (C. Swinyard, ed.), pp. 187–199, Charles C Thomas, Springfield, Ill.

Jurand, A., 1961, Further investigations on the cytotoxic and morphogenetic effects of some nitrogen mustard derivatives, *J. Embryol. Exp. Morphol.* **9:**492–506.

Jurand, A., 1968, The effect of hydrocortisone acetate on the development of mouse embryos, *J. Embryol. Exp. Morphol.* **20:**355–366.

Jurand, A., 1973, Teratogenic activity of methadone hydrochloride in mouse and chick embryos, *J. Embryol. Exp. Morphol.* **30:**449–458.

Källén, B., and Rüdeberg, S. I., 1966, Teratogenic effects of electric light on early chick embryos, *Acta Morphol. Neerl.-Scand.* **6:**95–99.

Kalter, H., 1968, *Teratology of the Central Nervous System,* Univ. of Chicago Press, Chicago, Ill.

Kauffman, S. L., 1968, Lengthening of the generation cycle during embryonic differentiation of the mouse neural tube, *Exp. Cell Res.* **49:**420–424.

Kauffman, S. L., 1969, Cell proliferation in embryonic mouse neural tube following urethane exposure, *Dev. Biol.* **20:**146–157.

Kauffman, S. L., and Herman, C., 1968, Ultrastructural changes in embryonic mouse neural tube cells after urethane exposure, *Dev. Biol.* **17:**55–74.

Kilham, L., and Margolis, G., 1974, Congenital infections due to reovirus type 3 in hamsters, *Teratology* **9:**51–63.

Kilham, L., Margolis, G., and Colby, E. D., 1967, Congenital infections of cats and ferrets by feline panleukopenia virus manifested by cerebellar hypoplasia, *Lab. Invest.* **17:**465–480.

Kochhar, D. M., 1968, Studies of vitamin A-induced teratogenesis: Effects on embryonic mesenchyme and epithelium, and on incorporation of H^3-thymidine, *Teratology* **1:**299–305.

Köhler, E., Merker, H.-J., Ehmke, W., and Wojnorwicz, F., 1972, Growth kinetics of mammalian embryos during the stage of differentiation, *Naunyn-Schmiedeberg's Arch. Pharmacol.* **272:**169–181.

Köhler, E., and Merker, H.-J., 1973, Effect of cyclophosphamide pretreatment of pregnant animals on the activity of nuclear DNA-dependent RNA-polymerases in different parts of rat embryos, *Naunyn-Schmiedeberg's Arch. Pharmacol.* **277:**71–88.

Kury, G., and Craig, J. M., 1966, Congenital malformations produced in chickens by fluorinated pyrimidines, *Arch. Pathol.* **81:**166–173.

Kury, G., and Craig, J. M., 1967, The effect of mitomycin C on developing chicken embryos, *J. Embryol. Exp. Morphol.* **17:**229–237.

Lash, J. W., and Saxén, L., 1972, Human teratogenesis: *In vitro* studies on thalidomide-inhibited chondrogenesis, *Dev. Biol.* **28:**61–70.

Lee, H., Cortés, J. L., and Levin, M. A., 1972, Teratogenic effects of phleomycin in early chick embryos, *Teratology,* **6:**201–203.

Lehmiller, D. J., 1971, Approaches to the study of molecular and cellular renewal, *in: Cellular and Molecular Renewal in the Mammalian Body* (I. L. Cameron and J. D. Thrasher, eds.), pp. 1–24, Academic Press, New York and London.

Lejour-Jeanty, M., 1966, Becs-de-lièvre provoqués chez le rat par un dérivé de la pénicilline, l'hadacidine, *J. Embryol. Exp. Morphol.* **15:**193–211.

Lockshin, R. A., and Beaulaton, J., 1974, Programmed cell death, *Life Sci.* **15:**1549–1566.

Marin-Padilla, M., and Benirschke, K., 1965, Thalidomide injury to the myocardium of armadillo embryos, *J. Embryol. Exp. Morphol.* **13:**235–241.

Marin-Padilla, M., and Ferm, V. H., 1965, Somite necrosis and developmental malformations induced by vitamin A in the golden hamster, *J. Embryol. Exp. Morphol.* **13:**1–8.

Maruyama, S., Chiga, M., and D'Agostino, A. N., 1968, Selective necrosis in the fetal rat central nervous system produced by 5-fluoro-2'-deoxyuridine. A morphologic study, *J. Neuropathol. Exp. Neurol.* **27:**96–107.

Menkes, B., Sandor, S., and Ilies, A., 1970, Cell death in teratogenesis, *in: Advances in Teratology* (D. H. M. Woollam, ed.), pp. 169–215, Academic Press, New York and London.

Merker, H. J., Pospisil, M., and Mewes, P., 1975, The effect of 6-mercaptopurine on the limb development of rat embryos, *Teratology* **11:**199–218.

Miller, S. A., 1974, Pattern of short term, *in vitro* ^{3}H-thymidine labeling in the tissues of the 8½ day mouse embryo, *Anat. Rec.* **178:**418–419.

Miller, S. A., and Runner, M. N., 1975, Differential permeability of murine visceral yolk sac to thymidine and to hydroxyurea, *Dev. Biol.* **45:**74–80.

Mott, W. J., Toto, P. D., and Hilgers, D. C., 1969, Labeling index and cellular density in palatine shelves of cleft-palate mice, *J. Dent. Res.* **48:**263–265.

Mottet, N. K., and Hammar, S. P., 1972, Ribosome crystals in necrotizing cells from the posterior necrotic zone of the developing chick limb, *J. Cell Sci.* **11:**403–414.

Naeye, R. L., 1967, Cytomegalic inclusion disease. The fetal disorder, *Am. J. Clin. Pathol.* **47:**738–744.

Neubert, D., Merker, H.-J., Köhler, E., Krowke, R., and Barrach, H. J., 1971, Biochemical aspects of teratology, *Adv. Biosci.* **6:**575–621.

O'Dell, D. S., and McKenzie, J., 1963, The action of aminopterin on the explanted early chick embryo, *J. Embryol. Exp. Morphol.* **11:**185–200.

Palludan, B., 1966, *A-Avitaminosis in Swine,* Munksgaard, Copenhagen.

Pasteels, J., 1943, Proliférations et croissance dans la gastrulation et la formation de la queue de vertébrates, *Arch Biol.* **54:**1–51.

Pexieder, T., 1975, Cell death in the morphogenesis and teratogenesis of the heart, *Adv. Anat. Embryol. Cell Biol.* **51:**1–100.

Preto, V., 1938, Analisi della distribuzione dell'attività mitotica in giovani embrioni di ratto, *Arch. Ital. Anat. Embriol.* **41:**165–206.

Råska, K., Zedeck, M. S., and Welch, A. D., 1966, Relationship between the metabolic effects and the pregnancy-interrupting property of 6-azauridine in mice, *Biochem. Pharmacol.* **15:**2136–2138.

Rickenbacher, J., 1968, The importance of the regulation for the normal and abnormal development. Experimental investigations on the limb buds of chicken embryos, *Biol. Neonate* **12:**65–87.

Ritter, E. J., Scott, W. J., and Wilson, J. G., 1971, Teratogenesis and inhibition of DNA synthesis induced in rat embryos by cytosine arabinoside, *Teratology* **4:**7–14.

Ritter, E. J., Scott, W. J., and Wilson, J. G., 1973, Relationship of temporal patterns of cell death and development to malformations in the rat limb. Possible mechanisms of teratogenesis with inhibitors of DNA synthesis, *Teratology* **7:**219–226.

Rubin, L., and Saunders, J. W., 1972, Ectodermal–mesodermal interactions in the growth of limb buds in the chick embryo: Constancy and temporal limits of the ectodermal induction, *Dev. Biol.* **28:**94–112.

Ruch, J. V., Robez-Kremes, G., Scheegans, O. E., and Rohmer, A., 1962, Sur les malformations cardiaques observées chez des monstres de poulets obtenues par le venin d'abbeille (Apivene), *C.R. Soc. Biol. Paris* **156:**379–382.

Sandor, S., 1968, The influence of ethyl alcohol on the developing chick embryo, *Rev. Roum. Embryol. Cytol. Ser. Embryol.* **5:**167–171.

Sandor, S., and Elias, S., 1968, Evolution of a circumscribed necrotic area in the axis of the chick embryo, induced by irradiation with ultraviolet rays, *Rev. Roum. Embryol. Cytol. Ser. Embryol.* **5:**173–180.

Sauer, M. E., and Duncan, D., 1962, Cytoplasmic inclusions containing deoxyrionucleic acid in the neural tube of chick embryos exposed to ionizing radiation, *in: Response of the Nervous System to Ionizing Radiation* (T. J. Haley and R. S. Snider, eds.), pp. 75–94, Academic Press, New York and London.

Saunders, J. W., 1966, Death in embryonic systems, *Science* **154:**604–612.

Schweichel, J.-U., 1970, Morphological studies on the problem of physiological cell death during embryonic development, *in: Metabolic Pathways in Mammalian Embryos during Organogenesis and its Modification by Drugs* (R. Bass, F. Beck, H.-J. Merker, D. Neubert, and B. Randhan, eds.), pp. 41–54, Freie Universität, Berlin.

Schweichel, J.-U., and Merker, H.-J., 1973, The morphology of various types of cell death in prenatal tissues, *Teratology* **7:**253–266.

Scott, W. J., Ritter, E. J., and Wilson, J. G., 1971, DNA synthesis inhibition and cell death associated with hydroxyurea teratogenesis in rat embryos, *Dev. Biol.* **26:**306–315.

Scott, W. J., Ritter, E. J., and Wilson, J. G., 1973a, DNA synthesis inhibition, cytotoxicity and their relationship to teratogenesis following administration of a nicotinamide antagonist, aminothiadiazole, to pregnant rats, *J. Embryol. Exp. Morphol.* **30:**257–266.

Scott, W. J., Ritter, E. J., and Wilson, J. G., 1973b, Polydactyly induced in rat fetuses by cytosine arabinoside (ara-C), *Teratology* **7:**A-26.

Scott, W. J., Ritter, E. J., and Wilson, J. G., 1975, Studies on induction of polydactyly in rats with cytosine arabinoside, *Dev. Biol.* **45:**103–111.

Seifertová, M., Čihák, A., and Vesely, J., 1973, Effect of 5-azacytidine and 6-azauridine on the synthesis of DNA in embryonic mouse brain mitotic activity and migration of ventricular cells, *Neoplasma* **20:**243–249.

Shapira, Y., and Shoshan, S., 1972, The effect of cortisone on collagen synthesis in the secondary palate of mice, *Arch. Oral Biol.* **17:**1699–1703.

Shepard, T. H., Lemire, R. J., Aksu, O., and Mackler, B., 1968, Studies of the development of congenital anomalies in embryos of riboflavin-deficient, galactoflavin fed rats. I. Growth and embryologic pathology, *Teratology* **1**:75–92.

Sinclair, J. G., 1950, A specific transplacental effect of urethane in mice, *Tex. Rep. Biol. Med.* **8**:623–632.

Singh, S., and Singh, G., 1973, Hemorrhages in the limbs of fetal rats after amniocentesis and their role in limb malformations, *Teratology* **8**:11–17.

Skalko, R. G., and Sax, R. D., 1974, Teratogenic and biochemical aspects of the interaction of 5-diazouracil and 5-fluorouracil in the mouse embryo, *Toxicol. Appl. Pharmacol.* **29**:124.

Škreb, N., and Frank, Z., 1963, Developmental abnormalities in the rat induced by heat shock, *J. Embryol. Exp. Morphol.* **11**:445–457.

Solter, D., Škreb, N., and Damjanov, I., 1971, Cell cycle analysis in the mouse egg-cylinder, *Exp. Cell Res.* **64**:331–334.

Stepanovich, V., and Gianelly, A., 1971, Preliminary studies of the lipids of normal and cleft palates of the rat, *J. Dent. Res.* **50**:1360.

Tahara, Y., and Kosin, I. L., 1967, Observations on congenital teratology in turkey embryos and its experimental transmission via transplantation, *J. Embryol. Exp. Morphol.* **18**:305–319.

Thalhammer, O., 1957, Die Vaccine-Virusembryopathie der weissen Maus, *Wien. Z. Inn. Med.* **38**:41.

Töndury, G., and Smith, D. W., 1968, Fetal rubella pathology, *J. Pediatr.* **68**:867–879.

Webster, D. A., and Gross, J., 1970, Studies on possible mechanisms of programmed cell death in the chick embryo, *Dev. Biol.* **22**:157–184.

Wechsler, W., 1973, Carcinogenic and teratogenic effects of ethylnitrosourea and methylnitrosourea during pregnancy in experimental rats, *in: Transplacental Carcinogenesis* (L. Tomatis and U. Mohr, eds.), Vol. 4, pp. 127–142, International Agency for Research on Cancer, Lyon.

Williamson, A. P., Blattner, R. J., and Robertson, G. C., 1965, The relationship of viral antigen to virus-induced defects in chick embryos. Newcastle disease virus, *Dev. Biol.* **12**:498–519.

Wilson, J. G., 1954, Differentiation and reaction of rat embryos to radiation, *J. Cell Comp. Physiol.* **43**(Suppl.):11–37.

Wolff, E., 1966, The experimental production and the explanation of phocomelia in the chick embryo, *in: International Workshop in Teratology*, pp. 84–94, Copenhagen.

Wolkowski, R., 1970, Effect of actinomycin D on early axial development in chick embryos, *Teratology* **3**:389–397.

Zwilling, E., 1959, Micromelia as a direct effect of insulin. Evidence from *in vitro* and *in vivo* experiments, *J. Morphol.* **104**:159–180.

Altered Biosynthesis   5

EDMOND J. RITTER

I. INTRODUCTION

Biosynthesis is essential to life. All living organisms require the synthesis of such macromolecules as DNA, the RNAs, and various structural and enzyme proteins. The synthesis of macromolecules is in turn dependent upon the synthesis or availability of precursors—the building blocks—and of ATP, the energy needed to put them together, as well as that required for the organism to function.

The embryo during organogenesis is particularly vulnerable to alterations of normal biosynthesis because it must maintain critical developmental schedules and because its cells are undergoing rapid proliferation. Such cells are in jeopardy if their normal cell cycle is interrupted for any reason, including altered patterns of biosynthesis. This vulnerability may lead to reduced cellular proliferation or outright necrosis and in turn to a lack of the correct number, type, or location of cells in the embryo. It is apparent that a paucity of cells is frequently associated with teratogenesis (see Chapter 4, this volume), although the steps leading from a cellular deficit to teratogenesis are poorly understood.

The embryo has regenerative capability and attempts to replenish the lost cells. If the repair is adequate and "in time," no gross malformations are seen. If not, teratogenesis occurs (Hicks *et al.*, 1957); that is, malformation is the outcome of interference with key developmental and growth processes, on the one hand, and of repair, on the other. By "in time" is meant not only the time in development, organogenesis, when embryo cells are particularly sensitive to damage, but also to times when certain structures are secondarily in

EDMOND J. RITTER • Children's Hospital Research Foundation and Department of Pediatrics, University of Cincinnati, Cincinnati, Ohio 45229.

jeopardy because of what might be called "architectural" relationships. For example, the neural tube must close on schedule otherwise spina bifida results. The palatal shelves must touch and fuse during a limited time span, and if they do not, they grow further apart resulting in cleft palate. A cytotoxic drug such as hydroxyurea kills mesodermal cells in the limb bud, but residual tissue attempts to replenish the cellular deficiency. If it does so by the critical time at which mesenchymal condensation lays down the digital rays, then the limb develops with a normal number of digits; if not, ectrodactyly occurs. The concept of interference with such temporal–spatial relationships is important to an understanding of teratogenic mechanisms, but definition is difficult since the controlling factors are not understood even for normal growth and development. One of the most intriguing questions is what triggers developmental events, why for example digital condensation occurs even though the limb bud is not ready, e.g., may have suffered cellular depletion. Subtle damage to organs and tissues may also occur that is less readily apparent than are gross malformations involving critical spatial relationships, for example, brain damage resulting in a performance deficit rather than a visible malformation (Butcher *et al.*, 1973).

There are various reasons for aberrations of normal biosynthesis. Many terata are genetic in origin, presumably due to genetically determined deficiencies of enzymes critical to development. Viruses which induce malformations may do so by taking over the cellular economy, leading to cytotoxicity and retardation of growth or differentiation. However, much of experimental teratology has been done by means of vitamin or other deficiencies or with chemical agents which interfere with embryonic development by inhibition of biosynthetic pathways; for example, cytosine arabinoside, an inhibitor of DNA synthesis, is a potent teratogen. 6-Aminonicotinamide may act, at least in part, by inhibiting the synthesis of adenosine triphosphate (Ritter *et al.*, 1975a). In fact, this chapter is concerned chiefly with studies of teratogenesis and teratogenic mechanisms involving such agents that appear to act by altering, usually reducing, biosynthesis.

Superficially this appears to be a rational enough approach, but several precautions are necessary. In only relatively few cases has the biochemical action of an agent been studied in the mammalian embryo. Mostly its action must be inferred from studies in other systems (cultured neoplastic cells, for example) and applied to mammals by inference. The primary inhibitory action attributed to an agent may depend on the particular test system in use, the concentration of the agent being used, and the endpoints employed. Thus, actinomycin D will inhibit only RNA in one circumstance, but may also affect DNA in another. To compound confusion, the primary inhibitory action may produce secondary effects. A protein-synthesis inhibitor, by inhibiting the synthesis of protein required for DNA synthesis, may secondarily inhibit DNA synthesis. Inhibition of messenger RNA synthesis precludes future protein synthesis. A DNA-synthesis inhibitor may kill cells, terminating all biosynthesis. The use of inhibitors of biosynthesis as tools to study teratogenic mechanisms leads to complexity, but this may still be less than is the case with spontaneous malformations. Metabolic inhibitors are only tools, and one must be aware of both their advantages and disadvantages.

Teratogenesis frequently involves cell death, although how cell death results from biochemical deviation is usually not known. In other cases necrosis may not be apparent, but there are other signs of reduced proliferation or abnormal differentiation. In most instances it seems that selective damage is required to produce teratogenesis, perhaps an organ that is growing faster than others at the time of insult is subject to greater damage. In other cases, as is presumed with some protein inhibitors, a low dose may produce generalized growth retardation and a larger dose may kill so many cells that the entire embryo dies without a teratogenic dose being found. As suggested by Warkany and Schraffenberger (1944), the teratogenic dosage is sometimes borderline. A higher dosage may be lethal by damaging too many cells and organs to be compatible with life, whereas the lesser damage produced by a lower dose may be within the ability of the embryo to repair so that no gross malformations are seen. It is important that alterations in biosynthesis caused by teratogenic agents, as well as the resulting effects such as cytotoxicity or reduced proliferation and differentiation, be viewed against the backdrop of an interaction between the embryos' reparative ability and the beginning of irreversible morphogenetic events.

II. INHIBITION OF DNA SYNTHESIS

Agents which inhibit DNA synthesis, widely used as cancer chemotherapy agents, are also powerful teratogens. This is expected since the embryo undergoes a rapid increase in cell number, requiring a doubling of DNA for each cellular proliferative cycle. Yet it is not clear that the inhibition of DNA synthesis per se is teratogenic. More likely the cell death which often accompanies inhibition of DNA synthesis is the actual cause of abnormality (see Chapter 4, this volume).

The pathways by which the inhibition of DNA synthesis leads to necrosis are little understood. Agents which inhibit the synthesis of a precursor required for DNA synthesis, for example, thymidine triphosphate, are thought to cause "unbalanced growth," and after a mitotic cycle or two the cells die. While the latter is probably true, it is not clear exactly how growth is unbalanced or precisely why the cells die. The mechanism of action of hydroxyurea is thought to involve inhibition of the enzyme ribonucleoside diphosphate reductase, thereby preventing the synthesis of deoxyribonucleotide precursors required for the synthesis of DNA. But hydroxyurea kills embryo cells as early as three hours after injection into the pregnant rat (Scott *et al.*, 1971), and there is no known mechanism by which the inhibition of DNA synthesis can cause necrosis so quickly.

In an effort to elucidate teratogenic mechanisms, a study was undertaken using the chemotherapeutic and teratogenic drug cytosine arabinoside (ara-C) administered to pregnant rats (Ritter *et al.*, 1971). This agent was chosen because it was thought to have only one action, the inhibition of DNA synthesis by interference with the enzyme DNA polymerase; thus it appeared to offer a simplified system for investigation. Ara-C was administered intraperitoneally to pregnant rats on day 12 of gestation using doses between 25 and 200 mg/kg. At various times thereafter the rats were injected with tritiated thymidine, and two hours later one uterine horn was removed and the embryos processed for scintillation counting to determine thymidine incorporation. The other horn was left *in situ* so that the fetuses could be examined at term. The purpose of this "split litter" technique was to permit comparison of the early depression of thymidine incorporation and later teratogenesis within the same litter, avoiding differences inherent in interlitter comparison. The results are seen in Figure 1 and Table 1. With 25 mg/kg there was no teratogenesis, despite a considerable reduction in DNA synthesis as indicated by decreased thymidine incorporation. At higher dosage thymidine incorporation was further decreased and teratogenesis became more severe: 100 and 200 mg/kg caused 95 and 100% of survivors to be malformed. The effects at higher doses were not surprising, but it was not expected to find that something as vital to embryonic growth and development as DNA synthesis could be inhibited as much as was the case with the 25 mg/kg dose and still produce no persistent abnormality. Further study suggested that the teratogenic factor was not inhibition of DNA synthesis per

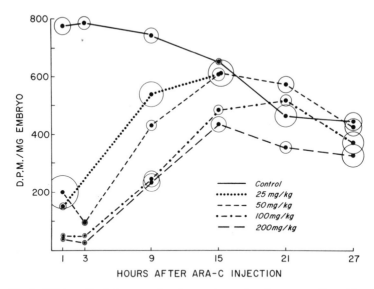

Fig. 1. DNA synthesis in control and experimental embryos as reflected by incorporation of [³H]thymidine beginning 1–27 hr after various dosages of ara-C given to 12-day pregnant rats. Each point represents the mean of scintillation counts on 3–5 embryos from each of 3 pregnant rats 2 hr after thymidine injection. Standard error of the mean is represented by the circles about each point (Ritter *et al.*, 1971).

**Table 1. Embryotoxicity Following Treatment
with Cytosine Arabinoside at Day 12 of Rat Gestation**

Dose of ara-C (mg/kg)	Total number implants	Embryotoxicity observed at day 20	
		Dead, resorbed (%)	Survivors malformed (%)
Without Thymidine and Surgery			
None[a]	477	4	2
50	74	4	16
100	101	4	67
200	82	3	85
Thymidine and One Uterine Horn Removed			
None	42	3	0
25	45	7	0
50	68	7	48
100	76	17	95
200	79	17	100

[a]Cumulative nontreated controls for our laboratory over last 4 years
(Ritter *et al.*, 1971).

se but rather the cytotoxicity which accompanied it. Histologic study of embryo sections showed little cell death with 25 mg/kg ara-C, whereas massive necrosis was seen with higher doses. It was apparent here, as reported earlier by Hicks *et al.* (1957) after X-irradiation, that necrotic damage was subject to repair and that the embryo thereby was able to compensate for moderate damage and prevent abnormality. They noted, as did Shepard *et al.* (1968) with teratogenesis induced by riboflavin deficiency, that cell death was predominant in particular areas, the locations corresponding with later sites of teratogenesis.

A similar study with hydroxyurea (Scott *et al.*, 1971) also showed a dose-related inhibition of thymidine incorporation: slight necrosis and mild teratogenesis with lower doses, and massive necrosis and severe teratogenesis with higher doses. Analysis of these and other studies suggests the primacy of cell death rather than inhibition of DNA synthesis per se as the teratogenic mechanism (Ritter *et al.*, 1973).

It is well established in teratology that the time during organogenesis at which sensitive tissues or organs are subjected to insult often determines the type of defect produced (see Chapter 2, Vol. 1). Both ara-C and hydroxyurea were observed to depress thymidine incorporation at about the same time, to roughly the same degree, and for about the same duration (Ritter *et al.*, 1973). Consequently, if teratogenesis depended mainly on the inhibition of DNA synthesis, then the same teratogenic manifestations should occur after both drugs. However, hydroxyurea was found to kill embryo cells very quickly, in about 3–5 hr, whereas after ara-C equivalent necrosis was not seen until 23–29 hr after drug administration. Using limb malformations as a basis of comparison, it was seen that hydroxyurea given on day 12 resulted in mainly forelimb ectrodactyly with only one digit missing per limb, whereas ara-C on day 12 gave ectrodactyly with usually more than one digit missing per forelimb and with some hindlimb ectrodactyly also. Taking into account the greater time available for repair after HU-induced necrosis, as compared with that resulting from ara-C, and in view of the schedule for digital development, it appeared likely that necrosis and degree of repair, rather than DNA-synthesis inhibition, were more closely associated with the teratogenic differences. Admittedly DNA-synthesis inhibition may have been directly or indirectly responsible for the necrosis, but the small difference in inhibition produced by the two drugs would make it difficult to account for the time difference in onset of necrosis.

Other agents which damage DNA or inhibit its synthesis are cytotoxic and teratogenic, including FUdR, methotrexate and other antifolates, 6-mercaptopurine, 6-thioguanine, and a wide variety of antineoplastic alkylating agents such as busulfan, cyclophosphamide, and chlorambucil (Shepard, 1973). It appears likely that any agent which reaches the embryo during organogenesis and manifests differential cytotoxicity will be teratogenic.

III. INHIBITION OF RNA SYNTHESIS

Few RNA-synthesis inhibitors have been studied as teratogens, but one, actinomycin D (act D) has been extensively investigated. It is a well-known inhibitor of DNA-dependent RNA synthesis, and consequently of protein synthesis as well. In higher dosage it can also inhibit DNA synthesis. The teratogenic mechanism is not clear, but since RNA synthesis, differentiation, and development appear causally related, it is reasonable to propose that act D acts as a teratogen by interfering with RNA synthesis, although this has not been established. In fact act D is known to possess marked cytotoxic properties, and necrosis must be considered a possible factor in teratogenesis. Again, one must look beyond the immediate effect, which may well involve inhibition of RNA synthesis, to the total effect, which may be biochemically complex and involve maternal as well as embryonic metabolism.

There is an extensive literature on the effect of act D in various organisms, but its teratogenic activity in the mammal was first described by Tuchmann-Duplessis and Mercier-Parot (1958, 1960). They administered act D intraperitoneally to Wistar rats at various doses between 25 and 100 μg/kg on various gestational days between 3 and 14 and found that it induced either resorptions or resorptions and malformations, depending on the dose and times of administration. Wilson (1966) studied the effects of 100, 200, and 300 μg/kg single intraperitoneal injections to Wistar rats on days 5–11. At all doses the greatest sensitivity was found on day 9. Chronic treatment was less effective than a single treatment on day 9 but accentuated the lethal and teratogenic effects when given in addition to administration on day 9.

Harvey and Srebnik (1968), using Long–Evans rats, found that act D, 200 μg/kg subcutaneously, on day 12 resulted in 63% living young of which 14% were malformed (club foot, gastroschisis) or severely stunted. They also found that treatment with thyroxine markedly reduced fetal death and prevented congenital malformation. The rationale for this effect is not apparent, but presumably thyroxine stimulates cellular growth and development as much as act D depresses it.

Jordan (1969) and Jordan and Wilson (1970) investigated the mechanism of act D teratogenesis in the rat by means of autoradiography. Following treatment of the dam with tritiated act D on either day 7, 8, 9, 10, or 11, small amounts of this agent were found in the 9-, 10-, and 11-day embryos. Treatment of histologic sections with deoxyribonuclease removed the label, indicating that the act D had been attached to the DNA as expected. The effect of act D on incorporation of tritiated uridine was determined on days 7 through 11. Uridine is largely incorporated into RNA. Little depression of uridine incorporation was seen on days 7 and 8, but on day 9 significant depression was seen in the visceral yolk sac and embryonic neural cells, with some depression in trophoblast giant cells and no significant decrease in maternal decidua or parietal yolk sac. On day 10 significant or moderate depression was found in all these tissues except the maternal decidua. On day 11 no significant depression of uridine incorporation was found in any of these tissues. Thus the decrease in embryonic uridine incorporation gives some degree of correlation with teratogenesis. It is interesting, however, that relatively few resorptions or malformations occur following treatment on day 11 nor was there a reduction in uridine incorporation, despite a positive finding of act D in embryonic tissue at that time. Of particular importance may be the fact that Jordan (1969) found cell damage (pyknotic nuclei, cytoplasmic vacuolization) on days 8, 9, and 10 when the embryos were particularly sensitive, but not at earlier times or on day 11. Thus teratogenesis showed positive correlation with necrosis as well as with inhibition of uridine incorporation. Why act D is cytotoxic is not apparent.

The lack of effect on the 11th day and the fact that resorptions and malformations occurred on days 5–10 led Köhler and Merker (1972) into further studies. They injected tritium-labeled act D into pregnant rats on days 11–14 of pregnancy, after which maternal and embryonic tissues were removed and radioactivity was measured by scintillation counting. The concentration of act D found in the 11-day embryo was far too low to effectively inhibit RNA polymerase but was much larger than the concentration on subsequent days. After treatment on day 11, embryos showed no reduction in RNA polymerase activity, whereas decidua, extraembryonic tissue, and maternal liver, kidney, and spleen showed large reductions. In other experiments pregnant rats were treated on day 8 with act D (300 μg/kg intraperitoneally), and implantation sites were removed at various times later for measurement of total RNA and DNA. At 36 and 48 hr both RNA and DNA were reduced by about 30–40%. Thus it appeared that a primary inhibition of RNA synthesis led secondarily to DNA synthesis inhibition, again pointing out the necessity of considering overall effects on the embryo when investigating teratogenic mechanisms. These workers also studied the cellular morphology of implantation sites on days 7, 8, and 9 at various times following treatment with act D by means of electron microscopy. The decidua showed pronounced degenerative nucleolar and other changes which were not seen in the embryo cells, but evidence of necrosis was observed in the embryonic and extraembryonic tissues. It was suggested that the lesion of the decidua is important to the embryotoxic effect of act D up to day 9 or 10 of gestation, possibly by inhibition of synthesis of proteins necessary for growth. Further discussions of the teratogenic mechanisms of act D are presented by Winfield and Bennett (1971).

In summary, while it is assumed that act D acts as a cytotoxic and teratogenic agent due to its well-known ability to complex with DNA and inhibit RNA synthesis, the mechanisms involved are not fully understood. Possibly act D does not act directly on the embryo, since the concentrations found in the embryo are quite small. Inhibition of RNA synthesis, embryonic cell death, and degenerative changes in decidual tissue have been temporally associated with teratogenesis. Possibly deranged decidual function is connected with necrotic or differentiative embryonic changes. The reason for the lack of teratogenesis at later gestational times may be either that act D does not reach the embryo at this time or that the embryo now receives adequate nutrition via the newly developed chorioallantoic placenta. These areas invite further research.

Daunomycin and chromomycin also bind to DNA and inhibit RNA synthesis (Balis, 1968, p. 137), and daunomycin inhibits DNA synthesis (Rusconi and Calendi, 1966). Under the name rubidomycin, daunomycin was found to be teratogenic in the rat by Roux and Taillemite (1969) and Roux *et al.* (1971). Shepard (1973) noted that M. Takaya reported chromomycin, an inhibitor of RNA synthesis, to be teratogenic in rats. T. Tanimura and H. Nishimura injected 0.5–2.0 mg/kg chromomycin into mice on day 10, 11, or 12 and found 36–51% of the fetuses had kinky tails, hydrops, or abnormal head shapes.

IV. INHIBITION OF PROTEIN SYNTHESIS

Inhibitors of DNA synthesis are virtually always teratogenic. By contrast, most inhibitors of protein synthesis show little or no teratogenic effects, if defined as structural defects. As to mechanisms, there are several points in protein synthesis which are amenable to inhibition by suitable agents. The probable site of inhibitory action is stated for the various agents reviewed below, but it should be noted that there is an extensive and sometimes controversial literature on this subject. Primary inhibition of protein synthesis will result in various secondary effects, including inhibition of nucleic acid synthesis since biosynthetic pathways require prior synthesis of particular enzymes.

The cytotoxic antibiotic chloramphenicol is thought to inhibit protein synthesis by blocking the attachment of messenger RNA to ribosomes (Armentrout and Weisberger, 1967; Ingall and Sherman, 1970). In the chick embryo chloramphenicol inhibited protein synthesis and growth but caused only infrequent malformations (Blackwood, 1962; Billet *et al.*, 1965). In the rabbit (Brown *et al.*, 1968) and monkey (Courtney and Valerio, 1968) no congenital defects were observed, and Fritz and Hess (1971) found little evidence of teratogenicity from chloramphenicol in the rat, mouse, or rabbit. In this laboratory preliminary results, following administration of chloramphenicol at 50–150 mg/kg to rats on day 12 of gestation, showed 96 normal surviving fetuses showing no significant weight reduction, 9 resorptions, and only one fetus with induced hematoma of the left forelimb (Wilson and Hallett, unpublished).

The tetracyclines apparently inhibit protein synthesis by interfering with ribosomal binding of aminoacyl tRNA (Carter and McCarty, 1966). They are also active chelating compounds, appear to interfere with the phosphorylation of glucose, and cause a derangement of cellular mechanisms responsible for nucleic acid synthesis (Kunin, 1970). Placental transfer occurs and tetracyclines are deposited and cause red discoloration during ossification of fetal cartilages (Filippi, 1967). The offspring of 20 rats treated from the 5th to the 20th day of pregnancy with oral tetracycline, 5 mg daily, showed cleft palate associated with hypoplasia in 30% and shortness of the distal segment of the limbs and syndactyly in 33%. Two other studies did not confirm these findings (Bevelander and Cohlan, 1962; Hurley and Tuchmann-Duplessis, 1963), and a third found only an increase in hydroureters following exposure to 500 mg daily (McColl *et al.*, 1965). Unpublished results in this laboratory showed no increase in embryotoxicity in offspring from 12 litters treated on day 12 of gestation with 40 mg/kg tetracycline.

Streptomycin and kanamycin produce misreadings in the genetic code at the level of the ribosome, resulting in synthesis of defective proteins according to Davies *et al.* (1964) and Sanders and Cluff (1970). Streptomycin administration at various times in gestation has been associated with deafness in children, although in one study 50 children who had been exposed *in utero* were found to have normal hearing (Varpela *et al.*, 1969). Boucher and Delost (1964) found growth retardation in the offspring of mice treated with streptomycin. Ericson-Strandvik and Gyllensten (1963) administered this drug to mice from days 9–13 of gestation and found at 1–6 hr after the last treatment significant amounts of streptomycin in the embryos, but there was no appreciable embryotoxicity. In unpublished studies in this laboratory, no significant effects were found following administration of streptomycin to rats on day 12 of pregnancy. Rats injected with kanamycin (100 mg/kg) from day 8 through 16 produced normal litters (Bevelander and Cohlan, 1962).

Puromycin, a compound that mimics the binding sites of tRNA, inhibits protein synthesis by aborting incomplete peptide chains. In an attempt to produce experimental congenital nephrosis, Hallman *et al.* (1960) administered puromycin to rats during the last days of pregnancy. The offspring died 1–4 days after birth; some of the young showed marked dilatation of the tubules, and in some a complete nephrolysis was in process. Bernstein *et al.* (1962), in a similar experiment, were unable to confirm these results. They administered various amounts of puromycin daily to 51 pregnant rats for periods of 7 days, beginning on the 5th to 14th day of gestation. The highest dosage used caused abortion in most animals. Examination of the 170 living offspring disclosed no gross malformations or microscopic abnormalities of the kidneys, and no other anomalies were observed. They suggested that the toxic effect of puromycin may have been mediated through interference with placental function since it either produced abortion or left the fetus unaffected. In view of this apparent disagreement, Alexander *et al.* (1966) studied the teratogenic effect of puromycin in rats and found that abnormal nephrogenic development occurred in 20% of newborn of mothers treated on days 8–10 with 3 mg puromycin per day. Except for the liver, which was normal, no attempt was made to study teratogenesis in other organs.

The rapid suppression of protein synthesis by cycloheximide may be followed by inhibition of DNA synthesis (Brown *et al.*, 1970). It is thought to inhibit protein synthesis by inhibiting ribosome function (Balis, 1968, p. 215). Zimmerman *et al.* (1970) found that 10 mg/kg in mice on day 11.5 resulted in resorptions but did not cause cleft palate. In unpublished experiments in this laboratory cycloheximide given to pregnant rats on day 12 of gestation produced a slight increase in malformations, some increase in intrauterine death and resorption, and a significant weight reduction.

Lincomycin, which inhibits protein synthesis by interfering with the binding of tRNA to the ribosome–messenger complex (Chang *et al.*, 1966), has also been studied in this laboratory (unpublished). Administration to rats on day 12 of gestation at doses of 500–1500 mg/kg produced no malformations or weight reduction but did cause increased resorptions at higher doses.

Kidd (1953) observed that certain mouse leukemias were suppressed by treatment with guinea pig serum, and Broome (1961) later found that the active factor in the serum was L-asparaginase (LA). L-Asparagine is necessary for protein synthesis and LA catalyzes the hydrolysis of L-asparagine to L-aspartic acid and ammonia, creating a lack of L-asparagine. Cells which are sensitive to LA are in general those which require exogenous L-asparagine for optimum growth, whereas cells which can synthesize sufficient L-asparagine are not sensitive (Becker and Broome, 1967). Sobin and Kidd (1966) showed that LA inhibited protein synthesis 25–50% within 15 min after the addition of LA to a culture of sensitive cells. Under similar conditions DNA synthesis and RNA synthesis were not significantly inhibited until after 2 and 4 hr of exposure, respectively. This shows that LA is primarily an inhibitor of protein synthesis but again illustrates that the syntheses of other macromolecules are inhibited as secondary effects.

L-Asparaginase (50 and 100 units/kg) was found to be teratogenic in the rabbit by Adamson and Fabro (1968), who administered it to does on days 8 and 9 of gestation. In addition to malformations, some increased resorptions and weight loss were also noted. In preliminary studies in this laboratory no malformations were found following a single intravenous administration of 1000 units/kg LA to pregnant Wistar rats on one day between days 8 and 12 of gestation; however, when this dose was given twice daily on each of days 8, 9, and 10 of gestation, 21 of 51 implantations resorbed and 7 fetuses were malformed. It has not yet been shown that some embryo cells are more sensitive to LA than others, but it is anticipated that this will be found to be the case. If so, this will be in agreement with the basic tenet in teratology that embryo cells must show a differential sensitivity in order to produce localized defects.

Embryonic cell death appears to lead to malformations if the number and locations of cells killed is less than lethal, but greater than the embryo's capacity for repair (see Chapter 4, this volume). For reasons not fully understood, embryo cells seem to show differential sensitivity to inhibitors of DNA synthesis. Logically this relates to the differing proliferative rates of various kinds of cells, although other factors may be involved (Ritter *et al.*, 1975b). Whatever the cause, there appear to be considerable differences in sensitivity of embryo cells to cytotoxic agents acting on DNA, such as alkylating agents, inhibitors of DNA polymerase, etc.

The infrequent incidence of teratogenesis with most inhibitors of protein synthesis suggests that they provoke very little differential cytotoxicity or other differential action. This is a reasonable assumption since all mammalian cells probably have a similar apparatus for protein synthesis. Presumably a large dose would kill so many cells that it would be embryolethal, whereas a smaller dose would spare most cells. Only infrequently would the dosage be so finely adjusted as to kill a relatively large number of cells in a moderate number of localized areas. An exception might be an agent such as L-asparaginase to which some cells, depending on their asparagine supply and requirement, do show a differential sensitivity. Although not proven, this is a likely explanation for the slight teratogenicity of most potent inhibitors of protein synthesis, in contrast to the positive effect noted in the studies with LA. Obviously more work is needed in this area.

V. ENERGY INHIBITION

Life is only a few calories away from death because it is maintained by a modest but continuous input of energy. The energy stored in organic compounds such as glucose is utilized by animals via the metabolic pathways of glycolysis, respiration, and the terminal electron transport system to make ATP. ATP provides the energy required for the biosynthesis of such macromolecules as RNA, DNA, and protein, and for the precursors of these macromolecules, as well as for membrane integrity, morphogenetic movement, osmotic regulation, in short for all of life, growth, and development. Consequently it could be anticipated that any sizable reduction in ATP level might be damaging to the rapidly growing embryo; as has been suggested by Landauer and Wakasugi (1968) and others, it appears likely that some genetic defects, nutritional deficiencies, and drugs may induce teratogenesis by mechanisms involving an insufficiency of ATP.

It has been suggested that protein–calorie malnutrition in infants and young children is the most serious health problem with which the world is faced (Alleyne *et al.*, 1972). It is clear that brain weight, cell number, myelination, etc., are reduced in children dying from malnutrition, as well as in experimental animals malnourished before and after birth. The time of myelination is a period of vulnerability for the nervous system. Sulfatides, the 3′-sulfate esters of galactocerebrosides which are markers of myelination, are synthesized at reduced rates in malnourished young rats (Chase *et al.*, 1967). While energy is only one factor, it is of interest that the synthesis of sulfatides requires "active sulfate," 3-phosphoadenosine-5′-phosphosulfate (PAPS), the synthesis of which in turn requires two molecules of ATP. To bring this closer to the human condition it may be noted that an ATP deficiency has been described in leukocytes of severely malnourished children (Yoshida *et al.*, 1968).

Malnutrition of both the embryo and the neonate is likely to be accompanied by vitamin deficiencies, including those involved in energy pathways such as riboflavin and nicotinamide. In experimental studies riboflavin deficiency was one of the early regimens found to be teratogenic (Warkany and Schraffenberger, 1943), and nicotinamide (NAM) deficiency induced by administration of the antimetabolite 6-aminonicotinamide has been extensively used in experimental teratogenic studies, with NAM found to be an effective protective agent (Landauer, 1957; Murphy *et al.*, 1957; Pinsky and Fraser, 1960; Chamberlain, 1967; and others). It is known that 6-AN acts by the formation of fraudulent nicotinamide adenine dinucleotide (NAD) and NADP (Johnson and McColl, 1955; Dietrich *et al.*, 1958). By interfering with these coenzymes, 6-AN must inhibit a great many different biochemical pathways, but prominent among these are glycolysis, respiration, and the terminal electron transport system, with probable inhibition of ATP synthesis. This possibility was investigated by placing pregnant rats on a NAM-deficient diet and administering 6-AN (6 mg/kg) on day 12 of gestation, with a protective dose of NAM administered at 1, 2, or 4 hr after 6-AN (Ritter, *et al.*, 1975a). When embryos were removed at various times later and assayed for ATP content, a significant decrease was found in treated animals (Figure 2). When 6-AN was used without NAM protection, ATP concentrations in the embryos were reduced to about 50% of control values. If NAM were given 1 hr after 6-AN, the reduction in ATP was only about 15%; given 2 or 4 hr later inter-

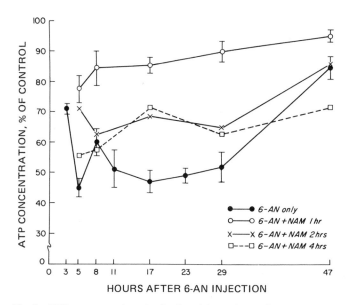

Fig. 2. ATP concentrations (μg/g tissue) in embryos from pregnant rats treated with 6 mg/kg 6-AN, intraperitoneal, on day 12 of gestation with or without 100 mg/kg NAM 1, 2, or 4 hr later, and removed for assay at the times indicated. SE of the mean is shown for the values obtained with 6-AN only and 6-AN + NAM at 1 hr; those obtained with 6-AN + NAM at 2 or 4 hr are omitted for clarity but they were of similar magnitude, 2–8% (Ritter *et al.*, 1975a).

mediate ATP concentrations resulted. If, instead, the young were allowed to go to term, all survivors following treatment with 6-AN alone were malformed, showing generally severe defects. Those given NAM 1 hr after 6-AN showed less severe defects in only 15% of survivors. Those given NAM at 2 or 4 hr after 6-AN showed defects intermediate in number and severity.

Thus, a close correlation was seen between teratogenesis and ATP concentration following 6-AN treatment. This does not prove the point, however, and it has been suggested that ATP content may not be involved and that damage to specific targets of 6-AN action may be causative (Chamberlain and Goldyne, 1970; Neubert *et al.*, 1970). This question cannot be answered at present. Probably both types of effects are involved, but the massive reduction in ATP concentration found in 6-AN-treated embryos must be kept in mind in mechanistic studies.

An important concept in teratogenesis is that the particular defects obtained with a given agent relate to the time in organogenesis at which the agent acts (see Chapter 2, Vol. 1). This was illustrated in a study by Curley *et al.* (1968), who administered 6-AN to mice on day 10.5 of gestation and observed such "early" (referring to time of induction) defects as vertebral anomalies and spina bifida, whereas 6-AN given on day 13.5 resulted in "late" defects such as micrognathia. This is consistent with the view that 6-AN via fraudulent NAD could reduce ATP synthesis by interfering with either glycolysis or respiration since NAD acts in both pathways. Studies by Tanimura and Shepard (1970), Cox and Gunberg (1972), and Mackler *et al.* (1971) indicate that glycolysis is the predominant energy pathway on day 10 of rat gestation, with respiration involving oxidative phosphorylation becoming increasingly important between days 10 and 14.

In contrast to the results with 6-AN, it would be expected that riboflavin deficiency would only depress ATP synthesis at the later times when respiration is predominant, since a major function of riboflavin is as a component in the flavoprotein coenzymes involved in respiration. In teratogenic studies where pregnant rats were maintained on riboflavin-deficient diets throughout gestation (with or without galactoflavin to intensify the deficiency state), only defects having inception late in organogenesis were seen (Warkany and Schraffenberger, 1944; Nelson *et al.*, 1956), consistent with the view that riboflavin deficiency reduced ATP synthesis at this stage of development. Based on this rationale, experiments similar to those above with 6-AN were conducted in this laboratory to determine whether there was a correlation between riboflavin deficiency, teratogenesis, and ATP concentrations. No such correlation was apparent (Ritter, Scott, and Wilson, unpublished results); concentrations of ATP in riboflavin-deficient embryos varied between about 80 and 120% of control values. This was surprising because no other mechanism for riboflavin-deficiency teratogenesis comes readily to mind. Another dissimilarity between the actions of riboflavin deficiency and 6-AN is that the former shows typical cell necrosis in areas subject to malformations (Shepard *et al.*, 1968), whereas very little such cell death is seen with 6-AN. 6-AN produces instead cellular vacuolization (Chamberlain, 1970; Caplan, 1972; Neubert *et al.*, 1970) which is not necessarily concentrated in areas destined to become grossly abnormal.

Interference with energy production has been suggested as the mechanism of action with arsenate-induced teratogenesis (Beaudoin, 1974), but preliminary investigation in this laboratory has failed to show diminished ATP synthesis in rat embryos at various times after a teratogenic dose of arsenate (unpublished results). Embryonic edema associated with hypoxia-induced teratogenesis may result from energy deficiency (Grabowski, 1970), although this investigator (1961) has also suggested that lactate accumulation initiates the edema syndrome. Seegmiller and Runner (1974) observed decreased $^{35}SO_4$ incorporation in chick embryos treated with 6-AN and suggested that inhibition of sulfation and failure to complete the synthesis of chondroitin sulfate may have been the specific defect causing micromelia. However, in this connection it should again be noted that PAPS, and therefore ATP, is required for the synthesis of chondroitin sulfate (Lash *et al.*, 1964), so that a 6-AN-induced deficiency of ATP might well result in a cartilage deficiency. In fact Seegmiller *et al.* (1972) postulated that the deficiency of matrix and the atypical appearance of chondrogenic cells in the chick embryonic limb following 6-AN treatment reflected reduced availability of high-energy phosphate.

In an investigation of the effect of organophosphorus and methyl carbamate insecticides on chick embryos, a correlation was found between teratogenesis and diminished NAD levels (Proctor and Casida, 1975). ATP levels were not determined.

It appears likely that an inadequate energy supply will lead to either death or abnormal embryonic development. In the latter event it is necessary to assume that energy needs vary throughout the embryo. Some cells must have a greater energy requirement than others, or a reduced ATP synthesis capability, so that under conditions of diminished supply they are more susceptible to injury. This would result in the differential damage that appears to be a prerequisite for teratogenesis.

REFERENCES

Adamson, R. H., and Fabro, S., 1968, Embryotoxic effect of L-asparaginase, *Nature* **218:**1164–1165.

Alexander, C. S., Swingle, K. F., and Nagasawa, H. T., 1966, Teratogenic effect of puromycin aminonucleoside on rat kidney, *Nephron* **3:**344–351.

Alleyne, G. A. O., Flores, H., Picou, D. I. M., and Waterlow, J. C., 1972, Metabolic changes in children with protein–calorie manutrition, *in: Nutrition and Development* (M. Winick, ed.), pp. 201–238, Wiley, New York.

Armentrout, S. A., and Weisberger, A. S., 1967, Inhibition of directed protein synthesis by chloramphenicol. Effect of magnesium concentration, *Biochem. Biophys. Res. Commun.* **26:**712–716.

Balis, M. E., 1968, *Antagonists and Nucleic Acids,* Wiley, New York.

Beaudoin, A. R., 1974, Teratogenicity of sodium arsenate in rats, *Teratology* **10:**153–158.

Becker, F. F., and Broome, J. D., 1967, L-Asparaginase: Inhibition of early mitosis in regenerating rat liver, *Science* **156:**1602–1603.

Bernstein, J., Meyer, R., and Pickman, D. S., 1962, Failure of amino-nucleoside to produce abnormalities in the fetus of the rat, *Nature* **195:**1209.

Bevelander, G., and Cohlan, S. Q., 1962, The effect on the rat fetus of transplacentally acquired tetracycline, *Biol. Neonate* **4:**365–370.

Billet, F. S., Collini, R., and Hamilton, L., 1965, The effects of D- and L-threochloramphenicol in the early development of the chick embryo, *J. Embryol. Exp. Morphol.* **13:**341–356.

Blackwood, U. B., 1962, The changing inhibition of early differentiaton and general development in the chick embryo by 2-ethyl-5-methylbenzimidazole and chloramphenicol, *J. Embryol. Exp. Morphol.* **10:**315–336.

Boucher, D., and Delost, P., 1964, Dévelopment post-natal des descendants issus de mères traitées par la streptomycine au cours de la gestation chez la souris, *C.R. Soc. Biol* **158:**2065–2069.

Broome, J. D., 1961, Evidence that the L-asparaginase activity of guinea pig serum is responsible for its antilymphoma effects, *Nature* **191:**1114–1115.

Brown, D. M., Harper, K. H., Palmer, A. K., and Tesh, S. A., 1968, Effect of antibiotics upon the rabbit, *Toxicol. Appl. Pharmacol.* **12:**295.

Brown, R. F., Umeda, T., Takai, S. I., and Lieberman, I., 1970, Effect of inhibitors of protein synthesis on DNA formation in liver, *Biochim. Biophys. Acta* **209:**49–53.

Butcher, R. E., Scott, W. J., Kazmaier, K., and Ritter, E. J., 1973, Postnatal effects in rats of prenatal treatment with hydroxyurea, *Teratology* **7:**161–166.

Caplan, A. I., 1972, Effects of a nicotinamide-sensitive teratogen 6-aminonicotinamide on chick limb cells in culture, *Exp. Cell Res.* **70:**185–195.

Carter, W., and McCarty, K. S., 1966, Molecular mechanisms of antibiotic action, *Ann. Intern. Med.* **64:**1087–1113.

Chamberlain, J. G., 1967, Effects of acute vitamin replacement therapy on 6-aminonicotinamide induced cleft palate late in rat pregnancy, *Proc. Soc. Exp. Biol. Med.* **124:**888–890.

Chamberlain, J. G., 1970, Early neurovascular abnormalities underlying 6-aminonicotinamide (6-AN)-induced congenital hydrocephalus in rats, *Teratology* **3:**377–388.

Chamberlain, J. G., and Goldyne, M. E., 1970, Intra-amniotic injection of pyridine nucleotides or adenosine triphosphate as counter therapy for 6-aminonicotinamide (6-AN) teratogenesis, *Teratology* **3:**11–16.

Chang, F. N., Sih, C. U., and Weisblum, B., 1966, Lincomycin, an inhibitor of aminoacyl sRNA binding to ribosomes, *Proc. Natl. Acad. Sci. U.S.A.* **55:**435–438.

Chase, H. P., Dorsey, J., and McKhann, G. M., 1967, The effect of malnutrition on the synthesis of a myelin lipid, *Pediatrics* **40:**551–559.

Courtney, K. D., and Valerio, D. A., 1968, Teratology in the *Macaca mulatta, Teratology* **1:**163–172.

Cox, S. J., and Gunberg, D. L., 1972, Metabolite utilization by isolated embryonic rat hearts *in vitro, J. Embryol. Exp. Morphol.* **28:**235–245.

Curley, F. J., Ingalls, T. H., and Zappasodi, P., 1968, 6-Aminonicotinamide induced skeletal malformations in mice, *Arch. Environ. Health* **16:**309–315.

Davies, J., Gilbert, W., and Gorini, L., 1964, Streptomycin, suppression, and the code, *Proc. Natl. Acad. Sci. U.S.A.* **51:**883–890.

Dietrich, L. S., Friedland, I. M., and Kaplan, L. A., 1958, Pyridine nucleotide metabolism: Mechanism of action of the niacin antagonist, 6-aminonicotinamide, *J. Biol. Chem.* **233:**964–968.

Ericson-Strandvik, B., and Gyllensten, L., 1963, The central nervous system of fetal mice after administration of streptomycin, *Acta Pathol. Microbiol. Scand.* **59:**292–300.

Fillipi, B., 1967, Antibiotics and congenital malformations. Evaluation of the teratogenicity of antibiotics, *Adv. Teratol.* **2:**239–256.

Fritz, H., and Hess, R. I., 1971, The effect of chloramphenicol on the prenatal development of rats, mice and rabbits, *Toxicol. Appl. Pharmacol.* **19:**667–674.

Grabowski, C. T., 1961, Lactic acid accumulation as a cause of hypoxia-induced malformations in the chick embryo, *Science* **134:**1359–1360.

Grabowski, C. T., 1970, Embryonic oxygen deficiency—a physiological approach to analysis of teratological mechanisms, *Adv. Teratol.* **4:**123–167.

Hallman, N., Hjelt, L., and Kouvalainen, K., 1960, Attempts to produce experimental congenital nephrosis, *Ann. Paediatr. Fenn.* **6:**289–298.

Harvey, J. E., and Srebnik, H. H., 1968, Fetal death and malformations produced by single injections of actinomycin D at midpregnancy and amelioration with thyroxine, *Anat. Rec.* **160:**362.

Hicks, S. P., Brown, B. L., and D'Amato, C. J., 1957, Regeneration and malformation in the nervous system, eye, and mesenchyme of the mammalian embryo after radiation injury, *Am. J. Pathol.* **33:**459–481.

Hurley, L. S., and Tuchmann-Duplessis, H., 1963, Influence de la tetracycline sur la développement pré- et post-natal du rat, *C.R. Acad. Sci.* **257:**302–304.

Ingall, D., and Sherman, J. D., 1970, Chloramphenicol, *in: Antimicrobial Therapy* (B. M. Kagan, ed.), pp. 61–77, W. B. Saunders, Philadelphia.

Johnson, W. J., and McColl, J. D., 1955, 6-Aminonicotinamide—a potent nicotinamide antagonist, *Science* **122:**834.

Jordan, R. L., 1969, Studies on the uptake and incorporation of tritiated uridine in the rat embryo and associated structures after treatment with actinomycin D, Thesis, University of Cincinnati.

Jordan, R. L., and Wilson, J. G., 1970, Radioautographic study of RNA synthesis in normal and actinomycin D treated rat embryos, *Anat. Rec.* **168:**549–564.

Kidd, J. G., 1953, Regression of transplanted lymphomas induced *in vivo* by means of normal guinea pig serum, *J. Exp. Med.* **98**:565–582.

Köhler, E., and Merker, H.-J., 1972, Studies of the teratogenic phase specificity of actinomycin D in the rat, *Naunyn Schmiedebergs Arch. Pharmakol.* **275**:31–44.

Kunin, C. M., 1970, The tetracyclines, *in: Antimicrobial Therapy* (B. M. Kagan, ed.), p. 47, W. B. Saunders, Philadelphia.

Landauer, W., 1957, Niacin antagonists and chick development, *J. Exp. Zool.* **136**:509–530.

Landauer, W., and Wakasugi, N., 1968, Teratological studies with sulfonamides and their implications, *J. Embryol. Exp. Morphol.* **20**:261–284.

Lash, J. W., Glick, M. C., and Madden, J. W., 1964, Cartilage induction *in vitro* and sulfate-activating enzymes, *Natl. Cancer Inst. Monogr.* **13**:39–49.

Mackler, B., Grace, R., and Duncan, H. M., 1971, Studies of mitochondrial development during embryogenesis in the rat, *Arch. Biochem. Biophys.* **144**:603–610.

McColl, J. D., Globus, M., and Robinson, S., 1965, Effect of some therapeutic agents on the developing rat fetus, *Toxicol. Appl. Pharmacol.* **7**:409–417.

Murphy, M. L., Dagg, C. P., and Karnofsky, D. A., 1957, Comparison of teratogenic chemicals in the rat and chick embryos, *Pediatrics* **19**:701–714.

Nelson, M. M., Baird, C. D. C., Wright, H. V., and Evans, H. M., 1956, Multiple congenital abnormalities in the rat resulting from riboflavin deficiency induced by the antimetabolite galactoflavin, *J. Nutr.* **58**:125–134.

Neubert, D., Merker, H.-J., Köhler, E., Krowke, R., and Barrach, H. J., 1970, Biochemical aspects of teratology, *in: Advances in the Biosciences* (G. Raspé, ed.), pp. 575–622, Pergamon Press, Oxford.

Pinsky, L., and Fraser, F. C., 1960, Congenital malformations after a two-hour inactivation of nicotinamide in pregnant mice, *Br. Med. J.* **2**:195–197.

Proctor, N. H., and Casida, J. E., 1975, Organophosphorus and methyl carbamate insecticide teratogenesis: Diminished NAD in chicken embryos, *Science* **190**:580–582.

Ritter, E. J., Scott, W. J., and Wilson, J. G., 1971, Teratogenesis and inhibition of DNA synthesis induced in rat embryos by cytosine arabinoside, *Teratology* **4**:7–14.

Ritter, E. J., Scott, W. J., and Wilson, J. G., 1973, Relationship of temporal patterns of cell death and development to malformations in the rat limb. Possible mechanisms of teratogenesis with inhibitors of DNA synthesis, *Teratology* **7**:219–226.

Ritter, E. J., Scott, W. J., and Wilson, J. G., 1975a, Inhibition of ATP synthesis associated with 6-aminonicotinamide (6-AN) teratogenesis in rat embryos, *Teratology* **12**:233–238.

Ritter, E. J., Scott, W. J., and Wilson, J. G., 1975b, Factors involved in differential teratogenic sensitivity to hydroxyurea in rats on days 10 and 12 of gestation, *Teratology* **11**:30A.

Roux, C., and Taillemite, J. L., 1969, Action tératogène de la rubidomycine chez le rat, *C.R. Soc. Biol.* **163**:1299–1302.

Roux, C., Emerit, I., and Taillemite, J. L., 1971, Chromosomal breakage and teratogenesis, *Teratology* **4**:303–316.

Rusconi, A., and Calendi, E., 1966, Action of daunomycin on nucleic acid metabolism in Hela cells, *Biochim. Biophys. Acta* **119**:413–415.

Sanders, E., and Cluff, L. E., 1970, Antimicrobial drugs: mechanisms and factors influencing their action, *in: Antimicrobial Therapy* (B. M. Kagan, ed.), p. 5, W. B. Saunders, Philadelphia.

Scott, W. J., Ritter, E. J., and Wilson, J. G., 1971, DNA synthesis inhibition and cell death associated with hydroxyurea teratogenesis in rat embryos, *Dev. Biol.* **26**:306–315.

Seegmiller, R. E., and Runner, M. N., 1974, Normal incorporation rates for precursors of collagen and mucopolysaccharide during expression of micromelia induced by 6-aminonicotinamide, *J. Embryol. Exp. Morphol.* **31**:305–312.

Seegmiller, R. E., Overman, D. O., and Runner, M. N., 1972, Histological and fine structural changes during chondrogenesis in micromelia induced by 6-aminonicotinamide, *Dev. Biol.* **28**:555–572.

Shepard, T. H., 1973, *Catalog of Teratogenic Agents,* The Johns Hopkins University Press, Baltimore.

Shepard, T. H., Lemire, R. J., Aksu, O., and Mackler, B., 1968, Studies of the development of congenital anomalies in embryos of riboflavin-deficient, galactoflavin fed rats. I. Growth and embryologic pathology, *Teratology* **1**:75–92.

Sobin, L. H., and Kidd, J. G., 1966, Alterations in protein and nucleic acid metabolism of lymphoma 6C3HED-OG cells in mice given guinea pig serum, *J. Exp. Med.* **123**:55–73.

Tanimura, T., and Shephard, T. H., 1970, Glucose metabolism by rat embryos *in vitro*, *Proc. Soc. Exp. Biol. Med.* **135**:51–54.

Tuchmann-Duplessis, H., and Mercier-Parot, L., 1958, Sur l'active tératogène chez le rat de l'actinomycine D, *C.R. Acad. Sci.* **247**:2200–2203.

Tuchmann-Duplessis, H., and Mercier-Parot, L., 1960, The teratogenic action of the antibiotic actinomycin D, *in: Ciba Foundation Symposium on Congenital Malformations* (G. E. W. Wolstenholme and C. M. O'Connor, eds.), pp. 115–128, Little, Brown, Boston.

Varpela, E., Hietalahti, J., and Aro, M. J. T., 1969, Streptomycin and dihydrostreptomycin medication during pregnancy and their effect on the childs inner ear, *Scand. J. Respir. Dis.* **50**:101–109.

Warkany, J., and Schraffenberger, E., 1943, Congenital malformations induced in rats by maternal nutritional deficiency. V. Effect of a purified diet lacking riboflavin, *Proc. Soc. Exp. Biol. Med.* **54**:92–94.

Warkany, J., and Schraffenberger, E., 1944, Congenital malformations induced in rats by maternal nutritional deficiency. VI. The preventive factor, *J. Nutr.* **27**:477–484.

Wilson, J. G., 1966, Effects of acute and chronic treatment with actinomycin D on pregnancy and the fetus in the rat, *Harper Hosp. Bull.* **24**:109–118.

Winfield, J. B., and Bennett, D., 1971, Gene–teratogen interaction: Potentiation of actinomycin D teratogenesis in the house mouse by the lethal gene brachyury, *Teratology* **4**:157–170.

Yoshida, J., Metcoff, J., and Frenk, S., 1968, Reduced pyruvic kinase activity, altered growth patterns of ATP in leukocytes, and protein–calorie malnutrition, *Am. J. Clin. Nutr.* **21**:162–166.

Zimmerman, E. F., Andrew, F., and Kalter, H., 1970, Glucocorticoid inhibition of RNA synthesis responsible for cleft palate in mice: A model, *Proc. Natl. Acad. Sci. U.S.A.* **67**:779–785.

Embryonic Intermediary Metabolism under Normal and Pathological Conditions

6

RALF KROWKE and DIETHER NEUBERT

I. INTRODUCTION

To elucidate the mode of action of drug-induced teratogenesis, the special susceptibility and sensitivity of the developing embryo to embryotoxic substances, compared to the toxicity seen in the adult organism, has to be considered. A number of parameters may be responsible for these toxicological differences including: (1) possible differences in drug metabolism and pharmacokinetics which may allow an accumulation of a given drug or its metabolites in embryonic cells; (2) the high rate of proliferation in certain stages of prenatal development; (3) typical induction and differentiation processes with characteristic transcriptional and translational regulations obligatory for a normal development; and (4) possible qualitative and quantitative differences in the intermediary metabolism of embryonic cells, for example, a metabolic pathway essential for normal prenatal development might produce little effect if inhibited in an adult organism.

Unfortunately, our knowledge of all of these matters is still rather fragmentary. Much more research using different animal species and human embryonic tissues is necessary before an understanding of the normal processes during prenatal development, not to mention interference with such

RALF KROWKE and DIETHER NEUBERT • Institut für Toxikologie und Embryonal-Pharmakologie der Freien Universität Berlin, Garystr. 1–9, 1000 Berlin 33.

processes, will be possible. In this chapter we will summarize and discuss some aspects of mammalian embryonic intermediary metabolism recently investigated. Since studies in the field of the biochemistry of mammalian embryonic tissues have been largely neglected, most of the data available have been collected within the last 5–6 years. Selected examples will be given for each point to be discussed and no attempt will be made in this brief survey to cover completely the subject or the literature.

II. SOME SPECIAL PROBLEMS CONCERNING METABOLIC PATHWAYS IN MAMMALIAN EMBRYONIC TISSUES

The mode of action of embryotoxic substances is much more difficult to explain than is a toxic event in an adult organism. It is well known that abnormal prenatal development can be triggered even when the teratogenic agent, for example, trypan blue, is largely excluded from the embryonic compartment (Wilson *et al.*, 1963; Beck *et al.*, 1967; Lloyd, 1970). Such indirect disturbances of prenatal development via interference with maternal or placental factors must be excluded before attempting to relate abnormal development to a direct interference with a particular metabolic pathway within the embryo.

A. The Two-Compartment System: Mother/Embryo

This two-compartment system is especially complicated since it does not represent a constant relationship, but one that is continuously changing during prenatal development. The means by which nutritional substances and drugs are transported into the embryonic compartment and by which metabolites are discharged into the maternal compartment are certainly different during the preimplantation period, the stage of histiotrophic nutrition, and the stage when the placenta is fully functional. However, it should be emphasized that none of these periods can be understood to represent constant conditions but must, from the nutritional as well as from the pharmacokinetic points of view, again be subdivided into various stages. It should also be recognized that for a well-balanced evaluation, toxicologically as well as biochemically, the embryonic compartment rarely can be analyzed separately, but generally must be dealt with as a part of this two-compartment system. It is fair to say that this approach has been even more neglected than has the study of isolated metabolic pathways in mammalian embryonic tissues.

Since induction and control of certain processes in postnatal organisms often are influenced by hormones, it is of interest to determine whether such hormones also play an integral part in the differentiations occurring during early mammalian embryogenesis. This topic has been discussed in a recent meeting (see Raspé, 1974). Two results argue against the usual hormones playing a major direct role in early organogenesis. First, New (1966) and other investigators have succeeded in growing mammalian postimplantation embryos *in vitro* over an extended period of organogenesis without addition of hormones. However, since the culture media contained serum, a certain amount of hormones may have been introduced by this route to the system. Second, pregnancy can be maintained in rodents after hypophysectomy on about day 6 of pregnancy if the animals are given estrogen and progesterone alone (Köhler *et al.*, 1974). Thus, normal mammalian organogenesis seems not to require hormones such as glucocorticoids, thyroid hormones, and growth hormone, all of which are typical "inducing" hormones in the postnatal organism. However, it cannot be ruled out that a certain amount of glucocorticoids was converted from the high doses of female sex hormones given in these studies.

There has been no previous indication that estrogens and progesterone exert a direct effect on the embryonic tissue during these early stages of development, and it has not been proven so far that receptors for these sex hormones are present in embryonic tissue. The data available are, therefore, consistent with the idea that most early mammalian embryonic inductions and differentiation processes can proceed without the aid of maternal hormones. This certainly does not mean, however, that an interference with maternal hormonal status has no effect on prenatal development. On the contrary, it is known from human pathology that prenatal development, especially at later stages of gestation, can be strongly influenced by an abnormal maternal hormonal status. To some extent this may be induced by the secondary changes occurring in the maternal compartment.

1. Some Aspects of the Pharmacokinetics

A special characteristic of the two-compartment system is that it dominates the pharmacokinetic parameters governing a drug-induced embryotoxic action. The term "embryotoxic" as defined earlier (Neubert *et al.*, 1973) denotes all transient or permanent toxic effects induced prenatally regardless of the mechanism of action. The terms "embryolethal" and "teratogenic," as well as "growth retardation," specify effects which may occur as a consequence of an embryotoxic action. Besides maternal factors, such as drug absorption, distribution, and elimination, the factors allowing a transfer of the drugs to the embryo and drug-metabolizing reactions occurring in the embryo are of utmost importance for the outcome of the drug action. Some of the toxicological aspects of teratology are discussed elsewhere (Neubert, 1976).

Little is known of the kinetics of the transfer of various compounds into the embryonic compartment at different stages of development. However, it is quite likely that the embryo behaves, under certain conditions, as what in pharmacokinetics is called a deep compartment, causing a compound to concentrate within the embryonic tissues at a level higher than in the maternal serum, especially after repeated applications. An example is a study performed in this laboratory using β-D-arabinofuranosylcytosine (cytosine arabinoside, ara-C). It is seen in the usual semilogarithmic plot (Figure 1) that the time-dependent decline of the drug concentration in the maternal serum is a straight line. Similar kinetics hold for the change in drug concentration in the maternal kidneys. In contrast, the concentration of the drug in the embryonic tissue as well as in the amniotic fluid steadily increases to reach a maximal value 2–3 hr after the injection. About 3 hr after application of the antimetabolite, the drug concentrations in maternal tissue and in the embryonic compartment are about equal under these experimental conditions. Subsequently the elimination of the drug from the embryonic compartment proceeds at a slower rate than from the maternal organism. It should be kept in mind that this example describes only one of many pharmacokinetic situations possible with various drugs and that with other embryotoxic compounds the concentration profile in the embryo might more closely resemble that seen in the maternal organism.

  Since it is well-known in pharmacology and toxicology that the pharmacokinetics for a given drug may differ greatly when studied in different animal species, including man, such species differences may also create varied toxicological situations in teratology. Chloramphenicol, for example, has a half-life in human serum of about 210 min, but this drug is eliminated from the rat at a considerably faster rate, giving a half-life of about 30 min (Voemel, personal communication, quoted in Oerter and Bass, 1975). This means that dose–response relationships as revealed in studies with experimental animals may be of little value in predicting teratogenic hazards in humans. When, therefore, teratogenic effects are quantitatively evaluated with a possible relevance for man, which is the most important aspect of teratological research, the pharmacokinetics of the drug must be known and taken into consideration in both the experimental animal as well as in human. If possible, the schedule of dosage in the animal model should be adjusted to match the pharmacokinetic situation in humans. So far these considerations have rarely been taken into account.

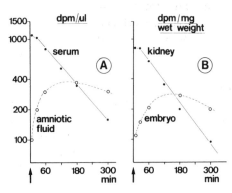

Fig. 1. Pharmacokinetics after a single intraperitoneal application of ara-C. Radioactivity in maternal serum and amniotic fluid (A) and in maternal kidneys and whole embryos (B) was measured after injection of [³H]ara-C in rats on day 12 of pregnancy.

2. Some Aspects of Drug Metabolism

Difficulties in predicting pharmacokinetics of a drug or its metabolites may result from an especially low or high rate of metabolism of the drug in a given species. It should be emphasized that metabolism can influence drug action either by inactivation or by activation. In most experimental animals typical cytochrome-P-450-containing drug-metabolizing enzymes are believed not to develop until the late prenatal or even the neonatal period (Fouts and Adamson, 1959; Fouts, 1962; Rane *et al.*, 1973). This would mean that the majority of changes, especially of an oxidative nature, on an exogenous molecule would not take place to an appreciable degree in embryonic tissue of experimental animals.

Two important points have to be taken into consideration. First, studies in recent years have shown that the above does not hold for the human fetus and that cytochrome-P-450-dependent drug-oxidizing enzymes are present in liver microsomes as early as the 8–15th week of gestation (Yaffe *et al.*, 1970; Pelkonen, 1973). In this respect most experimental animals are not good models for drug metabolism by human fetal tissues. Activation and inactivation reactions might, therefore, proceed at much faster rates in human fetuses or neonates than is likely at similar stages of experimental animals.

Second, discussions of drug metabolism tend mostly to be focused on such well-known reactions as oxidations in liver microsomes and conjugation reactions. But it should be realized that there exists a great variety of other metabolic pathways by which an inactive drug can be converted into a biologically active compound, without depending on cytochrome P-450. Thus, most antimetabolites are transformed into active drugs via pathways which are also used for the metabolism of naturally occurring metabolites. Many such metabolic pathways are operative in early embryonic tissues, as can be deduced from the effectiveness of antimetabolites on reactions of intermediary metabolism of embryonic cells. Antimetabolites of purine or pyrimidine metabolism or of pyridine nucleotides are well-known examples of such embryotoxic agents. Since the rates at which some of these metabolic pathways proceed are comparatively high in the embryo, as compared with many tissues in the adult organism, activation reactions involving antimetabolites may also proceed at a high rate in the embryo.

Because of differences in metabolic pathways some antimetabolites are activated in a different way in embryonic tissues than they are in tissues of an adult organism. This is apparently the case with 6-aminonicotinamide (6-AN) which in adult tissues is converted to the active antimetabolite (6-ANADP$^+$, the 6-AN analog of NADP$^+$) via an exchange reaction catalyzed by the enzyme NADP-glycohydrolase (EC 3.2.2.6), the activity of which has not been detected in embryonic tissues of rodents. Therefore, it appears that 6-AN is "activated" in the embryo via the pathway which is also used for the *de novo* synthesis of pyridine nucleotides (Figure 2).

When the activity of the "classical" drug-metabolizing enzymes is low in a given tissue, for example, in embryonic tissues of most experimental animals,

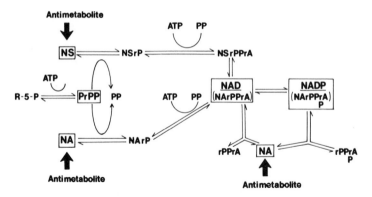

Fig. 2. Pathways which may give rise to the formation of an abnormal pyridine nucleotide from 6-AN. While in most tissues of an adult organism an exchange reaction between the nicotinamide moiety of a preformed pyridine nucleotide and the antimetabolite takes place [NAD(P)-glycohydrolase, right corner of the figure] in embryonic tissue, 6-AN can apparently be incorporated into the pyridine nucleotides via the pathways used for *de novo* synthesis (left side of the figure). NA = nicotinamide, NS = nicotinic acid. The antimetabolite replaces NA or NS in these reactions and is converted to 6-ANAD(P).

other drug-converting reactions may become significant, although they may not be considered important for adult tissues like liver. One should be aware of the possibility of such atypical enzymes catalyzing modifications in drugs reaching the embryo. Too little attention has been given to such possibilities.

A lack of drug-metabolizing enzymes could, on the other hand, lead to an accumulation of the parent compound in the embryo, if this drug is otherwise rapidly metabolized in tissue of the maternal organism. Modification of maternal drug metabolism, then, could lead to rather paradoxical findings. Such a situation was described by Gibson and Becker (1969, 1971) and has given rise to the suggestion that the well-known mechanism of alkylation produced by cyclophosphamide, a compound which has to be activated by metabolic oxidation, may not be the mode of teratogenic action of this cytostatic agent. According to data presented by Gibson and Becker (1969) using mice, the inhibition of maternal drug metabolism by SKF-525 A, a typical inhibitor of cytochrome-P-450-dependent mixed-function oxidations, *increased* the teratogenicity of cyclophosphamide, contrary to what might have been expected. Although this effect was not confirmed in rats (Engels, 1970), it was shown that the concentration of nonactivated cyclophosphamide can be greatly increased in embryonic tissue (Figure 3) by inhibiting maternal cytochrome-P-450-dependent drug-oxidizing enzymes by CFT 1201 (β-diethylaminoethylphenyldiallylacetate), another compound known to be a potent inhibitor of drug oxidations in liver microsomes (Neubert and Herken, 1955). Changes like these, i.e., altered pharmacokinetics within the embryonic

Fig. 3. Concentration of the activated metabolite in rat serum after a single subcutaneous injection of cyclophosphamide (Endoxan). In the controls only Endoxan was given. The experimental group was pretreated with an inhibitor (CFT 1201) of microsomal drug-metabolizing enzymes which are present predominantly in maternal liver (50 mg/kg orally given 5, 3, and 1 hr before Endoxan). The NBP reaction (Friedman and Boger, 1961) was used for measuring the alkylating form of Endoxan. It is seen that CFT 1201 pretreatment reduces the maximal concentration of the alkylating agent present in the serum. However, because of the slower rate of elimination, the activated metabolite is present much longer in the maternal serum.

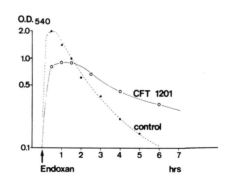

compartment, have to be taken into consideration if maternal metabolism of a given drug is changed under pathological conditions.

A pronounced increase in the teratogenicity, as a result of an interference with drug inactivation in maternal tissues, has been reported for 5-fluorouracil (5-FU). Simultaneous administration of a nonteratogenic dose of 5-FU and a natural pyrimidine base (uracil or thymine) or the corresponding nucleoside resulted in a clear-cut teratogenic effect, apparently because of a reduced metabolic degradation of 5-FU due to competition for the enzymes which catabolize both the physiological substances and the drug (Wilson *et al.*, 1969; Schumacher *et al.*, 1969). It is of interest that in these studies maternal serum concentrations of 5-FU, except for the first 60 min after the injection, were not found to be different in controls and uracil-treated animals, in contrast to the increase in 5-FU concentration in maternal liver and kidney and in embryonic tissue observed in uracil-treated rats.

B. Some Aspects of Replication, Transcription, and Translation

Replication of DNA and cell division, as well as reactions connected with gene expression, are among the metabolic functions particularly susceptible to attack by embryotoxic agents. Elucidation of these metabolic functions in embryonic tissues under normal and pathological conditions may help to explain the special vulnerability of embryonic tissues.

1. Replication and Cell Division

Mammalian embryonic tissue, especially during early organogenesis, is the most rapidly proliferating mammalian tissue known. This unusually high rate of replication and of cell divisions is very likely one reason for the special susceptibility of embryonic tissue to toxic substances. When measuring the increase in DNA content of the embryo in early organogenesis (days 8–10) it becomes obvious that the rat embryo increases its DNA content roughly 100-fold within a period of 2 days (Köhler, 1970; Neubert *et al.*, 1971a; Köhler *et al.*, 1972). This rate of DNA synthesis would correspond to an average of roughly 10 replications in 48 hr, or an average length of the cell cycle of less than 5 hr. Assuming an S phase of about 3–4 hr, this would mean that the majority of the cells would be in the replicating stage. Such a view is consistent with experimental data (Köhler, 1970; Köhler *et al.*, 1972) which indicate that the average DNA content of the embryonic cell nuclei lies roughly midway between that expected for a diploid and a tetraploid cell. Similar data for the mouse are given in Figure 4. The rate of proliferation was not so uniformly rapid during all of the gestational periods studied. Since it is fair to assume that not all cell types are proliferating at all stages of development, the cell cycle may even be shorter than 5 hr for some embryonic cells in the rat.

The study of DNA functions has been complicated by the discovery within the last 10 years of extrachromosomal DNA in mammalian mitochondria and the recognition of a special machinery for replication, transcription, and translation of this mitochondrial DNA. Although the significance of this extrachromosomal material is not yet fully understood, studies with substances like chloramphenicol, a typical inhibitor of mitochondrial protein syn-

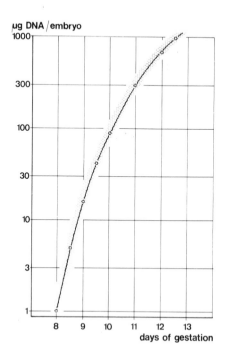

Fig. 4. Increase in DNA content of the mouse embryo during organogenesis. The DNA content was measured in at least 100 embryos of each developmental stage indicated, using the Burton (1956) method. DNA per embryo increases roughly 100-fold in a 48-hr period.

thesis, would suggest an important function of this DNA during embryonic development (Oerter and Bass, 1975). The present status of knowledge in this field and the possible significance of extrachromosomal DNA for animal development has been reviewed recently (Neubert *et al.*, 1975a).

2. Gene Expression

With the numerous inductions and differentiating processes occurring during organogenesis in embryonic tissues, transcription and its regulation must be among the most critical processes for normal embryonic development. The molecular biology of the regulation of gene expression is an area of knowledge that is just beginning to expand, and in the special case of embryonic development in mammals understanding of the biochemistry of the complicated regulatory processes is indeed limited. An excellent survey of knowledge of the biochemistry of differentiating processes in general was recently given by Rutter *et al.* (1973). Even when using comparatively simple model systems, such as pancreas differentiating in organ culture (Pictet and Rutter, 1972) or limb buds growing *in vitro* (Aydelotte and Kochhar, 1972; Neubert *et al.*, 1974a,b), the basic reactions governing changes at the chromatin level and regulating transcription are still poorly understood.

A drug-induced effect on RNA synthesis can be analyzed by two different approaches: (1) by studying the rate of synthesis of certain RNA species or the capacity of a given tissue to synthesize RNA, or (2) by characterizing more closely the RNA product formed under specified conditions.

The rate of synthesis depends on two principal factors (Figure 5): (1) the rate of the polymerizing reaction itself (DNA-dependent RNA polymerase), and (2) the availability of the nucleoside triphosphates which are required as precursors for a RNA polymerase. To analyze drug-induced teratogenic effects more closely and to gain insight into the complicated regulatory processes of embryonic differentiation during organogenesis and the possibilities of interfering with these reactions, this laboratory has developed methods to study both of the metabolic pathways above. The rate of DNA-dependent RNA synthesis, or the capacity to perform this transcription, is measured using isolated nuclei from embryonic tissues, or more recently, using homogenates of embryonic tissues. The initial studies had been performed with cell nuclei from whole rat (11–14 days) or mouse (9–13 days) embryos (Neubert, 1970a). Recently success in studying the different DNA-dependent RNA polymerase reactions with nuclei or homogenates from developing organs of the embryo in the late stage of organogenesis has been realized (Neubert and Rautenberg, 1976).

Two different DNA-dependent RNA polymerase reactions can be distinguished in nuclei from all mammalian tissues including embryonic tissues: (1) an α-amanitin-insensitive reaction localized mainly within the nucleolus and giving rise mostly to ribosomal RNA (rRNA) and possibly also to transfer

RNA (tRNA), and (2) an α-amanitin-sensitive reaction which apparently is responsible for the formation of messenger RNA (mRNA). Results obtained with embryonic tissues indicated that, in contrast to the activity of the DNA-dependent DNA polymerase (replicase), the activity of both types of RNA polymerase (based on DNA content) is not very high when compared with, for example, the activity found in adult liver tissue (Figure 6). This is surprising considering the numerous induction processes and the rapid growth rate in this tissue, but it agrees with earlier results obtained with rapidly growing tumor tissues (Neubert, 1966). The data suggest that the turnover of RNA might be reduced in this proliferating tissue. Results obtained so far indicate that greater emphasis should be put on the qualitative aspects of the RNA species formed, especially the mRNA. Success in isolating RNA fractions with mRNA properties, as characterized by hybridization to poly(U) and by stimulation of cell-free protein-synthesizing systems, has made possible quantitative and qualitative studies on this aspect of transcription after labeling the RNA *in vivo* or in organ culture systems (Rautenberg and Neubert, 1975).

Even less information is available on the stability of mRNA in embryonic tissues during organogenesis and on the possibilities of regulating differentiation at the translational level. Data recently obtained in several laboratories suggest that mRNA is not produced in a one-step procedure (Weinberg, 1973; Molloy *et al.*, 1974). The primary product of transcription seems to be a high-molecular-weight heterogeneous RNA (HnRNA) which then is specifically processed by being split into smaller fragments with the addition of a

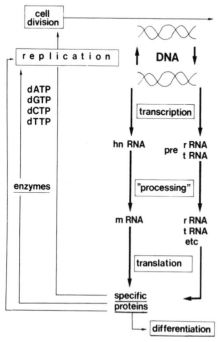

Fig. 5. Regulation of replication and of gene expression in a mammalian cell. The biochemical steps of replication, transcription, and "processing" occur predominantly in the cell nucleus. The direct product of transcription (HnRNA) as well as pre-rRNA and pre-tRNA are processed before being released into the cytoplasm by specific fragmentation processes and the addition of a poly(A) tail to some of the RNAs (mRNA). Although gene expression and replication are primarily independent processes, products of transcription and translation are required for replication and an interference with cell division and replication is possible via an alteration of gene expression.

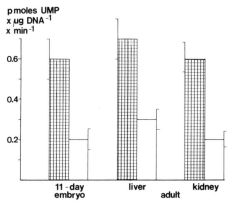

Fig. 6. Activity of DNA-dependent RNA polymerase in different mouse tissues. Incorporation of [³H]UTP into nuclear DNA was measured using nuclei isolated from adult and embryonic tissues. The hatched bars give the α-amanitine-insensitive reaction (3 min) in the presence of Mg^{2+} (mostly rRNA), the white bars show the rate of α-amanitine-sensitive RNA synthesis in the presence of Mn^{2+} (predominantly mRNA). It can be seen that the activity of the RNA polymerase in the rapidly growing embryonic tissue does not greatly exceed the activities measured in slowly proliferating adult tissues.

poly(A) tail to some of the mRNAs (Figure 5). It seems that roughly 50% of the initial HnRNA is lost during this processing and that only a fraction leaves the nucleus as mRNA which can be translated at the ribosomes. It may be that the processing of mRNA provides at least as powerful a mechanism for the regulation as does the process of transcription itself. The measurement of the total amount of HnRNA and of mRNA formed, as well as a characterization of HnRNA and of the different types of mRNAs (by hybridization or translation of isolated messenger and subsequent product analysis), seem to be the most promising approaches to increasing knowledge on the regulation of differentiation and induction during mammalian embryonic development. Such studies are in progress in this laboratory with a number of model systems.

C. Energy Sources and Uptake of Metabolic Substrates

With the high rates of proliferation and of synthetic reactions necessary to support rapid growth, efficient ATP-generating processes have to be operative. The available evidence seems to indicate that mammalian embryos, especially before placentation, possess the capacity of carrying out aerobic or anaerobic glycolysis and, in this respect, can be compared with many tumor tissues or rapidly proliferating tissue cultures (de Meyer and DePlaen, 1964; Neubert, 1970b; Neubert et al., 1971b; Shepard et al., 1970; Tanimura and Shepard, 1970). Under these conditions glucose is taken up by the embryo at a high rate and most of this carbohydrate (70–90%) accumulates in embryonic tissue as lactic acid. Contrary to the majority of adult tissues, a considerable portion of the remaining glucose fragments converted to acetyl-CoA is not oxidized aerobically to CO_2 and water, but is consumed in synthetic reactions (Neubert, 1970b). In accordance with this, mitochondria in embryonic tissues appear quite immature. Considerable mitochondrial biogenesis and maturation occurs around the early placentation period, at least in the rodent species studied (Bass, 1970). Some special aspects of this carbohydrate and energy metabolism in embryonic tissues, and the effects of interference by embryotoxic substances, are reviewed elsewhere (Bass *et al.*, 1977).

Energy metabolism in embryonic tissue, however, does not proceed entirely anaerobically. This can be deduced from extensive recent studies using techniques which allow the development of preimplantation as well as post-implantation embryos *in vitro* (New, 1966; Auerbach and Brinster, 1968). So far there is no evidence that it is possible to culture mammalian embryos under entirely anaerobic conditions. Although the oxygen tension in embryonic tissue at the different stages of development is not known, it may be deduced from these *in vitro* studies that a certain degree of aerobic metabolism is necessary for normal embryonic development, although anaerobic pathways of ATP formation may additionally be employed as in tumor tissue. Data from this laboratory (Neubert, 1970b) indicate that insulin is not necessary for glucose to enter the embryo and glucose uptake depends on the extraembryonic concentration; however, this proceeds at rates found in many adult tissues only in the presence of insulin. In this respect also, embryonic tissues resemble many tumor tissues in which the K_t for glucose is very low (Burk *et al.*, 1967).

Although there are some indications that during preimplantation development, as well as during organogenesis, carbohydrates are needed for embryonic development as an energy source, the question is not settled as to what degree other low-molecular-weight metabolites from the maternal compartment are utilized by the embryo. Fatty acids are taken up by embryonic tissue (Figure 7); however, the percentage of these substances being used as energy sources is not known. Also uncertain is which amino acids have to be taken up from the maternal compartment and to what degree amino acid metabolism can proceed in embryonic tissue. It should be realized that the term "essential amino acid" does not fully apply to embryonic tissue. It is possible that the rate of synthesis of a particular amino acid in rapidly proliferating tissue may not be high enough to guarantee a sufficient supply and additional amino acids, besides the usual ones, may become "essential." Embryotoxic actions of asparaginase (Adamson and Fabro, 1968) may be explained, at least partly, by this mechanism.

III. METHODS FOR STUDYING METABOLIC PATHWAYS IN MAMMALIAN EMBRYONIC TISSUES *IN VIVO* UNDER NORMAL AND PATHOLOGICAL CONDITIONS

To obtain information on metabolic pathways in embryonic tissues *in vivo* often requires isotopic tracer techniques. The two-compartment system, mother/embryo, combined with other complicating factors represent complex situations which demand utmost care in drawing conclusions from the data. However, such methods offer numerous advantages compared with experimental setups performed entirely *in vitro* for studying normal and abnormal embryonic development.

A. Usefulness of Radioactive Precursors

This laboratory has used different radioactively labeled precursors for the examination of metabolic pathways in rodent embryos *in vivo*. In addition to getting information on the rates at which several metabolites penetrate into the embryonic compartment, these studies were meant to provide clues on precursors suitable for screening metabolic pathways in the embryo relating to the action of teratogenic drugs. Figure 7 presents data from such an experimental series in the mouse on day 10 of gestation 3 hr after injection of labeled precursors. Similar experimental series have been performed in this laboratory for the rat on different days of gestation. Day 10 in the mouse and day 12 in the rat have proved to be especially suitable since this represents a period of high susceptibility to several teratogens and since enough embryonic tissue can be obtained from one experimental animal to do biochemi-

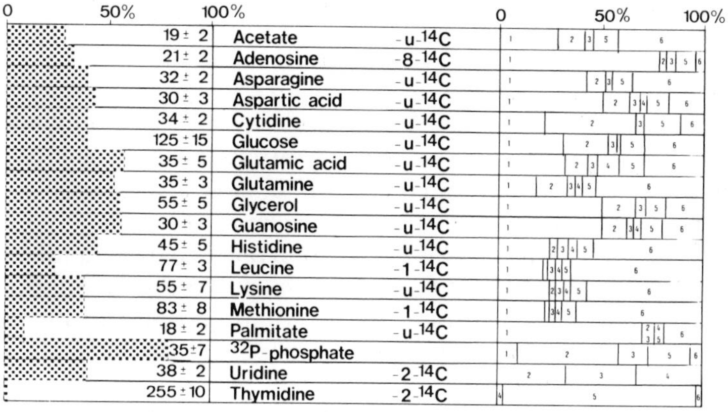

dpm/µg DNA	Precursor	Label
19 ± 2	Acetate	-u-^{14}C
21 ± 2	Adenosine	-8-^{14}C
32 ± 2	Asparagine	-u-^{14}C
30 ± 3	Aspartic acid	-u-^{14}C
34 ± 2	Cytidine	-u-^{14}C
125 ± 15	Glucose	-u-^{14}C
35 ± 5	Glutamic acid	-u-^{14}C
35 ± 3	Glutamine	-u-^{14}C
55 ± 5	Glycerol	-u-^{14}C
30 ± 3	Guanosine	-u-^{14}C
45 ± 5	Histidine	-u-^{14}C
77 ± 3	Leucine	-1-^{14}C
55 ± 7	Lysine	-u-^{14}C
83 ± 8	Methionine	-1-^{14}C
18 ± 2	Palmitate	-u-^{14}C
35 ± 7	^{32}P-phosphate	
38 ± 2	Uridine	-2-^{14}C
255 ± 10	Thymidine	-2-^{14}C

Fig. 7. Incorporation of different radioactively labeled metabolic precursors into acid-soluble and acid-insoluble cell components. Each of the tracers was injected intravenously into pregnant mice at a dose of 0.5 mCi/kg, and the radioactivity was measured after 90 min in the different cell components of the embryonic tissue (Krowke *et al.*, 1971c). On the left, the percentage of radioactivity found in the acid-soluble (shaded area) and in the acid-insoluble (white area) fractions is given. The figures give the dpm/µg DNA found in the acid-insoluble portion. On the right the distribution of radioactivity in the acid-insoluble fraction is given. The different fractions refer to: 1 = lipids; 2 = RNA; 3 = "glycosaminoglycans"; 4 = "carbohydrates"; 5 = DNA; 6 = proteins. In the case of nucleosides the radioactivity found in the lipid fraction may partly consist of the corresponding bases formed by metabolic degradation.

cal studies at least in triplicate (Krowke, 1970; Krowke *et al.*, 1971a,c). Experiments have also been performed on day 11 in rats and on day 9 in mice.

From the data given in Figure 7, it is seen that many amino acids, basic, neutral, and acidic ones, penetrate into the embryonic compartment at a rate high enough to allow measurement of incorporation into the main cellular components. The same holds for nucleosides which also give high labeling rates. Nucleic acid bases generally are poorer precursors for studies with embryonic tissues *in vivo*. Several other metabolites such as glucose, fatty acids, glycerol, acetate, and formate (Krowke, 1970), and inorganic tracers such as [^{32}P]orthophosphate and [^{35}S]sulfate (Kochhar *et al.*, 1968; Krowke *et al.*, 1971b; Schimmelpfennig, 1971; Schimmelpfennig *et al.*, 1971) also reach the embryo at a concentration high enough to allow incorporation studies. The suitability of other important precursors for labeling nucleic acid components, e.g., formate, glycine, and orotic acid, have been published elsewhere (Krowke, 1970). The possibility of labeling RNA moieties with some of these components during later stages of development in the rat fetus has been examined by Hayashi and Kazmierowski (1970). Bresnick *et al.* (1965) showed that orotic acid, widely used for labeling RNA in adult liver tissue, is not suited for embryonic tissues even in near-term stages of development.

B. "Specific" and "Nonspecific" Precursors

For different purposes it may be advisable to use either "specific" precursors which are incorporated into one cellular component only or more "nonspecific" ones which, in the embryo, are converted to several metabolites which can be incorporated into a variety of different cellular components. If the methods used are sophisticated enough to allow a separation of the different, labeled components, nonspecific precursors offer the advantage that several metabolic pathways can be evaluated in one experiment under identical experimental conditions. In routine examinations this laboratory has favored employing [U-^{14}C]glucose and ^{32}P inorganic phosphate, especially since these substances are comparatively inexpensive and since the physiological concentrations in the maternal organism and in the fetus are not altered appreciably by these compounds. This cannot be taken for granted with the more "specific" precursors which have to be given in unphysiologically high concentrations, e.g., thymidine (cf. Hughes *et al.*, 1973) in order to measure significant incorporation rates. A thymidine concentration of about 10^{-6} M was recently measured in mouse blood serum (Hughes *et al.*, 1973). Assuming a specific activity of 50 mCi/mmole and a dose of 0.5 mCi/kg, 10^{-5} moles are injected per kg body weight. The blood serum concentration of thymidine is raised about 100-fold. Several precursors are not very specific, as has clearly been shown in the case of thymidine, especially [^{3}H]thymidine, the label of which can be found in several cell components other than DNA (Schneider and Greco, 1970; Dobson and Cooper, 1971).

The degree of "specificity" can be judged from data given in Figure 7. It is seen that amino acids like asparagine and aspartic acid, as may be expected, are not only incorporated into proteins but also into nucleic acids. In the case of aspartic acid more radioactivity is found in nucleic acids than in proteins. A similar situation holds for glutamic acid while glutamine, histidine, leucine, glycine, and methionine are much more "specific" for proteins in embryonic tissues.

With regard to the radioactivity found in the acid-insoluble fractions, glucose, glycerole, leucine, glycine, methionine, and [2-^{14}C]thymidine were found to be convenient precursors for labeling experiments with embryonic tissues *in vivo*. It has, of course, to be taken into account that the pools of these different compounds in the maternal compartment as well as in the fetus are quite different, so that the figures given cannot be considered as true rates of incorporation of the different metabolites. On account of quite a few pitfalls the use of tritium-labeled substances should, if possible, be avoided in *in vivo* studies with embryonic tissues.

C. Screening Methods

Based on the data obtained with different radioactively labeled precursors, several years ago a screening system which provides some information on the effect of embryotoxic substances on several metabolic pathways in rodent embryos was designed in this laboratory (Krowke, 1970; Krowke *et al.*, 1971b,c).

1. Some General Considerations

This system has since been used successfully for the evaluation of numerous teratogens, including extensive studies with model substances. Employing a double-labeling technique, the labeling rates of lipids, RNA, DNA, and proteins, and occasionally of some minor components such as glycosaminoglycans, were followed over a suitable experimental period after the intravenous injection of [U-^{14}C]glucose and ^{32}P inorganic phosphate. Some of the pathways evaluated by this technique are schematically shown in Figures 8 and 9. In such studies it has to be taken into account that the two radioactive tracers, [^{14}C]glucose and [^{32}P]phosphate, are taken up in the embryonic compartment with quite different kinetics (Krowke *et al.*, 1971c); while with glucose a maximum incorporation is reached after about 2–3 hr, phosphate continues to be incorporated for at least 12 hr after intravenous injection.

Furthermore, it became evident that it was of utmost importance in such studies to follow the kinetics of the radioactivity in maternal blood, as well as in the acid-soluble fraction of the embryo which contains the total of the metabolites to be incorporated into the acid-insoluble components (Figure

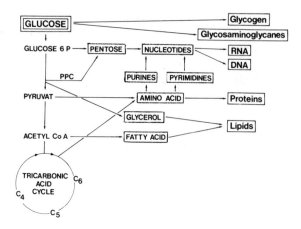

Fig. 8. Diagram showing the advantages of using a "non-specific" precursor. The pathways leading to the labeling of different cell components from radioactively labeled [U-^{14}C]glucose are shown.

10). It is thus possible also to evaluate effects of the teratogen on maternal metabolism, which effects occur more often than anticipated, as well as on transport of the radioactive precursor across the placenta. Some examples of the effects of teratogens on metabolism are given in the next paragraph. This method gives two kinds of clues: (1) information on the point of attack of the teratogen within the main pathways of intermediary metabolism, with a rough estimate of the extent of interference; and (2) information on the duration of block induced by a teratogen in a certain metabolic pathway of the embryo, provided an appropriate experimental design is used.

2. Analysis of Specific Metabolic Pathways

A few examples are given here of analyses of the action of teratogens on metabolic pathways in the embryo. Using inhibitors of DNA synthesis such as cytosine arabinoside (ara-C) or 5-fluorodeoxycytidine (FCdR), the data compiled in Figures 11 and 12 clearly show that it is possible to verify a biochemical lesion restricted to a single metabolic pathway in the embryo with this method. While a clear-cut inhibition of the incorporation of radioactively labeled precursors can be demonstrated via the steps of *de novo* synthesis of deoxynucleoside triphosphates (cf. Figure 9) and an inhibition of the synthesis of DNA molecules, the incorporation of radioactive precursors into other cell components such as lipids, RNA, and proteins proceeds in a completely normal way during the experimental period, especially in the case of ara-C. For other teratogens a more complex situation may exist wherein a number of metabolic

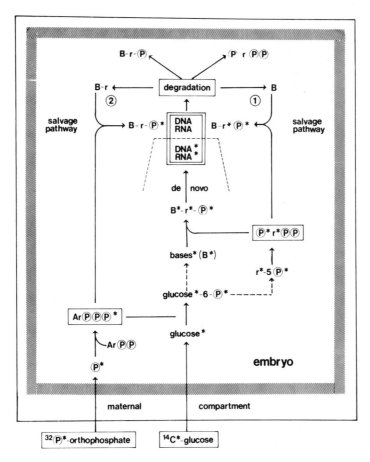

Fig. 9. Metabolic pathways (*de novo* synthesis as well as salvage pathway) giving rise to the labeling of nucleic acid components from the tracers [U−^{14}C]glucose and [^{32}P]orthophosphate. The extent of label appearing in the various nucleic acid moieties (base, sugar, and phosphate) obtained via the different pathways can be deduced.

pathways are affected in a direct or an indirect way following the inhibition of one primary pathway. Thus, not only is protein synthesis inhibited soon after the injection of the protein-synthesis inhibitor cycloheximide, but also RNA and DNA synthesis are affected (Krowke *et al.*, 1971a), because proteins necessary for replication and transcription become rate limiting. Contrary to the mode of action of cycloheximide, the cytostatic agent azaserine apparently blocks RNA and DNA synthesis simultaneously by a primary action on the *de novo* synthesis of the precursors of both metabolic pathways. Again no clear-cut block of the incorporation of precursors into lipids and protein is found during the period investigated (Figure 13).

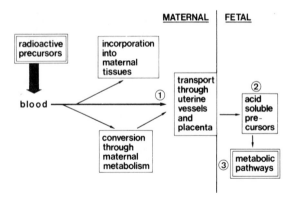

Fig. 10. Diagram of the distribution of radioactively labeled metabolites in the maternal and the fetal compartments after intravenous injection of a radioactive precursor. Drug-induced changes in the maternal organism can be observed by following the kinetics of the radioactivity in the maternal serum (1); changes in the transport through uterine vessels and the placenta can be detected by measuring the radioactivity in the acid-soluble components of the embryo (2); while metabolic pathways present in the embryo are detected by studying the incorporation of the label from the acid-soluble fraction into the various cell components (3).

A biochemical defect in embryonic metabolism can be detected with doses of FCdR as low as 0.1 mg/kg. This is about 1/30 of the minimal teratogenic dose. The biochemical screening system is thus a very sensitive method for predicting potential deviations in embryonic development.

It is not justifiable to deduce metabolic interference in embryonic metabolism from experiments performed with an antimetabolite in a different experimental system, e.g., tumors or tissue culture cells. This was shown in experiments from this laboratory performed with the cytostatic agent hydroxyurea (HU). Previous data in the literature shows that in a variety of systems this drug interferes with proliferation by inhibiting DNA synthesis. A study of DNA metabolism in rat embryos, using [^{3}H]thymidine incorporation as a criterion, clearly showed that an inhibition of replication by HU occurred in this tissue (Scott *et al.*, 1971). When applying the above-described technique and measuring the incorporation of [U-^{14}C]glucose fragments and of ^{32}P inorganic phosphate into different cell components, however, it became obvious that after a time the block was not restricted to DNA synthesis, but that the incorporation of these radioactive precursors into RNA was equally affected (Krowke *et al.*, 1971d; Krowke and Bochert, 1975). This broader effect, of course, will be overlooked if only the "specific" precursor thymidine is employed in similar studies (Table 1).

Initially the method was used with whole embryos, but recently it has been possible to adapt the technique to studies of various parts such as limb bud, brain vesicle, or other organ anlagen, of 10–12-day-old mouse embryos, or corresponding rat embryos (Krowke *et al.*, 1977). It has now been shown that it is possible with these biochemical methods to study some metabolic pathways in different cell types *in vivo*, for instance of a limb bud, after

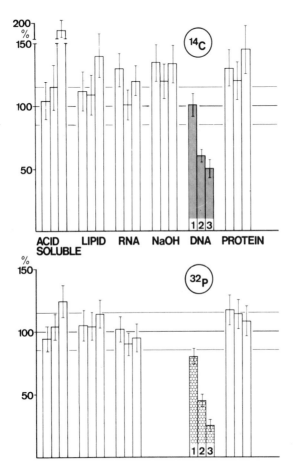

Fig. 11. Example of the effect of ara-C on rat embryos. 10–100 mg/kg ara-C were given i.p. on day 12 of pregnancy to rats followed by an intravenous injection of both [U-^{14}C]glucose and [^{32}P]-orthophosphate (1 mCi/kg each). The embryos were examined 3 hr after the injection of the radioactive tracers. A clear-cut reduction of the radioactive label was found in the DNA fraction while no decrease of the radioactive label incorporated into the other cell components was seen. 1 = 10 mg/kg ara-C; 2 = 30 mg/kg; 3 = 100 mg/kg. Average values of at least 4 experimental groups ± SD, expressed as percent of controls (control values, cf. Table 1).

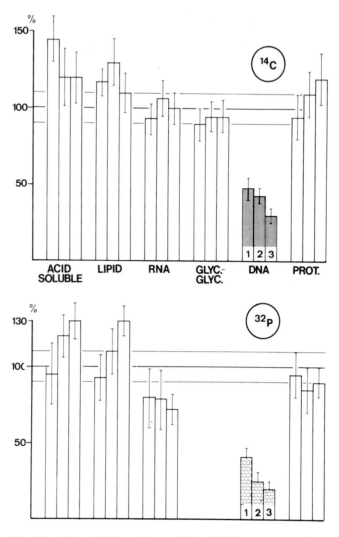

Fig. 12. Example of the effect of 5-fluorodeoxycytidine (FCdR) on mouse embryos. FCdR was given subcutaneously on day 10 of pregnancy to mice followed by an intravenous injection of both [U-14C]glucose and [32P]orthophosphate (1 mCi/kg each) 15 min later. The radioactivity was measured in the different cell components of the embryo 90 min after the injection of the radioactive precursors. A selective inhibition of the incorporation into DNA was seen when using the [14C]labeled precursor. With [32P] the radioactivity found in the DNA fraction was also drastically reduced. But a small decrease was also found for the radioactivity measured in the RNA fraction. This may reflect a secondary effect of the antimetabolite. 1 = 0.3 mg/kg; 2 = 1 mg/kg; 3 = 3 mg/kg. Average values of at least 4 experimental groups ± SD, expressed as percent of controls.

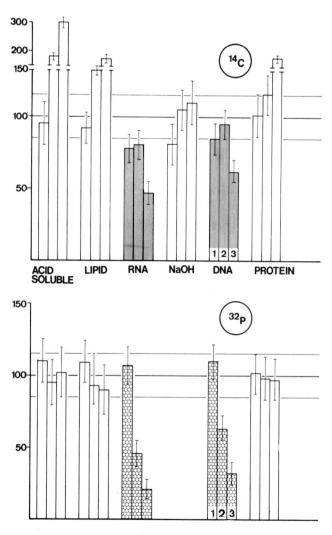

Fig. 13. Example of the effect of azaserine on mouse embryos. Experimental conditions same as in Figure 12. Azaserine was injected subcutaneously into mice on day 10 of pregnancy. A mixture of both radioactively labeled precursors was given intravenously 90 min later. The incorporation of the label into the cell components of the embryo was measured 90 min after the injection of the radioactive tracers. It can be seen that the incorporation of radioactivity into both RNA and DNA was reduced under these experimental conditions. However, the radioactivity found in lipids and proteins was not reduced, indicating some degree of selective action by the cytostatic agent. 1 = 1 mg/kg; 2 = 3 mg/kg; 3 = 10 mg/kg.

Table 1. Effect of Hydroxyurea (HU) on DNA and RNA Synthesis in Rat Embryos (Day 12)
(HU, 5 hr; ^{14}C + ^{32}P, 3 hr, 1 mCi/kg)[a]

	Acid soluble		Lipid		RNA		"NaOH"	DNA		Protein
	^{14}C	^{32}P	^{14}C	^{32}P	^{14}C	^{32}P	^{14}C	^{14}C	^{32}P	^{14}C
Control (dpm/µg DNA)	126 ± 35	420 ± 50	132 ± 21	11 ± 2.1	61 ± 8.1	59 ± 6.1	19 ± 2.3	64 ± 10	32 ± 3.3	87 ± 9.2
HU expressed as % of control										
250 mg	131 ± 11	99 ± 8	121 ± 10	92 ± 10	97 ± 3	77 ± 6.6	106 ± 8	77 ± 2	84 ± 8	123 ± 8
500 mg	144 ± 18	121 ± 10	118 ± 18	112 ± 8	66 ± 12	62 ± 8.5	96 ± 11	42 ± 6	20 ± 4.4	117 ± 11
750 mg	137 ± 22	104 ± 5	97 ± 10	92 ± 8	39 ± 12	39 ± 10	79 ± 8	30 ± 3	12 ± 3	99 ± 6

[a] At least 4 animals per group (controls = 30) received an i.p. injection of 250, 500, or 750 mg HU/kg and 2 hr later were given an i.v. injection of the isotopes. The embryos of each litter were collected and homogenized. An aliquot of the homogenate was analyzed as specified in Krowke et al. (1971c). The incorporation of both tracers was reduced in dose dependent fashion in the DNA *and* RNA fractions, indicating that the metabolic block caused by HU is not restricted to DNA synthesis only.

disassociating epithelial cells from blastema cells (Berg and Krowke, 1975). The techniques have also been applied successfully in studies of the effect of teratogens in organ culture systems (Gregg *et al.*, 1975).

3. Analysis of Nucleic Acid Moieties Labeled with [U-¹⁴C]Glucose

Many antimetabolites do not act on nucleic acid metabolism at the level of the polymerases but interfere with one of the many enzymic reactions necessary to ensure *de novo* synthesis of the purine and pyrimidine nucleoside triphosphates. Since in this case it is expected that primarily the synthesis of only a few of the 8 direct nucleic acid precursors is affected, more sophisticated methods can be used to obtain more precise information on the nature of metabolic block. A method has been developed in this laboratory which permits separate measurement of the incorporation rate into the base, sugar, and phosphate moieties of each of the 8 nucleic acid precursors after labeling by the metabolic pathways *in vivo* (Bochert *et al.*, 1973). This method, using a modified amino acid analyzer, allows the clean separation of the bases of nucleosides and permits measurement of the specific activity in each of the hydrolysis products from RNA or DNA (Table 2). However, as expected, the method gives detailed information only at the initial stages of the inhibition. After a block in one of the metabolic sequences has fully developed, the nucleoside triphosphates involved become rate limiting and an inhibition of the incorporation of all 4 nucleoside triphosphates into the nucleic acid component is the result. Nevertheless, the technique can give additional information, extending the data obtained by the simpler incorporation studies. More recently this separation of the DNA bases has proved to be useful for analyzing DNA after the administration of alkylating teratogens. It is now possible to separate alkylated bases like N-7-guanine or O-6-guanine from the natural

Table 2. Incorporation of [¹⁴C]Glucose Fragments into Deoxyribose and Base Moieties of DNA (dpm/nmol)[a]

	T	C	A	G
Deoxyribose	5.4 ± 0.1	9.6 ± 0.5	9.8 ± 0.2	13.5 ± 0.6
Base	2.0 ± 0.06	1.8 ± 0.01	1.6 ± 0.01	2.3 ± 0.06
¹⁴C in base as % of deoxynucleoside	29%	17%	14%	15%

[a]The table shows the distribution of radioactivity in sugar and bases of the separated deoxyribonucleotides from purified DNA of 12-day-old rat embryos. The DNA was labeled for 3 hr by 1 mCi [¹⁴C]glucose/kg applied intravenously. The nucleotides were obtained by phosphodiesterase and DNase hydrolysis and separated by high voltage electrophoresis or by a modified amino acid analyzer. The bases were prepared by formic acid hydrolysis and separated either on TLC cellulose plates or using the above mentioned instrument. The radioactivity was determined in Instagel (Packard Instruments) using a Packard Tricarb 3380. T = thymidine; C = cytidine; A = adenosine; G = guanosine.

bases and to correlate the rate of alkylation of different cell components with teratological data (Bochert, 1975).

D. Difficulties in Computing Metabolic Rates in Mammalian Embryonic Tissues *in Vivo* under Pathological Conditions

The data obtained with the experimental setup discussed have to be evaluated critically because of a number of possible pitfalls. The radioactivity incorporated into the final product (macromolecules) is a direct function of the specific activity reached in the last precursor of the metabolic pathway under experimental conditions (Figure 14). This specific activity, again, is a function of two parameters, one of which is the rate of the radioactively labeled compound passing through the sequence of biochemical reactions, and the second representing the pool sizes of the different metabolites of the given metabolic pathway. A decreased radioactivity in a final product (macromolecule) of the given metabolic pathway, therefore, could be the result of either a reduced availability of the radioactively labeled final precursor, the result of an increase in the pool size of any metabolite in the sequence, or the consequence of a reduced rate of polymerization.

A reduction in the pool size could, on the other hand, lead to an increase of the labeling rate of the final product. This would give rise to pronounced misinterpretations of the data, especially under circumstances when a radioactively labeled precursor is used which enters the metabolic pathway subsequent to the point of attack of the inhibitor to be evaluated. For these reasons thymidine is a tracer which should not be used for quantitative, or even semiquantitative, evaluations of the rate of DNA synthesis, unless the experimental conditions are under rigid control, for example, by measuring the pool size of TTP and the specific activity in this direct precursor of the replicase reaction. Furthermore, thymidine enters the TTP pool via a reaction (thymidine kinase) which in most tissues is not located in the main pathway of *de novo* synthesis, and it is well known that the activity of thymidine kinase can easily be changed under various experimental conditions.

The pitfalls mentioned become especially obvious if no steady-state kinetics are reached during the experimental period studied. With the method most often used, single injection of the radioactive precursor, a steady state certainly is not present and the supply of the radioactivity from the maternal serum, and the specific activity of the metabolites, change continuously. *For this reason it is not possible to compute reaction rates from the labeling data.* The results obtained represent average labeling rates achieved during the experimental period and can be used only for comparative purposes, e.g., controls vs. data from treated animals. Fortunately, a block within a metabolic pathway and a change in pool size are often dependent on each other. However, this may not necessarily be the case, especially if the time dependence of the

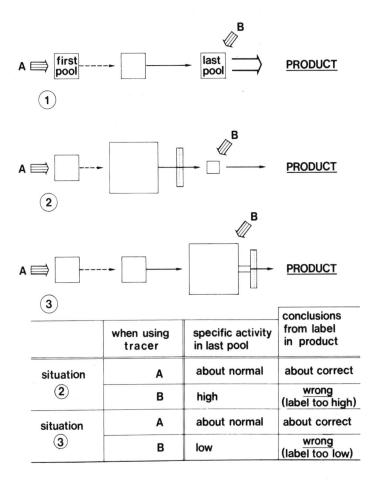

	when using tracer	specific activity in last pool	conclusions from label in product
situation ②	A	about normal	about correct
	B	high	wrong (label too high)
situation ③	A	about normal	about correct
	B	low	wrong (label too low)

Fig. 14. Schematic presentation of the action of an antimetabolite causing both an inhibition within a metabolic pathway and consequently a change in the pool size of some metabolites. (1) Normal conditions of the steady-state flow and pool sizes. (2) If a reaction in the middle of the sequence of the enzymic steps is inhibited, the pool of the preceding metabolite is expected to increase while the pool of the product of the inhibited reaction is expected to decrease. Different effects may be measured if a radioactively labeled precursor is given early (A) or late (B) in the sequence of the metabolic pathway. (3) If the last reaction of the metabolic pathway, that one leading to the formation of the final product, is inhibited, the pool of the direct precursor of the final product is expected to increase. Again radioactive precursors entering the pathway early (A) or late (B) will give different results. Changes in the feedback regulation by altering the pool sizes are not shown in this figure, but also have to be considered. The results possibly occurring during the inhibition of the metabolic pathway are summarized in the lower part of the figure. It should be considered that different results are to be expected during the onset of the inhibition and during the final stage when the inhibitor is being eliminated from embryonic tissues.

inhibition is studied. *Studies like those suggested should, therefore, be taken only as a hint to abnormalities existing in a pathway to be evaluated, and they do not necessarily allow a quantitative estimation of the degree of inhibition in a given tissue, or even in the different compartments within a single cell type.*

To obtain a true picture of the net increase in DNA in embryos during pathological conditions, the DNA content per embryo has to be measured directly in series of experimental animals under normal (cf. Figure 4) and pathological conditions. Pulse labeling and measuring the decline in the specific activity of the polydeoxynucleotide, due to the formation of new unlabeled DNA, can also represent a useful technique (Neubert *et al.*, 1968). These techniques cannot be used for evaluating the short-term effect of drugs. For studies on the onset of a metabolic interference in embryonic tissue, the administration of the drug *after* the injection of the labeled tracers gives more reliable results (Figure 15). Such a method is less susceptible to changes in the pool sizes of the metabolic precursors.

E. Direct Measurement of the Specific Activity Reached in the Last Precursor of the Metabolic Pathway to Be Evaluated

Some of the difficulties encountered in interpreting experimental data may be overcome if the specific activity present in the last precursor of the sequence is known, i.e., that of the metabolite directly involved in the polymerizing reaction giving rise to the formation of the macromolecule. This approach certainly is much more realistic than a simple incorporation study. With the techniques now available, such studies are possible and have already been performed with other mammalian tissues (Fridland, 1974). Using DNA precursors, Solter and Handschumacher (1969) described an enzymatic assay for deoxynucleoside triphosphates which may be very useful for studies with embryonic tissue (Skoog and Nordenskjold, 1971). Such an experimental approach does not, of course, completely rule out the difficulties in interpreting *in vivo* data. One could argue that the specific activity measured for a precursor does not reflect the actual specific activity at the site of action of the last enzyme in the metabolic pathway, for instance, in the cell nucleus. The specific activity of the substrates of the last enzymatic reaction would, furthermore, have to be measured for all of the metabolic pathways separately. This is certainly not easy in the case of amino acids or glycosaminoglycan or lipid precursors. For the nucleoside triphosphates this is further complicated by the fact that some of the metabolites may be rather labile, particularly under conditions of a hypoxia. Thus it may be almost impossible to obtain near *in vivo* conditions, i.e., to deep-freeze the embryo rapidly enough to guarantee an unchanged level of these nucleotides. For this reason the values reported for ATP in embryonic tissues (Chepenik *et al.*, 1970) become doubt-

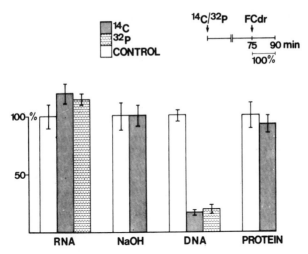

Fig. 15. Typical data from an experiment in which the metabolic inhibitor was administered after the injection of the radioactively labeled precursors. In order to avoid some of the difficulties mentioned in the text, the radioactively labeled tracers (1 mCi/kg each) were injected intravenously to achieve labeling of the metabolities in the embryo. After 75 min, 0.1 mg/kg FCdR was given to the mice (day 10 of pregnancy) intravenously, and the radioactive label in the different cell components of the embryo was measured, both immediately after injection of the inhibitor and 15 min later. Difficulties arising from changes in pool sizes are reduced by this procedure. It can be seen that the incorporation into DNA is greatly reduced during the 15-min period following the injection of the inhibitor, indicating a rapid onset of inhibition of the metabolic pathway in the mouse embryo. No deviations from controls are found in cell components such as RNA or protein. This experimental procedure is especially valuable when effects occurring shortly after the injection of the inhibitor are to be evaluated.

ful. Whether the same holds for the deoxynucleotide triphosphates is not known. They may be more stable. Furthermore, it is necessary to measure only the specific activities in these nucleotides, which is much easier than attempting to compute true concentrations in the tissue.

Since these rather time-consuming and complicated methods require that a complete time kinetics of the changes in deoxyribonucleotide triphosphates be done, such studies are only justified for the evaluation of special problems with selected antimetabolites. Thus, the authors feel that it is not worthwhile to attempt routine measurement of the specific activity of the low-molecular-weight precursors. However, the situation being as complex as it is, one must realize that without such data no real reaction rates can be deduced from incorporation studies *in vivo*, especially under pathological conditions. On the other hand, it is felt that the information gained from simple incorporation experiments is valuable in providing some insight into the principal changes occurring in intermediary metabolism of embryonic tissues during the action of teratogenic agents.

F. Incorporation of Antimetabolites into Nucleic Acids

There is another disadvantage in using radioactively labeled metabolites which may be of considerable importance. Only an inhibition of a main metabolic pathway can be recognized with the methods. If an antimetabolite is incorporated into the final product to a rather small extent, compared with the normal metabolite, this may not be detected. A reaction of this type, however, may be of great significance in teratology, and there are, in fact, some indications in the literature and from unpublished studies from this laboratory that the formation of such "false products" may represent an important embryotoxic mechanism. Therefore, it cannot be stressed too often that employing one method may jeopardize the evaluation of the mode of a teratogenic action and that for valid conclusions results should be obtained by as many biochemical methods as nearly concurrently as possible. Thus the incorporation techniques described here have to be used together with other methods in order to avoid wrong conclusions, especially if a negative result is obtained.

Data obtained with hydroxyurea (Section III.C.2), together with other results, make doubtful the assumption that the inhibition of DNA synthesis alone is the critical event which triggers abnormal development. By several means, such as administration of 6-aminonicotinamide (Köhler *et al.*, 1970; Barrach, 1970), 2,4,5-trichlorophenoxyacetic acid (Neubert and Dillmann, 1972), chloramphenicol (Oerter and Bass, 1975), or other drugs in an appropriate dose, it was possible to reversibly interfere with the proliferation of embryonic tissues. This resulted only in a growth retardation and in a limited number of malformations (in mice mostly cleft palate) and not in the numerous kinds of abnormalities seen after the administration of hydroxyurea, FCdR, or ara-C. At the present time the authors, therefore, hold the hypothesis that the teratogenic action of inhibitors of "DNA synthesis" is not triggered by their interference with DNA synthesis but is rather the result of an altered transcription. This might be initiated, as is the case with hydroxyurea, by the drug also interfering with RNA synthesis or, as might be stated for compounds such as FCdR, and ara-C, by being incorporated into the DNA molecule and thus giving rise to a *miscoding* during the transcription and differentiation processes.

If this hypothesis is correct, the critical point in the mode of action of these substances would not be the degree of inhibition of DNA synthesis but the extent of the incorporation of the drugs into nucleic acid molecules. In the case of FCdR such a view would be consistent with the findings of Sazaki and Ohmori (1973) that fractionated injections of an antimetabolite resulted in much stronger teratogenic effects than a single injection of an equivalent total dose, although the latter would be expected to interfere with the rate of DNA synthesis much more effectively.

A program of study is in progress on the late effects of the inhibitors of DNA synthesis on transcription and possibly on translation. A significant effect on the labeling of *RNA* after administration of FCdR was found (Figure 12). Similar results were obtained with 5-fluorodeoxyuridine (FUdR), as shown in Figure 16. In both experiments the incorporation of ^{32}P was the most sensitive indicator. Many more studies are needed to obtain information on the incorporation of such antimetabolites into nucleic acids and on the extent of replacement of natural nucleotides in the macromolecules. Reasonable data are published on the incorporation of 5-bromodeoxyuridine into DNA of mammalian embryos (Packard *et al.,* 1973), but here again the exact degree of replacement within the DNA molecules has not been evaluated. Moreover, this antimetabolite, which has been studied extensively in other differentiating mammalian systems (Stellwagen and Tomkins, 1971; Weintraub *et al.,* 1972), is incorporated into DNA to an unusually high degree. The demonstration of incorporation of an antimetabolite into DNA by itself does not explain the mode of a teratogenic action.

In the opinion of the authors the presence of a radioactively labeled inhibitor in an acid-insoluble fraction gives no evidence for incorporation of such a drug into nucleic acids. Only the clear-cut demonstration of the antimetabolite incorporated into highly purified nucleic acid can be accepted as evidence, preferably by quantitative characterization of the drug in a hydrolysate of the nucleic acid fraction. Experiments, especially those using tritium-labeled compounds, are of little value in studies measuring the incorporation

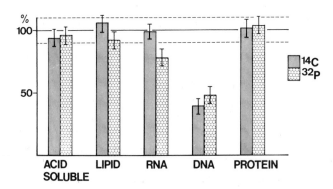

Fig. 16. Effect of 5-fluorodeoxyuridine (FUdR) on mouse embryos. FUdR was injected subcutaneously on day 10 of pregnancy; 15 min later the mixture of the radioactively labeled tracers (1 mCi/kg each) was injected intravenously into the mice. Incorporation into the cell components was measured 90 min after the injection of the tracers. Incorporation into the DNA fraction is found decreased for both the precursors used. However, during the period studied a small but significant effect on the incorporation into the RNA fraction is also obvious when [^{32}P] is used as a tracer. The effect of the antimetabolite can therefore not be considered to be entirely restricted to DNA metabolism (cf. Figure 12). 1 mg/kg FUdR was injected.

of radioactivity into an acid-insoluble fraction since degradation products of the initial drug can be expected to be incorporated into the nucleic acid molecule in an unspecific way and many tritiated compounds are acid labile. Therefore, although the mechanism of incorporation of antimetabolites into nucleic acids of embryonic tissue has been postulated, no evidence exists at present as to what extent this mechanism may be operative during a teratogenic action of such compounds. It would be of great help to find an inhibitor of DNA synthesis with an action restricted to interference with the incorporation of natural nucleoside triphosphates into DNA, and for which an incorporation into nucleic acids or a simultaneous effect on other metabolic pathways could be largely excluded. It is possible that such a restricted interference with replication for a period of time might not result in typical malformations.

As in many other mammalian tissues, 5-fluorouracil is incorporated into RNA of embryonic cells *in vivo* (Dagg *et al.*, 1966). Whether the comparatively small amount of radioactivity found in the DNA fraction represents an incorporation has been questioned by the authors on the basis of the limited data available. Data published more recently using a similar technique, radioactivity in the "DNA fraction" found by an Ogŭr–Rosen procedure (Ruddick and Runner, 1974), do not contribute to solving this problem.

Based on experiments from this laboratory and on recent data of experimental systems (Tidd and Paterson, 1974), the suggestion is made that 6-mercaptopurine (6-MP), a potent teratogen, may exert its action not by interfering with the *de novo* synthesis of nucleic acid precursors, generally assumed to be the mode of action of this drug, but by being incorporated into nucleic acids and giving rise to unphysiological templates. By using radioactively labeled 6-MP and measuring its incorporation into highly purified DNA, it was possible to show that roughly 1 out of 2×10^6 GMP moieties could be replaced by the analog when the antimetabolite was given *in vivo* in a teratogenic dose (Bochert, 1972, 1975).

These data focus attention on an effect of antimetabolites observed in several laboratories. Following the injection of substances such as FUdR (Webster *et al.*, 1973), hydroxyurea (Ritter *et al.*, 1973), and others, cell necroses can be demonstrated as early as 2–4 hr after the injection. It is hard to visualize why a selective inhibition of DNA synthesis should give rise to cell necroses after such a short period. An interference with mRNA synthesis and function may be a more likely event to explain this effect. It seems to the present authors that a better approach would be to determine which vital biochemical reactions become rate limiting in a portion of the rapidly proliferating cells in such a way as to be incompatible with cell survival under these conditions. Since the occurrence of necroses is a frequent event in early stages of teratogenesis (Scott *et al.*, 1971; Webster *et al.*, 1973), this may be an important key for understanding the special susceptibility of certain cells during embryonic development.

REFERENCES

Adamson, R. H., and Fabro, S., 1968, Embryotoxic effect of L-asparaginase, *Nature* **218:**1164–1165.

Auerbach, S., and Brinster, R. L., 1968, Effect of oxygen concentration on the development of two-cell mouse embryos, *Nature* **217:**465.

Aydelotte, M. B., and Kochhar, D. M., 1972, Development of mouse limb buds in organ culture: Chondrogenesis in the presence of a proline analog, L-azetidine-2-carboxylic acid, *Dev. Biol.* **28:**191–201.

Barrach, H.-J., 1970, Effect of 6-aminonicotinamide on the glucose metabolism of mammalian embryonic tissue, *in: Metabolic Pathways in Mammalian Embryos during Organogenesis and its Modification by Drugs* (R. Bass, F. Beck, H.-J. Merker, D. Neubert, and B. Randhahn, eds.), pp. 365–384, Free University Press, Berlin.

Bass, R., 1970, Respiration and oxicative phosphorylation of mitochondrial fractions isolated from rat embryos, *in: Metabolic Pathways in Mammalian Embryos during Organogenesis and its Modification by Drugs* (R. Bass, F. Beck, H.-J. Merker, D. Neubert, and B. Randhahn, eds.), pp. 309–319, Free University Berlin Press, Berlin.

Bass, R., Beck, F., Merker, H.-J., Neubert, D., and Randhahn, B., 1970, *Metabolic Pathways in Mammalian Embryos during Organogenesis and its Modification by Drugs*, Free University Berlin Press, Berlin.

Bass, R., Barrach, H.-J., and Neubert, D., 1977, Some aspects of carbohydrate and energy metabolism in mammalian embryos under normal and pathological conditions, *Ergeb. Physiol. Biol. Chem. Pharmacol.* (in press).

Beck, F., Lloyd, J. B., and Griffiths, A., 1967, Lysosomal enzyme inhibition by trypan blue: A theory of teratogenesis, *Science* **157:**1180–1182.

Berg, P., and Krowke, R., 1975, Biochemical studies with dissected blastema and epithelium of mouse embryo extremities after *in vivo* treatment with 6-mercaptopurine-riboside, *Naunyn Schmiedebergs Arch. Pharmakol.* **287:**R87(abst.).

Berg, P. L., Krowke, R., and Merker, H.-J., 1975, Studies on blastema and epithelium of the limb anlage of mouse embryos, *in: New Approaches to the Evaluation of Abnormal Embryonic Development* (D. Neubert and H.-J. Merker, eds.), pp. 151–160, Georg Thieme Publishers, Stuttgart.

Bochert, G., 1972, Eine Analyse der DNA-Synthese im embryonalen Säugetiergewebe unter dem Einfluss einiger embryotoxischer Pharmaka, Dissertation, Free University Berlin.

Bochert, G., 1975, unpublished results.

Bochert, G., Mewes, P., and Krowke, R., 1973, Labelling of RNA and DNA moieties of mammalian embryos *in vivo* by ^{14}C-U-glucose and ^{32}P-orthophosphate, *Naunyn Schmiedebergs Arch. Pharmakol.* **277:**413–428.

Bresnick, E., Lanclos, K., and Gonzales, E., 1965, The biosynthesis of ribonucleic acid in the liver of the rat fetus *in vivo*, *Biochim. Biophys. Acta* **108:**568–577.

Burk, D., Woods, M., and Hunter, J., 1967, On the significance of glycolysis for cancer growth, with special reference to Morris rat hepatomas, *J. Natl. Cancer Inst.* **38:**839–863.

Burton, K., 1956, A study of the conditions and mechanism of the reaction for the colorimetric estimation of deoxyribosenucleic acid, *Biochem. J.* **62:**315–323.

Chepenik, K. P., Johnson, E. M., and Kaplan, S., 1970, Effects of transitory maternal pteroylglutamic acid (PGA) deficiency on levels of adenosin phosphates in developing rat embryos, *Teratology* **3:**229–236.

Dagg, C. P., Doerr, A., and Offutt, C., 1966, Incorporation of 5-fluorouracil-2-C^{14} by mouse embryos, *Biol. Neonate* **10:**32–46.

Dobson, R. L., and Cooper, M. F., 1971, Incorporation of radioactivity from thymidine into mammalian glucose and glycogen, *Biochim. Biophys. Acta* **254:**393–401.

Engels, K., 1970, Biochemical studies on the embryotoxic action of cyclophosphamide, *in: Metabolic Pathways in Mammalian Embryos during Organogenesis and its Modification by Drugs* (R.

Bass, F. Beck, H.-J. Merker, D. Neubert, and B. Randhahn, eds.), pp. 123–136, Free University Berlin Press, Berlin.

Fouts, J. R., 1962, Metabolism of drugs by livers from fetal and newborn animals, in: Perinatal Pharmacology, Report of Forty-first Ross Conference on Pediatric Research (C. D. May, ed.), pp. 54–61, Ross Laboratories, Columbus, Ohio.

Fouts, J. R., and Adamson, R. H., 1959, Drug metabolism in the newborn rabbit, Science 129:897–898.

Friedman, O. M., and Boger, B. A., 1961, Colorimetric estimation of nitrogen mustards in aqueous media: Hydrolytic behavior of Bis(beta-chloroethyl)-amine, nor-HN_2, Anal. Chem. 33:906.

Fridland, A., 1974, Effect of methotrexate on deoxyribonucleotide pools and DNA synthesis in human lymphocytic cells, Cancer Res. 34:1883–1888.

Gibson, J. E., and Becker, B. A., 1969, Effect of phenobarbital and SKF 525-A on the teratogenicity of cyclophosphamide in mice, Teratoloty 1:393–398.

Gibson, J. E., and Becker, B. A., 1971, Teratogenicity of structural truncates of cyclophosphamide in mice, Teratology 4:141–151.

Gregg, C., Ott, D., Deaven, H., Spielmann, H., Krowke, R., and Neubert, D., 1975, The search for biological effects of ^{13}C-enrichment in developing mammalian systems, Proc. of the 2nd Intern. Conference on Stable Isotopes (E. R. Klein and P. D. Klein, eds.), pp. 61–70, United States Energy Research and Development Administration.

Hayashi, T. T., and Kazmierowski, D., 1970, Placental and fetal RNA metabolism. I. In vivo labelling of the fetal and placental RNA in the 14-day pregnant rat, Gynecol. Invest. 1:31–38.

Hughes, W. L., Christine, M., and Stollar, B. D., 1973, A radioimmunoassay for measurement of serum thymidine, Anal. Biochem. 55:468–478.

Kochhar, D. M., Larsson, K. S., and Boström, H., 1968, Embryonic uptake of ^{35}S-sulfate: Change in level following treatment with some teratogenic agents, Biol. Neonate 12:41–53.

Köhler, E., 1970, Growth kinetics of mammalian embryos during organogenesis, in: Metabolic Pathways in Mammalian Embryos during Organogenesis and its Modification by Drugs (R. Bass, F. Beck, H.-J. Merker, D. Neubert, and B. Randhahn, eds.), pp. 17–27, Free University Berlin Press, Berlin.

Köhler, E., Barrach, H.-J., and Neubert, D., 1970, Inhibition of NADP-dependent oxidoreductases by the 6-aminonicotinamide analogue of NADP, FEBS Lett. 6:225–228.

Köhler, E., Merker, H.-J., Ehmke, W., and Wojnorowicz, F., 1972, Growth kinetics of mammalian embryos during the stage of differentiation, Naunyn Schmiedebergs Arch. Pharmakol. 272:169–181.

Köhler, E., Wojnorowicz, F., and Borner, F., 1974, The maintenance of pregnancy by progesterone and estrone in rats fed a protein-free diet, in: Advances of the Biosciences 13. Hormones and Embryonic Development (G. Raspé, ed.), pp. 103–117, Pergamon Press, Oxford.

Krowke, R., 1970, Studies on the incorporation of ^{14}C-glucose and ^{32}P into nucleic acids and other cell components of rat embryos in vivo, in: Metabolic Pathways in Mammalian Embryos during Organogenesis and its Modification by Drugs (R. Bass, F. Beck, H.-J. Merker, D. Neubert, and B. Randhahn, eds.), pp. 99–120, Free University Berlin Press, Berlin.

Krowke, R., and Bochert, G., 1975, Inhibition of RNA synthesis, a possible mode of the embryotoxic action of hydroxyurea, Naunyn Schmiedebergs Arch. Pharmakol. 288:7–16.

Krowke, R., Siebert, G., and Neubert, D., 1971a, Effect of 5-fluoro-deoxycytidine, 5-fluoro-uracil and cycloheximide on the incorporation of ^{14}C-glucose and ^{32}P into different cell components of rat embryos in vivo, Naunyn Schmiedebergs Arch. Pharmakol. 269:229–234.

Krowke, R., Pielsticker, K., and Siebert, G., 1971b, A simple fractionation procedure for studying incorporation of radioactively labelled precursors into mammalian cell components, Naunyn Schmiedebergs Arch. Pharmakol. 271:121–124.

Krowke, R., Siebert, G., and Neubert, D., 1971c, A biochemical screening test with ^{14}C-glucose and ^{32}P-phosphate for the evaluation of embryotoxic effects in vivo, Naunyn Schmiedebergs Arch. Pharmakol. 271:274–288.

Krowke, R., Bochert, G., and Mewes, P., 1971d, Effects of some metabolic inhibitors on nucleic acid synthesis in early stages of mammalian embryonic development, *Acta Pharmacol. Toxicol.* **29**(Suppl. 4):27.

Krowke, R., Berg, P., and Merker, H.-J., 1977, Effect of cytosine arabinoside, 6-aminonicotinamide, and 6-mercaptopurine riboside on ectoderm and mesoderm of the mouse limb bud, *Teratology* (in press).

Lloyd, J. B., 1970, Histiotrophic nutrition—a target for "macromolecular" teratogens in rats, *in: Metabolic Pathways in Mammalian Embryos during Organogenesis and its Modification by Drugs* (R. Bass, F. Beck, H.-J. Merker, D. Neubert, and B. Randhahn, eds.), pp. 575–592, Free University Berlin Press, Berlin.

de Meyer, R., and DePlaen, J., 1964, An approach to the biochemical study of teratogenic substances on isolated rat embryos, *Life Sci.* **3**:709–713.

Molloy, G. R., Jellinek, W., Salditt, M., and Darnell, J. E., 1974, Arrangement of specific oligonucleotides within poly(A)terminated Hn RNA molecules, *Cell* **1**:43–53.

Neubert, D., 1966, Untersuchungen über den Nucleinsäurestoffwechsel in Zellkernen und Mitochondrien von Morris-Hepatomen, *in: Molekulare Biologie des malignen Wachstums, 17. Coll. Ges. Physiol. Chem. Mosbach* (H. Holzer and A. W. Holldorf, eds.), pp. 69–73, Springer-Verlag, Berlin.

Neubert, D., 1970a, Activity of DNA and RNA polymerases in isolated nuclei of mammalian embryos, *in: Metabolic Pathways in Mammalian Embryos during Organogenesis and its Modification by Drugs* (R. Bass, F. Beck, H.-J. Merker, D. Neubert, and B. Randhahn, eds.), pp. 79–95, Free University Berlin Press, Berlin.

Neubert, D., 1970b, Aerobic glycolysis in mammalian embryos, *in: Metabolic Pathways in Mammalian Embryos during Organogenesis and its Modification by Drugs* (R. Bass, F. Beck, H.-J. Merker, D. Neubert, and B. Randhahn, eds.), pp. 225–246, Free University Berlin Press, Berlin.

Neubert, D., 1977, An attempt of a molecular and multilateral approach towards developmental toxicology, *Arch. Toxicol.* (in press).

Neubert, D., and Dillmann, I., 1972, Embryotoxic effects in mice treated with 2,4,5-trichlorophenoxyacetic acid and 2,3,7,8-tetrachlorodibenzo-*p*-dioxin, *Naunyn Schmiedebergs Arch. Pharmakol.* **272**:243–264.

Neubert, D., and Herken, H., 1955, Wirkungssteigerung von Schlafmitteln durch den Phenyldiallylessigsäureester des Diäthylaminoathanols, *Naunyn Schmiedebergs Arch. Pharmakol.* **225**:453–462.

Neubert, D., and Rautenberg, M., 1976, Activity of nuclear DNA-dependent RNA polymerases in mouse limb buds differentiating *in vivo* or in organ culture, *Hoppe-Seyler's Z. Physiol. Chem.* **357**:1623–1635.

Neubert, D., Oberdisse, E., Merker, H.-J., Köhler, E., and Balda, B.-R., 1968, Biochemical studies on the nucleic acid metabolism of embryonic tissues and the effect of drugs, *Naunyn Schmiedebergs Arch. Pharmakol.* **259**:186–188.

Neubert, D., Merker, H.-J., Köhler, E., Krowke, R., and Barrach, H.-J., 1971a, Biochemical aspects of teratology, *in: Advances in the Biosciences 6, Schering Symposium on Intrinsic and Extrinsic Factors in Early Mammalian Development* (G. Raspé, ed.), pp. 575–621, Pergamon Press, Oxford.

Neubert, D., Peters, H., Teske, S., Köhler, E., and Barrach, H.-J., 1971b, Studies on the problem of "aerobic glycolysis" occurring in mammalian embryos, *Naunyn Schmiedebergs Arch. Pharmakol.* **268**:235–241.

Neubert, D., Zens, P., Rothenwallner, A., and Merker, H.-J., 1973, A survey of the embryotoxic effects of TCDD in mammalian species, *Environ. Health Perspect.* **5**:67–79.

Neubert, D., Merker, H.-J., and Tapken, S., 1974a, Comparative studies on the prenatal development of mouse extremities *in vivo* and in organ culture, *Naunyn Schmiedeberg Arch. Pharmakol.* **286**:251–270.

Neubert, D., Tapken, S., and Merker, H.-J., 1974b, Induction of skeletal malformations in organ cultures of mammalian embryonic tissues, *Naunyn Schmiedebergs Arch. Pharmakol.* **286**:271–282.

Neubert, D., Gregg, C. T., Bass, R., and Merker, H.-J., 1975, Occurrence and possible functions of mitochondrial DNA in animal development, *in: The Biochemistry of Animal Development* (R. Weber, ed.), Vol. III, Chapter 10, pp. 387–464, Academic Press, New York.

New, D. A. T., 1966, *The Culture of Vertebrate Embryos*, Logos Press Ltd., Elek Books, Ltd., London; Academic Press, New York, London.

Oerter, D., and Bass, R., 1975, Embryonic development and mitochondrial function. I. Effects of chloramphenicol infusion on the synthesis of cytochrome oxidase and DNA in rat embryos during organogenesis, *Naunyn Schmiedebergs Arch. Pharmakol.* **290:**175–189.

Packard, D. S., Menzies, R. A., and Skalko, R. G., 1973, Incorporation of thymidine and its analogue, bromodeoxyuridine, into embryos and maternal tissues of the mouse, *Differentiation* **1:**397–405.

Pelkonen, O., 1973, Drug metabolism in the human fetal liver. Relationship to fetal age, *Arch. Int. Pharmacodyn.* **202:**281–287.

Pictet, R., and Rutter, W. J., 1972, Development of the embryonic endocrine pancreas, *in: Handbook of Physiology—Endocrinology I* (D. F. Steiner and N. Freinkel, eds.), Chapter 2, pp. 25–66, American Physiology Society, Washington, D.C.

Rane, A., Berggren, M., Yaffe, S., and Ericsson, J. L. E., 1973, Oxidative drug metabolism in the perinatal rabbit liver and placenta. A biochemical and morphologic study, *Xenobiotica* **3:**37–48.

Raspé, G., ed., 1971, *Advances in the Biosciences 6, Schering Symposium on Intrinsic and Extrinsic Factors in Early Mammalian Development*, Pergamon Press, Oxford.

Raspé, G., ed., 1974, *Advances in the Biosciences 13, Hormones and Embryonic Development*, Pergamon Press, Oxford.

Rautenberg, P., and Neubert, D., 1975, Methods for the evaluation of mRNA fractions synthesized in limb buds differentiating in organ culture, *in: New Approaches to the Evaluation of Abnormal Embryonic Development* (D. Neubert and H.-J. Merker, eds.), pp. 133–144, Georg Thieme Publishers, Stuttgart.

Ritter, E. J., Scott, W. J., and Wilson, J. G., 1973, Relationship of temporal patterns of cell death and development to malformations in the rat limb. Possible mechanism of teratogenesis with inhibitors of DNA synthesis, *Teratology* **7:**219–226.

Ruddick, J. A., and Runner, M. N., 1974. 5-FU in chick embryos as a source of label for DNA and a depressant of protein synthesis, *Teratology* **10:**39–46.

Rutter, W. J., Pictet, R. L., and Morris, P. W., 1973, Toward molecular mechanisms of developmental processes, *Annu. Rev. Biochem.* **42:**601–645.

Sazaki, H., and Ohmori, K., 1973, The influence of single or fractionated doses of 5-fluoro-2'-deoxycytidine teratogenicity in mice, *Teratology* **8:**191–194.

Schimmelpfennig, K., 1971, Problems connected with *in vivo* labelling of embryonic glycosaminoglycans with $Na_2^{35}SO_4$ in teratological studies, *Naunyn Schmiedebergs Arch. Pharmakol.* **271:**320–324.

Schimmelpfennig, K., Baumann, I., and Kaufmann, C., 1971, Studies on glycosaminoglycans (GAG) in mammalian embryonic tissue. I. A micromethod for the fractionating and characterisation of GAG according to their anionic behaviour, *Naunyn Schmiedebergs Arch. Pharmakol.* **271:**430–456.

Schneider, W. C., and Greco, A. E., 1970, Incorporation of pyrimidine deoxynucleosides into liver lipids and other components, *Biochim. Biophys. Acta* **228:**610–626.

Schumacher, H.-J., Wilson, J. G., and Jordan, R. L., 1969, Potentiation of the teratogenic effects of 5-fluorouracil by natural pyrimidines. II. Biochemical aspects, *Teratology* **2:**99–106.

Scott, W. J., Ritter, E. J., and Wilson, J. G., 1971, DNA synthesis inhibition and cell death associated with hydroxyurea teratogenesis in rat embryos, *Dev. Biol.* **26:**306–315.

Shepard, T. H., Tanimura, T., and Robkin, M. A., 1970, Energy metabolism in early mammalian embryos, *Dev. Biol. Suppl.* **4:**42–58.

Skoog, L., and Nordenskjold, B. O., 1971, Effects of hydroxyurea and 1-β-D-arabinofuranosylcytosine on deoxyribonucleotide pools in mouse embryo cells, *Eur. J. Biochem.* **19:**81–89.

Solter, A. W., and Handschumacher, R. E., 1969, A rapid quantitative determination of deoxyribonucleoside triphosphates based on the enzymatic synthesis of DNA, *Biochim. Biophys. Acta* **174:**585–590.

Stellwagen, R. H., and Tomkins, G. M., 1971, Preferential inhibition by 6-bromodeoxyuridine of the synthesis of tyrosine aminotransferase in hepatoma cell cultures, *J. Mol. Biol.* **56:**167–182.

Tanimura, T., and Shepard, T. H., 1970, Glucose metabolism by rat embryos *in vitro*, *Proc. Soc. Exp. Biol. Med.* **135:**51–54.

Tidd, D. M., and Paterson, A. R. P., 1974, A biochemical mechanism for the delayed cytotoxic reaction of 6-mercaptopurine, *Cancer Res.* **34:**738–746.

Webster, W., Shimada, M., and Langman, J., 1973, Effect of fluorodeoxyuridine, colcemid, and bromodeoxyuridine on developing neocortex of the mouse, *Am. J. Anat.* **137:**67–86.

Weinberg, R. A., 1973, Nuclear RNA metabolism (review), *Annu. Rev. Biochem.* **42:**329–354.

Weintraub, H., Campbell, G. L. M., and Holtzer, H., 1972, Identification of a developmental program using bromodeoxyuridine, *J. Mol. Biol.* **70:**337–350.

Wilson, J. G., Shepard, T. H., and Gennaro, J. F., 1963, Studies on the site of teratogenic action of ^{14}C-labeled trypan blue, *Anat. Rec.* **145:**300.

Wilson, J. G., Jordan, R. L., and Schumacher, H., 1969, Potentiation of the teratogenic effect of 5-fluorouracil by natural pyrimidines. I. Biological aspects, *Teratology* **2:**91–98.

Yaffe, S. J., Rane, A., Sjöqvist, F., Boreus, L. O., and Orrenius, S., 1970, The presence of a monooxygenase system in human fetal liver microsomes, *Life Sci.* **9:**1189–1200.

Altered Electrolyte and Fluid Balance

7

CASIMER T. GRABOWSKI

I. INTRODUCTION

Every embryologist knows that there are many important fluid-filled spaces in and around the intrauterine individual, such as the ventricles of the brain, the canal of the spinal cord, the pelvis of the ureter, the lumens of various portions of the gut and glands, the entire cardiovascular system, the blastocoele, amnion, allantois, and yolk sac. Every teratologist also knows that many developmental abnormalities involve deficiencies or overenlargements of these spaces, e.g., hydronephrosis, hydroureter, hydrocephalus and hydrocephalus-like conditions, exencephaly, hydramnois, and the occlusion of various hollow organs. Less well known are details of the composition of the fluids that fill these spaces, how their volume is controlled, and their precise role in the normal and abnormal morphogenesis of the structures with which they are associated. Although the extracellular proteins (e.g., plasma proteins and some enzymes) of some developing fluids are reasonably well known, less is known of small organic molecules, inorganic constituents, and the means by which their concentrations are controlled. It is often assumed that, apart from that performed by the placenta, little or no regulation is possible on the part of the embryo and young fetus, due to physiological immaturity of organs (Howard, 1957) or cells (Widdowson, 1968). The primary aim of this chapter is to review reports which demonstrate that a study of the embryonic and fetal fluids, their composition, and the physiology of their regulation is a valuable approach to an analysis of abnormal development caused by known

CASIMER T. GRABOWSKI • Department of Biology, University of Miami, Coral Gables, Florida 33124.

teratogenic agents, as well as to provide insight into the role of these fluids in normal development.

Before discussing in detail several examples which illustrate these points, it is important to emphasize that most of the readily available information on this subject, especially in the books on fetal physiology such as those of Dawes (1968), Assali (1968), and Stave (1970), is primarily based on the perinatal physiology of large mammals (human and sheep). The late fetal and early postnatal periods constitute a unique, relatively short, transitional stage in the life cycle of the individual. Furthermore, the experimental work on the human is very limited in scope, and that done on sheep and other large mammals used a species of which little or nothing is known about teratogenic susceptibility. Most of this chapter will be concerned with preimplantation, embryonic, and early fetal stages, stages usually highly susceptible to the action of teratogenic agents.

There are several common misconceptions about this early period of development. For one thing, comparing the physiology of embryos and young fetuses to that of postnatal animals is unrealistic. They are not air-breathing, free-living animals. They are purely aquatic individuals dependent for gaseous interchange, water and ionic balance, waste removal, and metabolic control on radically different mechanisms than is the case after birth. Another is that much of the physiological regulation of fluids and molecules in mammalian embryos is attributed to an omnipotent placenta. A high degree of internal control over the composition and volume of these important fluid spaces by the embryo can be demonstrated to exist, presumably by mechanisms which are characteristic of successive stages of development.

Krogh, in his classic work, *Osmotic Problems of Aquatic Animals* (1939), devoted a long chapter to the fluid and ionic problems of animals developing in either fresh or salt water. He pointed out that a frog embryo, with an internal environment distinctly different from that of the fresh water surrounding it, has to exert some degree of control over its internal milieu from the earliest stages of development. Potts and Parry (1964), in their contemporary version of this classic, attempted to consider the fluid and ionic problems of embryos of aquatic and terrestrial animals but were limited, particularly in the case of mammalian embryos, by a paucity of information. Adolph (1967) comprehensively reviewed the available literature on fluid regulation in embryos, pointing out the necessity for both amniote and anamniote embryos to control the volume and composition of various fluid sacs, but he, too, concluded that the data on this subject were scanty. All of the above reviews have two things in common: (1) They were written by physiologists from the general physiological viewpoint, and (2) they agreed that many problems in this area are in need of attention. When the present writer became interested in the chemical composition of embryonic bloodstreams and the changes that might be induced by teratogenic agents, it was apparent that very little was known even about the most basic chemical constituents or ontogenetic changes in embryonic blood plasma. Extensive studies on normal embryos were required before changes induced by teratogenic agents could be studied (Grabowski, 1966b, 1967).

There are, however, several organ systems in which the fluids of hollow organs have been shown to be morphogenetically significant and in which interference with fluid volume, whether by mechanical or physiological means, has led to abnormal development. These are reviewed in Section II of this chapter, along with a brief discussion of the fluid imbalances in chick and mammalian embryos in general. The problems of ionic regulation in an individual cannot be totally separated from those of its fluids. In Section III are reviewed interesting data concerning extracellular potassium which indicates that the potassium physiology of the mammalian embryo and fetus is not only different from that of the adult, but also that this is a morphogenetically and teratogenetically significant ion.

II. FLUID REGULATION IN THE EMBRYO AND ITS MORPHOGENETIC SIGNIFICANCE

A. Function of the Vitreous Humor in Morphogenesis of the Eye

This example is one in which the alteration in the usual volume of a hollow organ was effected by mechanical means (see also Chapter 14, this volume). The eye of the chick undergoes a period of very rapid expansion between the 4th and 8th days of incubation, mostly due to increase in the volume of the fluid cavities (Coulombre, 1956). The volume of the vitreous humor as well as its pressure was reduced by placing glass capillaries in the right eye of chicks on day 4 of incubation, for several days. After removing the tubes, development of these eyes was compared with that of unoperated controls and controls in which a solid glass rod was placed in the eye. The loss of vitreous fluid via the inserted capillaries, and conversely the pressure exerted from its accumulation in normal eyes, was shown to affect total eye size, corneal size, corneal curvature, and the ciliary body (Coulombre, 1956, 1957; Coulombre and Coulombre 1957, 1958). Although the neural retina continued to develop at its own intrinsic rate, despite the loss of fluid and the folded configuration which resulted, the pigmented epithelium increased only in response to the tensile forces generated by the expansion of the normal eye, i.e., by the accumulation of vitreous humor (Coulombre et al., 1963). The mechanism of volume regulation of the vitreous humor of the eye is not known. The following example provides, however, a physiological basis for volume control as well as another demonstration of the morphogenetic significance of fluid pressure in a hollow organ.

B. Regulation of the Volume of Cerebrospinal Fluid and Its Role in Normal and Abnormal Development of the Brain

Between the time of closure of the anterior and posterior neuropores of the brain of the chick embryo (2nd day of incubation) and the opening in the roof of the fourth ventricle, about day 7 (Arey, 1954), the neural tube is a closed system. During this period (between days 3 and 8) the brain of the chick undergoes great expansion, largely caused by an increase in cerebrospinal fluid (CSF) rather than bulk of neural tissue. Weiss (1934) obtained evidence in tissue culture of secretory activity by the embryonic ependymal cells, and later he (Weiss, 1955) expanded on the concept that the secretion of CSF and the increased turgor caused by its accumulation had a significant morphogenetic role in the early stages of brain development.

1. Volume Regulation of CSF in the Chick Embryo

Freezing-point depression determinations on normal CSF between the 4th and 8th days of incubation (Browne, 1970a,b; Browne and Grabowski, 1977) have shown that osmotic pressure increases during the period of brain expansion, from a mean of 288 milliosmoles on day 4 to a maximum of 351 milliosmoles at day 7, and then drops abruptly to the same level as blood (290 milliosmoles) by the 8th day. This observation, together with that of Weiss (1934), suggests a two-stage intrinsic mechanism of control of the volume of CSF, namely: (1) secretion of fluid containing osmotically active substances by ependymal cells, and (2) an increase in the volume of the resulting hyperosmotic CSF by drawing in fluid from the surroundings by osmotic flux.

The hyperosmotic neural tube chamber would be expected to have some degree of turgor. This has been measured by Jelinek and Pexieder (1968), who found that the hydrostatic pressure of CSF in two-day chicks was equivalent to 2.18 mm H_2O, rises to 5.10 on day 5, and drops to 3.18 on day 6. This is clear evidence that there is an increased pressure in the CSF during the period of rapid expansion of the brain, as originally postulated by Weiss. The following experiments demonstrate that this increased turgor has a role both in normal and abnormal development.

2. Effects of Interfering with the Volume of Cerebrospinal Fluid

Although many neurologists and teratologists feel that malformations of the neural tube which involve persistent openings originate prior to closure of the neural folds, others feel that such dysraphic states could occur from overdistention and subsequent rupture of an already closed neural tube (Gardner, 1973). The following examples demonstrate that the latter is at least experimentally possible. Jelinek (1961) found that incisions in the wall of the brain of the young chick embryo caused collapse of the neural tube and subsequent maldevelopment. In a series of experiments with dimethyl sulfoxide (DMSO), it was found (Browne, 1970a,b; Grabowski, 1970; Browne and Grabowski 1977) that when it was injected into the air sac of 4-day chick eggs and the embryos were studied by photographic time-lapse techniques, the brain showed grossly visible distention (ca. 50% more in sagittal area than controls over an 8-hr period) followed occasionally by rupture of the brain. Eventually, obvious defects such as torn midbrains on day 5 and platyneuria in older embryos were found (Grabowski, 1970).

The swelling and rupture were attributed to a greater-than-normal increase in the osmotic pressure of CSF in treated embryos. The pressure in treated embryos rose to levels as high as 400 milliosmoles 2 hr after injection, compared to a mean level of about 280 milliosmoles in controls at 4 days (see p. 156). Thus it was demonstrated that (1) normal control of volume in the CSF is due, at least in part, to regulation of osmotic pressure of this fluid, and (2) an increase of normal pressure by DMSO injection can result in overdistention and rupture of the brain wall and subsequent abnormal development.

Exposure to another teratogenic agent, moderate hypoxia (i.e., 6% O_2 for 6 hr) has also been demonstrated to affect the volume of CSF in 3-day chick embryos and thereby temporarily affect the development of the cephalic and cervical flexures. This agent produces a generalized edema of the embryo (see below). During the period of observation, the angle between brain and trunk (i.e., cervical flexure) steadily decreases to about 60°, whereas in treated embryos there is an immediate increase in this angle to as much as 25° greater than that in controls (Grabowski and Schroeder, 1968). This temporary reversal of the development of the cephalic flexure is probably due to increased turgor within the neural tube, in turn caused by an increase in the volume of the CSF. Whether this subtle change is in itself capable of producing pathology is not known for certain, but brain abnormalities do result from such hypoxic treatment (Mushett, 1953; Grabowski and Paar, 1958; Grabowski, 1961).

3. Osmotic Changes and Water Uptake by the Rabbit Blastocyst

The above examples of control of the volume of hollow organs by osmotic changes are in the chick embryo. Can comparable mechanisms be demonstrated in developing mammals? Tuft and Böving (1968) found an increase in osmotic pressure in rabbit blastocyst fluid which corresponded with the period of rapid expansion of this cavity. The swelling occurs between the 4th and 8th day after fertilization, and the freezing-point depression of blastocyst fluid increases from 0.539°C at day 4 to a maximum of 0.554°C on day 6, while that of uterine fluid remains hypotonically constant at 0.536°C. The authors postulated that the swelling of the blastocyst, an important event at this stage because it helps to determine uterine spacing, is controlled by changes in osmotic pressure. Perhaps these examples, the chick CSF and rabbit blastocyst, may be indicative of a general mechanism for fluid volume control in other hollow organs during development.

C. The Role of Hemodynamics in Normal and Abnormal Development of the Cardiovascular System

1. Effects of Mechanical Interference with Hemodynamics

Significant effects of pressure and flow patterns in the sculpturing of the cardiovascular system have long been postulated on the basis of phylogenetic considerations (Spitzer, 1923) and mechanical models (Bremer, 1931) (see also Chapter 15, this volume). Within the last 20 years, however, a large number of experimental observations have clearly confirmed the importance of hemodynamics in normal and abnormal development of the cardiovascular system, whether in establishing a definitive artery or vein from a capillary bed, transformation of the aortic arches, or development of the heart. A variety of techniques have been employed including electrocautery of early vessels (Stephan, 1952), use of silver microclips to block aortic arches or constrict part of the heart (Rychter, 1962), cinematography to observe the role of bilaminar streams in partitioning of the heart (Jaffee, 1966), or measurements of blood pressure of the aortic arches in normal and abnormal circumstances (Pexieder, 1969).

The conclusion from these and many other studies, such as those of Jaffee (1965, 1967), Grabowski (1966a), and Harh *et al.* (1973), is that pressure and flow patterns of the blood are extremely important in the development of the circulatory system. Coincidentally, most of these workers have also found that almost any interference with aortic arch development results in ventricular septal defects (see especially review by Rychter, 1962). Although the experiments in which direct mechanical manipulation of the blood vessels of the embryo are, of necessity, confined to chick embryos, the nature of the data is such that the conclusions from them can and have been applied to the development of the mammalian circulatory system (Moffat, 1959; Rychter, 1962).

The really germane questions for the purpose of this chapter, however, are: (1) can teratogenic agents influence the hemodynamics of the circulatory system, and (2) can these disturbances result in abnormal development of the heart and blood vessels? The following experiments, done on both chicks and mammals, indicate strongly that the answers to both questions are affirmative.

2. Effects of Teratogenic Agents on Hemodynamics and the Circulatory System

The best documented example of hemodynamic changes induced by a teratogenic agent is that of moderate hypoxia on the chick. Exposure of 3- to 5-day chick embryos to moderate oxygen deficiency produces a marked swelling of the entire embryo, but for purposes of this chapter only the changes in volume of circulating fluids will be discussed. This moderate exposure (6–10% O_2 for 5–6 hr, depending on age of embryo) induces transient fluid volume changes which are evident from the first hour of exposure and last for several hours after return of the eggs to a normal atmosphere. The maximum increase in circulating fluids (hypervolemia) in 5-day chicks was estimated at 60% (Grabowski, 1966a). In a time-lapse photographic study of 3-day embryos, the effect of this hypervolemia on the heart and the diameter of blood vessels was measured. A gross enlargement of the heart was evident but could not be quantified because of cardiac contractions. The dorsal aorta increased up to 2½ times its normal diameter, and the anterior cardinal vein increased from 3 to 5 times normal size. Both of these vessels returned to normal size by 7 hr after treatment was discontinued (Grabowski and Schroeder, 1968). The blood pressure of treated 3- to 5-day embryos showed maximal increases about 1 hr after treatment (Grabowski et al., 1969; Grabowski, 1970). The mean blood pressure at this time, e.g., in 3-day embryos, was about 25% above normal; in some individuals it was as much as 2½ times above normal.

Do such obvious changes in blood volume and pressure directly lead to cardiovascular abnormalities? Abnormalities of the heart and blood vessels of the chick have been frequently described subsequent to hypoxic exposure (Byerly, 1926; Schellong, 1954; Grabowski and Paar, 1958; Jaffee, 1974). The latter, in particular, directly related the observed hypoxia-induced hypervolemia to the subsequent cardiovascular abnormalities (over 50%), and earlier this author (Jaffee, 1965) had correlated some hemodynamic changes in heart and in great vessel flow induced by trypan blue with known cardiovascular abnormalities in the chick.

In mammals the best known agent for inducing cardiovascular abnormalities is trypan blue (up to 60% in rats, Fox and Goss, 1956) which is one reason for its popularity in teratogenic research (Fox and Goss, 1956–1958; Beck and Lloyd, 1966). A conspicuous feature of trypan blue treatment of mice and rats is a generalized edema as well as a gross swelling of the pericardial cavity, heart, and major blood vessels (Waddington and Carter, 1953; Turbow, 1966; Grabowski et al., 1971). Fox and Goss attributed the cardiovascular anomalies they found to a displacement of the atrium during a critical stage of morphogenesis, a change which could have been caused by hypervolemia. In an attempt to correlate these cardiovascular abnormalities with the hemodynamic changes described, the present author's laboratory has measured the blood pressure of normal and trypan-blue-treated rats between

13½ and 18½ days. At all ages treated animals showed much more variability than did controls, and from day 14½ on, the mean blood pressure of treated embryos was significantly lower than that of controls (Grabowski *et al.*, 1971). The expected rise in blood pressure was not found; but there were possible explanations, e.g., technical inability to measure pressure at the earliest age at which hypervolemia was detectable. Nevertheless, the continued low pressure itself was indicative of prolonged hemodynamic changes produced by this cardiovascular teratogen.

D. General Aspects of the Fluid Problem in Chick and Mammalian Embryos under Normal and Abnormal Conditions

1. The Edema Syndrome in the Chick

Since this subject has been thoroughly reviewed elsewhere (Grabowski, 1970), only some highlights will be presented here. Among the fluids of the 5-day chick embryo, blood serum, cerebrospinal, chorionic, and amniotic fluids, all have a similar composition and total osmotic concentration. On the other hand, albumen, yolk-sac fluid, and allantoic fluid are quite different, containing less sodium and much more potassium and having a lower osmotic pressure (Grabowski, 1963, 1966a). It is evident that the young embryo is doing active osmotic work someplace within itself, or within its extraembryonic membranes, in order to maintain the gradients that exist between the embryonic bloodstream and the fluids of the yolk sac, albumen, and allantois. This, however, is based on deductive reasoning, and the route by which this conclusion was reached was somewhat circuitous.

It was earlier found that moderate degrees of hypoxia were more teratogenic than exposure to acute hypoxia and that those embryos exposed to moderate hypoxia did not show extensive cell death of a degree that would explain the abnormal development observed (Grabowski, 1964). However, immediately after exposure to moderate hypoxia, chick embryos displayed marked swelling, involving both the circulating fluids as well as the entire embryo (Grabowski, 1964). This demonstration of a gross increase in all embryonic fluids led to an analysis of the chemical composition of blood plasma in an attempt to understand the physiological basis for the changes (Grabowski, 1966a). The changes found in hypoxic embryos, however, did not agree with those expected on the basis of adult reactions to hypoxia. Changes in volume and composition of embryonic blood could be explained in terms of an osmoregulatory disturbance, resulting in an inrush of fluids from the hyposmotic and hyperkalemic yolk sac and albumen. These changes, in turn, lead to the generalized edema, hypervolemia, and consequent increase in blood pressure, as well as to the apparent increase in the pressure of cerebrospinal fluid mentioned above.

One important consequence of the hypervolemia and hypertension produced by moderate hypoxia was the formation of hematomas within a few hours after exposure. Most of the abnormalities eventually associated with this treatment could be correlated with the location of these hematomas immediately after treatment and the necrosis which they caused in surrounding tissues. The main cause of death was hemorrhage due to the rupture of major blood vessels during the period of distension and hypertension. The name given to this complex series of events was the *edema syndrome*. Moderate hypoxia induced osmoregulatory disturbances which were followed by hypertension, hypervolemia, and distension of the lumena of the embryo, as well as abnormal fluid accumulation in solid tissues. Although many of the embryos apparently recovered from such distortion, others showed damage from distension of hollow structures such as the neural tube and cardiovascular system or formation of persistent clear blisters and hematomas in or adjacent to rapidly developing structures causing necrosis and disruption in surrounding tissues. Thus it is not the lack of oxygen per se that causes abnormal development but a complex sequence of events (Figure 1).

Such edematous changes have been associated in the chick embryo with many teratogenic and hereditary agents (Sturkie, 1941; Cole, 1942; Ancel, 1950; Grabowski, 1970) and the sequence illustrated above may have widespread significance. The term edema syndrome to describe this sequence is also coming into more general usage (Turbow, 1966; Kaplan and Grabowski, 1967; Leist and Grauwiler, 1974; Jaffee, 1974).

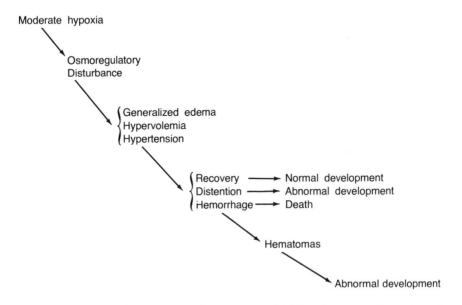

Fig. 1. The edema syndrome as seen after exposure of chick embryos to moderate hypoxia.

2. Fluid Imbalances in Mammalian Embryos

Do these considerations have any relationship to the development of mammalian embryos which are not likely to be bathed in hypotonic fluids or even fluids with compositions radically different from those of the mother? Frank edema has been induced by many teratogenic agents and mutant genes. Elements of the edema syndrome, namely hemorrhaging, blistering, and subsequent abnormal development of surrounding structures, have also been demonstrated in numerous instances. Since these relationships have been reviewed several times previously (Jost, 1953; Giroud *et al.,* 1955; Grabowski, 1970), there is no need to do so again except to add a few newer observations.

King *et al.* (1965) noted that administration of antihistamines of the cyclizine group to pregnant rats produced up to 100% cleft palate and induced severe edema. Posner and Darr (1970) confirmed these observations and suggested that the swollen tongue and edematous thorax which resulted from treatment kept the tongue wedged between the palatine shelves. Failure of the palatal shelves to close could then be considered a secondary result of the generalized edema. Why an antihistamine should induce edema is not clear, but the illustrations of edematous embryos produced in these studies are certainly striking. One cannot help but wonder what effects such swelling may have on circulation and the development of the cardiovascular system.

Khera (1975) noted that pregnant cats fed thalidomide during organogenesis produced near-term fetuses which displayed a very high frequency of cardiac malformations as well as conspicuous edema, although no attempt to correlate the two was made.

Leist and Grauwiler (1974) carefully traced the effects of transient uterine vessel clamping in pregnant rats on day 14, paying special attention to the limb abnormalities (up to 34%) after this treatment. They described generalized edema associated with this type of hypoxic treatment, including enlargement of blood vessels, blistering, and hemorrhaging, especially in the extremities and the snout, again providing evidence that the edema syndrome can be induced in mammals. It has been observed in the author's laboratory on numerous occasions that a generalized swelling occurs in rat embryos exposed to hypoxia, but this has been difficult to measure. For example, a 25% increase in volume is a considerable distension but is difficult to quantify. Regardless of what the basic physiological mechanism may be, edemic changes in mammalian embryos have been well documented in relation to a variety of causes; therefore, the edema syndrome must be recognized as a significant factor of mammalian teratology.

III. ELECTROLYTES IN NORMAL AND ABNORMAL DEVELOPMENT

There are significant differences in the mammal as to the relative amount of tissue water and the composition of intracellular and extracellular fluids and the ratio between intra- and extracellular concentrations of certain ions at different stages in development. Much data on late fetal and perinatal stages are available, including detailed information on gaseous transfer, acid–base balance, etc., but comparable data on the younger fetus and embryo are virtually nonexistent. This is most unfortunate because this is the time when abnormal development is most likely to be initiated.

A. Morphogenetic Effects of Extracellular Potassium

What evidence is there that changes in salt balance can adversely affect development? There are several bits of information which collectively indicate that a deficiency in extracellular potassium can cause abnormal development *in vitro* as well as *in vivo*. Crocker and Vernier (1970) and Crocker (1973) have demonstrated that human and mouse kidneys cultured in media containing about half the normal concentration of potassium develop abnormally, often forming cysts [although this contention has been questioned by Monie and Morgan (1975)]. Pantothenic acid deficiency in the rat has been clearly demonstrated (Giroud *et al.*, 1955) to result in abnormal kidney development and pronounced edema in the offspring. Studies in this laboratory on blood chemistry (Grabowski, 1970, and unpublished) have shown a significantly lowered glucose level in treated embryos and a lowered fetal serum potassium level, in some individuals to less than half of normal. Although further study has not been done to demonstrate that the lowered potassium values were directly related to the abnormalities of the kidney and the edema produced by this agent, there is a suggestive correlation.

Two other observations indicate a morphogenetic role for potassium. DeHaan (1967) found that lowered potassium in the medium in which embryonic chick heart cells were cultured increased the frequency with which beating cells were obtained. Hughes (1975) found the ratio of Na/K in normal whole chick embryos at 36–48 hr old to range from 4.0 to 2.0. In chicks exposed to several different teratogens, the ratio increased to 9.0 or more. He postulated that failure of the sodium pump to function properly could provoke, or at least contribute to, failure of neural tube closure.

B. Hypoxia, Hyperkalemia, and Fetal Cardiac Physiology

Although the above experiments indicate that mild hypokalemia may have a role in teratogenesis, the following studies demonstrate that embryonic and fetal blood plasma may be subject to considerable hyperkalemia. An excess of serum potassium is not an adult reaction to hypoxia and is, as a matter of fact, uncommon, difficult to induce, and a dangerous condition. Normal adult serum levels are maintained at 3–5 mEq/liter, with severe cardiac problems starting at 8 mEq/liter. Ventricular fibrillation and cardiac arrest occur when serum potassium levels reach 10–12 mEq/liter (Goldberger, 1965).

The normal level of serum potassium in mammalian embryos is a point of controversy. Although some workers have found it to be about the same as maternal levels (Fantel, 1975), others have reported it to be significantly higher in fetal humans, pigs, lambs and dogs (Widdowson, 1968; Kerpel-Fronius, 1970), and rats and rabbits (Adolph and Hoy, 1963; Chernoff and Grabowski, 1971; Grabowski, 1973). The author has also found that excised hearts from 15½-day rat fetuses continue to beat better and longer in solutions containing 9 mEq/liter of potassium than in those containing 5 mEq/liter. The relatively small differences in serum potassium in normal animals obtained by the above workers could be real or could be artifacts caused by variations in use of anesthetic and the speed and skill of the investigator in getting and centrifuging the blood samples. The intracellular potassium of erythrocytes is very high, and the dangers of contamination from damaged cells is appreciable. However, the hyperkalemia induced by hypoxia is of considerably higher magnitude than the variations found in normal sera by different workers. Furthermore it can be induced with such consistency and rapidity that the question of the precise level of serum potassium in normal fetuses is relatively unimportant in interpreting the following results.

Exposure of rat and rabbit fetuses to hypoxia induced by means of ogygen-deficient atmosphere, KCN, $NaNO_2$-induced methemoglobinemia, uterine clamping, or maternal injection of adrenaline and vasopressin (which induces acute hypoxia in embryos by constricting uterine blood vessels) uniformly raised the fetal serum potassium to levels of 10, 20, even 25 mEq/liter, depending on the degree of hypoxia attained (Grabowski and Chernoff, 1970; Chernoff and Grabowski, 1971; Grabowski, 1977). The rate of heartbeat was also affected. Bradycardia directly related to potassium concentration, rather than cardiac arrest, occurred at these high potassium levels. Thus a three-way correlation was found between degree of hypoxia, rate of heartbeat, and serum potassium. Regardless of the method of inducing hypoxia, age of embryo, or species, the rate of embryonic heartbeat decreased and serum potassium increased as the degree of hypoxia increased. For example, in 17½-day rat fetuses the normal rate of heartbeat is about 150–180 beats/min and serum potassium is 7 mEq/liter. During moderately acute hypoxia the rate decreased to 30 beats/min and K^+ increased to 20 mEq/liter. In these embryos the blood pressure was drastically reduced (Chernoff and Grabowski, 1971), but they could be maintained in such acute conditions of bradycardia and hyperkalemia for up to 40 min without death, and complete recovery could be obtained within a few minutes after normal oxygen levels were restored. These data show that fetal hypoxia can readily raise the concentration of serum potassium far beyond those tolerated by adult mammals. From the teratogenic standpoint, the concurrent bradycardia and hypotension could be significant in that they would tend to accentuate the oxygen deficiency and therefore increase the damage potential to localized regions of the fetus.

The striking tolerance of fetal hearts to high potassium has been confirmed by culturing isolated hearts from fetal and embryonic rats in solutions containing 4–30 mEq/liter K^+ (Grabowski, 1973, and unpublished). Solutions of 15 and even 19 mEq/liter of potassium were well tolerated by fetal hearts. Embryonic rat hearts (10.5–11.5 days) were even more tolerant to high potassium concentrations, continuing to beat overnight in solutions containing 20 mEq/liter. Excellent electrocardiograms of 15.5-day fetal rat hearts were obtained both *in vivo* and *in vitro* under normal and hyperkalemic conditions. Cardiac arrest from prolonged immersion in very high potassium solutions could be spontaneously reversed by replacement with a solution containing somewhat less potassium, a reversal which could be followed with a return of the normal ECG. These experiments were originally designed to study the initial responses of the mammalian fetus to hypoxia and to answer the question raised in Chapter 11, Vol. 1, namely, how much hypoxia is too much for a given organ at a given time in development, and consequently likely to induce damage.

The important points to be derived from this chapter on electrolyte balance are: (1) The physiological reactions of the mammalian embryo and fetus to teratogens may be quite different than expected on the basis of adult physiology. (2) The potassium regulatory mechanisms of mammalian fetuses are unstable, e.g., capable of going in either direction under the stimulus of different teratogenic agents. (3) The embryonic and fetal heart is surprisingly tolerant to very high concentrations of extracellular potassium. Such observations may be helpful in understanding teratogenic mechanisms, as well as in providing additional avenues for research.

C. Teratogenic Effects of Calcium

The following example further indicates how ions may serve to provoke or act as mediators of abnormal development. Grabowski (1966c) found that traces of calcium salts dropped on chick embryos produced a significant degree of malformations, especially of the CNS. A later observation, utilizing time-lapse photography (Grabowski, 1970), revealed a typical edema syndrome in a chick embryo which became grossly swollen, formed a caudal hematoma, and eventually developed rumplessness. However, the mode of action of this ion remains unknown.

IV. COMMENTS ON PHYSIOLOGICAL STUDIES IN TERATOLOGY

The major purpose of this chapter has been to demonstrate that embryonic and fetal physiology can be quite different from that of the perinatal or adult animal and that prenatal fluid and ionic physiology are often affected by teratogenic agents. The examples given indicate that physiological approaches can be useful in understanding teratogenic action as well as having significant heuristic value in studying normal morphogenesis. Teratologists use pharmacologically potent agents which disturb morphogenesis just as effectively as the classical embryologist's needle or hair loop. Their data should also be as useful in understanding normal development, but this goal will not be realized until the mode of action of teratogenic agents is thoroughly analyzed. One obstacle, in the writer's opinion, is limited application of the principles of physiology to studying changes in an embryo or early fetus affected by a teratogenic agent before any morphological effect—an end result—becomes apparent.

The experimental teratologist is usually confined to studying the embryos and fetuses of small laboratory animals. These are admittedly more difficult to study from the physiological standpoint than are lambs or humans. Nevertheless, equipment for studying physiological parameters such as embryonic blood pressure and ECG even in small animals is now available. Chemical analyses on microsamples can also be easily obtained. Among the major problems is obtaining suitable samples of embryonic tissues and fluids for quantitative analysis. The chemical testing of the components in blood and other body fluids has made possible great advances in clinical diagnosis and in better understanding the normal physiology of adults. Hopefully an equally valuable counterpart could be developed for the mammalian embryo.

Demonstrations that teratogenic agents can affect the chemistry of the bloodstream and other fluid compartments for prolonged periods of time raise the question as to whether such treatments may not also initiate metabolic disorders of long duration. For example, in addition to effects on serum potassium, a common effect of several teratogenic agents is reduced blood glucose. Is it possible that a prolonged period of hypoglycemia could affect the biochemical differentiation of the pancreas or liver by interfering with the orderly induction of a sequence of metabolic enzymes? Could the absence or excess of other small molecules at a critical period similarly have an effect on the enzymatic development of other organs? Such effects could be temporary, or they might be sufficiently prolonged to be responsible for metabolic disturbances in the newborn.

A matter of practical significance is the possibility of using some of these tests to facilitate teratogenic screening. The classical method of injecting a substance and looking for gross morphological abnormalities a week or two later is far from ideal as a means of identifying prenatal hazards. In addition to the uncertainties of extrapolating data from one species to another, there is also the problem that the morphological abnormalities looked for are simply the "tip of the iceberg." The subtler abnormalities of the nervous and endocrine system, and even such seemingly obvious problems as blindness or deafness, are not demonstrated by standard or presently recommended techniques. An analysis of embryonic and fetal blood chemistry after the administration of a teratogen could give some indication of whether or not an agent was affecting fetal physiology in a way which might lead to morphological abnormalities. A very simple test is the visual observation of the heart rate of a fetus shortly after experimental treatment. Certainly a marked bradycardia or cardiac arrest for a prolonged period of time could not be considered innocuous as far as embryonic or fetal health was concerned. Preliminary testing of this type is in progress in this and other laboratories. If the results of these efforts are to any degree predictive of permanent functional or structural disorder, they could add a new facet to the validity of testing. When more of the questions raised by these newer types of experiments are resolved, they may facilitate solving the serious and vexing problem of identifying possible teratogens. They will also have advanced knowledge in the areas of embryonic and fetal pathology, normal morphogenesis, and normal embryonic and fetal physiology.

REFERENCES

Adolph, E. F., 1967, Ontogeny of volume regulations in embryonic extracellular·fluids, *Q. Rev. Biol.* **42**:1–39.

Adolph, E. F., and Hoy, P. A., 1963, Regulation of electrolyte composition of fetal rat plasma, *Am. J. Physiol.* **204**:392–400.

Ancel, P., 1950, *La Chimiotératogenèse Chez les Vertébrés*, G. Doin, Paris.

Arey, L. B., 1954, *Developmental Anatomy*, 6th ed., W. B. Saunders, Philadelphia.

Assali, M. S., 1968, *Biology of Gestation*, Vols. I and II, Academic Press, New York.

Beck, F., and Lloyd, J. B., 1966, The teratogenic effects of azo dyes, *Adv. Teratol* **1**:131–193.

Bremer, J. L., 1931, The presence and influence of two spiral streams in the heart of the chick embryo, *Am. J. Anat.* **49**:409–440.

Browne, J. M., 1970a, The effect of embryonic fluids on the morphogenesis of the chick embryo, Ph.D. Dissertation, Univ. of Miami, Miami, Fla.

Browne, J. M., 1970b, Normal and abnormal changes in cerebrospinal fluid and their relation to morphogenesis of the chick brain, *Teratology* **3**:199.

Browne, J. M., and Grabowski, C. T., 1977, Cerebrospinal fluid of chick embryos: The osmoregulatory control of its volume and its role in brain morphogenesis (in press).

Byerly, T. C., 1926. Studies in growth—1. Suffocation effects in the chick embryo. *Anat. Rec.* **32**:249–270.

Chernoff, N., and Grabowski, C. T., 1971, Physiological responses of the rat fetus to maternal injections of epinephrine and vasopressin. *Br. J. Pharmacol.* **43**:270–278.

Cole, R. K., 1942, The "talpid lethal" in the domestic fowl, *J. Hered.* **33**:83–86.

Coulombre, A. J., 1956, The role of intraocular pressure in the development of the chick eye: I—Eye size, *J. Exp. Zool.* **133**:211–225.

Coulombre, A. J., 1957, The role of intraocular pressure in the development of the chick eye:II—Control of corneal size, *A.M.A. Arch Ophthalmol.* **57**:250–253.

Coulombre, A. J., and Coulombre, J. L., 1957, The role of intraocular pressure in the development of the chick eye: III—Ciliary body, *Am. J. Ophthalmol.* **44**:85–93.

Coulombre, A. J., and Coulombre, J. L., 1958, The role of intraocular pressure in the development of the chick eye: IV—Corneal curvature, *A.M.A. Arch. Ophthalmol.* **59**:502–506.

Coulombre, A. J., Steinberg, S. N., and Coulombre, J. L., 1963, The role of intraocular pressure in the development of the chick eye: V—Pigmented epithelium, *Invest. Ophthalmol.* **2**:83–89.

Coulter, D. B., and Small, L. L., Sodium and potassium concentrations of erythrocytes from perinatal, immature and adult dogs, *The Cornell Vet.* **63**:462–468.

Crocker, J. F. S., 1973, Human embryonic kidneys in organ culture: Abnormalities of development induced by decreased potassium, *Science* **181**:1178–1179.

Crocker, J. F. S., and Vernier, R. L., 1970, Fetal kidney in organ culture: Abnormalities of development induced by decreased amounts of potassium, *Science* **169:**485–487.

Dawes, G. S., 1968, *Foetal and Neonatal Physiology*, Year Book Medical Pub., Chicago, Ill.

DeHaan, R. L., 1967, Regulation of spontaneous activity and growth of embryonic chick heart cells in tissue culture. *Dev. Biol.* **16:**216–249.

Fantel, A. G., 1975, Fetomaternal potassium relations in the fetal rat on the twentieth day of gestation, *Pediatr. Res.* **9:**527–530.

Fox, M. H., and Goss, C. M., 1956, Experimental production of a syndrome of congenital cardiovascular defects in rats, *Anat. Rec.* **124:**189–208.

Fox, M. H., and Goss, C. M., 1957, Experimentally produced malformations of the heart and great vessels in rat fetuses. Atrial and caval abnormalities, *Anat. Rec.* **129:**309–332.

Fox, M. H., and Goss, C. M., 1958, Experimentally produced malformations of the heart and great vessels in rat fetuses. Transposition complexes and aortic arch abnormalities, *Am. J. Anat.* **102:**65–92.

Gardner, W. J., 1973, *The Dysraphic States, Excerpta Medica,* Amsterdam.

Gessner, I. H., 1966, Spectrum of congenital cardiac anomalies produced in chick embryos by mechanical interference with cardiogenesis, *Circ. Res.* **18:**625–633.

Giroud, A., Lefebvres, J., Prost, H., and Dupuis, R., 1955, Malformations des membres dues à des lésions vasculaires chez le foetus de rat déficient en acide pantothenique, *J. Embryol. Exp. Morphol.* **3:**1–12.

Goldberger, E., 1965, *A Primer of Water, Electrolyte and Acid–Base Syndromes,* 3rd ed., Lea and Febiger, Philadelphia.

Grabowski, C. T., 1961, A quantitative study of the lethal and teratogenic effects of hypoxia on the three-day chick embryo, *Am. J. Anat.* **109:**25–36.

Grabowski, C. T., 1963, Teratogenic significance of ionic and fluid imbalances, *Science* **142:**1064–1065.

Grabowski, C. T., 1964, The etiology of hypoxia-induced malformations in the chick embryo, *J. Exp. Zool.* **157:**307–326.

Grabowski, C. T., 1966a, Physiological changes in the bloodstream of chick embryos exposed to teratogenic doses of hypoxia, *Dev. Biol.* **13:**199–213.

Grabowski, C. T., 1966b, Ontogenetic changes in the concentration of serum proteins in chick and mammalian embryos, *J. Embryol. Exp. Morphol.* **16:**197–202.

Grabowski, C. T., 1966c, Teratogenic effects of calcium salts in chick embryos. *J. Embryol. Exp. Morphol.* **15:**113–118.

Grabowski, C. T., 1967, Ontogenetic changes in the osmotic pressure and sodium and potassium concentrations of chick embryo serum, *Comp. Biochem. Physiol.* **21:**345–350.

Grabowski, C. T., 1970, Embryonic oxygen deficiency—a physiological approach to analysis of teratological mechanisms, *Adv. Teratol.* **4:**125–169.

Grabowski, C. T., 1973, Fetal cardiac physiology and hypoxia-induced hyperkalemia, *Teratology* **7:**A-16.

Grabowski, C. T., 1977, The physiological effects of sodium nitrite-induced hypoxia in rat and rabbit fetuses (in prepration).

Grabowski, C. T., and Chernoff, N., 1970, Effects of hypoxia on the cardiovascular physiology of mammalian embryos, *Teratology* **3:**201.

Grabowski, C. T., and Paar, J. A., 1958, The teratogenic effects of graded doses of hypoxia on the chick embryo, *Am. J. Anat.* **103:**313–348.

Grabowski, C. T., and Schroeder, R. E., 1968, A time-lapse photographic study of chick embryos exposed to teratogenic doses of hypoxia, *J. Embryol. Exp. Morphol.* **19:**347–362.

Grabowski, C. T., Tsai, E. T., and Toben, H. R., 1969, The effects of teratogenic doses of hypoxia on the blood pressure of chick embryos, *Teratology* **2:**67–76.

Grabowski, C. T., Tsai, E. N. C., and Chernoff, N., 1971, The effects of Trypan blue on the blood pressure of rat embryos, *Teratology* **4:**69–74.

Harh, J. Y., Paul, M. H., Gallen, W. J., Friedberg, D. Z., and Kaplan, S., 1973, Experimental production of hypoplastic left heart syndrome in the chick embryo, *Am. J. Cardiol.* **31:**51–56.

Howard, E., 1957, Ontogenetic changes in the freezing point and sodium and potassium content of the subgerminal fluid and blood plasma of the chick embryo. *J. Cell. Comp. Physiol.* **50:**451–470.

Hughes, A., 1975, Teratogenesis and the movement of ions, *Dev. Med. Child Neurol.* **17:**111–114.

Jaffee, O. C., 1965, Hemodynamic factors in the development of the chick embryo heart, *Anat. Rec.* **151:**69–76.

Jaffee, O. C., 1966, Rheological aspects of the development of blood flow patterns in the chick embryo heart, *Biorheology* **3:**59–62.

Jaffee, O. C., 1967, The development of the arterial outflow tract in the chick embryo heart, *Anat. Rec.* **158:**35–42.

Jaffee, O. C., 1974, The effects of moderate hypoxia and moderate hypoxia plus hypercapnea on cardiac development in chick embryos, *Teratology* **10:**275–281.

Jelinek, R., 1961, K. otázce tzv. prerůstání neurální trubice, *Cs. Morfol.* **9:**151. (quoted in Jelínek and Pexieder, 1968).

Jelinek, R., and Pexieder, T., 1968, The pressure of encephalic fluid in chick embryos between the 2nd and 6th day of incubation, *Physiol. Bohem.* **17:**197–305.

Jost, A., 1953, La dégénérescence des extrémités du foetus de rat sous des actions hormonales (acroblapsie expérimentale) et la théorie des bulles myelencéphaliques de Bonnevie, *Arch. Fr. Pediatr.* **10:**865–870.

Kaplan, S., and Grabowski, C. T., 1967, Analysis of trypan blue-induced rumplessness in chick embryos, *J. Exp. Zool.* **165:**325–336.

Kerpel-Fronius, E., 1970, Electrolyte and water metabolism, in: *Physiology of the Perinatal Period* pp. 643–678, (U. Stave, ed.), Appleton-Century-Crofts, New York.

Khera, K. S., 1975, Fetal cardiovascular and other defects induced by Thalidomide in cats, *Teratology* **11:**65–72.

King, C. T. G., Weaver, S. A., and Narrod, S. A., 1965, Antihistamines and teratogenicity in the rat, *J. Pharmacol. Exp. Ther.* **147:**391–398.

Krogh, A., 1939, *Osmotic Regulation in Aquatic Animals,* Cambridge University Press, Cambridge.

Leist, K. H., and Grauwiler, J., 1974, Fetal pathology in rats following uterine-vessel clamping on day 14 of gestation, *Teratology* **10:**55–67.

Moffat, D. B., 1959, Developmental changes in the aortic arch system of the rat, *Am. J. Anat.* **105:**1–36.

Monie, I. W., and Morgan, J. R., 1975, Cysts in cultured fetal rat kidneys, *Teratology* **11:**143–151.

Mushett, C. W., 1953, Elektive Differenzierungsstörungen des Zentralnervensystems am Hühnchenkeim nach kurzfristigen Sauerstoffmangel, *Beitr. Pathol. Anat.* **113:**367–387.

Nelson, M. M., Wright, H. V., Baird, C. D. C., and Evans, H. M., 1957, Teratogenic effects of pantothenic acid deficiency in the rat, *J. Nutr.* **62:**395–402.

Pexieder, T., 1969, Blood pressure in the third and fourth aortic arch and morphogenetic influence of laminar blood streams in the development of the vascular system of the chick embryo, *Folia Morphol.* **17:**273–290.

Posner, H. S., and Darr, A., 1970, Fetal edema from benzhydrylpiperazines as a possible cause of oral–facial malformations in rats, *Toxicol. Appl. Pharmacol.* **17:**67–75.

Potts, W. T. W., and Parry, G., 1964, *Osmotic and Ionic Regulation in Animals,* MacMillan, New York.

Rychter, Z., 1962, Experimental morphology of the aortic arches and the heart loop in chick embryos, *Adv. Morphog.* **2:**333–371.

Schellong, G., 1954, Herz- und Gefaszbildungen beim Hünhchen durch kurzfristigen Sauerstoffmangel, *Beitr. Pathol. Anat.* **114:**212–243.

Spitzer, A., 1923, Uber den Bauplan des normal und missbildeten Herbens (Versuch einer phylogenetischen Theorie), *Virchows Arch. Pathol. Anat.* **243:**81–272.

Stave, U., 1970, *Physiology of the Perinatal Period,* 2 vols., Appleton-Century-Crofts, New York.

Stephan, F., 1952, Contribution expérimentelle à l étude du développement du système circulatoire chez l'embryon de poulet, *Bull. Biol. Fr. Belg.* **86:**217–309.

Sturkie, P. D., 1941, Studies on hereditary baldness in the domestic fowl. I. Embryological and physiological bases of the character, *J. Morphol.* **69:**517–535.

Tuft, P. H., and Böving, B. G., 1968, The uptake of water by the rabbit blastocyst, Annual Report of the Department of Embryology, Carnegie Institution, pp. 455–458.

Turbow, M. M., 1966, Trypan blue-induced teratogenesis of rat embryos cultivated *in vitro, J. Embryol. Exp. Morphol.* **15:**387–396.

Waddington, C. H., and Carter, T. C., 1953, A note on abnormalities induced in mouse embryos by trypan blue, *J. Embryol. Exp. Morphol.* **1:**167–180.

Weiss, P., 1934, Secretory activity of the inner layer of the embryonic mid-brain as revealed by tissue culture, *Anat. Rec.* **58:**299–302.

Weiss, P., 1955, Neurogenesis, *in: Analysis of Development* (B. H. Willier, P. Weiss, and V. Hamburger, eds.), pp. 346–401, W. B. Saunders, Philadelphia.

Widdowson, E. M., 1968, Growth and composition of the fetus and newborn, *in: Biology of Gestation* (N. S. Assali, ed.), pp. 1–51, Academic Press, New York.

Abnormal Cellular and Tissue Interactions

8

LAURI SAXÉN

I. INTRODUCTION

Morphogenesis in multicellular embryos involves proliferation, aggregation, migration, and organization of differentiating cells in a manner that is strictly controlled temporally and spatially. Most tissues comprise several cell types or tissue components with differing developmental histories. Consequently, such synchronism requires a precise morphogenetic building plan, which can be implemented only if cells are "aware" of each other and can communicate in a developmentally meaningful way. This morphogenetic communication, first demonstrated between the optic vesicle and the presumptive lens ectoderm by Spemann (1901), has long been known as *embryonic induction*. The classic term has more recently been replaced by the somewhat more defined expression "morphogenetic tissue interactions," defined by its proposer Grobstein (1956a) as follows: "Inductive or morphogenetic tissue interactions take place whenever in development two or more tissues of different history and properties become intimately associated and the alteration of the developmental course of the interactants results." In addition to such *heterotypic* tissue interactions, like cells also interact, as they must be capable of recognizing each other in order to aggregate selectively within the organism and form organ anlagen. This capacity of cells for *homotypic* interaction is gradually achieved during cell differentiation and constitutes another major prerequisite for normal development (Moscona, 1961; Moscona and Garber, 1968).

There should be no doubt today that such homotypic and heterotypic cell interactions constitute a central control mechanism for differentiation and

LAURI SAXÉN • Third Department of Pathology, University of Helsinki, 00290 Helsinki 29, Finland.

morphogenesis. Experimental interference with these processes has repeatedly been shown to lead to abnormal development and congenital defects. Furthermore, studies of various avian and mammalian strains have conclusively shown that certain mutant genes exert their deleterious effects through defective interactive processes (Zwilling, 1955; Saxén, 1973; Chapter 2, this volume). Hitherto, only scattered cases have been recognized of such exogenously produced or genetically determined abnormalities reflecting defects in interactive mechanisms. But now that the importance of morphogenetic tissue interactions in normal development is established, we may expect that many more congenital defects will be traced back to failures of interactive processes.

In this chapter, a short review of what we know of normal tissue interactions will be given, followed by a series of examples of defective control mechanisms that illustrate the various types of failure to complete such steps.

II. MORPHOGENETIC TISSUE INTERACTIONS

Morphogenetic interactions between like or unlike cells are the rule rather than the exception, and there are, in fact, no known instances where such communication is lacking between normal cells in close apposition during embryogenesis. The consequences of these interactive events may, however, vary greatly, and their interpretation may depend on the observer's special interest and the sensitivity of his method of detecting developmental changes. In addition to actual cytodifferentiation in biochemical, functional, or morphological terms, interactive events control such processes as cell aggregation, oriented migration, morphogenetic cell death, and proliferation. The term "morphogenetic tissue interaction" is usually restricted to cover only processes between cells and tissues in close proximity, up to 100 μm. This would exclude (from this presentation also) the events controlled by morphogenetically active hormones and various humoral factors such as erythropoietin, nerve-growth factor, chalones, etc.

A great number and variety of interactive processes contributing to embryogenesis have been explored with modern experimental and analytical techniques (see below), but as yet very few common denominators have been found. The topic must therefore be treated as an encyclopedia of examples rather than as a field of documented rules, laws, and generalizations. It has become increasingly evident that different interactive processes operate through different mechanisms, the only common feature being their importance in the control of normal development. Space permits only a few examples of the morphogenetic interactions already known to occur in vertebrate embryos, and the reader is referred to some more complete reviews of this complicated subject: Grobstein (1967), Fleischmajer and Billingham (1968), Saxén and Kohonen (1969), Kratochwil (1972), Saxén (1972).

A. Basic Techniques

For an understanding of the experimental results in this field, which will enable the reader to evaluate their significance in normal development and to consider critically their sources of error and limitations, a few comments on the basic methods are called for.

The classic procedure for demonstration of an interactive process with morphogenetic consequences is to *separate the interactants* and follow their subsequent development either in a separated state, after recombination, or after an experimental combination with a "new" partner. Recombinations can be performed either *in vivo*, as in the classic amphibian and avian experiments Zwilling, 1961; Saxén and Toivonen, 1962; Saunders and Gasseling, 1968), or *in vitro*, as in most of the recent experiments on mammalian embryos (Grobstein, 1956b; Wessells, 1967; Kratochwil, 1972). This basic technique affords possibilities for many experimental manipulations. Cells from embryos at different stages of development can be brought into experimental contact, tissues of mutant strains can be combined with their wild-type counterparts, normally interacting tissues can be pretreated separately or labeled, etc. An important improvement in the technique was introduced by Grobstein (1956b), who interposed a thin membrane filter between the interacting tissues. This technique, by preventing actual exchange of cells between the two tissues, allows the reseparation of the interactants at any time. Thus it is possible to study the kinetics of the process and the contact requirements of the interaction and also to make a separate biochemical or radiochemical analysis of the components at different stages of induction and thereafter. The basic method is illustrated in Figure 1.

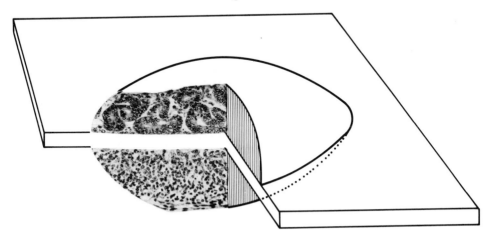

Fig. 1. Schematic illustration of the transfilter technique for studying morphogenetic tissue interactions. The metanephric mesenchyme responding with tubule formation is separated from the spinal cord inductor by a Nuclepore® membrane 12 μm thick.

Certain drawbacks and sources of errors inherent in this basic technique should be briefly outlined:

1. Separation and cleaning of the interacting tissues frequently involves *chemical treatment,* the active compounds most commonly used being various enzymes (trypsin, pancreatin, pronase, collagenase, and others). Such treatments are known to have a profound effect on the cell surface, removing surface-associated material that is presumably of importance both in interactive processes and in other cellular functions. After reassociation with the "inductor," such surface compounds are rapidly restored (Bernfield *et al.,* 1972), and it may even be questioned whether the interdependence so demonstrated reflects an interaction occurring in normal development—it might be merely a "healing" process. This possibility has naturally been taken into consideration by scientists being enzyme separation (e.g., Ellison and Lash, 1971), and proper controls can often be used.

2. Another problem in the separation of closely associated embryonic tissues of heterotypic origin is that *contaminant cells* (and cell fragments) tend to remain attached to one or both of the interacting tissues. This pitfall should be recognized, especially when "chemical cleaning" of the tissues is omitted or minimized in efforts to avoid surface changes. Some discrepant experimental results found in the literature might be explained by such incomplete separation (Saxén, 1970).

3. The possibility of artifacts resulting from mechanical manipulation of tissues, their separation from the organismal control mechanisms, and cultivation in artificial tissue culture conditions should always be borne in mind. While cell and tissue culture methods have proved most valuable in developmental biology and experimental teratology (Chapter 3, Volume 4), the many limitations of these techniques should not be forgotten (Saxén *et al.,* 1975).

B. General Features

1. Occurrence

Tissue interactions have been shown to guide cytodifferentiation and morphogenesis throughout development. From the early blastocyst stage through blastogenesis and organogenesis, they control the fate of the cells, and in the adult organism they are still responsible for the maintenance of the differentiated state. The decision determining the ultimate destiny of a mammalian blastocyst cell—whether it is to become a constituent of the inner cell mass or an outer trophoblast cell—is decided at the 8–16-cell stage, and seems to be determined by the position of the cell within this population (Hillman *et al.,* 1972). Soon afterward, the inner cell mass induces the proliferation of the polar trophoblast cells in its vicinity, whereas mural trophoblast cells further away cease to divide (Gardner and Papaionnau, 1975). During blastogenesis extensive work on amphibian embryos has shown a great number of interactive processes leading to segregation of the germinal layers and their early determination (Saxén and Toivonen, 1962). Subsequently, during actual organogenesis, interactive processes have been shown to control both cytodifferentiation and morphogenesis in every instance so far tested. For instance, epitheliomesenchymal interactions operate in the morphogenesis of glandular organs—lung, pancreas, thyroid, salivary, and mammary glands (Grobstein, 1953, 1967; Fleischmajer and Billingham, 1968; Kratochwil, 1972). Interdependence of ectoderm and mesoderm has also been shown in the developing limb and wing (Zwilling, 1956a, 1961; Saunders and Gasseling, 1968) and in the integument (Sengel, 1958; Wessells, 1962; Rawles, 1963). The latter is of special interest as it seems to continue into adult life, during which the regional characteristics of the epidermis are maintained through interaction with the dermal mesoderm (Billingham and Silvers, 1967, 1968).

These few examples seem to justify the conclusion that morphogenetic tissue interactions operate throughout life wherever morphogenetic movements bring two tissues with different developmental histories into close proximity.

2. Sequential Nature

From what has been said above, it also follows that the ultimate function and structure of every organ is the result of several interactive processes acting sequentially. The eye may serve as an example: During early gastrulation the presumptive neural ectoderm becomes "activated" in the neural direction (Nieuwkoop, 1973); this is followed by a second step when mesodermal cells are responsible for the regional segregation of the neural plate (Toivonen and Saxén, 1968). The optic cup developing in the forebrain region will subsequently induce lens from the epidermis (Spemann, 1901), and at an even later stage the various components of the eye are mutually interdependent for development (Coulombre, 1965; Chapter 14, this volume). The eye itself also has an inductive effect on its adnexa; for instance, the lens epithelium stimulates the development of the cornea (Meier and Hay, 1974), and the pigmented epithelium promotes scleral chondrogenesis (Newsome, 1972).

3. Reciprocity

Many, if not all, interactive processes guiding embryonic development operate in a two-way fashion, the two components being developmentally interdependent. For example, the metanephric kidney develops from two components in such a mutual interdependence: The ureter bud derived from the Wolffian duct branches into a metanephric blastema, giving rise to the excretory nephrons. These secretory tubules are induced by the ureter, and the branching of the ureter requires the morphogenetic action of the metanephric mesenchyme (Grobstein, 1955). Similar reciprocal actions have been demonstrated in the development of the liver between hepatic epithelium and mesenchyme (Croisille and LeDouarin, 1965), in the limb bud between the ectodermal ridge and the mesenchyme (Zwilling, 1956a, 1961), and in the tooth rudiment between the enamel epithelium and the mesenchyme of the pulp (Slavkin, 1972).

4. Specificity

During normal development *in vivo* every morphogenetic interaction is highly specific, for it is strictly limited both temporally and spatially, and in the event of its failure a certain developmental process will be thrown out of gear. Hence, the much-discussed question of the "specificity" of these processes is somewhat artificial and based entirely on experimental situations. In such instances, the specific requirement of the target tissue from its inductor seems to vary greatly, and frequently the normal inductor tissue can be replaced with a variety of "heterologous" tissues and in some instances even with cell-free extracts from embryos (Golosow and Grobstein, 1962; Fell and Grobstein, 1968; Rutter *et al.,* 1964). In such instances we are apparently analyzing a situation where the fate of the differentiating cells is already decided and their developmental options are strictly limited, as they may either express their narrow developmental potencies ("induced") or remain in their preinduction state. An extreme contrast would be a target tissue which could be converted into a variety of different phenotypes, in other words, where the target cells could be "directed" to choose one of many options. The vertebrate integument may serve as an example. Embryonic chick epidermis, when experimentally brought into contact with various heterologous mesenchymes, can be converted into keratinizing squamous epithelium, a mucus-secreting columnar epithelium, or a ciliated one (McLoughlin, 1961). Furthermore, mouse plantar epithelium combined with mesenchyme from a dental papilla will develop an enamel organ (Kollar and Baird, 1970). The chick epidermis shows a mammary-gland-like ingrowth when brought into close contact with mouse mammary mesenchyme (Propper, 1969). Apparently, the cells of the epidermal basal layer represent a reservoir of uncommitted cells able to express a variety of phenotypes when properly instructed.

In conclusion, the "specificity" of morphogenetic interactions varies greatly from true "directive" influences to supportive or "permissive" effects. Such stages of varying specificity can frequently be demonstrated during different steps of a sequential chain of inductions leading to the formation of an organ (Rutter *et al.,* 1964; LeDouarin, 1968; Spooner and Wessells, 1970).

C. Morphogenetic Signals

Much work and time has been wasted in efforts to isolate and characterize the "signal substances" supposed to be involved in morphogenetic tissue interactions. Only three successful approaches can be mentioned: A protein fraction which causes mesodermalization of amphibian gastrula ectoderm has been isolated from chick embryos and characterized (Tiedemann, 1971); another fraction from the same source acts specifically on the surface of mouse pancreatic epithelial cells, stimulating their proliferation and thus contributing to their morphogenesis (Rutter *et al.,* 1964); extracellular glycosaminoglycans derived from embryonic chick notochord have been shown to enhance somite chondrogenesis (Kosher *et al.,* 1973).

In addition to such transmissible factors, compounds of the extracellular matrix have been analyzed for their morphogenetic action. In epitheliomesenchymal interactions, the interactants have been reported to be separated by a basement membrane, and its collagenous components and proteoglycans have both been suggested to be involved in the interaction (Bernfield and Wessells, 1970; Bernfield *et al.,* 1972).

The classic "contact hypothesis" of Weiss (1947) suggested that complementary surface-associated molecules play an important role in morphogenetic interactions. Recently, the existence of such molecules has again been discussed, and an enzyme–substrate type of interaction between such molecules has been speculated upon (Roth, 1973). In homotypic, aggregative interactions such specific ligands have been demonstrated and characterized (Moscona, 1974), but so far not in heterotypic interactions. In certain interactive processes an intimate association of the cells seems to be required for transmission of inductive signals, which suggests that we may be dealing not with transmissible extracellular compounds but rather with molecules associated with the cell surface (Wartiovaara *et al.,* 1974; Saxén, 1975).

To conclude, not much is known about the chemistry and molecular biology of morphogenetic tissue interactions. Signal substances are practically unknown and so is their mode of action. Apparently, different interactive processes operate through different mechanisms, and until more is known about the active compounds and their transmission, each interactive event should be considered separately. This is especially true of failures of these processes, discussed next.

III. DEFECTIVE MORPHOGENETIC INTERACTIONS

The significance of morphogenetic tissue interactions for normal development, and the opposite, the role of disturbed interactions in teratogenesis, can be demonstrated in two ways: by experimental interference with these processes and by analysis of mutant strains with congenital defects.

The first approach has shown conclusively that physical and chemical inhibition of inductive interactions will lead to abnormal development; the second line of studies has brought to light a multitude of mutant genes whose action is manifested through defective interactive processes. In what follows, some work in which interactive processes have been experimentally prevented will be briefly discussed, and attention will be focused on the extensive evidence suggesting that mutant strains differ in possessing defective interactive mechanisms.

A. Experimental

The classic model system of experimental embryology, the amphibian gastrula, may also serve as a target for experimental interference with inductive processes. If gastrulation is totally or partially prevented (e.g., by exposure to hypertonic culture medium), the inductor tissue, the dorsal blastopore lip, does not invaginate properly and so fails to establish contact with the presumptive neural plate area. CNS defects of various degree result from this incomplete induction, the most common being microcephalia and anophthalmia, both suggesting insufficient contact in the anterior neural plate area (Saxén and Rapola, 1969). Contact between the interacting components of primary induction can be disrupted surgically or the contact area can be restricted (Figure 2). Careful dissection of the anterior part of the neural plate area exposes the inductor, the chordamesoderm, and leaves the way open for direct mechanical manipulations. A central fragment can be removed, the wound covered again with the neural plate, and the gastrula will heal perfectly except that the inductor tissue is now abnormally narrow. In response to this laterally narrowed inductor "template," the forebrain region of the neural plate becomes spatially restricted, and the eye anlagen approach each other and ultimately fuse partially or totally. The resulting malformation is either synophthalmia or cyclopia, known both from mythology and from clinical medicine.

Contact between interacting tissues can also be prevented experimentally by interposing a membrane. Certain types of membrane, if placed between the chordamesoderm and the presumptive neural plate, prevent neural induction and lead to a defect at the site of the membrane (Brahma, 1958). A porous membrane allowing diffusion of molecules but preventing actual cell contacts does not have this effect and induction is not inhibited (Saxén, 1961; Tarin et al., 1973).

Filter experiments have also been made in many interactive processes in mammalian embryos, and the conclusions seem to vary from system to system. In the interaction between the metanephric mesenchyme and its inductor, actual contact between cells or a close apposition of the cell membranes seems to be required, because induction is prevented by filters which allow free diffusion of large molecules but not cytoplasmic penetration (Wartiovaara *et*

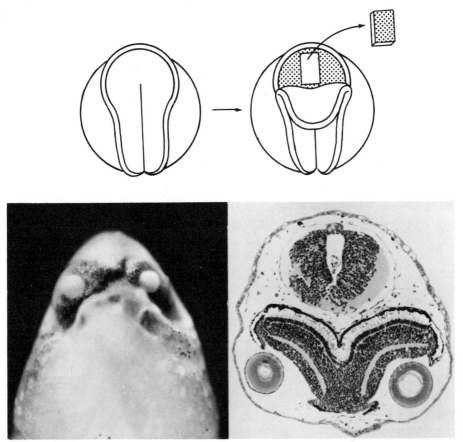

Fig. 2. Schematic representation of surgical narrowing of the mesodermal inductor and the consequence of the operation. The optic vesicles become fused, and the larva, photographed two weeks after the operation, has synophthalmia (Saxén and Kohonen, 1969).

al., 1974; Saxén, 1975). Whether this interactive situation should be considered an exception rather than the rule remains to be seen, but results with several of the model systems so far studied are consistent in showing that there regularly is a minimum pore size and a maximum distance for the transmission of inductive signals (Grobstein, 1967; Saxén, 1972). Hence, incomplete or disrupted contact between the interactants should always lead to a corresponding defect in the embryo. Long before the complex interactive situation in the developing limb bud was detected (p. 186), the significance of uninterrupted contact between the ectodermal ridge and the bud mesenchyme was recognized by Bagg (1929). In mice homozygous for the *myelencephalic bleb* mutant gene, various hereditary limb malformations were detected, and these were carefully examined during development (in cesarean sections) and again after birth. Hemorrhage of varying degree was seen at the tips of the limb buds early in the organogenetic period, and these blebs separated the ectoderm and mesoderm from each other at various sites and periods. When the embryos were followed to term, a clear correlation could be seen between the degree and time of separation and the various types of limb malformations (Figure 3).

Chemical interference with the induction process has given less constant and clear-cut results. Some potent teratogens like actinomycin D do prevent primary induction, and denaturation of the proteins in the inductor tissue specifically abolishes its mesodermalizing capacity (Toivonen, 1954; Toivonen *et al.*, 1964). Induction of cartilage in the avian and human nephric blastema under the influence of the limb bud is thalidomide-sensitive, but nothing is known about the target site of the teratogen in this interaction (Lash, 1964; Lash and Saxén, 1972). A thalidomide-type effect has also been produced with nitrogen mustard in chick embryos (Salzgeber, 1967). Sequential examination of the limb buds of treated embryos showed that the primary lesion

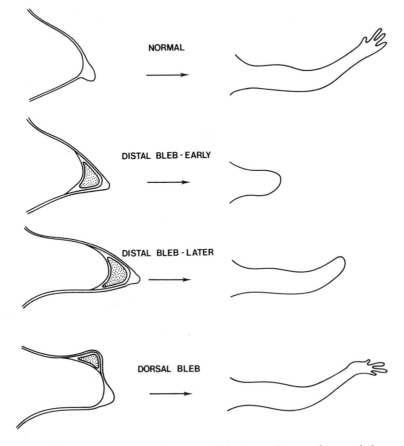

Fig. 3. Schematic illustration of Bagg's (1929) observations on the correlation between hemorrhagic blisters (stippled) in the limb bud and subsequent malformations. (After Zwilling, 1969.)

was located in the mesenchymal component of the young buds before they were induced by the ectodermal ridge. The mesenchyme then gradually recovered, and the restituted blastema became responsive to the inductive influence of the ectoderm. The explanation proposed was that the inductor tissue, which is known to program limb development in a proximal–distal sequence, had by then aged and lost its capacity to stimulate proximal differentiation and was able to promote development of the distal parts only. Further confirmation was obtained by treating the interacting components separately and examining the subsequent development of reciprocal combinations: treated mesenchyme with untreated ectodermal ridge, and intact mesenchyme with nitrogen-mustard-treated ectoderm. The former combination led to phocomelia and other defects, whereas in the latter set of experiments the inducing ectoderm resisted the treatment and supported normal limb development (Figure 4).

These examples show that mechanical or chemical interference with interactive processes can lead to abnormal development and congenital defects. How frequently exogenous teratogens exert their deleterious effect by interrupting tissue interactions is not known. On the other hand, many common defects in laboratory animals and human beings bear a clear resemblance to those produced by experimental inhibition of inductive processes through disrupted contact or chemical treatment. It is therefore conceivable that inductive tissue interactions are vulnerable to environmental agents, and may, in fact, be involved in teratogenesis.

B. Genetic Control of Interactive Processes

Three different ways in which a mutant gene might affect morphogenetic tissue interactions can be envisaged: growth disturbances affecting the contact relationships between the interacting tissues, an effect on the inducing potencies of one or both of the interactants, and an effect on their responsiveness or competence. Until more is known of the molecular basis of normal interactive events, very little can be said about the mode of action of a mutant gene in these processes, but examples are known of all three types of action listed above.

1. Disrupted Contact

Experimental prevention of contact between the optic vesicle and the overlying presumptive lens ectoderm inhibits lens formation (McKeehan, 1951). A similar mechanism seems to be the explanation for anophthalmia (or microphthalmia) in a mutant mouse strain extensively studied by Chase and Chase (1941). In these embryos, the optic cup is initially formed in the same way as in embryos of a wild-type mouse at the same stage of development.

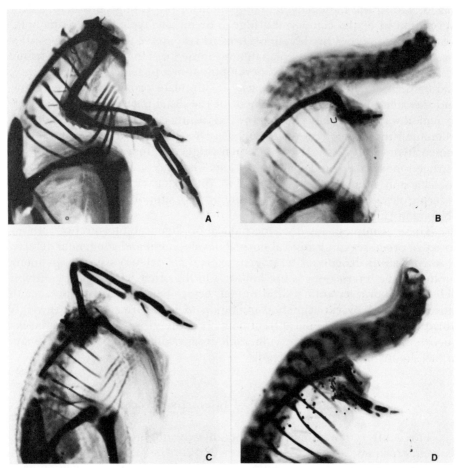

Fig. 4. Photographs of cleared and alizarin-stained chick wings illustrating the effect of nitrogen mustard treatment of separated wing-bud components (Salzgeber, 1968). A: A wing bud treated with Tyrode solution and transplanted to an embryo at stage 20; normal development. B: A wing bud treated with nitrogen mustard and similarly grafted; typical phocomelic defect. C: A wing bud developed from dissociated and reassociated ectodermal and mesodermal components treated with Tyrode solution; normal appearance. D: Wing bud obtained after operations similar to those in C, but after treatment of the mesodermal component with nitrogen mustard; phocomelic defect.

Subsequently, however, it becomes retarded in growth and either never reaches the ectoderm or establishes only spatially limited contact with it (Figure 5). Consequently, lens induction is impaired and an anophthalmic or microphthalmic embryo develops.

In mutant chick embryos with severe kidney malformations or kidney aplasia, Gruenwald (1952) described another genetic defect resulting from lack of proper contact between interacting tissue components. He was able to show that the actual failure was a growth retardation of the ureter, which did not make proper contact with the nephric blastema. As shown by subsequent experimental work (p. 176), the ureter induces the formation of secretory tubules in the mesenchymal nephric blastema, and its impaired growth in these mutants prevents normal interaction. A block in this interactive system has also been postulated to be the causal mechanism of kidney defects in the *Sd* strain of mouse (Gluecksohn-Waelsch and Rota, 1963). Homozygous *Sd/Sd* embryos show no kidney differentiation, but microscopic analysis of early embryos demonstrated the presence of both the ureter and the metanephric blastema. Hence, the failure to develop a kidney was suggested to be due to inability of the components to interact properly. The authors then performed a series of recombination experiments *in vitro*. These demonstrated tubule formation both in the mutant mesenchyme when exposed to the influence of the ureter bud from a wild-type embryo, and in a wild-type metanephric mesenchyme when induced by the mutant ureter bud. Also when a piece of dorsal spinal cord, a potent tubule inductor, was dissected from mutant embryos and brought into contact with the metanephric mesenchyme of the *Sd/Sd* mutant, tubule formation occurred. The authors concluded that both the inductive properties of the ureter bud and the responsiveness of the mesenchymal metanephric blastema were retained, and that the mutation

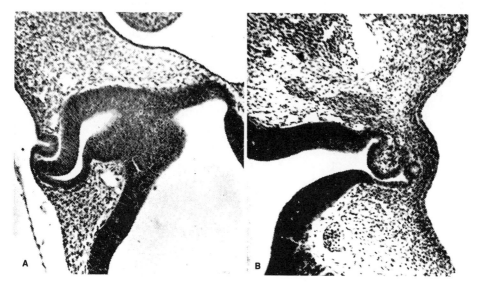

Fig. 5. Micrographs of the optic vesicle of a 10-day wild-type mouse embryo (A) and the corresponding stage of an anophthalmic strain (B). In the normal embryo the vesicle is in contact with the developing lens, whereas retarded growth of the mutant has prevented contact and lens induction (Chase and Chase, 1941).

does not exert its effect by suppressing these potencies. The differences between development *in vivo* and the observations *in vitro* might be explained by retarded growth of the ureter, with consequent delay in the establishment of contact between the interactants *in vivo*. *In vitro*, the metanephric mesenchyme loses its competence to respond to the tubule inductor in about 24 hr (Nordling *et al.*, 1971). Hence, if contact were delayed *in vivo*, failure to form tubules would be expected, but *in vitro* the mutant tissue components, freed of the temporal restriction, would ultimately establish contact, as in fact they did.

As a further example of disrupted contacts preventing normal differentiation, a mutant gene interfering with cell migration may be mentioned. As suggested by experiments concerning the migration of precardiac mesenchyme cells and neural crest cells, oriented cell migration is achieved through preformed "paths" presumably provided by heterotypic cells in close association with migrating ones (DeHaan, 1964; Weston, 1970). Experimental disruption of this intimate association between the migratory cells and their substrate leads to cessation of migration, as shown by DeHaan (1958), who produced double-hearted chick embryos by treating them with chelating agents. Apparently, removal of Ca^{2+} from the milieu affected cell contacts and stopped oriented movements of the precardiac myoblasts.

Recent electron-microscopic analysis of the postnatal development of the cerebellum in mouse and monkey has revealed an interesting migratory pattern of the granule cells (Rakic, 1971, 1972). These cells proliferate on the external surface of the cerebellum, and the daughter cells, while undergoing maturation, migrate inward to their adult position in the granular layer. This oriented migration seems to be guided by the Bergman glial cell fibers, which extend through the molecular layer and to which the granule cell neurons are directly apposed. In the weaver (*wv*) mutant mouse, this migration is severely impaired, as illustrated in Figure 6 (Rakic and Sidman, 1973a,b; Sidman, 1974). Examination of the heterozygous +/*wv* mouse during early postnatal life shows that the number of normal Bergman glial fibers is reduced, but

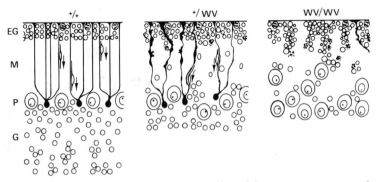

Fig. 6A. A schematic representation of the effect of the *weaver* (*wv*) gene on the cerebellum in a 10-day-old mouse. EG: external granular layer; M: molecular layer; G: granular layer. Arrows indicate the granule cells migrating along the dark Bergman glial cells (see text) (Rakic and Sidman, 1973a).

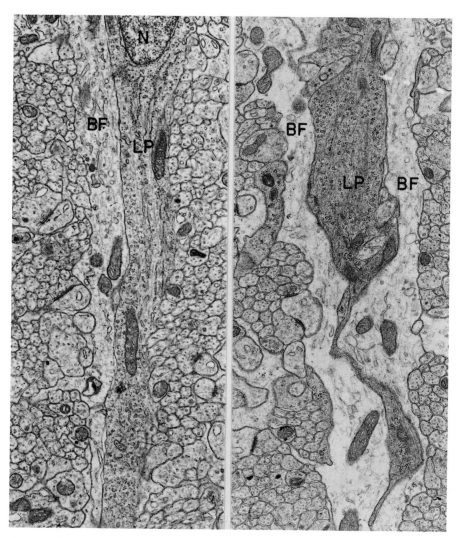

Fig. 6B. Electron micrographs of the vertically oriented granule cell nucleus (N) and the leading process (LP) opposed to a Bergman fiber (BF) in the cerebellum of a normal wild-type mouse (left), and of a +/wv mutant. In the latter, the Bergman fibers fail to show the normal elongated shape and the contact area between these and the leading processes becomes irregular (×16,000 reduced 5% for reproduction) (Rakic and Sidman, 1973b).

these are aligned with normal, migrating granule cells. In addition, granule cells which have lost contact with the glial fibers can be detected. The migration of these is prevented, and they seem to degenerate rapidly. In the *wv/wv* homozygous mice the changes are similar but more marked. Only a few Bergman glial fibers can be found, and most of the granule cells have remained in the external granular layer. Here they do not form axons but die, with the result that the cerebellum is drastically reduced in size from the third week of life on. While the primary site of action of the *weaver* gene remains unknown, detailed electron microscopic analysis suggests that destruction of the Bergman glial fibers precedes the cessation of granule cell migration. Hence, the sequence of events can be taken as an example of an abnormal intercellular relationship, where loss of the guiding contact between two interacting cell types ultimately leads to blocked migration and death of the granule cells at the site of their genesis.

2. Defective Inductor

As has repeatedly been stressed, during early limb-bud development the apical ectodermal ridge (AER) and the mesenchyme are in contact and developmentally interdependent. Experimental breaking of this contact prevents limb outgrowth, as was noted on p. 180, in connection with defects resulting from an obstacle placed between the interactants. According to the present, simplified model, the AER promotes outgrowth of the limb and is responsible for its sequential proximal–distal "induction." The mesoderm produces a "maintenance factor" (MF) which is responsible for the preservation of the AER. This factor is still undefined, but seems to be transmissible through porous membranes and through the mesenchymal blastema itself (Saunders, 1948; Zwilling, 1956a, 1961, 1974; Saunders and Gasseling, 1963).

Much of our present knowledge of the interactive processes guiding limb development has been gained from experiments on mutant chick strains. Studies of reciprocal combinations between the interacting tissue components from mutant and wild-type embryos can indicate where the primary failure lies. Especially illustrative is the series of experiments on a wingless mutant chick by Zwilling (1956a,b). In the wingless mutant an AER initially develops but soon regresses, giving rise to a shortened wing where only the girdle region has differentiated. This defect is not due to unresponsiveness of the mutant mesoderm, for the latter, in combination with a wild-type AER, does differentiate beyond the stage reached by the mutant. Again, however, the transplanted normal AER regresses, and in consequence the distal development of the wing ceases. Zwilling inferred that the mutation primarily impairs the "maintaining" properties of the mesoderm and the defect results in an inductively inactive and regressing AER.

This conclusion gained support from a later series of experiments (Zwilling, 1974). If the maintenance factor (MF) lacking from the wingless mesoderm is transmissible, this should eventually become available to the normal AER on mutant mesoderm if the recombined tissues are transplanted to a normal wing bud. This hypothesis was tested. Mutant mesoderm carrying a wild-type AER was transplanted to the dorsal surface of the limb bud of a normal embryo. The AER then survived for longer periods, and outgrowth of the wing bud was stimulated beyond that of the controls.

Thus, the wingless mutant has revealed the existence of a gene acting primarily by production (or release) of a morphogenetically significant, transmissible factor.

Experiments on another mutant, the polydactylous strain of chick, also yielded results suggestive of abnormalities in the production or distribution of MF. Combination of AER from this strain and wild-type mesenchyme regularly led to the development of abnormal limb, whereas the reciprocal combination (wild-type AER to mutant mesenchyme) produced polydactylous limbs (Figure 7) (Zwilling and Hansborough, 1956). The authors concluded that excess of or abnormally distributed MF is responsible for the development of an elongated, oversized AER which, in turn, induces supernumerary structures in the mesoderm. The final shape of a limb is now known to be "carved" by zones of programmed cell death at the margins of the bud and between the developing digits. The setting of the "death clock" in cells of strictly limited areas has been suggested to be due to interactive processes (Saunders and Fallon, 1966), and treatment with suitable exogenous chemicals prevents interdigital necrosis, the result being soft-part syndactyly (Menkes and Delanu, 1964; Saunders and Fallon, 1966). Recent work on the polydactylous *talpid*[3]

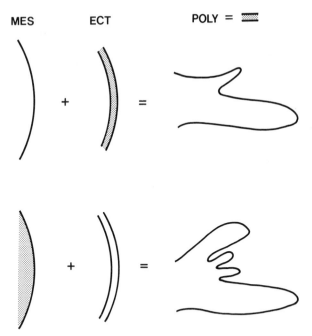

Fig. 7. Scheme of the results of reciprocal interchanges of ectoderm and mesoderm from limb buds of wild-type embryos and from a polydactylous strain of chick. (After Zwilling, 1956a.)

mutant chick has indicated that the necrotic zones shaping the limb bud are missing in this strain (Hincliffe and Ede, 1967; Hincliffe and Thorogood, 1974). The absence of the marginal zones is thought to account for the abnormal size and shape of the *talpid*[3] mesoblast, which in turn is responsible for the excessive development of the AER, because the mesenchymal MF is now distributed over a wider-than-normal area. These observations are thus in good agreement with the original hypothesis advanced by Zwilling and Hansborough (1956). What site, in this complicated and only partially understood series of interactive processes, is the primary target of the mutant gene still remains to be shown.

3. Defective Responding Tissue

The capacity of embryonic cells and tissues to respond to inductive influences is restricted both spatially and temporally and can be affected by exogenous agents, as shown in experiments on amphibian gastrulas (Leikola, 1963; Gebhardt and Nieuwkoop, 1964; Grunz, 1972). Examples of mutant genes affecting this responsiveness (competence) are known in amphibians, birds, and mammals. They all lead to prevention of morphogenetic interactions.

In the eyeless (*e/e*) mutant axolotl, no optic vesicle develops, and this defective morphogenesis has recently been studied by the technique of reciprocal combinations (Van Deusen, 1973). Presumptive neural ectoderm of the forebrain region was interchanged between mutant and normal embryos at an early gastrula stage, and similar exchanges were made with the inductor of the eye region, the precordal mesoderm. No optic vesicles developed in the series where the mutant *e/e* ectoderm was exposed to normal mesoderm inductor, whereas the reciprocal exchange, normal ectoderm on mutant inductor, regularly gave rise to fully developed eyes (Figure 8). The author concludes that the mutant gene affects the competence of the ectoderm to respond to the inductive stimulus of the mesoderm and thus leads to complete failure of eye development.

The ectodermal scales on avian legs also result from an ectodermal–mesodermal interaction, as shown by Sengel (1958). Failure of this interactive process is known to lead to complete absence of scales in the "scaleless" chick, a mutant studied extensively by Abbott and her collaborators (Goetinck and Abbott, 1960, 1963; Sengel and Abbott, 1962; Sawyer and Abbott, 1972). Reciprocal exchanges of the ectoderm and mesoderm between the mutant strain and normal embryos yielded clear-cut results: A limb constituted from mesoderm of the scaleless strain and ectoderm from a wild-type chick regularly showed normal scale formation. In contrast to this, limbs developed from mutant ectoderm and normal wild-type mesoderm never developed scales (Figure 9). Microscopic examination of the histogenesis of the scaleless skin

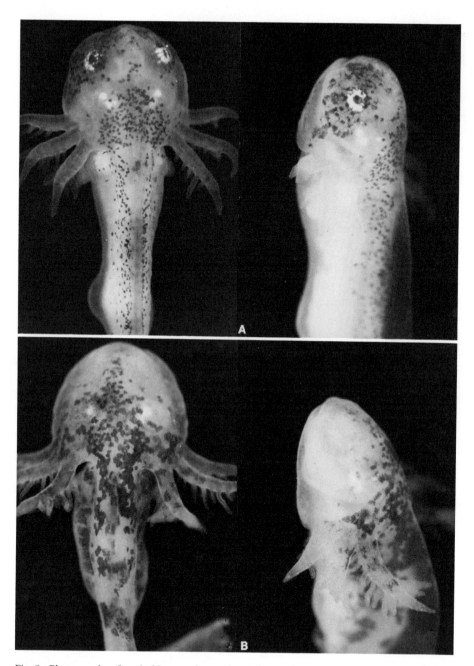

Fig. 8. Photographs of axolotl larvae after reciprocal transplantation of the anterior neural plate region and the precordal mesenchyme from wild-type gastrulas and from those of an eyeless mutant. A: wild-type graft (pigmented) develops an eye in mutant host; B: mutant ectodermal graft (white) fails to develop in a wild-type host (Van Deusen, 1973).

Fig. 9. Photographs of chicken legs produced by reciprocal combinations of ectoderm and mesoderm from buds derived from wild-type embryos and a scaleless strain. A: mutant ectoderm plus wild-type mesoderm; B: wild-type ectoderm plus mutant mesoderm (Goetinck and Abbott, 1963).

indicated that the trait is not due to secondary degeneration but that even the initial stages of scale formation are defective. These observations strongly suggest that the mutant gene acts primarily upon the ectoderm and affects its competence to form scales in response to the mesodermal inductor.

Mouse embryos carrying a mutant gene at the *T* locus are severely malformed and often die during early embryonic development. Surviving mutants show a great variety of congenital defects, among which defective development of the vertebral cartilage is a prominent feature. This cartilage is believed to be induced *in vivo* by factors emanating from the notochord. In experimental conditions early somite tissue can be induced to form cartilage by both notochord and spinal cord (Grobstein and Holtzer, 1955, Lash, 1963). In the light of these observations, it was of interest to test these interactive tissues from the *T* locus mutant (Bennett, 1964). Ventral spinal cord of the *T/T* mutant embryo stimulated chondrogenesis in wild-type somite tissue, but mutant mesenchyme showed no response when brought in contact with the inductor tissue from a normal mouse embryo. This observation strongly suggests that the mutation at the T locus reduces the responsiveness of the tissue to a chondrogenic stimulus.

4. Homotypic Interactions

As emphasized in the introduction, embryogenesis involves interactive processes not only between cells of different origin and developmental history, but also between like cells. When different cell types (e.g., from the liver and the neural retina) are experimentally disaggregated and subsequently mixed and cultured *in vitro,* like cells recognize each other, aggregate, and "sort out" into homotypic islands (Townes and Holtfreter, 1955; Moscona, 1961, 1974; Steinberg, 1964). Experimental or genetic interference with such basic properties of embryonic cells would be expected to have profound consequences for differentiation and morphogenesis. In fact, years ago Gluecksohn-Waelsch (1963) suggested that altered adhesive properties might be responsible for multiple defects in the "phocomelic" strain of mouse. More recently, Bennett (1975) has made the same suggestion concerning the T-locus mutants. Indirect evidence for alterations of the cell surface in these mutants was obtained by showing antigenic differences in the surface of the sperm between different mutants of the T locus. Methods are now available for *in vitro* studies of both animal and human cells and their aggregative and morphogenetic behavior, and it will be extremely interesting to expose cells from defective human embryos to such tests (Cassiman and Bernfield, 1974).

The *talpid*[3] mutant already mentioned is characterized by a great variety of defects, especially in the skeletal mesoderm. Ede and his collaborators, in their detailed study of development in this strain, started from the assumption that the mutant gene may primarily interfere with the adhesive properties of the mesenchymal cells and so secondarily affect their aggregation, migration, and differentiation. Experiments with disaggregated wing-bud mesenchyme cells suggested that reaggregation was enhanced in the mutant cells as compared with cells from wild-type embryos at the same stage of development. Moreover, the aggregates formed by the *talpid*[3] cells were only half of the size of those formed by the control cultures (Ede and Agerbak, 1968; Ede and Flint, 1972). Recent ultrastructural studies on cells of this strain have added some most interesting structural differences to the comparison between *talpid*[3] and normal cells (Ede *et al.,* 1974): In a normal embryo transmission and scanning electron micrographs reveal abundant, thin cytoplasmic processes in the ectodermal and mesodermal cells of the limb bud. Mesenchymal filopodia approach the epithelial basal lamina, leaving only a narrow gap between them. In the *talpid*[3] limb bud, the gap between the interacting ectoderm and mesoderm is wider, and there are statistically fewer mesenchymal processes closely approaching the basal lamina (Figure 10). How these differences are causally related to the observed differences in the adhesive properties be-

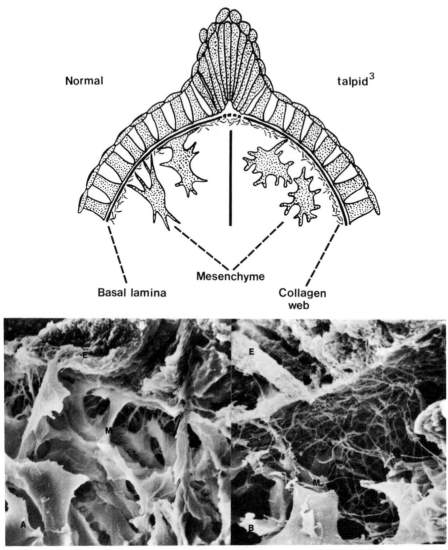

Fig. 10. Scanning electron micrographs and a scheme of the limb bud of a normal chick embryo (A) and of a *talpid*[3] mutant (B). AER: apical ectodermal ridge; M: mesenchyme; E: ectoderm. The scanning electron microscope pictures were taken from the proximal region of the limb buds (Ede *et al.*, 1974).

tween mutant and normal cells and to the abnormalities in the development of the *talpid*[3] embryos is not yet clear, but certain speculations may be offered. Only recently the old "contact hypothesis" of inductive tissue interactions has been revived, with emphasis on the possible role of intimate associations between the interacting cells (Saxén, 1975). In some instances where tissues are known to be in interactive relationship, cytoplasmic processes very much like those described above have been demonstrated and have been suggested to act as mediators of intercellular communication (Slavkin *et al.,* 1969; Saxén, 1972; Slavkin, 1972; Wartiovaara *et al.,* 1974). The findings of Ede and his group seems to fit well with such a suggestion and might be taken as evidence for distortion of the mechanism of intercellular communication due to the action of a mutant gene on the cell surface and its configuration. If so, this might be the beginning of a molecular approach to the understanding of both normal and abnormal cell interactions during embryonic development.

IV. CONCLUDING REMARKS

In the introduction to this chapter the importance of morphogenetic tissue interactions as a basic control mechanism in embryogenesis was stressed. Examples of abnormal development have been presented in support of this statement and to show how defective interactive processes may be the underlying cause of severe congenital malformations. It might be postulated that, in fact, inductive interactions are the critical events in development and are highly sensitive to both exogenous and genetic teratogenic factors. Failures of these mechanisms are possibly much more common than is recognized today, but our insufficient and fragmentary knowledge of the normal interactive processes greatly limits our ability to evaluate their significance as mediators of teratogenic effects. On the other hand, studies on mutant genes affecting interactive processes have contributed greatly to our understanding of "embryonic induction." Hence, intimate "interaction" between researchers in the fields of developmental biology and experimental teratology might afford a key to a better understanding of the unsolved problems in the field of morphogenetic tissue interactions.

REFERENCES

Bagg, H. J., 1929, Hereditary abnormalities of the limbs, their origin and transmission. II. A morphological study with special reference to the etiology of clubfeet, syndactylism, hypodactylism, and congenital amputation of X-rayed mice, *Am. J. Anat.* **43**:167.

Bennett, D., 1964, Abnormalities associated with a chromosome region in the mouse. II. Embryological effects of lethal alleles in the *t*-region, *Science* **144**:263.

Bennett, D., 1975, *T*-locus mutants: Suggestions for the control of early embryonic organisation through cell surface components, *in: The Early Development of Mammals* (M. Balls and A. E. Wild, eds.), pp. 207–218, Cambridge University Press, London.

Bernfield, M. R., and Wessells, N. K., 1970, Intra- and extra-cellular control of epithelial morphogenesis, *Dev. Biol. Suppl.* **4**:195.

Bernfield, M. R., Banerjee, S. D., and Cohn, R. H., 1972, Dependence of salivary epithelial morphology and branching morphogenesis upon acid mucopolysaccharideprotein (proteoglycan) at the epithelial surface, *J. Cell Biol.* **52**:674.

Billingham, R. E., and Silvers, W. K., 1967, Studies on the conservation of epidermal specificities of skin and certain mucosas in adult mammals, *J. Exp. Zool.* **125**:429.

Billingham, R. E., and Silvers, W. K., 1968, Dermoepidermal interactions and epithelial specificity, *in: Epithelial–Mesenchymal Interaction* (R. Fleischmajer and R. E. Billingham, eds.), pp. 252–266, Williams & Wilkins, Baltimore.

Brahma, S. K., 1958, Experiments on the diffusibility of the amphibian evocator, *J. Embryol. Exp. Morphol.* **6:**418.

Cassiman, J. J., and Bernfield, M. R., 1974, Morphogenetic properties of human embryonic cells: Aggregation of dissociated cells and histogenesis in cultured aggregates, *Pediatr. Res.* **8:**184.

Chase, H. B., and Chase, E. B., 1941, Studies on an anophthalmic strain of mice. I. Embryology of the eye region, *J. Morphol.* **68:**279.

Coulombre, A. J., 1965, The eye, *in Organogenesis* (R. L. DeHaan and H. Ursprung, eds.), pp. 219–251, Holt, New York.

Croisille, Y., and LeDouarin, N. M., 1965, Development and regeneration of the liver, *in: Organogenesis* (R. L. DeHaan and H. Ursprung, eds.), pp. 421–466, Holt, New York.

DeHaan, R. L., 1958, Cell migration and morphogenetic movements, *in: The Chemical Basis of Development* (W. D. McElroy and B. Glass, eds.), p. 339, Johns Hopkins Press, Baltimore.

DeHaan, R., 1964, Cell interactions and oriented movements during development, *J. Exp. Zool.* **157:**127.

Ede, D. A., and Agerbak, G. S., 1968, Cell adhesion and movement in relation to the developing limb pattern in normal and talpid[3] mutant chick embryos, *J. Embryol. Exp. Morphol.* **20:**81.

Ede, D. A., and Flint, O. P., 1972, Patterns of cell division, cell death and chondrogenesis in cultured aggregates of normal and talpid[3] mutant chick limb mesenchyme cells, *J. Embryol. Exp. Morphol.* **27:**245.

Ede, D. A., Bellairs, R., and Bancroft, M., 1974, A scanning electron microscope study of the early limb-bud in normal and talpid[3] mutant chick embryos, *J. Embryol. Exp. Morphol.* **31:**761.

Ellison, M. L., and Lash, J. W., 1971, Environmental enhancement of *in vitro* chondrogenesis, *Dev. Biol.* **26:**486.

Fell, P. E., and Grobstein, C., 1968, The influence of extra-epithelial factors on the growth of embryonic mouse pancreatic epithelium, *Exp. Cell Res.* **53:**301.

Fleischmajer, R., and Billingham, R. E., 1968, *Epithelial–Mesenchymal Interactions,* Williams & Wilkins, Baltimore.

Gardner, R. I., and Papaionnau, Y. E., 1975, Determination in the trophoblast and inner cell mass and their derivatives, *in: Mammalian Early Development* (M. Balls and A. Wild, eds.), pp. 107–132, Cambridge University Press, Cambridge.

Gebhardt, D. O. E., and Nieuwkoop, P. D., 1964, The influence of lithium on the competence of the ectoderm in *Ambystoma mexicanum, J. Embryol. Exp. Morphol.* **12:**317.

Glücksohn-Waelsch, S., 1963, Lethal genes and analysis of differentiation, *Science* **142:**1269.

Glücksohn-Waelsch, S., and Rota, T. R., 1963, Development in organ tissue culture of kidney rudiments from mutant mouse embryos, *Dev. Biol.* **7:**432.

Goetinck, P. F., and Abbott, U. K., 1960, Transplantation studies with the scaleless mutant, *Poult. Sci.* **39:**1252.

Goetinck, P. F., and Abbott, U. K., 1963, Tissue interaction in the scaleless mutant and the use of scaleless as an ectodermal marker in studies of normal limb differentiation, *J. Exp. Zool.* **154:**7.

Golosow, N., and Grobstein, C., 1962, Epitheliomesenchymal interaction in pancreatic morphogenesis, *Dev. Biol.* **4:**242.

Grobstein, C., 1953, Epithelio-mesenchymal specificity in the morphogenesis of mouse submandibular rudiments *in vitro, J. Exp. Zool.* **124:**383.

Grobstein, C., 1955, Inductive interaction in the development of the mouse metanephros, *J. Exp. Zool.* **130:**319.

Grobstein, C., 1956a, Inductive tissue interaction in development, *Adv. Cancer Res.* **4:**187–236.

Grobstein, C., 1956b, Trans-filter induction of tubules in mouse metanephrogenic mesenchyme, *Exp. Cell Res.* **10:**424.

Grobstein, C., 1967, Mechanisms of organogenetic tissue interactions, *Natl. Cancer Inst. Monogr.* **26:**279.

Grobstein, C., and Holtzer, H., 1955, *In vitro* studies of cartilage induction in mouse somite mesoderm, *J. Exp. Zool.* **128:**333.

Gruenwald, P., 1952, Development of the embryo excretory system, *Ann. N.Y. Acad. Sci.* **55:**142.

Grunz, H., 1972, Einfluss von Inhibitoren der RNS- und Protein-Synthese und Induktoren auf die Zellaffinität von Amphibiengewebe, *Roux Arch. Entw Mech. Org.* **169:**41.

Hillman, N., Sherman, M. I., and Graham, C., 1972, The effect of spatial arrangement on cell determination during mouse development, *J. Embryol. Exp. Morphol.* **28:**263.

Hincliffe, J. R., and Ede, D. A., 1967, Limb development in the polydactylous talpid mutant of the fowl, *J. Embryol. Exp. Morphol.* **17:**385.

Hincliffe, J. R., and Thorogood, P. V., 1974, Genetic inhibition of mesenchymal cell death and the development of form and skeletal pattern in the limbs of talpid³ (ta³) mutant chick embryos, *J. Embryol. Exp. Morphol.* **31:**747.

Kollar, E. J., and Baird, G. R., 1970, Tissue interaction in embryonic mouse tooth germs. II. The inductive role of the dental papilla, *J. Embryol. Exp. Morphol.* **24:**173.

Kosher, R. A., Lach, J. W., and Minor, R. R., 1973, Environmental enhancement of *in vitro* chondrogenesis. IV. Stimulation of somite chondrogenesis by exogenous chondromucoprotein, *Dev. Biol.* **35:**210.

Kratochwil, K., 1972, Tissue interaction during embryonic development. General properties, *in: Inductive Tissue Interactions and Carcinogenesis* (D. Tarin, ed.), pp. 1–47, Academic Press, London.

Lash, J. W., 1963, Tissue interaction and specific metabolic response: Chondrogenic induction and differentiation, *in: Cytodifferentiation and Macromolecular Synthesis* (M. Locke, ed.), pp. 235–260, Academic Press, New York.

Lash, J. W., 1964, Normal embryology and teratogenesis, *Am. J. Obstet. Gynecol.* **90:**1193.

Lash, J. W., and Saxén, L., 1972, Human teratogenesis: *In vitro* studies on thalidomide inhibited chondrogenesis, *Dev. Biol.* **28:**61.

LeDouarin, N., 1968, Synthèse du glycogène dans les hépatocytes en voie de différenciation: Rôle des mésenchymes homologue et hétérologues, *Dev. Biol.* **17:**101.

Leikola, A., 1963, The mesodermal and neural competence of isolated gastrula ectoderm studied by heterogenous inductors, *Ann. Zool. Soc. Fenn.* **25:**(2):1.

McKeehan, M. S., 1951, Cytological aspects of embryonic lens induction in the chick, *J. Exp. Zool.* **117:**31.

McLoughlin, C. B., 1961, The importance of mesenchymal factors in the differentiation in chick epidermis. II. Modification of epidermal differentiation by contact with different types of mesenchyme, *J. Embryol. Exp. Morphol.* **9:**385.

Meier, S., and Hay, E. D., 1974, Control of corneal differentiation by extracellular materials. Collagen as a promoter and stabilizer of epithelial stroma production, *Dev. Biol.* **38:**249.

Menkes, B., and Delanu, M., 1964, Leg differentiation and experimental syndactyly in the chick embryo. Part II. Experimental syndactyly in chick embryo, *Rev. Roum. Embryol. Cytol. Ser. Embryol.* **1:**(2):193.

Moscona, A. A., 1961, Rotation-mediated histogenetic aggregation of dissociated cells, *Exp. Cell Res.* **22:**455.

Moscona, A. A., 1974, Surface specification of embryonic cells: Lectin receptors, cell recognition, and specific cell ligands, *in: The Cell Surface in Development* (A. A. Moscona, ed.), p. 67, John Wiley, New York.

Moscona, A. A., and Garber, B. B., 1968, Reconstruction of skin from single cells and integumental differentiation in cell aggregates, *in: Epithelial–Mesenchymal Interactions* (R. Fleischmajer and R. E. Billingham, eds.), pp. 230–243, William & Wilkins, Baltimore.

Newsome, D. A., 1972, Cartilage induction by the retinal pigmented epithelium of the chick embryo, *Dev. Biol.* **27:**575.

Nieuwkoop, P. D., 1973, The "organization center" of the amphibian embryos: Its origin, spatial organization and morphogenetic action, *in: Advances in Morphogenesis* (M. Abercrombie, J. Brachet, and T. J. King, eds.), Vol. 10, pp. 1–39, Academic Press, New York.

Nordling, S., Miettinen, H., Wartiovaara, J., and Saxén, L., 1971, Transmission and spread of embryonic induction. I. Temporal relationships in transfilter induction of kidney tubules *in vitro, J. Embryol. Exp. Morphol.* **26**:231.

Propper, A., 1969, Compétence de l'épiderme embryonnaire d'oiseau vis-à-vis de l'inducteur mammaire mésenchymateu, *C. R. Acad. Sci.* **268**:1423.

Rakic, P., 1971, Neuron–glia relationship during granule cell migration in developing cerebellar cortex. A Golgi and electronmicroscopic study in *Macacus rhesus, J. Comp. Neurol.* **141**:283.

Rakic, P., 1972, Mode of cell migration to the superficial layers of fetal monkey neocortex, *J. Comp. Neurol.* **145**:61.

Rakic, P., and Sidman, R. L., 1973a, Weaver mutant mouse cerebellum: Defective neuronal migration secondary to abnormality of Bergmann glia, *Proc. Natl. Acad. Sci. U.S.A.* **70**:240.

Rakic, P., and Sidman, R. L., 1973b, Sequence of developmental abnormalities leading to granule cell deficit in cerebellar cortex of weaver mutant mice, *J. Comp. Neurol.* **152**:103.

Rawles, M. E., 1963, Tissue interactions in scale and feather development as studied in dermal–epidermal recombinations, *J. Embryol. Exp. Morphol.* **11**:49.

Roth, S., 1973, A molecular model for cell interactions, *Q. Rev. Biol.* **48**:541.

Rutter, W. J., Wessells, N. K., and Grobstein, C., 1964, Control of specific synthesis in the developing pancreas, *Natl. Cancer Inst. Monogr.* **13**:51.

Salzgeber, B., 1967, Sur l'étude expérimentale de la genèse de la phocomélie chez l'embryon de poulet, *C.R. Acad. Sci. Paris* **264**:395.

Salzgeber, B., 1968, Étude des malformations distales obtenues après traitement à l'ypérite azotée du constituant ectodermique des bourgeons de membres, *C.R. Acad. Sci. Paris* **267**:90.

Saunders, J. W., Jr., 1948, The proximo-distal sequence of origin of the parts of the chick wing and the role of the ectoderm, *J. Exp. Zool.* **108**:363.

Saunders, J. W., Jr., and Fallon, J. F., 1966, Cell death in morphogenesis, *in: Major Problems in Developmental Biology* (M. Locke ed.), pp. 289–314, Academic Press, New York.

Saunders, J. W., Jr., and Gasseling, M. T., 1963, Trans-filter propagation of apical ectoderm maintenance factor in the chick embryo wing bud, *Dev. Biol.* **7**:64.

Saunders, J. W., and Gasseling, M. T., 1968, Ectodermal–mesenchymal interactions in the origin of limb symmetry, *in: Epithelial–Mesenchymal Interactions* (R. Fleischmajer and R. E. Billingham, eds.), pp. 78–97, William & Wilkins, Baltimore.

Sawyer, R. H., and Abbott, U. K., 1972, Defective histogenesis and morphogenesis in the anterior shank skin of the scaleless mutant, *J. Exp. Zool.* **181**:99.

Saxén, L., 1961, Transfilter neural induction of amphibian ectoderm, *Dev. Biol.* **3**:140.

Saxén, L., 1970, Failure to demonstrate tubule induction in a heterologous mesenchyme, *Dev. Biol.* **23**:511.

Saxén, L., 1972, Interactive mechanisms in morphogenesis, *in: Inductive Tissue Interactions and Carcinogenesis* (D. Tarin, ed.), pp. 49–80, Academic Press, London.

Saxén, L., 1973, Tissue interaction and teratogenesis, *in: Pathobiology of Development* (E. V. D. Perrin and M. Finegold, eds.), pp. 31–51, Williams and Wilkins, Baltimore.

Saxén, L., 1975, Transmission and spread of kidney tubule induction, *in: Extracellular Matrix Influences on Gene Action* (H. Slavkin, ed.) pp. 523–529, Academic Press, New York.

Saxén, L., and Kohonen, J., 1969, Inductive tissue interactions in vertebrate morphogenesis, *Int. Rev. Exp. Pathol.* **8**:57–128.

Saxén, L., and Rapola, J., 1969, *Congenital Defects,* Holt, Rinehart and Winston, New York.

Saxén, L., and Toivonen, S., 1962, *Primary Embryonic Induction,* Academic Press, London.

Saxén, L., Karkinen-Jääskeläinen, M., and Saxén, I., 1975, Organ culture in teratology, *in: Current Topics in Pathology* (A. Gropp and K. Benirschke, eds.), pp. 124–143, Springer-Verlag, Berlin.

Sengel, P., 1958, Recherches expérimentales sur la différenciation des germes plumaires et des pigment de la peau de l'embryon de poulet en culture *in vitro, Ann. Sci. Nat. Zool.* **20**:431.

Sengel, P., and Abbott, U. K., 1962, Comportement *in vitro* de l'épiderme et du derme d'embryon de poulet mutant "scaleless" en association avec le derme et l'épiderme d'embryon normal, *C.R.Acad. Sci.* **255**:1999.

Sidman, R. L., 1974, Contact interactions among developing mammalian brain cells, *in: The Cell Surface in Development* (A. A. Moscona, ed.), p. 221, John Wiley, New York.

Slavkin, H. C., 1972, Intercellular communication during odontogenesis, *in: Developmental Aspects of Oral Biology* (H. C. Slavkin and L. A. Bavetta, eds.), pp. 165–199, Academic Press, London.

Slavkin, H. C., Bringas, P., Cameron, J., LeBaron, R., and Bavetta, L. A., 1969, Epithelial and mesenchymal cell interactions with extracellular matrix material *in vitro, J. Embryol. Exp. Morphol.* **22:**395.

Spemann, H., 1901, Entwicklungsphysiologische Studien am Triton-Ei. I, *Arch. Entw. Mech. Org.* **12:**224.

Spooner, B. S., and Wessells, N. K., 1970, Mammalian lung development: Interactions in primordium formation and bronchial morphogenesis, *J. Exp. Zool.* **175:**445.

Steinberg, M. S., 1964, The problem of adhesive selectivity in cellular interactions, *in: Cellular Membranes in Development* (M. Locke, ed.), pp. 321–366, Academic Press, New York.

Tarin, D., Toivonen, S., and Saxén, L., 1973, Studies in ectodermal–mesodermal relationships in neural induction. II. Intercellular contacts, *J. Anat.* **115:**147.

Tiedemann, H., 1971, Extrinsic and intrinsic information transfer in early differentiation of amphibian cells, *in: Control Mechanisms of Growth and Differentiation* (D. D. Davies and M. Balls, eds.), pp. 223–234, Cambridge University Press, Cambridge.

Toivonen, S., 1954, The inducing action of the bone-marrow of the guinea-pig after alcohol and heat treatment in implantation and explanation experiments with embryos of *Triturus, J. Embryol. Exp. Morphol.* **2:**239.

Toivonen, S., and Saxén, L., 1968, Morphogenetic interaction of presumptive neural and mesodermal cells mixed in different ratios, *Science* **159:**539.

Toivonen, S., Vainio, T., and Saxén, L., 1964, The effect of actinomycin D on primary embryonic induction, *Rev. Suisse Zool.* **71:**139.

Townes, P. I., and Holtfreter, J., 1955, Directed movements and selective adhesion of embryonic amphibian cells, *J. Exp. Zool.* **128:**53.

Van Deusen, E., 1973, Experimental studies on a mutant gene (e) preventing the differentiation of eye and normal hypothalamus primordia in the axolotl, *Dev. Biol.* **34:**135.

Wartiovaara, J., Nordling, S., Lehtonen, E., and Saxén, L., 1974, Transfilter induction of kidney tubules: Correlation with cytoplasmic penetration into nucleopore filters, *J. Embryol. Exp. Morphol.* **31:**667.

Weiss, P., 1947, The problem of specificity in growth and development, *Yale J. Biol. Med.* **19:**235.

Wessells, N. K., 1962, Tissue interactions during skin histodifferentiation, *Dev. Biol.* **4:**87.

Wessells, N. K., 1967, Avian and mammalian organ culture, *in: Methods in Developmental Biology* (F. H. Wilt and N. K. Wessells, eds.), pp. 445–456, Thomas Y. Crowell Co., New York.

Weston, J. A., 1970, The migration and differentiation of neural crest cells, *Adv. Morphog.* **8:**41–114.

Zwilling, E., 1955, Teratogenesis, *in: Analysis of Development* (R. H. Willier, P. Weiss, and V. Hamburger, eds.), pp. 699–719, Saunders, Philadelphia.

Zwilling, E., 1956a, Genetic mechanism in limb development, *Cold Spring Harbor Symp. Quant. Biol.* **21:**349.

Zwilling, E., 1956b, Interaction between limb bud ectoderm and mesoderm in the chick embryo. IV. Experiments with a wingless mutant, *J. Exp. Zool.* **132:**241.

Zwilling, E., 1961, Limb morphogenesis, *Adv. Morphog.* **1:**301–330.

Zwilling, E., 1969, Abnormal morphogenesis in limb development, *in: Limb Development and Deformity: Problems of Evaluation and Rehabilitation* (C. A. Swinyard, ed.), pp. 100–118, Charles C Thomas, Springfield, Ill.

Zwilling, E., 1974, Effects of contact between mutant (wingless) limb buds and those of genetically normal chick embryos: Confirmation of a hypothesis, *Dev. Biol.* **39:**37.

Zwilling, E., and Hansborough, L. A., 1956, Interaction between limb bud ectoderm and mesoderm in the chick embryo. III. Experiments with polydactylous limbs, *J. Exp. Zool.* **132:**219.

Cell Morphogenetic Movements

9

KENNETH MANAO YAMADA

I. CELL MORPHOGENETIC MOVEMENTS DURING DEVELOPMENT

During embryological development cells undergo carefully coordinated movements which, if carried out correctly, will result in the formation of tissues and organs. These morphogenetic movements can be classified as migratory, elongating, folding, or "passive" movements.*

A. Migration

Many cells actively migrate to new locations in the embryo. For example, neural crest cells leave the dorsum of the neural tube and travel to an impressive number of distant sites to form facial structures, skin pigment, nerve ganglia, and numerous other tissues (Hörstadius, 1950; Weston, 1970; Johnston and Listgarten, 1972). Axons grow outward from stationary nerve cell bodies to make connections in distant tissues. Precardiac mesodermal cells migrate medially to form the heart (DeHaan, 1965). These cells or their processes migrate in groups or singly. Cells in epithelia, however, tend to

---

*Many of the underlying principles and some of the best examples of concepts discussed in this chapter were initially developed using invertebrates or amphibians. Much of that work is regretably omitted for reasons of space and because the examples were chosen for their ease of extrapolation to human development. The reviews listed in the References should provide access to this literature.

KENNETH MANAO YAMADA • Laboratory of Molecular Biology, National Cancer Institute, National Institutes of Health, Bethesda, Maryland.

adhere tightly to each other and to migrate as flat sheets of cells (e.g., see Vaughan and Trinkaus, 1966; Middleton, 1973). It seems likely that the underlying mechanisms of these various forms of cell motility will be found to be similar (see Section V.B).

B. Thickening or Elongating

Some tissues, such as neural tube, lens, and inner ear, begin as thickenings in epithelia. These thickenings form relatively rapidly from localized regions of the ectoderm to establish the neural plate, the lens placode, and the otic placode. They appear to form by the lengthening of individual cells, which undergo active extension at the expense of cell width.

C. Invagination and Evagination

Flat epithelial sheets must often form infoldings or outpocketings during the formation of organ rudiments. Pancreas and lung buds form by evaginations of epithelia, while lens vesicles and neural tubes form by invagination or tube formation. These processes are quite similar mechanically, and when each of these events is analyzed in detail, it appears that the mechanisms of movement are also similar.

D. Passive Movements

Groups of cells may move relative to other cells simply because of differential mitotic rates or by localized cell death, as discussed in Chapter 4 in this text. In addition, it is likely that new tissue configurations resulting from active cell movements can be stabilized by subsequent patterns of cell division. (Each class of active cell movement will be analyzed in detail in later sections.)

II. MECHANISMS OF CELL MOVEMENT

Microfilaments and microtubules are two recently discovered subcellular structures which are probably central to cell movement (Figure 1). Intermediate-sized 100-Å filaments constitute a third class of filamentous structure which, in contrast, has not yet been shown to play a role in movement. Such intermediate filaments include neurofilaments, whose functions remain uncertain, and tonofilaments, which loop through desmosomes and possibly help to anchor cells at junctions with other cells (Wuerker and Kirkpatrick, 1972; Kelly, 1966).

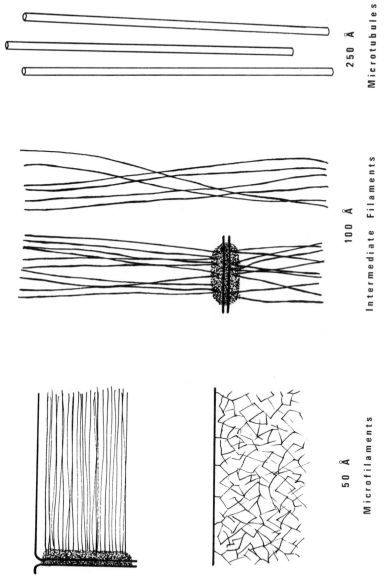

Fig. 1. Comparison of three classes of cytoplasmic organelles (see text). Microfilaments occur in linearly oriented arrays or in the form of a network or meshwork. Intermediate filaments include tonofilaments inserting into desmosomes at cell–cell junctions and the neurofilaments of nerve cells. Microtubules can even be several times longer than depicted here.

A. Microfilaments

Filaments approximately 60 Å in diameter, but ranging from 40 to 70 Å, are found immediately adjacent to the inner surface of the plasma membrane in many cells. They are found in oriented, parallel arrays or as a loose network (Figures 1 and 2). In general, tissues undergoing active, folding types of morphogenetic movements display the linearly oriented type of microfilament pattern. Migrating cells have both types, while the growing tips of nerve axons have only the network pattern.

Microfilaments are similar in diameter to muscle actin. Recent work has provided further evidence that both types of microfilaments represent polymerized cytoplasmic actin. Linearly oriented microfilaments will specifically bind heavy meromyosin, which consists of the active heads of myosin molecules. This binding is reversed by ATP or pyrophosphate, as has been previously established for muscle actin. In addition, the heavy meromyosin is found to bind in characteristic "arrowhead" arrays, which suggests that the underlying filament substructure is similar to that of actin filaments (Ishikawa et al., 1969; Behnke et al., 1971; Spooner et al., 1973; Ludueña and Wessells, 1973; McNutt et al., 1973; Goldman and Knipe, 1973; Perdue, 1973; see also Huxley, 1973). Microfilaments in the network pattern also apparently bind heavy meromyosin (Burton and Kirkland, 1972; Chang and Goldman, 1973). Moreover, antibodies against electrophoretically purified fibroblast cytoplasmic actin bind to cells in patterns strikingly similar to the known patterns of microfilaments (Lazarides and Weber, 1974; see also Perdue, 1973; Gabbiani et al., 1973). Fluorescein-labeled heavy meromyosin can also be used to demonstrate the overall distribution of polymerized actin in cells in vitro (Sanger, 1975a,b).

Cytoplasmic actin has been demonstrated to be present by electrophoretic and other criteria in a large variety of tissues (Bray, 1973) and has been isolated from fibroblasts (Yang and Perdue, 1972) and platelets (Bettex-Galland and Lüscher, 1965; Adelstein and Conti, 1973). As might be expected, myosin and the regulatory protein tropomyosin have also been isolated from mammalian or avian nonmuscle cells (Adelstein and Conti, 1973; Booyse et al., 1973; Stossel and Pollard, 1973; Ostlund et al., 1974; Ash, 1975; Fine et al., 1973; Cohen and Cohen, 1972). Antibodies prepared against smooth muscle myosin and against skeletal muscle tropomyosin (which apparently cross-react with cellular myosin or tropomyosin) also appear to bind to bundles of linearly oriented microfilaments (Weber and Groeschel-Stewart, 1974; Lazarides, 1975).

How these contractile proteins interact to produce cell movement is still uncertain. A popular hypothesis is that cytoplasmic actin or myosin is inserted into selected regions of the plasma membrane and that contraction by a sliding-filament mechanism, similar to that of muscle, pulls key portions of cell membrane together, resulting in movement (Huxley, 1973; Pollard and Weihing, 1974; and specific examples in Section V).

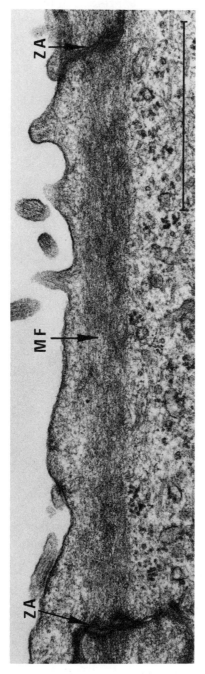

Fig. 2. Band of microfilaments across the apex of an epithelial cell undergoing morphogenesis. These microfilaments (MF) terminate in amorphous, electron-dense material at the zonula adherens region (ZA) of the junctions between this cell and its neighbors. Note the protrusions of the plasma membrane and how the band of microfilaments excludes other organelles such as ribosomes and vesicles. This cell is from chick oviduct epithelium invaginating to form glands in response to estrogen. Similar microfilament arrays are found in other invaginating or evaginating epithelia (see text). Bar indicates 1 μm. (From Wrenn, 1971.)

The patterns of microfilaments seen by electron microscopy probably reflect the organization of such contractile proteins in the cell. For example, the parallel arrays of microfilaments probably represent highly oriented actin filaments, and the microfilament networks less-organized actin. Even lower concentrations of polymerized actin would be expected to appear faintly fibrillar or amorphous.

That thick filaments corresponding to aggregated myosin are not seen readily in nonmuscle cells is not surprising for several reasons: Even smooth muscle myosin can be seen only under certain conditions (see Somlyo et al., 1971; Cooke and Fay, 1972); cell movements may not require the highly structured actomyosin arrays of contracting muscle; and the concentration of cytoplasmic myosin is much lower than in muscle (Pollard and Weihing, 1974).

Many examples of insertion of microfilaments into localized plaques or zones adjacent to plasma membranes have been reported (see Section V.B. and V.D. and Wessells et al., 1971) and even cases of possible insertion of microfilaments directly into the plasma membrane (Yamada et al., 1971; McNutt et al., 1971). However, there is no proof that either actin or myosin itself actually inserts into the lipid bilayer or onto a specific intramembranous protein. Nevertheless, actin remains attached to isolated plasma membranes from several cell types (Pollard and Korn, 1973; Spudich, 1974; Wickus et al., 1975; Gruenstein et al., 1975). Myosin can be demonstrated immunologically on the inner surface of the plasma membrane of cultured human fibroblasts (Painter et al., 1975). Myosin has also been detected on the outer surfaces of mammalian and avian fibroblasts (Gwynn et al., 1974; Willingham et al., 1974). Whether the "myosin" in these two locations represents myosin inserted completely through the plasma membrane, myosin molecules sticking to both sides of the membrane, or cross-reacting nonmyosin membrane materials, is not yet known.

B. Microtubules

Microtubules are slender, unbranched, apparently rigid organelles roughly 250 Å in diameter and up to several microns long (see reviews by Porter, 1966; Olmsted and Borisy, 1973). They are distinguished from microfilaments and intermediate filaments by their larger diameter and their apparently hollow 150-Å centers when viewed in cross section (Figure 3). Microtubular protein, called tubulin, is a dimer with subunits of molecular weight 55,000–60,000 (50,000 in fibroblasts; Ostlund and Pastan, 1975). Isolated tubulin can be reassembled into microtubules if the Ca^{2+} concentration is sufficiently low (Weisenberg, 1972) and if the appropriate starter or cofactor is present, e.g., microtubule fragments or "tau factor" (Borisy and Olmsted, 1972; Kirschner et al., 1974; Weingarten et al., 1975). Microtubule assembly

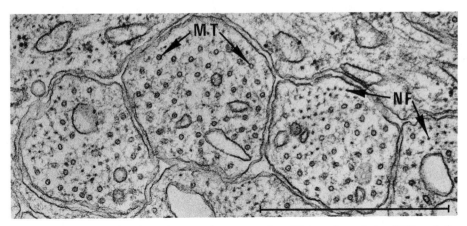

Fig. 3. Microtubules of embryonic axons in cross-section. These microtubules (MT) and the smaller neurofilaments (NF) are prominent in elongating axons. Bar indicates 1 μm. (From Yamada *et al.,* 1971.)

can be directed *in vitro* by adding basal bodies or mitotic spindles to the mixture of polymerizing tubulin (Snell *et al.,* 1974; Cande *et al.,* 1974).

Although electron miscroscopy provides the most detailed information on the structure and orientation of microtubules in cells, immunofluorescence using antibodies against purified tubulin has recently provided a simple means of determining the overall distribution of microtubules in cells cultured *in vitro* (Fuller *et al.,* 1975; Weber *et al.,* 1975).

Microtubules appear to provide structural support for cellular asymmetries, particularly long protrusions from cells such as embryonic axons (Porter, 1966; Tilney, 1968; Yamada *et al.,* 1970; Daniels, 1972). However, they may also be associated with active cell movements, either by transporting materials to selected locations in the cell (e.g., Freed and Lebowitz, 1970; see also Olmsted and Borisy, 1973), by lengthening due to addition of more tubulin (Inoué and Sato, 1967), or even possibly by actively sliding past neighbors (McIntosh *et al.,* 1969).

C. Adhesion and the Cell Surface

Tissue or individual cell morphogenetic movements take place in complex environments. Cells complete morphogenesis in altered relationships to other cells, and even at times in association with changed extracellular materials (Toole *et al.,* 1972; Bernfield *et al.,* 1972). How such cells can undergo and complete morphogenesis in appropriate environments is a major puzzle, particularly in the case of neural crest cell migratory behavior, where an initially localized group of cells must migrate to a whole host of new environments. Changes in cell adhesion could be part of the answer since cells might locomote over, or adhere preferentially to, only certain other cells or extracellular matrices (see Moscona, 1973; Toole *et al.*, 1972). Adhesion alone could theoretically provide developmental information by establishing pathways of migration on preferred substrata (Carter, 1967; Harris, 1973a; Letourneau, 1975) or by resulting in a local sorting out of cells according to a general hierarchy of adhesive strengths (Steinberg, 1970). More information could be provided if such adhesions were also selective, requiring recognition of specific adhesive sites (e.g., Barbera *et al.*, 1973). Alternatively, cell movement might also be directed by changing glycoprotein–glycosyltransferase interactions (Roth, 1973), by the sequential production and removal of an extracellular material such as hyaluronic acid which affects locomotion (Toole *et al.*, 1972; Pratt *et al.*, 1975) or, less likely, because of local production of a diffusable morphogenetic substance [e.g., a chemotactic substance such as cyclic AMP (Bonner, 1971; Grimes and Barnes, 1973)].

Obviously, the "cell surface," consisting of the outer portion of the plasma membrane lipid bilayer plus the glycocalyx (see Rambourg, 1971), will be the first site to receive any such information. Adhesion also takes place at the cell surface. Consequently, the isolation and characterization of molecules responsible for adhesion and other cell surface events (Lilien, 1969; McClay and Moscona, 1974; Yamada *et al.*, 1975) will probably be crucial to understanding morphogenetic cell movements.

III. DESCRIPTIVE AND GENETIC ANALYSIS OF CELL MOVEMENT

Detailed histological and ultrastructural studies of the morphogenesis of many organs and tissues have been performed in hopes of determining the actual mechanisms of morphogenetic movements (see DeHaan and Ursprung, 1965; Trinkaus, 1969; and examples in Section V). Migrating cells from a particular source have been identified by morphologic or histochemical criteria. They have also been marked by radioactive thymidine (Weston, 1970) or by using xenografts with distinctive morphology, such as quail cells grafted into chicken embryos (LeDouarin, 1973).

In addition, correlations have been made between changes in numbers and patterns of organelles as seen by electron microscopy and the stages of cell morphogenetic movements. Certain procedures are useful in preserving such highly labile structures as microtubules and newly assembled membranes. For instance, rapid fixation with glutaraldehyde–paraformaldehyde and optimal osmotic strength of fixatives can help prevent distortions of ultrastructure (see Hayat, 1970).

Genetically abnormal morphogenesis may provide insight into the normal mechanisms of movement, as well as providing models for congenital defects. For example, the weaver mutant in mice has been shown by Rakic and Sidman (1973) to involve defective nerve cell migration in the cerebellum, probably due to a failure of glial interaction with the migrating cells (see

Chapter 8, this volume). Other single-gene mutations may directly affect migratory cells, for instance, one of the *T* locus mutations in mice (Spiegelman and Bennett, 1974) and the *talpid* mutation in the chick (Ede *et al.*, 1974). Another mouse mutation called *steel* adversely affects several of the final environments which receive migratory cells, resulting in faulty pigmentation, hematopoiesis, and germ cell function (see Mayer, 1973). Other mutants with defects pertinent to particular classes of morphogenetic movements (e.g., epithelial folding) may already exist among the large number of mutant mice now available (Green, 1966; see also *Mouse News Letter*).

IV. DRUGS AS SPECIFIC PROBES

A. General Approaches

Another potentially powerful technique is to apply a drug which specifically disrupts only one type of organelle and then to determine the resulting morphogenetic defect. If the drug's action is reversible by removal or antidote, then correlations can be established between the reappearance of the structure and resumption of morphogenesis. Since morphogenetic movements are often programmed for a short time period, the drug's effects should vary depending on the embryo's stage of development.

Drugs used as this type of probe must either be without side effects or used with careful controls. All the drugs to be described have potentially major side effects, and the interpretation of their actions on some experimental systems in the literature is clouded by failure to perform important controls.

Several of the drugs which have proven useful in analyzing cell movement are cytochalasin, colchicine, and agents affecting Ca^{2+} or cyclic AMP metabolism.

B. Cytochalasin

A large number of cell movements are inhibited by the mold metabolite cytochalasin B (types A, B, C, and D differ in side groups, potency, and in some of their morphological and biochemical effects). Removal of the drug usually results in a resumption of morphogenesis. By electron microscopy, the microfilament arrays associated with these movements are generally observed to be reversibly disrupted or altered. No other subcellular organelles besides microfilaments appear to be altered in structure by cytochalasin (Carter, 1967; reviewed by Wessells *et al.*, 1971).

It has been suggested that cytochalasin may act by affecting actin or myosin directly. Evidence for such direct effects on purified contractile proteins from skeletal muscle or from nonmuscle cells exists (Spudich and Lin, 1972; Spudich, 1973; Puszkin *et al.*, 1973; Nicklas and Berl, 1974). However, muscle protein preparations which were less highly purified and which may have also contained troponin–tropomyosin were not substantially affected (Wessells *et al.*, 1971; Forer *et al.*, 1972; Pollard and Weihing, 1974).

The cytochalasins are highly hydrophobic and must be dissolved in organic solvents such as dimethylsulfoxide before use. As would be expected, radioactively labeled cytochalasin binds to lipid-rich cell structures such as the plasma and microsomal membranes (Lin and Spudich, 1974; Mayhew *et al.*, 1974; Tannenbaum *et al.*, 1975). Thus cytochalasin could also directly affect membrane assembly (Bluemink and deLaat, 1973), membrane mobility (Huestis and McConnell, 1974), or intramembranous proteins and enzymes (see below).

Interestingly, two classes of microfilaments may not be disrupted by cytochalasin. Microfilaments in epithelial microvilli, e.g., in avian oviduct cells (Wrenn, 1971), are not affected or may even lengthen in cytochalasin B [avian intestine (Burgess and Grey, 1974)]. The highly oriented sheath filaments of migratory cells are also variably resistant to cytochalasin (see Section V.B.1). Such resistant filaments may represent a biochemically distinct class, such as another form of actin, or they may be more highly stabilized than cytochalasin-sensitive filaments.

The cytochalasins have major side effects. The most prominent is a dose-dependent inhibition of the transport of hexoses, such as glucose and glucosamine, in concentration ranges even lower than those used for ultrastructural studies (see references in Bloch, 1973). This inhibition is reversible after removal of the drug, and shows noncompetitive kinetics in most of the cell types studied. The inhibition does not affect cellular ATP levels for over 30 min in fibroblasts (Warner and Perdue, 1972; no later time points were reported). A similar abrupt glucose deprivation induced by transfer of cell cultures to glucose-free medium does not mimic the effects of cytochalasin B on morphogenesis or on ultrastructure in several systems nor does it interfere with the drug's action (Yamada and Wessells, 1973; Taylor and Wessells, 1973). Cytochalasin D can apparently cause morphological effects on some cell lines without altering hexose transport (Miranda *et al.*, 1974a). In some cell types, other side effects of cytochalasin B may include moderately altered uridine and orotic acid transport. It has not been shown to affect protein, RNA, or DNA synthesis; amino acid transport; Ca^{2+} fluxes; or mucopolysaccharide synthesis measured by other than hexose incorporation (see Cohn *et al.*, 1972; Spooner, 1973).

Several authors have suggested that cytochalasin's effects on transport may be related to its effects on microfilament structure because the drug affects a common structure in the plasma membrane, for example, an intramembranous protein serving as a microfilament insertion site. However, recent binding studies with radioactively labeled cytochalasin B suggest that cells contain two classes of binding sites. The high-affinity sites are located in

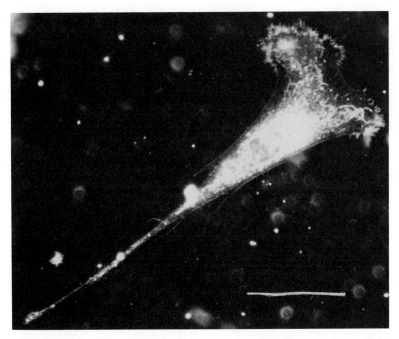

Fig. 4. Human fetal lung fibroblast migrating in culture. The anterior of this fibroblastic cell consists of broad, fan-shaped lamellae which undergo ruffling as the cell advances. Bar indicates 20 μm; dark field microscopy with the aid of Dr. M. Willingham.

the plasma membrane and are associated with sugar transport, while the low-affinity sites could be associated with actin or other proteins (Lin *et al.*, 1974; Lin and Spudich, 1974). It is not known whether the effects of cytochalasin B on a large number of secretory events (Allison, 1973) are due to effects on microfilament function, to direct action on the plasma membrane, or to effects on a common intermediate.

C. Colchicine and Vinblastine

Microtubules can be disrupted by several agents including colchicine and its reversible analog colcemid, vinblastine and its analog vincristine, high hydrostatic pressure, and cold (see review by Olmsted and Borisy, 1973). Colchicine binds tightly to tubulin, forming a complex which cannot assemble into microtubules, while vinblastine binds at a separate site and apparently aggregates microtubular protein into crystalline intracellular masses. No other organelle appears to be altered by these drugs, although large numbers of intermediate-sized filaments which could be microtubule breakdown products are found in cells treated with colchicine or vinblastine. Colchicine-

and cold-resistant microtubules are found in cilia and flagella. As suggested for the cytochalasin-resistant filaments, such resistant microtubules have been postulated to contain the usual subunit proteins, which are stabilized against depolymerization by as yet undetermined means (Tilney and Gibbins, 1968).

These antimicrotubular drugs may have nonspecific side effects. Colchicine can bind to membrane and chromatin preparations (Stadler and Franke, 1972), although at least part of the binding to membrane preparations may be due to trapping of tubulin (Ostlund and Pastan, 1975). Colchicine has been reported to affect intramembranous particle mobility (Wunderlich *et al.*, 1973); colchicine and vinblastine can also affect membrane stability (Seeman *et al.*, 1973). However, the latter effects were demonstrated at concentrations higher than those usually used to disrupt microtubules.

Other membrane phenomena, such as effects on cap formation in anti-immunoglobulin-treated lymphocytes (Edidin and Weiss, 1972; Yahara and Edelman, 1973), aggregation of fibroblasts (Waddell *et al.*, 1974), and many secretory events (Allison, 1973) can be inhibited by colchicine, colcemid, or vinblastine, even though there is no evidence for substantial microtubule–membrane associations. These results could reflect either direct effects on the membrane or effects on microtubules secondarily transmitted to the membrane through some unknown mediator. Colchicine may also affect membrane transport of nucleosides (Mizel and Wilson, 1972). Vinblastine, under some conditions (Wilson *et al.*, 1970), but apparently not others (Marantz *et al.*, 1969; Olmsted *et al.*, 1970), can precipitate actin-like substances other than tubulin. Antimicrotubular drugs have also been reported to affect RNA and protein synthesis (Creasy and Markiw, 1965).

D. Other Drugs

Papaverine, lanthanum, local anesthetics, and EGTA have been utilized to disrupt calcium metabolism and to halt various cell migratory and morphogenetic movements. Although their effects on movement are all consistent with calcium control of contraction, the fact that these drugs can have other major side effects makes this conclusion tentative (Ash *et al.*, 1973; Letourneau and Wessells, 1974; Gail *et al.*, 1973). Directly injecting calcium immediately under the plasma membrane of amphibian eggs results in a local area of contraction. Since network-type microfilaments are located in this region, this result suggests that calcium concentrations could control contractility (Gingell, 1970; Wessells *et al.*, 1971).

The role of cyclic nucleotides, particularly cyclic AMP and cyclic GMP, in cell movement is coming under investigation (see reviews by Pastan and Johnson, 1974; Willingham, 1975). Both dibutyryl cyclic AMP, a cyclic AMP analog, as well as prostaglandin E_1, which elevates cyclic AMP levels, can decrease the motility of cells (Johnson *et al.*, 1972; Pick, 1972). These drugs have also been claimed to stimulate axonal production by neuroblastoma cells *in vitro* (Prasad, 1972). In addition, both cyclic AMP and dibutyryl cyclic AMP were reported to stimulate axonal outgrowth from cultured sensory ganglia (Roisen *et al.*, 1972; Haas *et al.*, 1972). Such experiments must be interpreted with caution until the naturally occurring levels of cyclic nucleotides are determined. For example, although ganglion cells stimulated to produce axons by nerve growth factor might be expected to show an elevation in cyclic AMP levels, no effects on cyclic AMP or adenylate cyclase levels could be found in these cells (Hier *et al.*, 1973; Frazier *et al.*, 1973).

A highly toxic drug isolated from the *Amanita* mushroom, phalloidin, has recently been shown to polymerize actin irreversibly (Lengsfeld *et al.*, 1974). This polymerization is inhibited by cytochalasin B both in preparations of hepatocyte membranes and of purified muscle actin. The authors suggest that its toxicity to liver cells may result from an inhibition of normal microfilament depolymerization. If cells other than hepatocytes can be shown to be sensitive to this drug, it may prove to be useful for studies on cell movement.

E. Experimental Precautions

Many of the drugs discussed have specific and reversible ultrastructural effects and concomitant effects on cell movements. Each may also have accompanying side effects. Several classical approaches have been used to decrease the chances of artifacts secondary to drug side effects *in vitro*: (1) demonstration of ultrastructural correlation between the presence of a structure and cell movement, drug-induced disruption and cessation of the movement, and restoration of structure and movement, where possible on time scales (decreased movement should accompany or follow progressive decreases in microfilament order); (2) absence of gross effects on cell viability or on general processes such as protein synthesis; (3) identification of specific side effects and the use of appropriate controls (e.g., cytochalasin and glucose deprivation or comparisons of colchicine and its inactive analog lumicolchicine; (4) use of multiple drugs with different mechanisms of action to disrupt an organelle (e.g., colchicine, its reversible analog colcemid, vinblastine, and podophyllotoxin for microtubules); and (5) comparison of drugs of different specificity in the same cell (e.g., cytochalasin rapidly disrupts axonal tip movements and the microfilaments there, while colchicine has effects on the microtubules of the axon but does not halt movements of the microtubule-lacking tip).

Drugs have often been used as preliminary probes to evaluate possible roles of microfilaments, microtubules, or Ca^{2+} fluxes in particular morphogenetic movements. Unless at least some of the controls listed above are performed, it seems risky to attribute the resultant developmental defects to action on a single organelle.

V. ANALYSIS OF SPECIFIC MORPHOGENETIC EVENTS

A. Cytokinesis

Two important types of cell movement occur during cell division. The mitotic spindle separates the chromosomes in karyokinesis, and the cleavage furrow pinches the cell in two. Although these movements are not strictly morphogenetic, they demonstrate important principles. The mitotic spindle is composed of arrays of microtubules, and antimicrotubular agents which disrupt the spindle will halt karyokinesis (see Olmsted and Borisy, 1973). Many theories of mitosis exist; recently a proposal has been made that spindle microtubules move relative to each other by means of side arms, in analogy to the sliding-filament model of muscle contraction (McIntosh *et al.*, 1969).

In contrast, cytokinesis apparently involves a "contractile ring" of microfilaments, which form a band of parallel filaments lining the bottom of the cleavage furrow and encircle the center of the cell (see Schroeder, 1970, 1973). Such microfilaments probably contain cellular actin, since they bind heavy meromyosin. They are disrupted by cytochalasin, with a loss of the cleavage furrow. Interestingly, karyokinesis continues, so that the chromosomes are separated, but in a single cell which is then binucleate (Carter, 1967).

B. Cell Migration

1. Locomotion of Cells *in Vitro*

Cell migration has been studied most extensively in fibroblasts (Figure 4) and in two neural crest derivatives, glial and nerve cells. Although cell migration can be observed *in vivo* (Speidel, 1933; see Trinkaus, 1973), the ease of experimental manipulation of cells *in vitro* has made cells cultured on artificial substrata a favorite of experimenters. Recent evidence suggests that cell migration involves both the addition of new membrane to the anterior end of the cell (Abercrombie *et al.*, 1972; Harris and Dunn, 1972), as well as some mechanism of forming and breaking adhesions to the substratum. Although it is generally assumed that locomotion involves contractile events, their nature is not known.

Fibroblastic and glial cells are similar in ultrastructure (Buckley and Porter, 1967; Spooner *et al.*, 1971; Abercrombie *et al.*, 1971; McNutt *et al.*, 1971, 1973; Goldman and Knipe, 1973; Perdue, 1973; Ludueña and Wessells, 1973; see also Reaven and Axline, 1973; Miranda *et al.*, 1974). Both cell types contain bundles of microfilaments parallel to the lower surface of the cell, as well as a loose network of microfilaments immediately under the plasma membrane (see Figures 5 and 6). The network is most prominent in the portions of the cell periphery which are the sites of active movement, the leading lamellae and the ruffled membranes. Striking arrays of parallel microfilaments form a sheath along the bottom of the cell, and others can be found along the dorsal

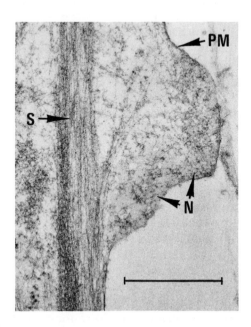

Fig. 5A. A bundle of sheath-type microfilaments (S) adjacent to network-type microfilaments (N). PM, plasma membrane. Bar indicates 0.5 μm. (From Ludueña and Wessells, 1973.)

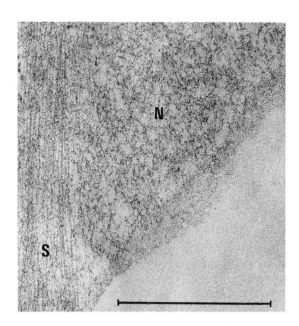

Fig. 5B. Similar region at higher magnification. S, linearly oriented sheath-type microfilaments; N, network-type microfilaments. Bar indicates 0.5 μm. (From Spooner et al., 1971.)

surface of the cell. Some of these linearly oriented microfilaments may insert into "plaques" on the plasma membrane at points where the cell is adhering to the substratum or to other cells (Abercrombie *et al.*, 1971; McNutt *et al.*, 1971; Brunk *et al.*, 1971; Heaysman and Pegrum, 1973). In contrast, microtubules are more centrally located in the cell, generally lying along the long axis of the cell and extending into processes.

Cytochalasin B treatment of migrating cells results in a rapid inhibition of ruffling and a cessation of locomotion (Carter, 1967; Spooner *et al.*, 1971). The network-type microfilaments are rapidly affected by this drug. Gaps appear in the network, permitting endoplasm to extend out to the plasma membrane, while other regions of the microfilament network apparently condense into masses of dense material. In contrast, the linearly oriented arrays of microfilaments often remain at least partially intact (Spooner *et al.*, 1971; Wessells *et al.*, 1971; Goldman, 1972; Miranda *et al.*, 1974b; cf. Axline and Reaven, 1974, on the linear microfilaments of macrophages). Microtubules and other organelles are not altered in ultrastructure. Over the next several hours in culture, large portions of the cytoplasm gradually collapse or contract inward, finally resulting in a stellate shape. The ultrastructure is similar to that described previously, except that the remaining linearly oriented microfilaments are concentrated into the long cell processes which have not collapsed inward, and the masses of dense material are larger. Removal of the drug allows a rapid resumption of motility. Cytochalasin D has generally similar effects on the ultrastructure of several established cell lines, although its morphological effects appear to be more drastic. Nevertheless, removal of this drug permits a rapid restoration of normal cell shape and ultrastructure (Miranda *et al.*, 1974a,b).

The bundles of linearly arranged microfilaments in these cells are probably cellular actin (with associated myosin and tropomyosin) as noted earlier. Although at points these bundles appear to merge into the network filaments, convincing proof that the latter are also actin is not yet available in these cells, since network microfilaments appear to be disrupted during the preparation of samples for heavy meromyosin binding (Spooner *et al.*, 1973). The cytochalasin-induced masses of material are probably compacted aggregates of actin, since those resulting from cytochalasin D treatment can bind heavy meromyosin (Miranda *et al.*, 1974b).

Colchicine treatment results in a breakdown of microtubules, but unlike cytochalasin, does not halt ruffled membrane activity or cell locomotion. However, more subtle effects have been reported: Net migration in a single direction is inhibited, and ruffled membrane activity can be spread over the entire perimeter of the cell, rather than remaining confined to one region. The impression is of a loss of the capacity to continue migrating in one direction and a loss of stabilization of the nonruffling portions of the cell membrane (Vasiliev *et al.*, 1970; Goldman, 1971; Gail and Boone, 1971).

It is informative to compare these results with studies on the elongating processes of nerve cells. The migratory tips of axons, the growth cones, are

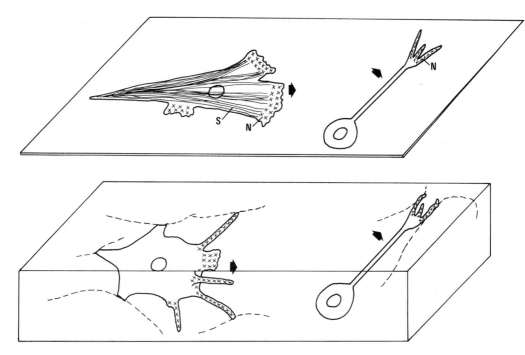

Fig. 6. Schematic comparison of cells migrating *in vitro* on a flat substratum (top) vs. *in vivo* (bottom). The cells are migrating, and the axons are elongating, in the directions indicated by the arrows. N, network-type microfilaments; S, sheath-type linearly oriented microfilaments. Dashed lines indicate other cells or extracellular materials which the cells use as substrata for locomotion *in vivo*. It is not yet known whether cells *in vivo* also contain sheath-type microfilaments.

the major site for the addition of new surface material (Bray, 1970). Growth cones and their filopodia contain the network type of microfilament in varying degrees of orientation (Yamada *et al.*, 1970, 1971; Bunge, 1973; Privat *et al.*, 1973). Neither the tip nor the rest of the cell contain the highly organized, linear microfilaments found in glial and fibroblastic cells (Ludueña and Wessells, 1973). Such a microfilament network in neuroblastoma cells reportedly binds heavy meromyosin (Burton and Kirkland, 1972; Chang and Goldman, 1973). Cytochalasin treatment of elongating axons causes a rapid distortion of the microfilament network, seemingly collapsing or contracting it. Axonal migration ceases immediately, and axons can remain extended but immobile for over 18 hr. This effect is reversed on removal of the drug (Yamada *et al.*, 1970, 1971).

Colchicine, in contrast, causes a slower disruption of microtubules and results in a shortening axon, even though the motile tip remains active. The axons eventually retract back into the nerve cell body. Removal of these drugs permits new axons to form (Yamada *et al.*, 1970, 1971; Daniels, 1972).

The actual mechanisms of all these types of cell migration are not known. However, fibroblasts and glial cells do contain substantial quantities of myosin

(0.5–3% of total cell protein) (Ostlund *et al.*, 1974; Ash, 1975). Large quantities of actin are also present by electrophoretic criteria and constitute as much as 10–20% of total cell protein (Orkin *et al.*, 1973; Fine and Bray, 1971). The regulatory protein tropomyosin has been isolated from embryonic nerve cells and from platelets (Fine *et al.*, 1973; Cohen and Cohen, 1972).

How these contractile proteins interact to produce locomotion is under intensive investigation. Both migrating cells and elongating axons apparently add new membrane to actively extending portions of the cell (Bray, 1970; Abercrombie *et al.*, 1970, 1972; Harris and Dunn, 1972) which could involve some form of contractile activity. However, indirect evidence suggests that cell migration may differ from axonal elongation in the need for another, more obviously contractile type of activity. Axons can elongate away from a cell body which remains stationary, while a migrating cell must translocate its cell body. Thus migrating cells must break adhesions at the posterior end of the cell (Harris, 1973b) in order to achieve net translocation along the substratum (Ingram, 1969; Abercrombie *et al.*, 1970).

The breaking of posterior adhesions may be mediated via the sheath-type microfilaments, which are present in migratory cells and absent from axons (Wessells *et al.*, 1973; Ludueña and Wessells, 1973). Consistent with this hypothesis is the finding that altering such adhesions by treatment with lanthanum or by culturing cells in agar will inhibit cell locomotion but not axonal elongation (Strassman *et al.*, 1973; Letourneau and Wessells, 1974). Moreover, treatment of cultured sensory ganglia with dibutyryl cyclic AMP, which can increase the adhesion of cells to the substratum (Johnson and Pastan, 1972; Gazdar *et al.*, 1972; Grinnell *et al.*, 1973), also inhibits the outgrowth of migratory cells while actually stimulating axonal outgrowth (Haas *et al.*, 1972).

2. Migration of cells *in Vivo*

The migration of neural crest cells, of cells ingressing through the primitive streak to form the mesoderm, and of precardiac mesodermal cells, to cite only a few examples, probably involves groups of variably touching cells rather than isolated cells. There is currently no reason to believe that they differ from single cells *in vitro* in their basic methods of locomotion, since microfilament structures are found in such cells *in vivo*. However, the actual shapes of cells may differ due to migration in a three-dimensional matrix rather than on a flat substratum (see Figure 6; Trelstad *et al.*, 1967; Tennyson, 1970; Johnston and Listgarten, 1972; Elsdale and Bard, 1972; Trinkaus, 1973). In particular, the leading edge of migrating cells *in vitro* often consists of broad lamellae, while this region of cells migrating *in vivo* is more often occupied by long filopodial processes. Axonal growth cones and their filopodia are surprisingly similar *in vitro* and *in vivo*, although *in vivo* the filopodia become twisted as they interdigitate between cells along their paths.

C. Cell Elongation

Lens placodes are formed by a thickening of the ectoderm overlying the optic cup. This thickening is due to a rapid elongation of individual ectodermal cells (see Figure 7). Byers and Porter (1964) demonstrated that such cells contain a circumferential row of microtubules oriented parallel to the long axis of the cells. Treatment of such placodes with colcemid blocked further cell elongation, but did not inhibit subsequent formation of the lens vesicle from the placode (Pearce and Zwann, 1970).

After lens vesicle formation (see Section V.D), a further elongation of cells occurs in the posterior lens epithelium. These cells also contain many oriented microtubules, and this elongation is also prevented by colchicine (Byers and Porter, 1964; Piatigorsky *et al.*, 1972). Initially, elongation and microtubule accretion can proceed in the absence of protein synthesis, indicating that these cells can utilize a pool of microtubule precursors for microtubule assembly.

Similarly, neurulation commences with the formation of a thickened structure, the neural plate, which is also composed of elongating ectodermal cells (see review by Karfunkel, 1974). These cells also contain many microtubules oriented along the axis of elongation. Colchicine and vinblastine disrupt the microtubules and result in a loss of the elongated cell shape and an inhibition of morphogenesis. Interestingly, cytochalasin treatment does not disrupt these microtubules, and elongation can continue in its presence.

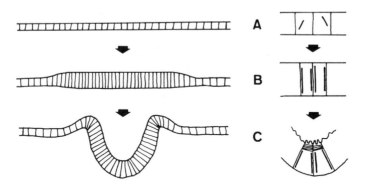

Fig. 7. Digrammatic representation of lens vesicle formation or neurulation viewed in cross section (left). Two typical cells of the epithelium are depicted on the right. A. Ectoderm prior to morphogenesis. B. Thickening lens placode or neural plate. The elongating cells contain microtubules oriented along the axis of elongation. C. Invaginating lens vesicle or neural tube. The epithelial cells contain bands of microfilaments across their apices and oriented microtubules along the long axis, and the plasma membranes are thrown into folds. Structures are *not* drawn to scale.

These examples, based on mouse and chick embryos, imply that cells which actively elongate do so by polymerizing new microtubules parallel to the axis of elongation. It is not known whether microtubules can produce this type of elongation directly, or whether they may act by transporting cytoplasm or other morphogenetic materials asymmetrically in the cell.

D. Invagination and Evagination

Flat or hollow epithelia give rise to many organ rudiments via infoldings and outpocketings. In the cases which have been examined by electron microscopy, bands of microfilaments in the regions of folding have been implicated in the morphogenetic event (Figures 2 and 7).

After the lens placode is formed, its cells invaginate to form the lens vesicle. The apical ends of these cells contain bands of circumferentially arrayed microfilaments parallel to the ends of the cells. The microfilaments apparently insert into electron-dense material at the zonula adherens region of the junctional complex. This end of the cells is narrower than the basal end, and the plasma membrane has an unusual number of bulges, protrusions, and microvilli; these findings are suggestive of contraction at this end of the cells (Wrenn and Wessells, 1969).

The process of neurulation is similar. The thickened neural plate forms a trough, which deepens until the neural folds fuse to form the neural tube. The invaginating cells contain arrays of microfilaments at their apical ends; these ends of the cell also display numerous microvilli and other protusions (Karfunkel, 1974; Freeman, 1972). Cytochalasin treatment disrupts these microfilaments, the cells lose their wedge shape, and the forming neural folds fall back into the ectoderm. Such cytochalasin treatment can result in embryos with neural tube closure defects (Karfunkel, 1972; Linville and Shepard, 1972).

Pancreatic and lung buds arise as outpocketings from the primitive gut. Pancreatic rudiment formation cannot be ascribed to local changes in mitotic rate, but instead appears to result from changes in the shapes and relative positions of a small number of cells (Wessells and Cohen, 1967; Wessells and Evans, 1968). Microfilament bundles are particularly abundant at the apical ends of such cells and may be involved in the shape changes. Lung rudiments also contain microfilament bundles at both apical and basal ends of the cells. Cytochalasin inhibits lung morphogenesis, although it could do so by effects on either mitosis or microfilaments (Bernfield and Wessells, 1970). Lung rudiment microfilaments have recently been shown to bind heavy meromyosin, suggesting that they are cytoplasmic actin (Spooner et al., 1973). Thus pancreatic and lung morphogenesis may involve contractile events similar to those described for other forms of folding morphogenesis.

Salivary gland development has been extensively analyzed as a model of organ morphogenesis (reviewed by Bernfield and Wessells, 1970; Spooner,

1974). The rudiment begins a distinctive multilobed branching pattern by the formation of clefts in a single bulb. Each new lobe is in turn divided into more lobes by more clefts. Each of these clefts contains highly oriented microfilament bands parallel to the axis of apparent contraction, with the adjacent plasma membrane and basal lamina thrown into folds. These ultrastructural findings suggest that cleft formation and lens and neural tube formation occur by similar mechanisms. However, microfilaments are also found at the opposite end of these salivary gland cells, suggesting that the presence of microfilaments alone may not be diagnostic of a contractile event. Both sets of microfilaments bind heavy meromyosin. These microfilaments, and the process of salivary cleft formation, are disrupted by cytochalasin B, and removal of the drug results in a return of the microfilament bands and a resumption of morphogenesis (Spooner and Wessells, 1972). After they are formed, clefts may become stabilized by mucopolysaccharides and polmerized collagen (Bernfield and Wessells, 1970; Bernfield et al., 1972), and they are then no longer sensitive to cytochalasin B.

Morphogenesis of glands in the chick oviduct will occur in response to the injection of estrogen. After estrogen treatment, increasingly larger bundles of microfilaments appear at the apical end of oviduct cells, this end of the cells narrows, and the epithelium begins to invaginate to form glands. Cytochalasin B does not affect either gland formation or microfilament structure when used at the usual concentrations. However, doses approximately fivefold higher are effective in disrupting microfilaments and in inhibiting morphogenesis (Wrenn, 1971). These results demonstrate the importance of monitoring drug effects by electron microscopy, since negative results may merely be due to inadequate local intracellular concentrations of the drug.

VI. GENERALITIES AND HYPOTHESES

It is not currently possible to present a definitive formulation of the mechanisms of mammalian cell morphogenetic movement. Nevertheless, certain generalities have emerged, and *tentative* conclusions can be suggested subject to later revision as more information becomes available:

1. Active cell movement depends primarily on interactions of cytoplasmic actin, myosin, and the appropriate control molecules (Jahn and Bovee, 1969; Pollard and Weihing, 1974). Other forms of motility could be mediated by microtubules.

2. Microtubules probably provide structural support for elongated cell processes, such as axons, and may also help maintain elongated cell shapes, for example, columnar epithelial regions. Microtubules may also stabilize cell membranes, limiting regions of membrane activity to localized areas, for example, confining ruffling to the anterior of migrating cells.

3. Microfilaments probably represent highly oriented actin–myosin arrays, which contract to form localized foldings of epithelial sheets. The in-

tracellular location of the plasma membrane insertion sites for such contractile proteins would determine the orientation of contraction and consequently the direction of movement. For example, microfilament insertion sites in an epithelial cell could be limited to the zonula adherens region around the apical end of the cell, and contraction would narrow only that end of the cell; the epithelium would then invaginate. Sites could be arranged even more asymmetrically, such as at only two sides of that end of the cell, so that contraction would flatten that end; the epithelium would then curl up to form a tube.

4. Cell extension can involve microfilaments or microtubules. Extension of a cell process can result from the activity of a microfilament network system as is seen in migratory cells and axonal tips. Theoretically, elongation of a cell could also occur by contraction of cytoplasmic actomyosin in microfilaments linking circumferential attachment sites at both ends of a cell; if its neighbors prevented the cell from bulging at its center, it would be squeezed into a more elongated shape. Alternatively, microtubules could be responsible for elongation by means of controlled tubulin polymerization, microtubule-mediated transport of materials to peripheral portions of the cell, or sliding of microtubules relative to each other.

5. Failures of morphogenesis can result from drug treatments at the appropriate time in development. Correlations can be made between disruption of an organelle and cessation of morphogenesis, as well as restoration of structure and function after drug removal. However, it is difficult to prove that a drug has affected *only* that organelle, since the drug probably exerts other biochemical effects of varying significance.

VII. CONTROL OF CELL MOVEMENTS AND CONGENITAL DEFECTS

Table 1 lists a number of recently suggested possible control points for cell morphogenetic movements. Although none have yet been conclusively proven to be important *in vivo*, a change in any of these parameters could result in a modulation of cell movements. Likewise, a defect at any level could have disastrous effects on morphogenesis. Some defects would be expected to be lethal very early in development, for example, a significant mutation in a binding or regulatory site might even disrupt cleavage. Other defects, e.g., an altered cell surface protein in a single cell type, could lead to organ agenesis or dysgenesis. In contrast, a less drastic alteration in a more basic control element could lead to pleiotropic defects.

Chemical or ionizing agents could easily interfere with a regulatory step. An important point to remember is that morphogenetic movements take place in a very limited period of time. If the agent results in late or aberrant epithelial folding or cell migration, the defect in morphogenesis will be highly specific to the dose and duration of the teratogen, as well as to the gestational age of the embryo.

Many human congenital defects could theoretically be due to defects in cell morphogenetic movements. To cite only a few examples, lung agenesis or

Table 1. Possible Control Points of Cell Movement

Microfilaments/Actin–Myosin	Microtubules	General
1. Rates of synthesis and turnover of contractile proteins	1. Rates of synthesis and turnover of tubulin	1. Synthesis of regulatory proteins, e.g., tropomyosin
2. Number or arrangement of insertion points in plasma membrane	2. Number and arrangement of microtubule organizing centers	2. Levels of cyclic AMP and cyclic GMP
3. Local Ca^{2+} concentration effects on troponin–tropomyosin (increase activates)	3. Local Ca^{2+} effects on tubulin polymerization (decrease activates)	3. Changes in cell–cell or cell–substratum adhesion
4. Phosphorylation of myosin	4. Phosphorylation of tubulin or associated proteins	4. Cell surface alterations, e.g., in protein glycosylation or in membrane mobility
	5. Local GTP concentration	5. "Induction"
	6. Rates of synthesis and turnover of tau factor	

ectopic lung lobes probably result from absent or additional lung buds forming from the foregut, and rachischisis from failure of the neural tube to complete closure. Similarly, various cardiac defects could result from faulty cell migration or folding during the complex early phases of heart development. Common congenital defects such as cleft palate and cardiac septal defects could also result from abberrant active cell movements. The recent demonstration of substantial quantities of cytoplasmic actin and myosin in the developing palate is consistent with this possibility (Lessard *et al.*, 1974). Application of the three-pronged approach of electron microscopy, measurements of contractile proteins, and treatments with drugs known to disrupt morphogenesis may prove to be valuable in analyzing these defects, both by elucidating the normal mechanisms of development and by providing models for abnormal development.

VIII. CONCLUSION

Many morphogenetic events in development have been analyzed at the ultrastructural level. In particular, microfilaments and microtubules have been implicated in cell morphogenetic movements. Drugs which disrupt these organelles, but which may be accompanied by side effects, will produce specific defects in morphogenetic cell movements and in organogenesis in tissue culture. These experiments provide models for defects in development. Similar *in vitro* and ultrastructural analyses of the effects of known teratogenic drugs and of known genetic defects in animals could provide information directly pertinent to some human congenital defects.

REFERENCES*

Abercrombie, M., Heaysman, J. E. M., and Pegrum, S. M., 1970, The locomotion of fibroblasts in culture. III. Movements of particles on the dorsal surface of the leading lamella, *Exp. Cell Res.* **62**:389.

Abercrombie, M., Heaysman, J. E. M., and Pegrum, S. M., 1971, The locomotion of fibroblasts in culture. IV. Electron microscopy of the leading lamella, *Exp. Cell. Res.* **67**:359.

Abercrombie, M., Heaysman, J. E. M., and Pegrum, S. M., 1972, Locomotion of fibroblasts in culture. V. Surface marking with concanavalin A, *Exp. Cell Res.* **73**:536.

Adelstein, R. S., and Conti, M. A., 1973, The characterization of contractile proteins from platelets and fibroblasts, *Cold Spring Harbor Symp. Quant. Biol.* **37**:599.

Allison, A. C., 1973, The role of microfilaments and microtubules in cell movement, endocytosis and exocytosis, *in: Ciba Foundation Symposium 14. Locomotion of Tissue Cells*, pp. 109–148, Elsevier, New York.

Ash, J. F., 1975, Purification and charactization of myosin from the clonal rat glial cell strain C-6, *J. Biol. Chem.* **250**:3560.

Ash, J. F., Spooner, B. S., and Wessells, N. K., 1973, Effects of papaverine and calcium-free medium on salivary gland morphogenesis, *Dev. Biol.* **33**:463.

Axline, S. G., and Reaven, E. P., 1974, Inhibition of phagocytosis and plasma membrane mobility

*Selected recent reviews are indicated by asterisks.

of the cultivated macrophage by cytochalasin B. Role of subplasmalemmal microfilaments, *J. Cell Biol.* **62:**647.

Barbera, A. J., Marchase, R. B., and Roth, S., 1973, Adhesive recognition and retinotectal specificity, *Proc. Natl. Acad. Sci. U.S.A.* **70:**2482.

Behnke, O., Kristensen, B. I., and Nielsen, L. E., 1971, Electron microscopical observations on actinoid and myosinoid filaments in blood platelets, *J. Ultrastruct. Res.* **37:**351.

Bernfield, M. R., and Wessells, N. K., 1970, Intra- and extracellular control of epithelial morphogenesis, *Dev. Biol. Suppl.* **4:**195.

Bernfield, M. R., Banerjee, S. D., and Cohn, R. H., 1972, Dependence of salivary epithelial morphology and branching morphogenesis upon acid mucopolysaccharide–protein (proteoglycan) at the epithelial surface, *J. Cell Biol.* **52:**674.

Bettex-Galland, M., and Lüscher, E. F., 1965, Thrombosthenin, the contractile protein from blood platelets and its relation to other contractile proteins, *Adv. Protein Chem.* **20:**1.

Bloch, R., 1973, Inhibition of glucose transport in the human erythrocyte by cytochalasin B, *Biochemistry* **12:**4799.

Bluemink, J. G., and deLaat, S. W., 1973, New membrane formation during cytokinesis in normal and cytochalasin B-treated eggs of *Xenopus laevis*. I. Electron microscope observations, *J. Cell Biol.* **59:**89.

Bonner, J. T., 1971, Aggregation and differentiation in the cellular slime molds, *Annu. Rev. Microbiol.* **25:**75.

Booyse, F. M., Hoveke, T. P., and Rafelson, M. E., 1973, Human platelet actin. Isolation and properties, *J. Biol. Chem.* **248:**4083.

Borisy, G. G., and Olmsted, J. B., 1972, Nucleated assembly of microtubules in porcine brain extracts, *Science* **177:**1196.

Bray, D., 1970, Surface movements during the growth of a single explained neuron, *Proc. Natl. Acad. Sci. U.S.A.* **65:**905.

Bray, D., 1973, Cytoplasmic actin: A comparative study, *Cold Spring Harbor Symp. Quant. Biol.* **37:**567.

Brunk, U., Ericsson, J. L. E., Pontén, J., and Westermark, B., 1971, Specialization of cell surfaces in contact-inhibited human glia-like cells *in vitro. Exp. Cell Res.* **67:**407.

Buckley, I. K., and Porter, K. R., 1967, Cytoplasmic fibrils in living cultured cells. A light and electron microscope study, *Protoplasma* **64:**349.

Bunge, M. B., 1973, Fine structure of nerve fibers and growth cones of isolated sympathetic neurons in culture, *J. Cell Biol.* **56:**713.

Burgess, D. R., and Grey, R. D., 1974, Alterations in morphology of developing microvilli elicited by cytochalasin B. Studies of embryonic chick intestine in organ culture, *J. Cell Biol.* **62:**566.

Burton, P. R., and Kirkland, W. L., 1972, Actin detected in mouse neuroblastoma cells by binding of heavy meromyosin, *Nature (London) New Biol.* **239:**244.

Byers, B., and Porter, K. R., 1964, Oriented microtubules in elongating cells of the developing lens rudiment after induction, *Proc. Natl. Acad. Sci. U.S.A.* **52:**1091.

Cande, W. Z., Snyder, J., Smith, D., Summers, K., and McIntosh, J. R., 1974, A functional mitotic spindle prepared from mammalian cells in culture, *Proc. Natl. Acad. Sci. U.S.A.* **71:**1559.

Carter, S. B., 1967, Effects of cytochalasins on mammalian cells, *Nature* **213:**261.

Chang, C.-M., and Goldman, R. D., 1973, The localization of actin-like fibers in cultured neuroblastoma cells as revealed by heavy meromyosin binding, *J. Cell Biol.* **57:**867.

Cohen, I., and Cohen, C., 1972, A tropomyosin-like protein from human platelets, *J. Mol. Biol.* **68:**383.

Cohn, R. H., Banerjee, S. D., Shelton, E. R., and Bernfield, M. R., 1972, Cytochalasin B: lack of effect on mucopolysaccharide synthesis and selective alterations in precursor uptake, *Proc. Natl. Acad. Sci. U.S.A.* **69:**2865.

Cooke, P. H., and Fay, F. S., 1972, Thick myofilaments in contracted and relaxed mammalian smooth muscle cells, *Exp. Cell Res.* **71:**265.

Creasy, W. A., and Markiw, M. E., 1965, Biochemical effects of the vinca alkaloids. IV. The synthesis of ribonucleic acid and the incorporation of amino acids in Ehrlich ascites cells *in vitro, Biochim. Biophys. Acta* **103:**635.

Daniels, M. P., 1972, Colchicine inhibition of nerve fiber formation *in vitro*, *J. Cell Biol.* **53:**164.

DeHaan, R. L., 1965, Morphogenesis of the vertebrate heart, *in: Organogenesis* (R. L. DeHaan and H. Ursprung, eds.), pp. 377–419, Holt, Rinehart and Winston, New York.

*DeHaan, R. L., and Ursprung, H., eds., 1965, *Organogenesis,* Holt, Rinehart and Winston, New York.

Ede, D. A., Bellairs, R., and Bancroft, M., 1974, A scanning electron microscope study of the early limb-bud in normal and *talpid*³ mutant chick embryos, *J. Embryol. Exp. Morphol.* **31:**761.

Edidin, M., and Weiss, A., 1972, Antigen cap formation in cultured fibroblasts: A reflection of membrane fluidity and of cell motility, *Proc. Natl. Acad. Sci. U.S.A.* **69:**2456.

Elsdale, T., and Bard, J., 1972, Cellular interactions in mass cultures of human diploid fibroblasts, *Nature* **236:**152.

Fine, R. E., and Bray, D., 1971, Actin in growing nerve cells, *Nature (London) New Biol.* **234:**115.

Fine, R. E., Blitz, A. L., Hitchcock, S. E., and Kaminer, B., 1973, Tropomyosin in brain and growing neurons, *Nature (London), New Biol.* **245:**182.

Forer, A., Emmersen, J., and Behnke, O., 1972, Cytochalasin B: Does it affect actin-like filaments? *Science* **175:**774.

Frazier, W. A., Ohlendorf, C. E., Boyd, L. F., Aloe, L., Johnson, E. M., Ferrendelli, J. A., and Bradshaw, R. A., 1973, Mechanism of action of nerver growth factor and cyclic AMP on neurite outgrowth in embryonic chick sensory ganglia: Demonstration of independent pathways of stimulation, *Proc. Natl. Acad. Sci. U.S.A.* **70:**2448.

Freed, J. J., and Lebowtiz, M. M., 1970, The association of a class of saltatory movements with microtubules in cultured cells, *J. Cell Biol.* **45:**334.

Freeman, B. G., 1972, Surface modifications of neural epithelial cells during formation of the neural tube in the rat embryo, *J. Embryol. Exp. Morphol.* **28:**437.

Fuller, G. M., Brinkley, B. R., and Boughter, J. M., 1975, Immunofluorescence of mitotic spindles by using monospecific antibody against bovine brain tubulin, *Science* **187:**948.

Gabbiani, G., Ryan, G. B., Lamelin, J.-P., Vassalli, P., Majno, G., Bouvier, C. A., Cruchaud, A., and Luscher, E. F., 1973, Human smooth muscle antibody. Its identification as antiactin antibody and a study of its binding to "nonmuscular" cells, *Am. J. Pathol.* **72:**473.

Gail, M. H., and Boone, C. W., 1971, Effect of Colcemid on fibroblast motility, *Exp. Cell Res.* **65:**221

Gail, M. H., Boone, C. W., and Thompson, C. S., 1973, A calcium requirement for fibroblast motility and proliferation, *Exp. Cell Res.* **79:**386.

Gazdar, A., Hatanaka, M., Herberman, R., Russell, E., and Ikawa, Y., 1972, Effects of dibutyryl cyclic adenosine phosphate plus theophylline on murine sarcoma virus transformed nonproducer cells, *Proc. Soc. Exp. Biol. Med.* **141:**1044.

Gingell, D., 1970, Contractile responses at the surface of an amphibian egg, *J. Embryol. Exp. Morphol.* **23:**583.

Goldman, R. D., 1971, The role of three cytoplasmic fibers in BHK-21 cell motility. I. Microtubules and the effects of colchicine, *J. Cell Biol.* **51:**752.

Goldman, R. D., 1972, The effects of cytochalasin B on the microfilaments of baby hamster kidney (BHK-21) cells, *J. Cell Biol.* **52:**246.

Goldman, R. D., and Knipe, D. M., 1973, The functions of cytoplasmic fibers in nonmuscle cell motility, *Cold Spring Harbor Symp. Quant. Biol.* **37:**523.

Green, E. L., ed., 1966, *Biology of the Laboratory Mouse,* 2nd ed., Mc-Graw-Hill, New York.

Grimes, G. J., and Barnes, F. S., 1973, A technique for studying chemotaxis of leucocytes in well-defined chemotactic fields, *Exp. Cell Res.* **79:**375.

Grinnell, F., Milam, M., and Srere, P. A., 1973, Cyclic AMP does not affect the rate at which cells attach to a substratum, *Nature (London), New Biol.* **241:**82.

Gruenstein, E., Rich, A., and Weihing, R. R., 1975, Actin associated with membranes from 3T3 mouse fibroblast and HeLa cells, *J. Cell Biol.* **64:**223.

Gwynn, I., Kemp, R. B., Jones, B. M., and Groschel-Stewart, V. 1974, Ultrastructural evidence for myosin of the smooth muscle type at the surface of trypsin-dissociated embryonic chick cells, *J. Cell Sci.* **15:**279.

Haas, D. C., Hier, D. B., Arnason, G. W., and Young, M., 1972, On a possible relationship of cyclic AMP to the mechanism of action of nerve growth factor, *Proc. Soc. Exp. Biol. Med.* **140:**45.

Harris, A., 1973a, Behavior of cultured cells on substrata of variable adhesiveness, *Exp. Cell Res.* **77:**285.

Harris, A., 1973b, Location of cellular adhesions to solid substrata, *Dev. Biol.* **35:**83.

Harris, A., and Dunn, G., 1972, Centripetal transport of attached particles on both surfaces of moving fibroblasts, *Exp. Cell Res.* **73:**519.

Hayat, M. A., 1970, *Principles and Techniques of Electron Microscopy. Biological Applications,* Vol. I, Van Nostrand Reinhold, New York.

Heaysman, J. E. M., and Pegrum, S. M., 1973, Early contacts between fibroblasts. An ultrastructural study, *Exp. Cell Res.* **78:**71.

Hier, D. B., Arnason, B. G. W., and Young, M., 1973, Nerve growth factor: Relationship to the cyclic AMP system of sensory ganglia, *Science* **182:**79.

Hörstadius, S., 1950, *The Neural Crest,* Oxford Univ. Press, New York.

Huestis, W. H., and McConnell, H. M., 1974, A functional acetylcholine receptor in the human erythrocyte, *Biochem. Biophys. Res. Commun.* **57:**726.

Huxley, H. E., 1973, Muscular contraction and cell motility, *Nature* **243:**445.

Ingram, V. M., 1969, A side view of moving fibroblasts, *Nature* **222:**641.

Inoué, S., and Sato, H., 1967, Cell motility by labile association of molecules. The nature of mitotic spindle fibers and their role in chromosome movement, *J. Gen. Physiol. Suppl.* **50:**259.

Ishikawa, H., Bischoff, R., and Holtzer, H., 1969, Formation of arrowhead complexes with heavy meromyosin in a variety of cell types, *J. Cell Biol.* **43:**312.

*Jahn, T. L., and Bovee, E. C., 1969, Protoplasmic movements within cells, *Physiol. Rev.* **49:**793.

Johnson, G. S., and Pastan, I., 1972, Cyclic AMP increases the adhesion of fibroblasts to the substratum, *Nature (London), New Biol.* **236:**247.

Johnson, G. S., Morgan, W. D., and Pastan, I., 1972, Regulation of cell motility by cyclic AMP, *Nature* **235:**54.

Johnston, M. C., and Listgarten, M. A., 1972, Observations on the migration, interaction, and early differentiation of orofacial tissues, in: *Developmental Aspects of Oral Biology* (H. C. Slavkin and L. A. Bavetta, eds.), pp. 53–80, Academic Press, New York.

Karfunkel, P., 1972, The activity of microtubules and microfilaments in neurulation in the chick, *J. Exp. Zool.* **181:**289.

Karfunkel, P., 1974, The mechanisms of neural tube formation, *Int. Rev. Cytol.* **38:**245.

Kelley, D. E., 1966, Fine structure of desmosomes, hemidesmosomes, and an adepidermal globular layer in developing newt epidermis, *J. Cell Biol.* **28:**51.

Kirschner, M. W., Williams, R. C., Weingarten, M., and Gerhart, J. C., 1974, Microtubules from mammalian brain: Some properties of their depolymerization products and a proposed mechanism of assembly and disassembly, *Proc. Natl. Acad. Sci. U.S.A.* **71:**1159.

Lazarides, E., 1975, Tropomyosin antibody: The specific localization of tropomyosin in nonmuscle cells, *J. Cell Biol.* **65:**549.

Lazarides, E., and Weber, K., 1974, Actin antibody: The specific visualization of actin filaments in non-muscle cells, *Proc. Natl. Acad. Sci. U.S.A.* **71:**2268.

LeDouarin, N., 1973, A biological cell labeling technique and its use in experimental embryology, *Dev. Biol.* **30:**217.

Lengsfeld, A. M., Löw, I., Wieland, T., Dancker, P., and Hasselbach, W., 1974, Interaction of phalloidin with actin, *Proc. Natl. Acad. Sci. U.S.A.* **71:**2803.

Lessard, J. L., Wee, E. L., and Zimmerman, E. F., 1974, Presence of contractile proteins in mouse fetal palate prior to shelf elevation, *Teratology* **9:**113.

Letourneau, P. C., 1975, Cell-to-substratum adhesion and guidance of axonal elongation, *Dev. Biol.* **44:**92.

Letourneau, P. C., and Wessells, N. K., 1974, Migratory cell locomotion versus nerve axon elongation. Differences based on the effects of lanthanum ion, *J. Cell Biol.* **61:**56.

Lilien, J. E., 1969, Toward a molecular explanation for specific cell adhesion, *Curr. Top. Dev. Biol.* **4:**169.

Lin, S., and Spudich, J. A., 1974, Biochemical studies on the mode of action of cytochalasin B. Cytochalasin B binding to red cell membrane in relation to glucose transport, *J. Biol. Chem.* **249:**5778.

Lin, S., Santi, D. V., and Spudich, J. A., 1974, Biochemical studies on the mode of action of cytochalasin B. Preparation of [³H]cytochalasin B and studies on its binding to cells, *J. Biol. Chem.* **249:**2268.

Linville, G. P., and Shepard, T. H., 1972, Neural tube closure defects caused by cytochalasin B, *Nature (London), New Biol.* **236:**246.

Ludueña, M. A., and Wessells, N. K., 1973, Cell locomotion, nerve elongation, and microfilaments, *Dev. Biol.* **30:**427.

Marantz, R., Ventilla, M., and Shelanski, M., 1969, Vinblastine-induced precipitation of microtubule protein, *Science* **165:**498.

Mayer, T. C., 1973, Site of gene action in steel mice: Analysis of the pigment defect by mesoderm–ectoderm recombinations, *J. Exp. Zool.* **184:**345.

Mayhew, E., Poste, G., Cowden, M., Tolson, N., and Maslow, D., 1974, Cellular binding of ³H-cytochalasin B, *J. Cell Physiol.* **84:**373.

McClay, D. R., and Moscona, A. A., 1974, Purification of the specific cell-aggregating factor from embryonic neural retina cells, *Exp. Cell Res.* **87:**438.

McIntosh, J. R., Hepler, P. K., and van Wie, D. G., 1969, Model for mitosis, *Nature* **224:**659.

McNutt, N. S., Culp, L. A., and Black, P. H., 1971, Contact-inhibited revertant cell lines isolated from SV40-transformed cells. II. Ultrastructural study, *J. Cell Biol.* **50:**691.

McNutt, N. S., Culp, L. A., and Black, P. H., 1973, Contact-inhibited revertant cell lines isolated from SV40-transformed cells. IV. Microfilament distribution and cell shape in untransformed, transformed, and revertant Balb/c 3T3 cells, *J. Cell Biol.* **56:**412.

Middleton, C. A., 1973, The control of epithelial cell locomotion in tissue culture, *in: Ciba Foundation Symposium 14. Locomotion of Tissue Cells*, pp. 251–270, Elsevier, New York.

Miranda, A. F., Godman, G. C., Deitch, A. D., and Tanenbaum, S. W., 1974a, Action of cytochalasin D on cells of established lines. I. Early events, *J. Cell Biol.* **61:**481.

Miranda, A. F., Godman, G. C., and Tanenbaum, S. W., 1974b, Action of cytochalasin D on cells of established lines. II. Cortex and microfilaments, *J. Cell Biol.* **62:**406.

Mizel, S. B., and Wilson, L., 1972, Nucleoside transport in mammalian cells. Inhibition by colchicine, *Biochemistry* **11:**2573.

Moscona, A. A., 1973, Cell aggregation, *in: Cell Biology in Medicine* (E. E. Bittar, ed.), pp. 571–591, John Wiley, New York.

Nicklas, W. J., and Berl, S., 1974, Effects of cytochalasin B on uptake and release of putative transmitters by synaptosomes and on brain actomyosin-like protein, *Nature* **247:**471.

*Olmsted, J. B., and Borisy, G. G., 1973, Microtubules, *Annu. Rev. Biochem.* **42:**507.

Olmsted, J. B., Carlson, K., Klebe, R., Ruddle, F., and Rosenbaum, J., 1970, Isolation of microtubule protein from cultured mouse neuroblastoma cells, *Proc. Natl. Acad. Sci. U.S.A.* **65:**129.

Orkin, R. W., Pollard, T. D., and Hay, E. D., 1973, SDS gel analysis of muscle proteins in embryonic cells, *Dev. Biol.* **35:**388.

Ostlund, R., and Pastan, I., 1975, Fibroblast tubulin, *Biochemistry* **14:**4064–4068.

Ostlund, R. E., Pastan, I., and Adelstein, R. S., 1974, Myosin in cultured fibroblasts, *J. Biol. Chem.* **249:**3903.

Painter, R. G., Sheetz, M., and Singer, S. J., 1975, Detection and ultrastructural localization of human smooth muscle myosin-like molecules in human non-muscle cells by specific antibodies, *Proc. Natl. Acad. Sci. U.S.A.* **72:**1359.

Pastan, I., and Johnson, G. S., 1974, Cyclic AMP and the transformation of fibroblasts, *Adv. Cancer Res.* **19:**303.

Pearce, T. L., and Zwaan, J., 1970, A light and electron microscopic study of cell behavior and microtubules in the embryonic chicken lens using Colcemid, *J. Embryol. Exp. Morphol.* **23:**491.

Perdue, J. F., 1973, The distribution, ultrastructure, and chemistry of microfilaments in cultured chick embryo fibroblasts, *J. Cell Biol.* **58:**265.

Piatigorsky, J., Webster, H. D., and Wollberg, M., 1972, Cell elongation in the cultured embryonic chick lens epithelium with and without protein synthesis, *J. Cell Biol.* **55**:82.

Pick, E., 1972, Cyclic AMP affects macrophage migration, *Nature (London), New Biol.* **238**:176.

Pollard, T. D., and Korn, E. D., 1973, Electron microscopic identification of actin associated with isolated amoeba plasma membranes, *J. Biol. Chem.* **248**:448.

*Pollard, T. D., and Weihing, R. R., 1974, Actin and myosin and cell movement, *CRC Crit. Rev. Biochem.* **2**:1.

*Porter, K. R., 1966, Cytoplasmic microtubules and their functions, *in: Ciba Foundation Symposium. Principles of Biomolecular Organization,* pp. 308–356, Little, Brown, Boston.

Prasad, K. N., 1972, Morphological differentiation induced by prostaglandin in mouse neuroblastoma cells in culture, *Nature (London), New Biol.* **236**:49.

Pratt, R. M., Larsen, M. A., and Johnston, M. C., 1975, Migration of cranial neural crest cells in a cell-free hyaluronate-rich matrix, *Dev. Biol.* **44**:298.

Privat, A., Drian, M. J., and Mandon, P., 1973, The outgrowth of rat cerebellum in organized culture, *Z. Zellforsch. Mikrosk. Anat.* **146**:45.

Puszkin, E., Puszkin, S., Lo, L. W., and Tanenbaum, S. W., 1973, Binding of cytochalasin D to platelet and muscle myosin, *J. Biol. Chem.* **248**:7754.

Rakic, P., and Sidman, R. L., 1973, Weaver mutant mouse cerebellum: Defective neuronal migration secondary to abnormality of Bergmann glia, *Proc. Natl. Acad. Sci. U.S.A.* **70**:240.

Rambourg, A., 1971, Morphological and histochemical aspects of glycoproteins at the surface of animal cells, *Int. Rev. Cytol.* **31**:57.

Reaven, E. P., and Axline, S. G., 1973, Subplasmalemmal microfilaments and microtubules in resting and phagocytizing cultivated macrophages, *J. Cell Biol.* **59**:12.

Roisen, F. J., Murphy, R. A., Pichichero, M. E., and Braden, W. G., 1972, Cyclic adenosine monophosphate stimulation of axonal elongation, *Science* **175**:73.

Roth, S., 1973, A molecular model for cell interactions, *Q. Rev. Biol.* **48**:541.

Sanger, J. W., 1975a, Changing patterns of actin localization during cell division, *Proc. Natl. Acad. U.S.A.* **72**:1913.

Sanger, J. W., 1975b, Presence of actin during chromosomal movement, *Proc. Natl. Acad. Sci. U.S.A.* **72**:2451.

Schroeder, T. E., 1970, The contractile ring. I. Fine structure of dividing mammalian (HeLa) cells and the effects of cytochalasin B. *Z. Zellforsch. Mikrosk. Anat.* **109**:431.

Schroeder, T. E., 1973, Actin in dividing cells: Contractile ring filaments bind heavy meromyosin, *Proc. Natl. Acad. Sci. U.S.A.* **70**:1688.

Seeman, P., Chau-Wong, M., and Moyyen, S., 1973, Membrane expansion by vinblastine and strychnine, *Nature (London), New Biol.* **241**:22.

Snell, W. J., Dentler, W. L., Haimo, L. T., Binder, L. I., and Rosenbaum, J. L., 1974, Assembly of chick brain tubulin onto isolated basal bodies of *Chlamydomonas reinhardi, Science* **185**:357.

Somlyo, A. P., Devine, C. E., and Somlyo, A. V., 1971, Thick filaments in unstretched mammalian smooth muscle, *Nature (London), New Biol.* **233**:218.

Speidel, C. C., 1933, Studies of living nerves. II. Activities of ameboid growth cones, sheath cells, and myelin segments, as revealed by prolonged observation of individual nerve fibers in frog tadpoles, *Am. J. Anat.* **52**:1.

Spiegelman, M., and Bennett, D., 1974, Fine structural study of cell migration in the early mesoderm of normal and mutant mouse embryo (*T*-locus: t^9/t^9), *J. Embryol. Exp. Morphol.* **32**:723.

Spooner, B. S., 1973, Cytochalasin B: Toward an understanding of its mode of action, *Dev. Biol.* **35**:f-13.

Spooner, B. S., 1974, Morphogenesis of vertebrate organs, *in Concepts of Development* (J. Lash and J. R. Whittaker, eds.), pp. 213–240, Sinauer Associates, Stamford, Conn.

Spooner, B. S., and Wessells, N. K., 1972, An analysis of salivary gland morphogenesis: Role of cytoplasmic microfilaments and microtubules, *Dev. Biol.* **27**:38.

Spooner, B. S., Yamada, K. M., and Wessells, N. K., 1971, Microfilaments and cell locomotion, *J. Cell Biol.* **49**:595.

Spooner, B. S., Ash, J. F., Wrenn, J. T., Frater, R. B., and Wessells, N. K., 1973, Heavy meromyosin binding to microfilaments involved in cell and morphogenetic movements, *Tissue Cell* **5**:37.

Spudich, J. A., 1973, Effects of cytochalasin B on actin filaments, *Cold Spring Harbor Symp. Quant. Biol.* **37**:585.

Spudich, J. A., 1974, Biochemical and structural studies of actomyosin-like proteins from non-muscle cells. II. Purification, properties, and membrane association of actin from amoebae of *Dictyostelium discoideum, J. Biol. Chem.* **249**:6013.

Spudich, J. A., and Lin, S., 1972, Cytochalasin B, its interaction with actin and actomyosin from muscle, *Proc. Natl. Acad. Sci. U.S.A.* **69**:442.

Stadler, J., and Franke, W. W., 1972, Colchicine-binding proteins in chromatin and membranes, *Nature (London), New Biol.* **237**:237.

Steinberg, M. S., 1970, Does differential adhesion govern self-assembly processes in histogenesis? Equilibrium configurations and the emergence of a hierarchy among populations of embryonic cells, *J. Exp. Zool.* **173**:395.

Stossel, T. P., and Pollard, T. D., 1973, Myosin in polymorphonuclear leukocytes, *J. Biol. Chem.* **248**:8288.

Strassman, R. J., Letourneau, P. C., and Wessells, N. K., 1973, Elongation of axons in an agar matrix that does not support cell locomotion, *Exp. Cell Res.* **81**:482.

Tannenbaum, J., Tanenbaum, S. W., Lo, L. W., Godman, G. C., and Miranda, A. F., 1975, Binding and subcellular localization of tritiated cytochalasin D, *Exp. Cell Res.* **91**:47.

Taylor, E. L., and Wessells, N. K., 1973, Cytochalasin B: Alterations in salivary gland morphogenesis not due to glucose depletion, *Dev. Biol.* **31**:421.

Tennyson, V. M., 1970, The fine structure of the axon and growth cone of the dorsal root neuroblast of the rabbit embryo, *J. Cell Biol.* **44**:62.

Tilney, L. G., 1968, The assembly of microtubules and their role in the development of cell form, *Dev. Biol. Suppl.* **2**:63.

Tilney, L. G., and Gibbins, J. R., 1968, Differential effects of antimitotic agents on the stability and behavior of cytoplasmic and ciliary microtubules, *Protoplasma* **65**:167.

Toole, B. P., Jackson, G., and Gross, J., 1972, Hyaluronate in morphogenesis: Inhibition of chondrogenesis *in vitro, Proc. Natl. Acad. Sci. U.S.A.* **69**:1384.

Trelstad, R. L., Hay, E. D., and Revel, J. P., 1967, Cell contact during early morphogenesis in the chick embryo, *Dev. Biol.* **16**:78.

*Trinkaus, J. P., 1969, *Cells into Organs. The Forces that Shape the Embryo,* Prentice-Hall, Englewood Cliffs, N. J.

Trinkaus, J. P., 1973, Modes of cell locomotion *in vivo,* in: *Ciba Foundation Symposium 14. Locomotion of Tissue Cells,* pp. 233–249, Elsevier, New York.

Vasiliev, Ju. M., Gelfand, I. M., Domnina, L. V., Ivanova, O. Y., Komm, S. G., and Olshevskaja, L. V., 1970, Effect of Colcemid on the locomotory behavior of fibroblasts, *J. Embryol. Exp. Morphol.* **24**:625.

Vaughan, R. B., and Trinkaus, J. P., 1966, Movements of epithelial cell sheets *in vitro, J. Cell Sci.* **1**:407.

Waddell, A. W., Robson, R. T., and Edwards, J. G., 1974, Colchicine and vinblastine inhibit fibroblast aggregation, *Nature* **248**:239.

Warner, D. A., and Perdue, J. F., 1972, Cytochalasin B and the adenosine triphosphate content of treated fibroblasts, *J. Cell Biol.* **55**:242.

Weber, K., and Groeschel-Stewart, U., 1974, Antibody to myosin: The specific visualization of myosin-containing filaments in nonmuscle cells, *Proc. Natl. Acad. Sci. U.S.A.* **71**:4561.

Weber, K., Pollack, R., and Bibring, T., 1975, Antibody against tubulin: The specific visualization of cytoplasmic microtubules in tissue culture cells, *Proc. Natl. Acad. Sci. U.S.A.* **72**:459.

Weingarten, M. D., Lockwood, A. H., Hwo, S.-Y., and Kirschner, M. W., 1975, A protein factor essential for microtubule assembly, *Proc. Natl. Acad. Sci. U.S.A.* **72**:1858.

Weisenberg, R. C., 1972, Microtubule formation *in vitro* in solutions containing low calcium concentrations, *Science* **177**:1104.

Wessells, N. K., and Cohen, J. H., 1967, Early pancrease organogenesis: Morphogenesis, tissue interactions, and mass effects, *Dev. Biol.* **15**:237.

Wessells, N. K., and Evans, J., 1968, Ultrastructural studies of early morphogenesis and cytodifferentiation in the embryonic mammalian pancreas, *Dev. Biol.* **17**:413.

*Wessells, N. K., Spooner, B. S., Ash, J. F., Bradley, M. O., Ludueña, M. A., Taylor, E. L., Wrenn, J. T., and Yamada, K. M., 1971, Microfilaments in cellular and developmental processes, *Science* **171**:135.

Wessells, N. K., Spooner, B. S., and Ludueña, M. A., 1973, Surface movements, microfilaments and cell locomotion, *in: Ciba Foundation Symposium 14. Locomotion of Tissue Cells,* pp. 53–82, Elsevier, New York.

Weston, J. A., 1970, The migration and differentiation of neural crest cells, *Adv. Morphog.* **8**:41.

Wickus, G., Gruenstein, E., Robbins, P. W., and Rich, A., 1975, Decrease in membrane-associated actin of fibroblasts after transformation by Rous sarcoma virus, *Proc. Natl. Acad. Sci. U.S.A.* **72**:746.

Willingham, M. C., 1975, Cyclic AMP and cell behavior, *Int. Rev. Cytol.* **44**:319.

Willingham, M. C., Ostlund, R. E., and Pastan, I., 1974, Myosin is a component of the cell surface of cultured cells, *Proc. Natl. Acad. Sci. U.S.A.* **71**:4144.

Wilson, L., Bryan, J., Ruby, A., and Mazia, D., 1970, Precipitation of proteins by vinblastine and calcium ions, *Proc. Natl. Acad. Sci. U.S.A.* **66**:807.

Wrenn, J. T., 1971, An analysis of tubular gland morphogenesis in chick oviduct, *Dev. Biol.* **26**:400.

Wrenn, J. T., and Wessells, N. K., 1969, An ultrastructural study of lens invagination in the mouse, *J. Exp. Zool.* **171**:359.

Wuerker, R. B., and Kirkpatrick, J. B., 1972, Neuronal microtubules, neurofilaments, and microfilaments, *Int. Rev. Cytol.* **33**:45.

Wunderlich, F., Müller, R., and Speth, V., 1973, Direct evidence for a colchicine-induced impairment in the mobility of membrane components, *Science* **182**:1136.

Yahara, I., and Edelman, G. M., 1973, Modulation of lymphocyte receptor redistribution by concanavalin A, anti-mitotic agents and alterations of pH, *Nature (London), New Biol.-* **236**:152.

Yamada, K. M., and Wessells, N. K., 1973, Cytochalasin B: Effects on membrane ruffling, growth cone and microspike activity, and microfilament structure not due to altered glucose transport, *Devel. Biol.* **31**:413.

Yamada, K. M., Spooner, B. S., and Wessells, N. K., 1970, Axon growth: Roles of microfilaments and microtubules, *Proc. Natl. Acad. Sci. U.S.A.* **66**:1206.

Yamada, K. M., Spooner, B. S., and Wessells, N. K., 1971, Ultrastructure and function of growth cones and axons of cultured nerve cells, *J. Cell Biol.* **49**:614.

Yamada, K. M., Yamada, S. S., and Pastan, I., 1975, The major cell surface glycoprotein of chick fibroblasts is an agglutinin, *Proc. Natl. Acad. Sci. U.S.A.* **72**:3158.

Yang, Y.-Z., and Perdue, J. F., 1972, Contractile proteins of cultured cells. 1. The isolation and characterization of an actin-like protein from cultured chick embryo fibroblasts, *J. Biol. Chem.* **247**:4503.

NOTE ADDED IN PROOF

Several publications which have appeared since the completion of this chapter provide new insight into the biochemical basis of cell movement.

Actin: At least one cellular actin, human platelet actin, has an amino acid sequence different from cardiac muscle actin, and is therefore synthesized by a different gene. This type of diversification would permit separate regulation of actins during development. (Elzinga, M., Maron, B. J., and Adelstein, R. S., 1976, *Science* **191**:94.

Myosin: Myosins also differ in composition, and their expression is altered during myoblast development. (Chi, J. C., Fellini, S. A., and Holtzer, H., 1975,

Proc. Natl. Acad. Sci. U.S.A. **72**:4999). The activity of myosin ATPase can be modulated by the extent of phosphorylation of one of the myosin light chain subunits. (Adelstein, R. S., and Conti, M. A., 1975, *Nature* **256**:597.

Microtubules: Beside the major structural protein, tubulin, isolated microtubules also contain at least two proteins of high molecular weight called microtubule-associated proteins (MAPs). These proteins exist as slender filaments projecting from microtubules, and they may help regulate microtubular assembly (Dentler, W. L., Granett, S., and Rosenbaum, J. L., 1975, *J. Cell Biol.* **65**:237; Murphy, D. B., and Borisy, G. G., 1975, *Proc. Natl. Acad. Sci. U.S.A.* **72**:2696). A new procedure for determining the percentage of tubulin incorporated into microtubules vs. that still soluble in the cytoplasm may permit studies of the changes in microtubular assembly during development (Rubin, R. W., and Weiss, G. D., 1975, *J. Cell Biol.* **64**:42).

Cell-surface proteins: The aggregation-promoting protein originally isolated from cultured neural retina cells can be isolated directly from intact neural retinas, but only at a specific stage in development, strongly suggesting that it plays a major role in development (Hausman, R. E., and Moscona, A. A., 1976, *Proc. Natl. Acad. Sci. U.S.A.* **73**:3594). The major cell surface protein of cultured chick fibroblasts (CSP or LETS protein) increases adhesion of cells to substrata. Reconstitution of the glycoprotein on the surface of neoplastically transformed cells partially reverts morphology, alignment at confluence, and cell surface architecture to a more normal phenotype (Yamada, K. M., Yamada, S. S., and Pastan, I., 1976, *Proc. Natl. Acad. Sci. U.S.A.* **73**:1217; Yamada, K. M., Ohanian, S. H., and Pastan, I., 1976, *Cell* **9**:241).

Other proteins: A troponin-like protein has been isolated from brain, nearly completing the inventory of muscle proteins found in nonmuscle cells. (Fine, R., Lehman, W., Head, J., and Blitz, A., 1975, *Nature* **258**:260). A major cell protein has recently been discovered in association with actin in leukocytes, smooth muscle, and other cell types. "Actin-binding protein" or "filamin" may bind to actin to result in cytoplasmic sol-to-gel transformations and cell movements (Stossel, T. P., and Hartwig, J. H., 1976, *J. Cell Biol.* **68**:602; Wang, K., Ash, J. F., and Singer, S. J., 1975, *Proc. Natl. Acad. Sci. U.S.A.* **72**:4483).

Cytochalasin B: Effects of this drug on actin-containing microfilaments of fibroblasts and other cells were confirmed by immunofluorescence staining using anti-actin antibodies. (Norberg, R., Lidman, K., and Fagraeus, A., 1975, *Cell* **6**:507). Cytochalasin B inhibited the gelation of cell extracts, as well as inhibiting gel formation by purified actin plus actin-binding protein. Cytochalasin may therefore produce its effects on microfilament organization and on cell movement by disrupting the interaction of actin with actin-binding protein. (Hartwig, J. H., and Stossel, T. P., 1976, *J. Cell Biol.* **71**:295; Weihing, R. R., 1976, *J. Cell Biol.* **71**:303).

Alterations in the Metabolism of Glycosaminoglycans and Collagen

10

D. M. KOCHHAR and K. SUNE LARSSON

I. INTRODUCTION

Glycosaminoglycans (GAG, previously acid mucopolysaccharides) and collagen are components of extracellular materials which surround the cells of all multicellular organisms. Extracellular matrix is an integral part of tissues and organs and provides the supportive framework on which the structural and functional organization of the body is based. The study of extracellular macromolecules, and the cells that synthesize them, is important to the developmental biologist who finds intriguing problems in cell differentiation, regulation of biosynthesis and secretion, interaction and functional organization of the secreted products, self-assembly, and pattern formation. Abnormality in any of these events or others, such as degradation, through the interference of intrinsic or extrinsic factors, would be expected to lead to faulty development of an organ.

Glycosaminoglycans of connective tissues are a family of high-molecular-weight compounds distributed in such a manner that specific

D. M. KOCHHAR • Department of Anatomy, School of Medicine, University of Virginia, Charlottesville, Virginia. Present address: Department of Anatomy, Jefferson Medical College, Philadelphia, Pennsylvania.   K. SUNE LARSSON • Laboratory of Teratology, Karolinska Institutet, Stockholm, Sweden.

members predominate in certain locations in the body. Diverse connective tissues and organs such as dermis, cornea, tendons, aortic wall, bone, and cartilage have been investigated as to the chemical structure of GAG. However, cartilage, uniquely endowed with discrete anatomic localization, intracellular specialization, and extracellular matrix, has been a choice tissue for study by developmental biologists, biochemists, and clinicians. Since skeletal defects are among the most commonly described, teratologists should be alert to the consequences of intervention in the development of cartilage, which provides models for most of the skeletal system in the body.

This chapter will attempt to summarize current information about the chemistry and biosynthesis of main groups of GAG, with particular reference to cartilage. Only those aspects are mentioned which are well established and relevant to investigation in teratological studies; more detailed information can be obtained from reviews mentioned in the text. For lack of concrete information about the functional role of GAG, it is instructive to include a discussion of a group of genetic diseases, mucopolysaccharidoses, with disorders of GAG metabolism and storage. A few examples of other hereditary disorders in man and animals are also mentioned. A number of teratogenic agents in other systems such as cell and tissue culture are able to alter the biosynthesis and morphogenesis of extracellular matrix. The role played by these in the origin of malformations, whether through their effect on embryonic GAG metabolism or some other means, is still conjectural. Experimental studies summarized in the last section of this chapter will deal with interference with GAG metabolism by some teratogenic agents and with the possible mechanisms of their action.

Recent evidence on the molecular diversity of collagen molecules among different connective tissues has given new impetus to the study of this fibrillar protein. Since collagen and GAG are intimately associated in the extracellular matrices, and since factors that influence the metabolism of one may alter the chemical or physical properties of the other, some information about collagen is also included.

II. GLYCOSAMINOGLYCANS

A. Structure

The glycosaminoglycans (GAG) consist of linear carbohydrate chains covalently linked to a protein core to form macromolecules termed proteoglycans. In proteoglycans, the carbohydrate moiety is the dominant feature of the macromolecule. In contrast, glycoproteins (including collagen) are molecules in which only monosaccharides or oligosaccharides are covalently bound to protein chains, hence the protein moiety dominates.

The names, general distribution, and component moieties of various GAG are summarized in Table 1. The principal GAG of cartilage are chondroitin 4-sulfate, chondroitin 6-sulfate, and keratan sulfate. Space-filling tissue fluids such as vitreous humor and synovial fluid are rich in hyaluronic acid. Embryonic tissues including skin and umbilical cord also have a high content of hyaluronic acid, and some hyaluronic acid is also present in cartilage. The carbohydrate chains of all GAG, with the exception of keratan sulfate, consist of alternating units of hexosamines and uronic acids (Figure 1). The hexosamine moieties are either glucosamine or galactosamine, and the uronic acid moieties are either glucuronic acid or iduronic acid. In keratan sulfate the repeating disaccharide unit contains galactose instead of uronic acid. The GAG are polyanions because of numerous carboxylate and/or sulfate groups present in the carbohydrate chains. The polyanionic character of proteoglycans is important in their interactions with other molecules and ions of the extracellular matrix (Ogston, 1970). With the exception of hyaluronic acid, all GAG are sulfated; the sulfate group is present in ester linkage at carbon 4 (C-4) or C-6 of hexosamine residues, hence the designation such as chondroitin 4-sulfate or chondroitin 6-sulfate. Some varieties of GAG may

Table 1. Glycosaminoglycans of Vertebrate Connective Tissues

Name[a]	Distribution	Hexosamine	Uronic acid	Sulfate
Hyaluronic acid	Vitreous body, synovial fluid, skin, umbilical cord	Acetylglucosamine	Glucuronic	−
Chondroitin	Cornea	Acetylgalactosamine	Glucuronic	−
Chondroitin 4-sulfate (chondroitin sulfate A)	Cartilage, bone, skin, cornea	Acetylgalactosamine	Glucuronic	+
Chondroitin 6-sulfate (chondroitin sulfate C)	Cartilage, skin, umbilical cord	Acetylgalactosamine	Glucuronic	+
Dermatan sulfate (chondroitin sulfate B)	Skin, aorta, tendons, umbilical cord	Acetylgalactosamine	Iduronic or glucuronic	+
Heparan sulfate (heparitin sulfate)	Aorta, lung	Acetylglucosamine	Glucuronic or induronic	+
Heparin	Liver, skin, lung	Acetylglucosamine	Glucuronic or induronic	+
Keratan sulfate I (keratosulfate)	Cornea	Acetylglucosamine	Galactose	+
Keratan sulfate II	Cartilage, nucleus pulposus	Acetylglucosamine	Galactose	+

[a]The nomenclature of Jeanloz (1960); older names in parentheses.

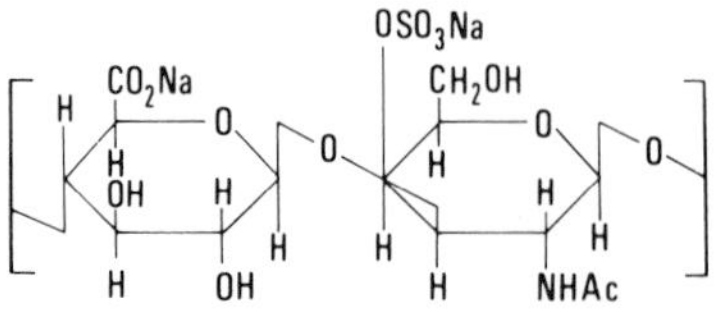

Fig. 1. Repeating unit of sodium salt of chondroitin 4-sulfate. D-Glucuronic acid is linked to N-acetylgalactosamine via β1-3 glycosidic linkage.

bear more than one sulfate group per hexosamine residue. In keratan sulfate a sulfate group is present at C-6 of both monosaccharide residues.

A varying number of polysaccharide chains of GAG are attached to a single protein core to give rise to a complex proteoglycan macromolecule with a molecular weight as high as 4 million (Mathews, 1965, 1967a). The average cartilage proteoglycan molecule contains about 100 chondroitin sulfate chains and 30–60 keratan sulfate chains linked the same core protein (Mathews and Lozaityte, 1958; Partridge *et al.*, 1961). Rodén (1970a) and his colleagues have established that in cartilage chondroitin sulfate chains are linked to the core protein by means of a sequence of galactose–galactose–xylose; the xylose residue is covalently linked to serine residues of the core protein (Figure 2). A

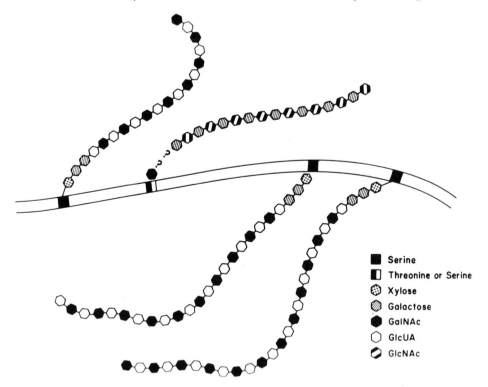

Serine
Threonine or Serine
Xylose
Galactose
GalNAc
GlcUA
GlcNAc

Fig. 2. Structure of cartilage proteoglycan. The protein backbone carries both chondroitin sulfate and keratan sulfate chains. GalNAc, N-acetylgalactosamine; GlcUA, glucuronic acid; GlcNAc, N-acetylglucosamine. (Reproduced by permission from Levitt and Dorfman, 1974.)

similar linkage region is found in all other sulfated GAG except keratan sulfate (Figure 2); here, the linkage occurs to asparagine in the case of keratan sulfate I (in cornea), and to serine, threonine, or glutamic acid in keratan sulfate II (in skeletal tissue).

To understand the structure of proteoglycans, the core protein needs special attention since it provides a backbone from which polysaccharide chains extend in all directions like bristles on a brush (Hascall and Sajdera, 1970). The core protein is noncollagenous in nature and has a molecular weight of 200,000. Mathews (1971) has proposed that it is a single peptide of 2000 amino acids and that chondroitin sulfate chains are distributed along the core protein in closely spaced groups, primarily as pairs, with fewer than 10 amino acid residues between individual chains (Figure 2). A segment of unsubstituted 35 amino acid residues intervenes between these groups. Complete analysis of amino acid sequence is not yet available and the validity of the 45-unit amino acid sequence as the true repeating unit is not yet established. More recent evidence has revealed a further substructure of the core protein (Rosenberg et al., 1975b; Hascall and Heinegård, 1975). In addition to the region where the polysaccharide chains are attached, the core protein has a binding site at one of its ends where it is attached to hyaluronic acid of the cartilage matrix. The linkage between the globular end of the core protein and the hyaluronic acid molecule is noncovalent and is stabilized by two low-molecular-weight proteins called link proteins (Gregory, 1973; Hascall and Heinegård, 1975). Chemical evidence has been augmented by electron-microscopic studies in formulating a molecular model of proteoglycan aggregate of native cartilage (Figure 3). It is characterized by a filamentous backbone provided by hyaluronic acid with proteoglycan subunits arising laterally at regular intervals from the opposite sides of the hyaluronic acid chain (Rosenberg et al., 1975a,b).

B. Biosynthesis and Degradation

Advances in modern methodology have yielded rather precise information about GAG metabolism of embryonic and adult tissues which has been the basis for considerable progress in unraveling some aspects of processes of cell differentiation and in understanding the nature of cellular changes associated with some pathologic states. The biosynthetic pathways for most GAG have been summarized in several review articles (Thorp and Dorfman, 1967; Stoolmiller and Dorfman, 1969; Rodén, 1970b; Levitt and Dorfman, 1974; Lamberg and Stoolmiller, 1974). Glucose is the precursor of all glycosyl residues of GAG by way of uridine nucleotide sugars. Several monosaccharide interconversion reactions yield the appropriate UDP derivatives from which the sugars are sequentially transferred to the growing GAG chains (Figure 4). The sequential order of sugars, unlike that of proteins and nucleic acids coded by a template, is specified by individual enzymes termed glycosyl trans-

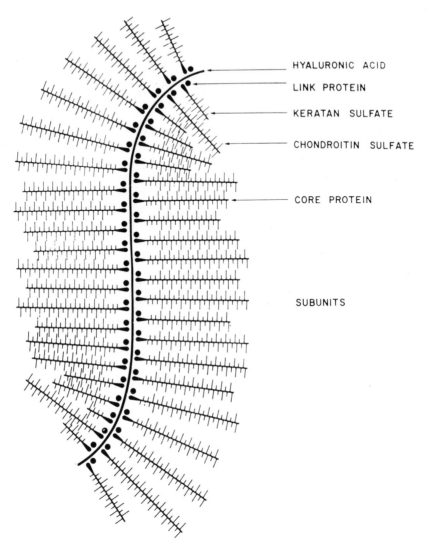

HYALURONIC ACID

LINK PROTEIN

KERATAN SULFATE

CHONDROITIN SULFATE

CORE PROTEIN

SUBUNITS

Fig. 3. Diagrammatic representation of the structure of proteoglycan aggregate. (Reproduced by permission from Rosenberg *et al.*, 1975b.)

ferases. During biosynthesis of chondroitin sulfate in cartilage, coordinated catalytic action of six glycosyl transferases and a sulfotransferase is required for the initiation and elongation of the GAG chains (Figure 4). Studies on embryonic chick cartilage have shown that the synthesis of the protein backbone must precede the initiation of GAG chain formation (Telser *et al.,* 1965). Horwitz and Dorfman (1968) have reported that the glycosyl trans-ferases are usually found as multienzyme complexes bound to membranes. This and other investigations on the intracellular sites of GAG synthesis

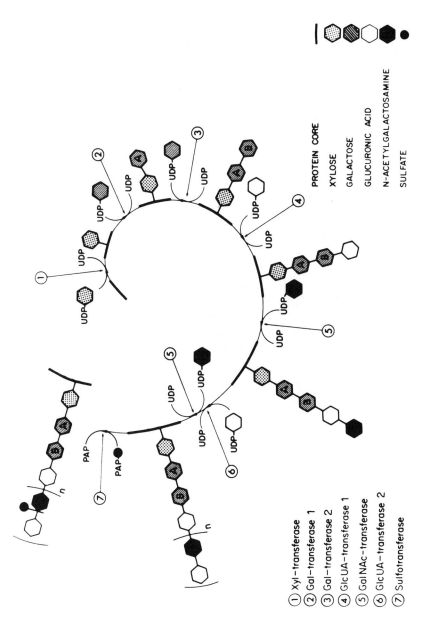

Fig. 4. Biosynthesis of chondroitin sulfate. UDP, uridine-5'-diphosphate; GlcUA, glucuronic acid; GalNAc, N-acetylgalactosamine; PAP, 5'-adenosine-3'-phosphate. (Reproduced by permission from Levitt and Dorfman, 1974.)

suggest that the protein backbone and glycosyl transferases are synthesized by membrane-bound ribosomes in the rough endoplasmic reticulum. The synthesis of the linkage region and the GAG chain formation is initiated in the cisternae of the rough endoplasmic reticulum and continues while the macromolecule progresses into the smooth endoplasmic reticulum and Golgi apparatus. Chain elongation and sulfation occur largely in the Golgi vesicles and vacuoles (Godman and Lane, 1964; Neutra and Leblond, 1966; Revel, 1970; Kimata *et al.*, 1971). Proteoglycan is released by the secretion vacuoles into the extracellular matrix. Although several factors are known to affect the synthesis and release of GAG in experimental systems (Bosmann, 1968; Nevo and Dorfman, 1972), including the matrix macromolecules themselves, the precise regulatory mechanisms operating under *in vivo* situations are only beginning to be elucidated.

The total GAG content of developing tissues is regulated not only by synthesis but also by degradatory processes. The process of degradation of proteoglycan complexes, as judged by the degradatory components in the urine, may occur by the action of lysosomal enzymes such as proteases, glycosidases, and sulfatases; only some of these have been known to be operative in physiological conditions.

From the current concepts of the structure of proteoglycan complexes, the digestion of protein backbone by proteases may be expected to lead to marked changes in the structure of the molecule. Studies on the effects of papain on rabbit ear cartilage (Thomas, 1956) and the effect of excess retinol on the release of cathepsin D from chick embryo cartilage and consequent dissolution of the cartilage matrix (Weston *et al.*, 1969) are certainly suggestive of such a mechanism. However, the role of these and other proteases in actual degradation of proteoglycan is still speculative. Similarly, it is known that sulfatase activity is present in human skin fibroblasts *in vitro* (Dorfman and Matalon, 1969) and that desulfation of GAG is involved in the degradatory processes; however, from current evidence it appears such enzymes do not attack the intact molecule but degrade partially degraded products.

Glycosidases, lysosomal hydrolytic enzymes that degrade GAG chains, are of particular interest since a lack or inactivity of some of them is associated with a number of mucopolysaccharide storage disorders (see below). Hyaluronidase, historically known as the "spreading factor," is widely distributed in the tissues of vertebrate and invertebrate animals, including mammals. Testicular hyaluronidase is an endoglycosidase, and it extensively degrades hyaluronic acid and chondroitin 4- and 6-sulfate by splitting bonds within the polysaccharide chains adjacent to glucosamindic residues. Dermatan sulfate is only partially degraded since it is a hybrid molecule containing both iduronic acid and glucuronic acid. In contrast to a broad specificity expressed by testicular hyaluronidase, leech head hyaluronidase specifically degrades only hyaluronic acid (Weissmann, 1955). Other glycosidases, termed exoglycosidases, further degrade the oligosaccharides released through the action of hyaluronidase. The hydrolases reportedly detected so far are: βglucuronidase, *N*-acetylhexosaminidase, α-L-iduronidase, β-galactosidase, and β-xylosidase (see review by Lamberg and Stoolmiller, 1974).

Recent discoveries of two degradative enzymes, chondroitinase-ABC isolated from *Proteus vulgaris* and chondroitinase-AC from *Flavobacterium heparium* (Yamagata *et al.*, 1968), have greatly facilitated the detection and identification of trace amounts of individual GAG components in mixtures of native complexes as present in most tissues. Chondroitinase-ABC readily cleaves the acetylgalactosaminyl bonds of chondroitin 4-sulfate, chondroitin 6-sulfate, and dermatan sulfate but degrades hyaluronic acid and chondroitin at an extremely reduced rate. It does not attack keratan sulfate, heparin, or heparan sulfate. Chondroitinase-AC exhibits similar action but leaves dermatan sulfate intact. Further, the disaccharides released through the action of these enzymes are identified by chromatographic and electrophoretic analyses and reflect the repeating units of the parent molecules.

III. COLLAGEN

In the developing embryo the importance of collagen is reflected by its presence at very early stages of differentiation. Its participation in the macromolecular organization of the extracellular matrix through its association with glycosaminoglycans and its role in cellular interactions and tissue morphogenesis are some of the aspects which are being investigated now. Recent investigations have shown that vertebrate collagens are a heterogeneous population of molecules representing at least five different gene products. The current information is reviewed by Nigra *et al.* (1972), Trelstad (1973), Miller and Matukas (1974), Serafini-Fracassini and Smith (1974), and Martin *et al.* (1975).

In the formation of microfibrils, the collagen molecules are polymerized in parallel alignment with a quarter stagger pattern resulting in the characteristic axial repeating period of 650–700Å. A considerable amount of information is available which correlates the morphological and biochemical evidence regarding synthesis and assembly of the collagen molecule. It is now known that besides the cells commonly associated with collagen synthesis such as fibroblasts, chondrocytes, and osteocytes, other cell types are capable of synthesis under certain conditions, including embryonic neural tube (Trelstad *et al.*, 1973), notochord (Linsenmeyer *et al.*, 1973), and embryonic corneal epithelium (Trelstad *et al.*, 1974). Current evidence indicates that the molecular assembly and helix formation of the nascent α-chains occur on the polyribosomes.

A number of chemical modifications which occur in the polypeptide chains of collagen subsequent to assembly of the primary structure and prior to deposition of the collagen fiber in the extracellular matrix are summarized in Figure 5. There are at least six enzymes involved in the performance of post-translational modifications of the collagen α-chains after formation of the

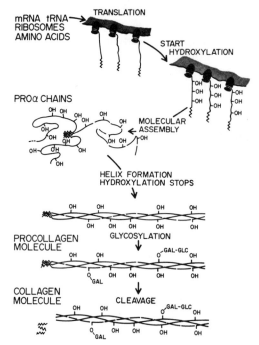

Fig. 5. Biosynthesis of collagen. (Reproduced by permission from Martin *et al.*, 1975.)

peptide bonds (Table 2). It has been demonstrated that the biosynthetic precursor of the α-chain contains a nonhelical extension at either end of the peptide so that the precursor, termed pro-α-chain, has a molecular weight approximately 15–20% greater than that of α-chain (see review by Martin *et al.*, 1975). The modifications undergone by pro-α-chains prior to their release from the ribosomes include hydroxylation of selected proline residues by

Table 2. Enzymes and Events in Postsynthetic Modification of Collagen

Event	Enzyme
Hydroxylation	Prolyl hydroxylase, lysyl hydroxylase
Molecular assembly	—
Helix formation	—
Glycosylation	Glucosyl transferase, galactosyl transferase
Procollagen to collagen conversion	Procollagen peptidase
Cross-linking	Lysyl oxidase

prolyl hydroxylase and hydroxylation of selected lysine residue by lysyl hydroxylase. The glycosylation of some of the hydroxylysyl residues by glucosyl and galactosyl transferases is an event which occurs subsequent to hydroxylation. The three pro-α-chains are assembled into triple helical procollagen, probably immediately following hydroxylation, through linkage by interchain disulfide bonds. It is postulated that the nonhelical extensions of procollagen α-chains may play a key role in alignment of the chains for helix formation. It is considered that the hydroxylation of proline residues lends stability to the helicity of the molecule. Although the molecular mechanism by which hydroxyproline stabilizes α-chains is not known, evidence now indicates a role of prolyl hydroxylase in controlling cellular biosynthesis and transport of collagen (Olsen *et al.*, 1975). Using immunoferritin techniques, these authors have shown that procollagen is found in both the cisternae of the endoplasmic reticulum and the large Golgi vacuoles of embryonic tendon fibroblasts. The current view of the manner in which collagen is transported by Golgi vacuoles and discharged into the extracellular space by exocytosis is similar to that held for the transport and secretion of GAG (Weinstock and Leblond, 1974).

Two additional enzymes are required before collagen attains its final fibrillar form. Since the nonhelical polypeptide extensions are not present in normal collagen fibrils, they must be removed to give rise to the definitive monomeric molecule. It is now established that this cleavage is performed by a neutral protease, procollagen peptidase (Bornstein *et al.*, 1972), several minutes after the extrusion of procollagen from the cell. The monomeric molecules are progressively assembled into primary filaments and fibrils through intermolecular covalent cross-links. Although several types of cross-links have been detected in collagen, the initial cross-linking involves lysyl residues located in the amino terminal region of α-chains. The ε-amino group of specific lysyl residues is oxidatively deaminated by lysyl oxidase to α-aminoadipic-δ semialdehyde (allysine), and the allysyl residues on adjacent molecules spontaneously condense with one another to form cross-links (Piez, 1968; Pinnell and Martin, 1968). The cross-links of newly formed collagen are acid-labile, but as the fibril formation and maturation progresses these cross-links become more acid-stable. Pinnell and Martin (1968) have shown that lathyrogenic agents, such as β-aminopropionitrile (β-APN), irreversibly inhibit lysyl oxidase and thus prevent aldehyde formation required for cross-linking.

As a component of extracellular matrix, collagen is synthesized simultaneously by cells that produce GAG. It is beginning to be appreciated that the differences between diverse tissues such as tendon, bone, cartilage, basement membrane, and dermis must be based not only on qualitative and quantitative differences in collagen and GAG but also on their interaction. A small number of hereditary disorders involving collagen metabolism are reportedly associated with congenital bony and other abnormalities in the infant; these are briefly described in a later section.

IV. HEREDITARY DISEASES

A. Mucopolysaccharidoses

These are a group of hereditary diseases characterized by increased intralysosomal storage and excessive urinary excretion of certain GAG. An appreciation of these mucopolysaccharidoses is important for teratologists since they may hold the key to future knowledge necessary to study cause-and-effect relationships between GAG metabolism and structural/functional developmental abnormalities. In the Hurler's syndrome the newborn shows normal development for some months, after which symptoms begin to appear and deterioration becomes progressive (Neufeld and Fratantoni, 1970). The head becomes peculiarly large with wide set eyes, large lips, and flat nose bridge. Other defects such as opacity of the cornea, deterioration of hearing, and growth and mental retardation gradually become prominent. There is widespread skeletal involvement with wide ribs, stiff joints, and anomalous development of the vertebrae and long bones. There is enlargement of liver and spleen, and cardiovascular abnormalities in the heart valves and walls of major blood vessels. The affected children usually die in the second decade. Although a discussion of the clinical findings in each of these diseases is beyond the scope of this chapter, a list of the known mucopolysaccharidoses and an enzyme deficiency associated with each is given in Table 3 (see summary by Rennert, 1975). Certain other hydrolytic enzymes are found to be elevated in the livers of some patients. It should be noted that the deficiencies of enzymes listed in Table 3 do not alone explain the failure of degradation of dermatan sulfate and heparan sulfate which are found in the more common mucopolysaccharidoses (such as Hurler, Hunter, and Sanfilippo), and in fact some of the clinical manifestations may not be a result of the primary defect but of some far removed interlinked metabolic disturbance.

B. Other Hereditary Diseases

1. Man

a. Niemann-Pick Disease.   This is a genetically determined disorder characterized by an extensive involvement of the nervous system. The primary lesion is a deficiency of the enzyme phospholipase C that specifically cleaves sphingomyelin (Brady, 1968). Hence, the tissue concentration of sphingomyelin is raised. Ganglioside content is found to be elevated in both white and gray matter (review by Schettler and Kahlke, 1967). It has now been reported (Brunngraber et al., 1973) that brain GAG also shows a twofold elevation.

Table 3. Classification of Genetic Mucopolysaccharidoses and Associated Defects[a]

Type	Designation	Major clinical features	Excessive GAG accumulated or excreted	Enzymatic defect
I	Hurler	Skeletal, visceral and CNS involvement, corneal clouding	Dermatan sulfate/heparan sulfate (70:30)	α-Iduronidase
	Scheie	Skeletal involvement, particularly joints, corneal clouding, cardiac involvement, no mental retardation	Dermatan sulfate/heparan sulfate (70:30)	α-Iduronidase
II	Hunter	Milder but similar to Hurler except corneal clouding	Dermatan sulfate, heparan sulfate	Sulfoiduronate sulfatase
III	Sanfilippo	CNS involvement, minimal skeletal involvement	Heparan sulfate	Type A: Heparan sulfate sulfatase, type B: N-acetylglucosaminidase
IV	Morquio	Severe skeletal deformity, corneal clouding, cardiac involvement	Keratan sulfate	N-acetylhexosamine sulfate sulfatase
V	Maroteaux–Lamy	Skeletal deformity, corneal involvement, no mental retardation	Dermatan sulfate	Arylsulfatase B
VI	β-Glucuronidase deficiency	Hepatosplenomegaly, skeletal involvement, mental retardation	Dermatan sulfate	β-Glucuronidase

[a]Adapted from Rennert (1975).

b. **Marfan's Disease.** This is an autosomal dominant disease character-ized by defects in connective tissue including loose-jointedness, long extrem-ities, dislocated lenses, aortic aneurism, and mitral regurgitation (McKusick, 1972). Increased amounts of salt-soluble collagen are found in the skin (Lait-inen *et al.*, 1968) and fibroblast cultures (Macek *et al.*, 1966) of Marfan patients. It has been reported that the increased amounts of hyaluronic acid which accumulate in cultures of Marfan fibroblasts are due to a greater rate of synthesis as opposed to a decreased rate of breakdown (Matalon and Dorfman, 1968; Lamberg and Dorfman, 1973).

Metabolic changes of urinary GAG have been reported in two other genetic diseases, Weber Christian disease (Murata *et al.*, 1973) and pachyder-moperiostosis (Yamato *et al.*, 1974). A beginning is also being made in the ultrastructural characterization of generalized human skeletal dysplasias. In a patient with pseudoachondroplastic dwarfism, Cooper *et al.* (1973) reported a storage disorder in the rough endoplasmic reticulum. In the chondrocytes of epiphyseal cartilage, these authors observed inclusions and whorls of alternate electron-dense and electron-lucent layers within the rough endoplasmic re-ticulum. Similarly, in a case of metaphyseal dysostosis, Cooper and Ponseti (1973) revealed the cisternae of the endoplasmic reticulum of chondrocytes to be markedly dilated by a granular precipitate. Further characterization of these inclusions by cytochemical and fractionation methods is needed.

2. Animals

There are a number of hereditary defects of skeleton in experimental animals which provide good material for studying altered GAG metabolism (Grüneberg, 1965). However, only a few of these have been investigated in metabolic studies. Mathews (1967b) analyzed the GAG content in sternal and limb cartilages of micromelic chick embryos belonging to three types of recess-ive lethal mutations. Only one type, nanomelia (first described by Landauer, 1965) showed clear-cut differences in the amount of GAG from that present in normal embryos. It was shown that nanomelic cartilage contained about one tenth as much chondroitin sulfate of normal cartilage. Collagen content seemed a little greater than normal. GAG content of other tissues, such as skin and vitreous humor, was normal. Recently, Palmoski and Goetinck (1972) have reported that the biosynthesis of the major proteochondroitin sulfate component of normal chick chondrocytes in culture is absent from cultures of nanomelic chondrocytes.

In mutant mouse embryos homozygous for the gene congenital hy-drocephalus $(ch+/ch+)$, a 60% reduction in the total GAG has been reported in the sternocostal cartilage (Breen *et al.*, 1973). It was interesting that the rela-tive proportion of individual GAG was not altered, and the heterozygous embryos had a higher (up to 114% of control) GAG content. While the homozygous embryos showed a defective sternum, in heterozygous embryos calcification was only delayed in the sternum.

3. Collagen-Related Diseases

Although several heritable disorders of connective tissue are on record, such as Ehlers–Danlos and Marfan's syndromes (see review by McKusick and Martin, 1975), the first evidence of a molecular defect in collagen metabolism associated with a clinical syndrome was reported by Pinnell *et al.* (1972) in two sisters who showed severe scoliosis, hyperelastic skin, and recurrent dislocations and hyperextensibility of joints. Amino acid analysis of collagen from dermis and several other locations revealed a marked decrease in hydroxylysine content. Since the content of lysine (which is the only precursor of peptidyl hydroxylysine) was found to be increased in collagen, a low activity of lysyl hydroxylase was postulated as the most likely explanation. It is not yet known precisely how a deficiency of hydroxylysine leads to the observed clinical manifestations. It is possible that the lack of hydroxylysine-containing cross-links would make collagen less stable and may result in structural abnormalities.

Another congenital abnormality, termed dermatosparaxis, has been detected in cattle and is characterized by a lack of procollagen peptide (Lenaers *et al.*, 1971). The afflicted calves have poor dermal tensile strength; the individual collagen filaments appear to be normal, but their packing within the fibrils is impaired. The calves show no other gross congenital anomalies. It is probable that the retention of polypeptide extensions on the collagen molecules may interfere with precise molecular alignment during fiber formation.

V. FUNCTIONAL ROLE OF EXTRACELLULAR MATRIX IN DEVELOPING SYSTEMS

The detailed information about the molecular structure of matrix molecules has not revealed the physiologic significance of the different varieties of GAG and collagens in developing or adult tissues, although some information is available which may be important for deducing the functional role of these polymers in development. Studies on skin from developing human (Breen *et al.*, 1970, 1972), pig (Loewl and Meyer, 1958), and frog tadpole (Lipson and Silbert, 1968) have generally indicated that the concentration of hyaluronic acid predominated in early stages and decreased as development progressed. In human and other species the total content of both hyaluronic acid and chondroitin sulfate decreased during growth and aging in the skin (Schiller and Dorfman, 1960; Smith *et al.*, 1965). Robinson and Dorfman (1969) have reported that as embryonic cartilage matures, its content of chondroitin 4-sulfate gradually diminishes while the synthesis of chrondroitin 6-sulfate increases. The increased chondroitin 6-sulfate content of aging cartilage is accompanied by a similar increase in keratan sulfate. A functional role of these components, however, is difficult to ascertain from purely biochemical determinations. From physiological and biophysical studies some evidence has suggested that, while chondroitin sulfate and dermatan sulfate play a part in the formation of collagen fibrils from collagen monomer (Lowther and Toole, 1968; Lowther and Natarajan, 1972), hyaluronic acid is associated with the dispersal of collagen fibrils (Flint, 1972). Special arrangement in the cornea of collagen fibrils in a three-dimensional matrix may be regulated by the associated proteoglycans, thus resulting in transparency of the cornea (Maurice, 1957).

The following discussion points out some instances wherein developmental significance of matrix macromolecules and their interactions is under investigation.

A. Epithelial–Mesenchymal Interactions

Cellular interactions have been extensively studied in organs composed mainly of epithelia and mesenchyme, such as skin, glands, lung, urogenital organs, tooth, and limb. Extracellular matrix materials at the epithelial–mesenchymal interface were implicated in organogenetic tissue interactions after studies suggested that direct tissue contact was not required for the interaction to take place (Grobstein, 1967). Although it is probable that informational molecules from one tissue may be transferred to the other through the matrix (Slavkin, 1972), the evidence to date points to the occurrence of a permissive interaction in which one tissue merely stimulates the other to attain its intrinsic developmental potentialities. Connective tissue macromolecules have been proposed as mediators of such tissue interactions (Konigsberg, 1970; Cohen and Hay, 1971; Bernfield and Banerjee, 1972; Trelstad *et al.*, 1973). A number of investigators have studied the histochemistry and fine structure of interactants and their interface during morphogenesis of glands (Bernfield *et al.*, 1973), limb (Kelley, 1973; Ede *et al.*, 1974), and somite (Cohen and Hay, 1971; O'Hare, 1973). Biochemical characterization of morphogenetically active interface materials has only just begun (Trelstad *et al.*, 1973, 1974), and promising results are expected which may throw light on enigmatic tissue interactions. The role of teratogenic agents in disturbing epithelial–mesenchymal interactions is still unstudied and needs immediate attention.

B. Hyaluronate, Hyaluronidase, and Morphogenesis

Toole and co-workers (see review, 1972) have demonstrated a unique sequence of biochemical events, involving GAG, which occurs during morphogenesis in several embryonic regions such as limb, cornea, axial skeleton, and brain. These studies show that during the early morphogenetic phase of development of an organ, such as in limb bud of chick embryo, the mesodermal cells synthesize hyaluronic acid. This phase is characterized by proliferation and absence of aggregation of cells in the blastema. It is followed by phases during which the cells aggregate and become differentiated into cartilage cells. It is shown that cell aggregation begins upon the removal of hyaluronate by endogenous hyaluronidase, followed by accelerated synthesis of chondroitin sulfate. An analogous series of steps involving hyaluronate and hyaluronidase accompanies migration of corneal fibroblasts and neuronal cells and their subsequent differentiation in the chick cornea and brain, respectively. The evidence thus derived has lead to the postulate that hyaluronate inhibits certain cell interactions and encourages cell migration. These processes are reversed upon enzymatic removal of hyaluronate, and this step may be operating under hormonal (or other) regulation so that temporal control over developmental events is exercised. It has now been confirmed by several authors that nanogram quantities of purified hyaluronate block chondrogenesis in cultures of chondrogenic cells (Toole *et al.*, 1972; Solursh *et al.*, 1974). The evidence that addition of exogenous chondromucoprotein to cultures of chondrocytes stimulates production of more chondromucoprotein (Nevo and Dorfman, 1972) also supports the postulate. The precise mechanism by which these GAG influence cellular behavior is still obscure, but an active role for them in morphogenesis is certainly evident.

C. GAG–Collagen Interactions

The extracellular matrix of cartilage has been a favorite site to visually resolve the macromolecular relationships between collagen and GAG. It has long been recognized that the collagen fibrils differ in size in cartilages of different species; larger fibrils, 500–600 Å diameter, are seen in ox epiphyseal plate while rabbit epiphyseal plate has smaller fibrils of less than 200 Å diameter (Smith, 1970). While larger fibrils show the characteristic band pattern of collagen, smaller fibrils lack this pattern and show a periodicity of about 210 Å. Matukas *et al.* (1967) observed that in chick embryo there is an abrupt change from large-fibril articular cartilage to small-fibril epiphyseal cartilage in the same cartilage model. Since differences in GAG composition and metabolism also exist between these two regions of chick embryo cartilage (Oohira *et al.*, 1974; Kimata *et al.*, 1974), the size of collagen fibrils may bear a relationship to their interactions with proteoglycan macromolecules.

Most of the data obtained thus far support the view that the organization of collagen fibrils in tissues is influenced by the amounts and kinds of proteoglycan present in the tissues (Toole and Lowther, 1968; Oegema *et al.*, 1975); small amounts of proteoglycan monomer or proteoglycan aggregate markedly retard the precipitation of collagen fibrils from solutions. Since such interactions have been carried out in tissue extracts only, it is not yet possible to determine if similar interactions occur between native macromolecules *in situ*.

Routine preparative techniques for electron microscopy using uranyl acetate and lead citrate stains reveal a considerable amount of material in the interfibrillar spaces of the matrix (Figure 6). The most prominent components are particles termed matrix granules, about 700 Å diameter, many of which are joined to neighboring particles by 30-Å thread-like filaments. The matrix granules and filaments are suspended in the matrix and show frequent alignment along collagen fibrils. Cytochemical and other studies employing enzymatic digestion have shown that the matrix granules consist principally of proteoglycan complexes (Matukas *et al.,* 1967; Smith, 1970). Serafini-Fracassini and Smith (1974) have proposed that proteoglycan molecules, which are perhaps dispersed evenly *in situ,* attain their clumped particulate image in electron micrographs artifactually by shrinkage of the noncollagenous proteins of the matrix during fixation and tissue preparation; thus the proteoglycan complexes are molded into matrix granules by entrapment of GAG chains within a stellate reticulum. Bismuth nitrate staining of the matrix granules reveals that each is an aggregate of 20–25 smaller granules, about 30 Å in diameter (Serafini-Fracassini and Smith, 1966). In accordance with the model proposed by Mathews and Lozaityte (1958) and Mathews (1965), one 30-Å granule probably represents a doublet composed of a pair of chondroitin sulfate macromolecules. The filaments observed in association with the matrix granules may be hyaluronic acid chains (Rosenberg *et al.,* 1975a,b).

An interesting observation concerning the orientation of matrix granules at the surface of developing chondrocytes was recently reported in the embryonic mouse limb bud (Kochhar *et al.,* 1974). The matrix granules were seen as beads strung along short, individual threads, mostly oriented perpendicular to the cell surface projecting into the lacuna (Figure 7). This configuration was also observed within the cell in large vacuoles. At higher magnification a fibrous component which strings these granules together was observed (Figure 8). The granules were 200–300 Å apart from each other. A particulate component that was revealed within the large matrix granule could be a precipitation artifact; however, it was found only in association with the granule. The beaded-string formations were not observed away from the cell surface, and the granules associated with these formations were the largest in the matrix, each measuring about 600 Å in diameter. It was interesting that the size of the matrix granules decreased as one viewed the matrix further from the cells. Recently Campo and Phillips (1973) and Eisenstein *et al.* (1973) have also shown that the matrix granules are largest in close proximity to the chondrocytes, and that they are partially resistant to dissociative extraction by 4 M guanidinium chloride—an extraction that removed 85% of the total proteoglycans of the cartilage. It has been suggested that these large, newly extruded, pericellular proteoglycan granules are structurally or metabolically different from more mature, smaller granules of the interterritorial matrix.

Although most electron-microscopic observations reveal that a definite physical association between proteoglycan aggregates and collagen fibrils exists in the extracellular matrix, the chemical nature of this association is far

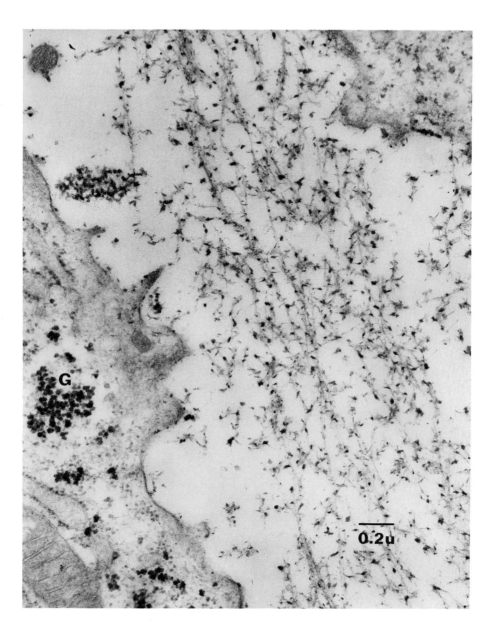

Fig. 6. Cartilage matrix of organ cultured mouse limb bud. Collagen fibrils, matrix granules, and threadlike filaments are suspended in the intercellular space (see text). Intracellular large glycogen particles (G) are prominent. Preparation fixed in glutaraldehyde and osmium tetroxide, stained with uranyl acetate and lead citrate (×72,700, reduced 36% for reproduction).

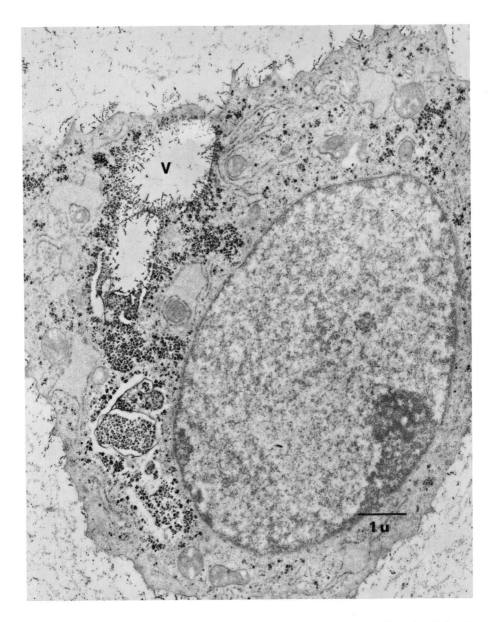

Fig. 7. Developing chondrocyte of embryonic mouse limb bud. Matrix granules in beaded string formation are seen on the cell surface as well as within large vacuoles (V) of the cell. Glycogen granules and dilated cisternae of rough endoplasmic reticulum are prominent. Preparation similar to that for figure 6 (×21,000, reduced 36% for reproduction).

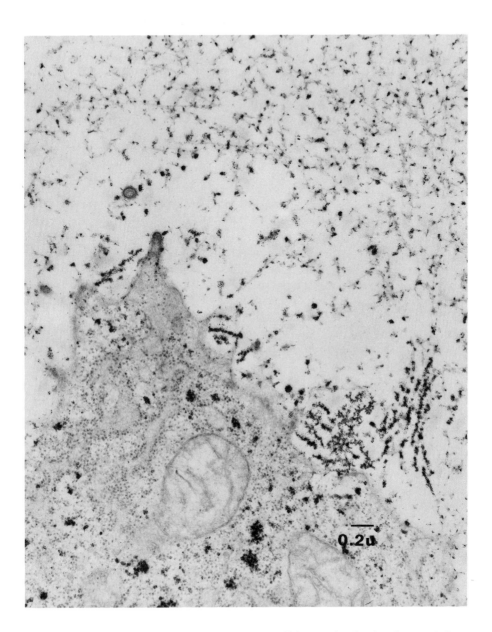

Fig. 8. Higher magnification of cell surface and extracellular matrix of a chondrocyte similar to one shown in Figure 7. See text for explanation (×49,000, reduced 36% for reproduction).

from clear. Greenwald *et al.* (1975) have recently demonstrated a strong ionic interaction between collagen and the core protein of the proteoglycan.

From the standpoint of embryogenesis and teratogenesis, more important than the above considerations is the actual role that GAG and collagen play in the three-dimensional organization of organs and tissues. It is here that a lack of information in the literature is startling. Two organogenetic systems are briefly described below where correlations between morphogenesis of an embryonic region and extracellular materials have been sought.

1. Limb Development

A large literature has accumulated on the initiation of GAG and collagen synthesis and its temporal relationship to chondrogenesis in embryonic limb bud and somites (see review by Levitt and Dorfman, 1974). For the present purpose, however, emphasis is in the response of overall morphogenesis of the limb to factors which alter the metabolism of GAG and collagen, and the possibility that the experiments mentioned may clarify the action of teratogenic agents in the origin of congenital defects. When limb buds of mouse embryos are excised at early stages of development and explanted in culture in media to which appropriate inhibitory agents are added, the developmental alteration reflects to a degree the morphogenetic role of the missing component (Kochhar, 1975). In two such experiments the authors compared the effects on limb morphogenesis of a proline analog, L-azetidine, and of a glutamine analog, DON (6-diazo-5-oxo-L-norleucine) which inhibits GAG synthesis (Aydelotte and Kochhar, 1972, 1975). It was found that both agents retarded limb development in culture, but the structure and shape of chondrified skeletal elements were different in each case (Figure 9). L-Azetidine delayed chondrogenesis, and the cartilage which subsequently appeared failed to elongate and became swollen, soft, and malformed. No collagen fibrils were observed in the cartilage matrix, but in alcian-blue-stained histological preparations and in electron micrographs some proteoglycans were demonstrated. On the other hand DON-treated limb buds acquired elongated cartilage elements but the matrix remained unstained after alcian blue (Aydelotte and Kochhar, 1975); the chondrocytes remained close together and electron microscopy revealed abundant collagen fibrils in the intercellular spaces (Kochhar *et al.*, 1976). Although conclusions from these experiments are tentative since L-azetidine and DON are also capable of interfering, respectively, with protein and purine metabolism in other systems, it can be argued that collagen and GAG played different roles in the attainment of form by the limbs during morphogenesis.

2. Orofacial Development

The extracellular components of mesenchyme in the early morphogenesis of the face have been studied as to their role in neural crest

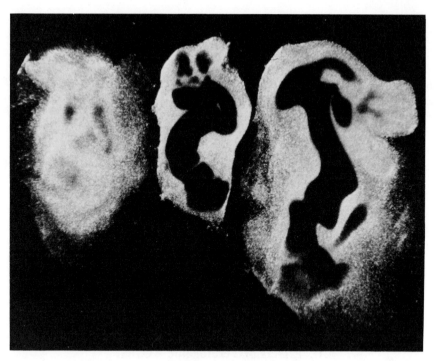

Fig. 9. Forelimb buds of 12-day mouse embryos organ cultured for 6 days in a medium containing DON (left), L-azetidine (middle), or no additives (right). See text for explanation. Whole mounts stained with toluidine blue (×25).

cell migration. These cells migrate extensively and give rise to numerous derivatives including skeletal and connective tissues. Cranial neural crest cells in the chick embryo migrate ventrally to the facial region in a relatively cell-free space beneath the surface ectoderm (Johnston, 1966; Johnston and Listgarten, 1972). A biochemical and radioautographic study after *in ovo* administrations of labeled glucosamine, fucose, and sulfate has shown that the space contains mainly hyaluronate (Pratt *et al.*, 1975), which may partially create a substratum suitable for crest cell migration. Removal of selected areas of the hyaluronate matrix could mediate cell aggregations after migration is completed. The impairment of the crest cell migration and its possible significance as a pathogenic mechanism in humans in otocephaly and familial facial osteodysplasia has been proposed (Johnston *et al.*, 1973).

The secondary palate has been shown to contain both GAG and collagen during developmental stages including the time of movement of the palatine shelves from the vertical to horizontal position. Histochemical studies in conjunction with enzyme digestion have indicated the presence of hyaluronic acid and chondroitin sulfates in the palatine shelves (Walker and Fraser, 1956; Larsson *et al.*, 1959; Larsson, 1962c; Kochhar and Johnson, 1965; Andersen and Matthiessen, 1967). Radioautographic studies on the incorporation of

[^{35}S]sulfate revealed a high synthetic rate in the mesenchyme at the time of palatal closure in the mouse (Larsson, 1960, 1962a; Walker, 1961; Jacobs, 1964) and in the rat (Kochhar and Johnson, 1965; Nanda, 1970). Chemical evidence has been given for the presence of chondroitin 6-sulfate, chondroitin 4-sulfate, and hyaluronic acid (Larsson, 1960; Pratt et al., 1973). Hyaluronic acid accounted for 60% of total GAG content in the rat palate (Pratt et al., 1973). It was shown that during the critical stages of palatal closure there was a twofold increase in the amount of hexosamines per dry weight of palate tissue.

Collagen has been demonstrated histochemically and electron microscopically in the palate of rat embryos (Hassell and Orkin, 1976). The amount of collagen per unit dry weight in rat embryonic palate has been shown to increase significantly just prior to and during palatal closure (Pratt and King, 1971). A much greater rate of increase was, however, noted after closure during the period of bone formation. This has been confirmed by measuring the incorporation of [^{3}H]proline in hydroxyproline in the shelves of rat embryos (Shapira and Shoshan, 1972).

D. Effects of Teratogenic Agents

A small number of teratogenic agents have been linked directly or indirectly with the metabolism of GAG and collagen. Information from some of these is summarized below to illustrate possible connections between effects on extracellular matrix and on morphogenesis.

1. BUdR (5-Bromo-2'-deoxyuridine)

This halogenated thymidine analogue is capable of interfering with the differentiation of developing tissues by inhibiting the synthesis of specialized cell products (Abbott and Holtzer, 1968). Among others, Lasher and Cahn (1969) have shown that low levels of BUdR can reversibly abolish the chondrogenic expression of chondrocytes in clonal cultures. A similar effect was also observed on myogenesis (Coleman et al., 1970). Although such effects were earlier reported to be reversible, more recently it has been found that the effects are irreversible if the cells are treated at stages of development before "determination" has been stabilized (Levitt and Dorfman, 1972; Abbott et al., 1972; Agnish and Kochhar, 1976).

Recent reports (Skalko et al., 1971; Kochhar, 1975) have confirmed previous observations on the teratogenicity of BUdR and have further shown that the production in mouse embryos of malformations such as exencephaly, micrognathia, limb defects, and cleft palate is dependent on the stage of development when the drug is injected. Skalko *et al.* (1971) and Agnish and Kochhar (1976) have further reported that the label from [³H]BUdR appeared in embryonic DNA during the susceptible stages. No report has linked alterations in GAG or collagen metabolism to BUdR-induced skeletal or other defects, but the following studies in cell culture could serve as a basis for further work in developing embryos.

In chondrocyte cultures of embryonic chick sterna, Daniel *et al.* (1973) found that BUdR depressed but did not completely abolish chondroitin sulfate synthesis. They also found that in the presence of BUdR, chondroitin 6-sulfate was the predominant isomer synthesized, instead of chondroitin 4-sulfate of the untreated cultures. In BUdR cultures the synthesis of other sulfated GAG (predominantly heparan sulfate) and glycoprotein components was greatly stimulated, while no effect was seen on the synthesis of hyaluronic acid and collagen. Since this biosynthetic pattern of treated cultures simulated that of prechondrogenic cells, the authors believed that BUdR rendered ineffective the normal control mechanisms for the regulation of GAG production by chondrocytes during differentiation.

Levitt and Dorfman (1973) have argued that the reduced content of chondroitin sulfate proteoglycan in limb bud cultures after BUdR treatment could result from: (1) deficient synthesis of nucleotide sugar precursors; (2) deficient glycosyl transferase activity; (3) augmented rate of degradation; (4) decreased synthesis of a specific core protein; or (5) synthesis of an aberrant core protein that is not glycosylated. Based on an observation that inhibition of chondroitin sulfate synthesis in cultured chick chondrocytes was reversed by D-xylose, these authors demonstrated a similar reversal for the BUdR effect by D-xylose in limb-bud cells. In contrast to the findings of Schulte-Holthausen *et al.* (1969) and Marzullo (1972) of strikingly decreased activities of UDP-glucose dehydrogenase, UDP-*N*-acetylgalactosamine-4-epimerase and UDP-glucose epimerase in BUdR-treated chondrocytes, Levitt and Dorfman (1973) found only a slight reduction in four enzymes involved in chondroitin sulfate production, not enough to account for the decrease in GAG synthesis. A slight increase in the turnover rate of sulfated GAG also was not sufficient to explain the marked decrease in chondroitin sulfate content. The authors concluded that the basic effect of BUdR treatment was in preventing cells from acquiring the ability to form a specific core protein. Since the synthesis of chondroitin sulfate proteoglycan is initiated with the formation of core protein, its absence or chemical alteration would prevent the synthesis of GAG chains.

Additional evidence (Levitt *et al.,* unpublished observations) indicates BUdR treatment also prevents the limb-bud cells from acquiring the capacity to synthesize cartilage-specific collagen with proper $\alpha1:\alpha2$ chain ratio. A likely mechanism of action of BUdR on cells would be its preferential incorporation into the DNA controlling synthesis of specialized products. Britten and Davidson (1969) have proposed the existence of such distinctive types of DNA that regulate the genes responsible for differentiated functions.

2. DON (6-Diazo-5-oxo-L-norleucine)

The teratogenic glutamine analogues, DON and azaserine, are of considerable importance in studying GAG metabolism in teratogenesis. Glutamine serves as the amino donor in the conversion of fructose-6-phosphate to glucosamine-6-phosphate, an important step in the biosynthesis of GAG. DON is known to prevent formation of glucosamine from glucose (Ghosh *et al.*, 1960) and to inhibit the synthesis of chondroitin sulfate in preparations of minced embryonic cartilage (Telser *et al.*, 1965). The teratogenic effects produced by DON in rats and mice include cleft palate and several types of skeletal defects (Greene and Kochhar, 1975a; see review by Chaube and Murphy, 1968). Since limb defects produced by DON were not mitigated by simultaneous administration to mice of glutamine and glucosamine, it is probable that interruption of some other glutamine-dependent metabolic step, rather than GAG metabolism, is operative in limb defects (Greene and Kochhar, 1975a). However, a role in inhibition of GAG and surface glycoprotein synthesis in the developing rat palate was indicated by the observations of Pratt *et al.* (1973) and Greene and Pratt (1975).

3. 6-AN (6-Aminonicotinamide)

6-AN has been shown to be teratogenic in chick embryos by Landauer (1957), and this has been confirmed by Pinsky and Fraser (1959) in the mouse and by Chamberlain and Nelson (1963) in the rat. Cartilage and skeletal abnormalities in 6-AN-treated chick embryos were correlated with degenerative changes seen in the chondrogenic cells (Caplan, 1972a). Seegmiller *et al.* (1972) also observed unusually distended cisternae of the rough endoplasmic reticulum and an absence of Golgi vacuoles in chondroblasts, implying that there was an interference in the transport of materials to the Golgi apparatus. The effect could either be ascribed to a defect at the Golgi, such as an impairment of sulfation or packaging of GAG, or to a primary defect in synthesis on the rough endoplasmic reticulum. It is known that the nicotinamide analogues interfere with energy requirements. Köhler *et al.* (1970) have shown that 6-amino NAD impairs utilization of glucose by interference with dehydrogenase-linked reactions. Landauer and Sopher (1970) have provided evidence that the teratogenic effects of 6-AN in the chick are prevented by the use of energy sources such as glycerophosphate. The necessity of ATP in activation of sulfate and the requirement of energy metabolites in GAG synthesis have been demonstrated (Lash *et al.,* 1964; Marzullo and Lash, 1970).

Based on teratogenic studies with nicotinamide analogues and on mitigating effects of nicotinamide against the teratogenic effects of a large number of teratogens, Caplan (1972b) has suggested that nicotinamide plays a central or controlling role in mesodermal cell differentiation into myogenic or chondrogenic phenotypes. Further studies with 6-AN and another nicotinamide analog, 3-acetylpyridine, in cell differentiation may implicate cyclic nucleotides as mediators in determining whether an undifferentiated mesenchymal cell will become a cartilage cell or a muscle cell (McMahon, 1974).

4. β-APN (β-Aminopropionitrile)

When lathyrogenic compounds (β-APN, D-penicillamine, aminoacetonitrile, and semicarbazide) are administered at critical stages of gestation in the rat and mouse, cleft palate is produced in a high frequency (Abramovich and DeVoto, 1968; Steffek *et al.,* 1971). The sensitive period of cleft palate production by β-APN corresponded to the time when the palatal shelves were rotating from a vertical to a horizontal position. The action of β-APN according to Pinnell and Martin (1968) is in preventing the formation of cross-links in newly synthesized collagen molecules by directly inhibiting the enzyme lysyl oxidase. It should be noted that the synthesis per se of collagen is not affected, but a more soluble form is produced. Pratt and King (1972) have observed a positive correlation between the inhibition of collagen cross-linking and cleft palate formation in the rat when β-APN was given in various concentrations on day 15. β-APN did not affect the synthesis of GAG or proteins, including collagen, in the palate. In other studies, however, β-APN did decrease sulfate incorporation into chondroitin sulfate in rat cartilage *in vivo* and in cartilage slices *in vitro* (Karnovsky and Karnovsky, 1961). Recently Elders *et al.* (1973) have reported that β-APN treatment of the chick embryo at 7 days of incubation decreased the amount of N-acetylhexosamines and of hexosamines esterified by nucleotides.

5. Cortisone and Other Corticosteroids

Since the discovery by Baxter and Fraser (1950) that cortisone produces isolated cleft palate in the mouse, corticosteroids have been widely used in experimental production of this congenital malformation. Several theories have been proposed as possible mechanisms by which these hormones may produce cleft palate. In addition to disruption of GAG or collagen synthesis or both, stabilization of intracellular lysosomal membranes, myopathy, weakened midline fusion, and loss of amniotic fluid have been suggested as feasible pathogenic mechanisms (Larsson, 1974; Greene and Kochhar, 1975b).

Corticosteroids are known to interfere with the synthesis of GAG and collagen in embryonic as well as in adult tissues (Schiller and Dorfman, 1957; Larsson, 1962b; Cutroneo *et al.*, 1971; Shapira and Shoshan, 1972). A decreased sulfation or a diminished production of sulfated GAG during a specified period of the palatal closure process in the mouse after cortisone treatment was demonstrated by Larsson (1962b).

Semiquantitative radioautographic techniques, in contrast to histochemical methods, were sensitive enough to demonstrate an interference in the palatal GAG. Alterations in the metabolism of GAG were interpreted to result in insufficient development of the "internal force" in the palatine shelves. Quantitative measurements of inhibition of GAG synthesis have been made in an attempt to correlate them with the frequency of cleft palate induced with cortisone or triamcinolone (Andrew and Zimmerman, 1971). These authors showed in the C3H mice that triamcinolone could induce 80% clefts concomitant with a 26% inhibition of labeled sulfate incorporation while cortisol produced only 2% clefts in conjunction with a 32% inhibition of labeled sulfate incorporation. However, since the two teratogens were given as single injections on day 11—much before the time when shelves change position—and the effect on [^{35}S]sulfate incorporation was not studied until day 14, such a lack of correlation is not very informative. It has not been possible so far to demonstrate a strain difference in GAG metabolism in sensitive (A/Jax) and resistant (CBA) mice with respect to frequencies of corticosteroid-induced cleft palate. Even if such a strain difference exists, however, isotope and histochemical methods might not be sensitive enough to demonstrate it. Also species differences exist concerning teratogenic susceptibility to cortisone and other corticosteroids, which are highly potent teratogens in the mouse and the rabbit but not in the rat. It has, moreover, been shown that in the rat methylprednisolone is not teratogenic, in contrast to triamcinolone, betamethasone, and dexamethasone (Walker, 1971). Nevertheless, impairment by corticosteroids of the synthesis of GAG or collagen could possibly cause alterations in the physical properties of the extracellular matrix in the palatal shelves, thereby leading to abnormal cell and tissue movements.

6. Salicylates

Warkany and Takacs (1959) demonstrated the teratogenic effect of salicylates in the rat and found mainly craniorachischisis, exencephaly, hydrocephaly, facial clefts, and other skeletal anomalies. Later they reported that both methyl and sodium salicylate could also induce cardiovascular malformations (Takacs and Warkany, 1968). The teratogenic effect of various salicylates has since been reported for many species (review by Larsson, 1970).

Since salicylates are very active in producing depression of GAG synthesis, their effects on several enzyme systems have been investigated. It has been suggested that salicylate inhibition of GAG formation is due to the interference with oxidative phosphorylation (Böstrom *et al.*, 1964). However, findings by Kalbhen and Domenjoz (1967) on the failure of salicylate to reduce the ATP level in rat paw edema seem to contradict this theory. Moreover, it has been shown that the synthesis of glucosamine-6-phosphate by L-glutamine-D-fructose-6-phosphate amino transferase prepared from mouse fetal tissue is inhibited by sodium salicylate (Jacobson *et al.*, 1964). The results of *in vitro* studies on [^{35}S]sulfate incorporation in calf cartilage also point to an inhibition by salicylates in formation of glucosamine-6-phosphate from fructose-6-phosphate since a stimulatory effect by 0.1 mM glutamine was abolished by salicylate at 6 mM concentration (Larsson *et al.*, 1968; Beaudoin *et al.*, 1969). Salicylates have a much smaller effect on the alternative pathway for glucosamine-6-phosphate formation from glucosamine and ATP since they did not markedly interfere with the stimulatory effect of glucosamine (Beaudoin *et al.*, 1969). The synthesis of GAG can also be affected by chelating agents (Marsh and Fraser, 1973). Salicylates have been shown to inhibit protocollagen proline hydroxylase in a cell-free system, counteracted by addition of Fe^{2+} (Nakagawa and Bentley, 1971; Liu and Bhatnagar, 1973). Thus, the mode of teratogenic action of salicylates, as well as of other anti-inflammatory agents, might be through a disturbed production of GAG or collagen, but direct proof is lacking (Kimmel *et al.*, 1971).

7. Vitamin A

Studies with excess vitamin A (retinol) have not yielded a definitive mechanism of teratogenic action. The role of retinol in cellular metabolism is reported to be diverse. The tissues from animals deficient in retinol show inhibition in the synthesis of protein, RNA, glycoprotein, and mucopolysaccharide; the latter metabolic effects have been reported to occur through inhibition of enzymes for sulfate-activation such as ATP-sulfurylase (Sundaresan, 1972).

It has long been known from clinical studies that the integrity of epithelial tissues is dependent on the availability of vitamin A. Experimental studies have revealed that under conditions of excess retinol the normal keratinization of chick embryo skin is suppressed and the ectoderm differentiates into mucus-secreting epithelium (Fell and Mellanby, 1953). Further, retinol deficiency has been shown to cause a reduction in the number of mucus-secreting goblet cells in the small intestine (DeLuca *et al.*, 1970). The control of differentiation in the latter system by retinol may depend on enzymatic effects on biosynthesis of specialized products such as glycoproteins and GAG. It has been shown that in vitamin A-deficient rats the incorporation of labeled glucosamine into a specific fucose-containing glycopeptide in intestinal goblet cells is greatly impaired (DeLuca *et al.*, 1970). This vitamin A-sensitive glycoprotein is characterized by Kleinman and Wolf (1974a,b) as being nonsulfated and found within the membranes of rough and smooth endoplasmic reticulum and Golgi apparatus.

Solursh and Meier (1973) have suggested that vitamin A may interfere in chondrogenesis in ways other than through enhanced degradation of proteoglycan in established cartilage, as proposed by Dingle *et al.* (1966) and Goodman *et al.* (1974). In cultures of chick embryo chondrocytes, vitamin A produced a concentration-dependent inhibition of GAG synthesis (Solursh and Meier, 1973). Vitamin A treatment inhibited to a similar extent the incorporation of [^{14}C]glucose and [^{35}S]sulfate into isolated GAG, indicating that the effect was not merely due to altered sulfate pools or to undersulfation of the molecule. This inhibition was accompanied by no effect on collagen or general protein synthesis.

Only a few studies have been performed on the influence of teratogenic doses of vitamin A on GAG metabolism in abnormally developing embryos. Incorporation of [^{35}S]sulfate into polymeric material of vitamin A-treated rat embryos was reported to be greatly enhanced (Kochhar and Johnson, 1965; Kochhar *et al.*, 1968b; Nanda, 1970; Schimmelpfennig *et al.*, 1972). Schimmelpfennig (1971) attributed this increase to slower removal of [^{35}S]sulfate from the bloodstream of treated mothers rather than to an accelerated rate of GAG synthesis. However, in other systems, such as cultured 3T6 mouse fibroblasts, retinol enhances 2–4 times the incorporation of [^{14}C]glucosamine into polymeric material which the cells release into the medium (Kochhar *et al.*, 1968a). Since the glucosamine-labeled material synthesized under the influence of retinol has not been characterized, it is not possible to explain the results obtained by Solursh and Meier (1973) in cultured chondrocytes. Schimmelpfennig *et al.* (1972) have reported that in vitamin A-treated embryos the ratio of fractions with lower anionic properties to those with higher anionic properties shows an increase. If this observation implies that synthesis of hyaluronic acid is stimulated in treated embryos, this may explain the inhibition of chondrogenesis observed in the malformed limbs of such embryos (Kochhar, 1970). However, the role of GAG metabolism in vitamin A-induced malformations must be more complex. Heterotopic chondrogenesis was observed in the region of presumptive maxillary membrane bone of treated embryos and may play a part in cleft palate formation (Kochhar and Johnson, 1965).

8. Thalidomide

Among many investigations on the metabolic basis for thalidomide-induced skeletal defects, only two point to the possible involvement of extracellular materials. Neubert (1970) showed that the serum from rats treated with high doses of thalidomide contained a factor which inhibited protocollagen prolyl hydroxylase, a key enzyme in collagen synthesis. This effect was pronounced at low concentrations of ascorbic acid in a cell-free system. The addition of thalidomide to the reaction mixture instead of serum, however, produced an inconsistent response. These studies have not been confirmed. Lash and Saxén (1972) have reported that thalidomide inhibits chondrogenesis *in vitro* only when mesonephric mesenchyme was associated with limb bud in the explant. Chondrogenesis in the isolated limb bud was not inhibited by thalidomide in either chick or human embryonic tissues. It was proposed that mesonephric mesenchyme at a specific stage of development has a positive influence upon limb chondrogenesis and that thalidomide inhibits this interaction thus inhibiting chondrogenesis.

VI. CONCLUSIONS

Biochemical information on glycosaminoglycans has undergone remarkable expansion within the past 20 years. The chain composition of GAG, the linkage to protein, and microheterogeneity within the molecules have been elucidated. Isolation and characterization procedures have facilitated the search for mechanisms of teratogenesis. Further references, however, are necessary for the application of microchemical procedures to the small tissue samples available to embryologists. Similar expansion in the field of collagen has occurred more recently. Referring to current work on GAG, Ogston (1970) has commented ". . . I see the main interest in the immediate future as likely to lie in the study of systems rather than substances, and by physical, physicochemical and engineering (and of course, biological), rather than by chemical, methods."

The functional role of these extracellular macromolecules as mediators of tissue interaction, cell movement, cell migration, morphogenesis of organs, and as markers of cell differentiation is fast emerging. A number of teratogens which may influence these developmental processes through alterations of GAG and collagen metabolism are discussed. It seems likely that the genesis of malformations may be correlated with these alterations. Further information from developmental biology should stimulate teratologists to design appropriate experiments to clarify the role of alterations in the extracellular materials in maldevelopment.

ACKNOWLEDGMENTS

Financial assistance of Swedish Medical Research Council (grant 14X-993-10) and Stiftelsen Sigurd och Elsa Golges Minne is gratefully acknowledged. Original work mentioned in this chapter was supported by USPHS grant HD-06550 from the National Institute of Health. We are indebted to Miss Suzanne Riebe and Miss Pamela Baker for assistance.

REFERENCES

Abbott, J., and Holtzer, H., 1968, The loss of phenotypic traits by differentiated cells. V. The effect of 5-bromodeoxyuridine on cloned chondrocytes, *Proc. Natl. Acad. Sci. U.S.A.* **59**:1144.

Abbott, J., Mayne, R., and Holtzer, H., 1972, Inhibition of cartilage development in organ cultures of chick somite by the thymidine analog, 5-bromo-2'-deoxyuridine, *Dev. Biol.* **28**:430.

Abramovich, A., and DeVoto, F. C. H., 1968, Anomalous maxillofacial patterns produced by maternal lathyrism in rat fetuses, *Arch. Oral Biol.* **13**:823.

Agnish, N. D., and Kochhar, D. M., 1976, Direct exposure of postimplantation mouse embryos to 5-bromodeoxyuridine *in vitro* and its effects on subsequent chondrogenesis in the limbs, *J. Embryol. Exp. Morphol.* **36** (in press).

Andersen, H., and Matthiessen, M., 1967, Histochemistry of the early development of the human central face and nasal cavity with special reference to the movements and fusion of the palatine processes, *Acta Anat.* **68**:473.

Andrew, F. D., and Zimmerman, E. F., 1971, Glucocorticoid induction of cleft palate in mice; no correlation with inhibition of mucopolysaccharide synthesis, *Teratology* **4**:31.

Aydelotte, M. B., and Kochhar, D. M., 1972, Development of mouse limb buds in organ culture: Chondrogenesis in the presence of a proline analog, L-azetidine-2-carboxylic acid, *Dev. Biol.* **28**:191–201.

Aydelotte, M. B., and Kochhar, D. M., 1975, Influence of 6-diazo-5-oxonorleucine (DON), a glutamine analogue, on cartilagenous differentiation in mouse limb buds *in vitro*, *Differentiation* **4**:73.

Baxter, H., and Fraser, F. C., 1950, Production of congenital defects in the offspring of female mice treated with cortisone, *McGill Med. J.* **19**:245.

Beaudoin, A. R., Boström, H., Friberg, V., and Larsson, K. S., 1969, The effect of sodium salicylate on the glutamine- and glucosamine-induced stimulation of ^{35}S-sulfate incorporation *in vitro*, *Ark. Kemi.* **30**:523.

Bernfield, M. R., and Banerjee, S. D., 1972, Acid mucopolysaccharide (glycosaminoglycans) at the epitheliomesenchymal interface of mouse embryo salivary glands, *J. Cell Biol.* **52**:664.

Bernfield, M. R., Cohn, R. H., and Banerjee, S. D., 1973, Glycosaminoglycans and epithelial organ formation, *Am. Zool.* **13**:1067.

Bornstein, P., Ehrlich, H. P., and Wyke, A. W., 1972, Procollagen: Conversion of the precursor to collagen by a neutral protease, *Science* **175**:544.

Bosmann, H. B., 1968, Cellular control of macromolecular synthesis: Rates of synthesis of extracellular macromolecules during and after depletion by papian, *Proc. R. Soc. London Ser. B* **169**:399.

Boström, H., Bernsten, K., and Whitehouse, M. W., 1964, Biochemical properties of antiinflammatory drugs. II. Some effects on sulfate-^{35}S metabolism *in vivo*, *Biochem. Pharmacol.* **13**:413.

Brady, R. O., 1968, Enzymatic defects in the sphingolipidoses, *Adv. Clin. Chem.* **11**:1.

Breen, M., Weinstein, H. G., Johnson, R. L., Veis, A., and Marshall, R. T., 1970, Acidic glycosaminoglycans in human skin during fetal development and adult life, *Biochim. Biophys. Acta* **201**:54.

Breen, M., Johnson, R. L., Sittig, R. A., Weinstein, H. G., and Veis, A., 1972, The acidic glycosaminoglycans in human fetal development and adult life: Cornea, sclera, and skin, *Connect. Tissue* **1**:291.

Breen, M., Richardson, R., Bondareff, W., and Weinstein, H. G., 1973, Acidic glycosaminoglycans in developing sterno-costal cartilage of the hydrocephalic ($ch+ch+$) mouse, *Biochim. Biophys. Acta* **304**:828.

Britten, R. J., and Davidson, E. H., 1969, Gene regulation for higher cells: A theory, *Science* **165**:349.

Brunngraber, E. G., Berra, B., and Zambotti, V., 1973, Altered levels of tissue glycoproteins, gangliosides, glycosaminoglycans and lipids in Niemann–Pick's disease, *Clin. Chim. Acta* **48:**173.

Campo, R. D., and Phillips, S. J., 1973, Electron microscopic visualization of proteoglycans and collagen in bovine costal cartilage, *Calcif. Tissue Res.* **13:**83.

Caplan, A. I., 1972a, The site and sequence of action of 6-aminonicotinamide in causing bone malformations of embryonic chick limb and its relationship to normal development, *Dev. Biol.* **28:**71.

Caplan, A. I., 1972b, Comparison of the capacity of nicotinamide and nicotinic acid to relieve the effects of muscle and cartilage teratogens in developing chick embryos, *Dev. Biol.* **28:**344.

Chamberlain, J. G., and Nelson, M. M., 1963, Congenital abnormalities in the rat resulting from single injections of 6-aminonicotinamide during pregnancy, *J. Exp. Zool.* **153:**285.

Chaube, S., and Murphy, M. L., 1968, The teratogenic effects of the recent drugs active in cancer chemotherapy, *Adv. Teratol.* **3:**181.

Cohen, A. M., and Hay, E. D., 1971, Secretion of collagen by embryonic neuroepithelium at the time of spinal cord-somite interaction, *Dev. Biol.* **26:**578.

Coleman, A. W., Coleman, J. R., Kantel, D., and Werner, I., 1970, The reversible control of animal cell differentiation by the thymidine analog, 5-bromodeoxyuridine, *Exp. Cell Res.* **59:**319.

Cooper, R. R., and Ponseti, I. V., 1973, Metaphyseal dysostosis: Description of an ultrastructural defect in the epiphyseal plate chondrocytes, *J. Bone Joint Surg.* **55A:**485.

Cooper, R. R., Ponseti, I. V., and Maynard, J. A., 1973, Pseudoachondroplastic dwarfism. A rough-surfaced endoplasmic reticulum storage disorder, *J. Bone Joint Surg.* **55A:**475.

Cutroneo, K. R., Costello, D., and Fuller, G. C., 1971, Alteration of proline hydroxylase activity by glucocorticoids, *Biochem. Pharmacol.* **20:**2797.

Daniel, J. C., Kosher, R. A., Lash, J. W., and Hertz, J., 1973, The synthesis of matrix components by chondrocytes in vitro in the presence of 5-bromodeoxyuridine, *Cell Differ.* **2:**285.

DeLuca, L., Schumacher, M., and Wolf, G., 1970, Biosynthesis of a fucose-containing glycopeptide from rat small intestine in normal and vitamin A-deficient conditions, *J. Biol. Chem.* **245:**4551.

Dingle, J. T., Fell, H. B., and Lucy, J. A., 1966, Synthesis of connective tissue components. The effect of retinol and hydrocortisone on cultured limb bone rudiments. *Biochem. J.* **98:**173.

Dorfman, A., and Matalon, R., 1969, The Hurler and Hunter syndromes, *Am. J. Med.* **47:**691.

Ede, D. A., Bellairs, R., and Bancroft, M., 1974, A scanning electron microscope study of the early limb-bud in normal and talpid[3] mutant chick embryos, *J. Embryol. Exp. Morphol.* **31:**761.

Eisenstein, R., Larsson, S. E., Sorgente, N. and Kuettner, K. E., 1973, Collagen–proteoglycan relationships in epiphyseal cartilage, *Am. J. Pathol.* **73:**443.

Elders, M. J., Smith, J. D., Smith, W. G., and Hughes, E. R., 1973, Alterations in glycosaminoglycan metabolism in β-aminopropionitrile-treated chick embryos, *Biochem. J.* **136:**985.

Fell, H. B., and Mellanby, E., 1953, Metaphasia produced in cultures of chick ectoderm by high vitamin A, *J. Physiol.* **119:**470.

Flint, M., 1972, Interrelationships of mucopolysaccharide and collagen in connective tissue remodeling, *J. Embryol. Exp. Morphol.* **27:**481.

Ghosh, S., Blumenthal, H. J., Davidson, E., and Roseman, S., 1960, Glucosamine metabolism. V. Enzymatic synthesis of glucosamine-6-phosphate, *J. Biol. Chem.* **235:**1265.

Godman, G. C., and Lane, N., 1964, On the site of sulfation in the chondrocyte, *J. Cell Biol.* **21:**353.

Goodman, D. S., Smith, J. E., Hembry, R. M., and Dingle, J. T., 1974, Comparison of the effects of vitamin A and its analogs upon rabbit ear cartilage in organ culture and upon growth of the vitamin A-deficient rat, *J. Lipid Res.* **15:**406.

Greene, R. M., and Kochhar, D. M., 1975a, Limb development in mouse embryos. Protection against teratogenic effects of 6-diazo-5-oxonorleucine (DON) *in vivo* and *in vitro, J. Embryol. Exp. Morphol.* **33:**355.

Greene, R. M., and Kochhar, D. M., 1975b, Some aspects of corticosteroid-induced cleft palate: A review, *Teratology* **11:**47.

Greene, R. M., and Pratt, R. M., 1975, Inhibition of palatal epithelial cell adhesion *in vitro* by diazo-oxo-norleucine, *Teratology* **11**:19A.

Greenwald, R. A., Schwartz, C. E., and Cantor, J. O., 1975, Interaction of cartilage proteoglycans with collagen-substituted agarose gels, *Biochem. J.* **145**:601.

Gregory, J. D., 1973, Multiple aggregation factors in cartilage proteoglycan, *Biochem. J.* **133**:383.

Grobstein, C., 1967, Mechanism of organogenetic tissue interaction, *Natl. Cancer Inst. Monogr.* **26**:279–299.

Grüneberg, H., 1965, *The Pathology of Development,* John Wiley, New York.

Hascall, V. C., and Heinegård, D., 1975, The structure of cartilage proteoglycans, *in: Extracellular Matrix Influences on Gene Expression* (H. C. Slavkin and R. C. Greulich, eds.), pp. 423–433, Academic Press, New York.

Hascall, V. C., and Sajdera, S. W., 1970, Physical properties and polydispersity of proteoglycan from bovine nasal septum, *J. Biol. Chem.* **245**:4920.

Hassell, J. R., and Orkin, R. W., 1976, Synthesis and distribution of collagen in the rat palate during shelf elevation, *Dev. Biol.* **49**:80.

Horwitz, A. L., and Dorfman, A., 1968, Subcellular sites for synthesis of chondromucoprotein of cartilage, *J. Cell Biol.* **38**:358.

Jacobs, R. M., 1964, S^{35}-liquid scintillation count analysis of morphogenesis and teratogenesis of the palate in mouse embryos, *Anat. Rec.* **150**:271.

Jacobson, B., Boström, H., and Larsson, K. S., 1964, The effect of sodium salicylate on hexosamine synthesis in eviscerated mouse fetuses, *Acta Chem. Scand.* **18**:818.

Jeanloz, R. W., 1960, The nomenclature of acid mucopolysaccharides, *Arthritis Rheum.* **3**:323–327.

Johnston, M. C., 1966, A radioautographic study of the migration and fate of cranial neural crest cells in the chick embryo, *Anat. Rec.* **156**:143.

Johnston, M. C., and Listgarten, M. A., 1972, Observations on the migration interaction, and early differentiation of orofacial tissues, *in: Developmental Aspects of Oral Biology* (H. C. Slavkin and L. A. Bavetta, eds.), pp. 55–80, Academic Press, New York.

Johnston, M. C., Bhakdinaronk, A., and Reid, Y. C., 1973, An expanded role of the neural crest in oral and pharyngeal development, *in: Fourth Symposium on Oral Sensation and Perception, Development in the Fetus and Infant* (J. F. Bosma, ed.), pp. 37–52, DHEW Publication No. (NIH) 73-546, Washington, D.C.

Kalbhen, D. A., and Domenjoz, R., 1967, The effect of phenylbutazone and sodium salicylate on the adenosinetriphosphate level in rat paw edema, *Br. Chem. Ther.* **5**:375.

Karnovsky, M. J., and Karnovsky, M., 1961, Metabolic effects of lathyrogenic agents on cartilage *in vivo* and *in vitro*, *J. Exp. Med.* **113**:381.

Kelley, R. O., 1973, Fine structure of the apical rim mesenchyme complex during limb morphogenesis in man, *J. Embryol. Exp. Morphol.* **29**:117.

Kimata, K., Okayama, M., Suzuki, S., Suzuki, I., and Hoshino, M., 1971, Nascent mucopolysaccharides attached to the Golgi membrane of chondrocytes, *Biochim. Biophys. Acta* **237**:606.

Kimata, K., Okayama, M., Oohira, A., and Suzuki, S., 1974, Heterogeneity of proteochondroitin sulfate produced by chondrocytes at different stages of cytodifferentiation, *J. Biol. Chem.* **249**:1646.

Kimmel, C. A., Wilson, J. G., and Schumacher, H. J., 1971, Studies on metabolism and identification of the causatine agent in aspirin teratogenesis in rats, *Teratology* **4**:15.

Kleinman, H. K., and Wolf, G., 1974a, The biosynthesis of a fucose-containing glycoprotein from intestinal mucosa of normal and vitamin A-deficient rats, *Biochim. Biophys. Acta* **354**:17.

Kleinman, H. K., and Wolf, G., 1974b, Extraction and characterization of a "native" vitamin A-sensitive glycoprotein from rat intestine, *Biochim. Biophys. Acta* **359**:90.

Kochhar, D. M., 1970, The role of altered metabolism of glycosaminoglycans in the production of experimentally induced congenital malformations, *in: Symposium on Metabolic Pathways in Mammalian Embryos during Organogenesis and their Modification by Drugs* (R. Bass, F. Beck, H. J. Merker, D. Neubert, and B. Randhahn, eds.), SFB 29, pp. 421–440, Free University Press, Berlin.

Kochhar, D. M., 1975, The use of *in vitro* procedures in teratology, *Teratology* **11**:273.

Kochhar, D. M., and Johnson, E. M., 1965, Morphological and autoradiographic studies of cleft palate induced in rat embryos by maternal hypervitaminosis A, *J. Embryol. Exp. Morphol.* **14**:223.

Kochhar, D. M., Dingle, J. T., and Lucy, J. A., 1968a, The effects of vitamin A (retinol) on cell growth and incorporation of labeled glucosamine and proline by mouse fibroblasts in culture, *Exp. Cell Res.* **52**:591.

Kochhar, D. M., Larsson, K. S., and Broström, H., 1968b, Embryonic uptake of ^{35}S-sulfate: Change in level following treatment with some teratogenic agents, *Biol. Neonate* **12**:41.

Kochhar, D. M., Kochhar, O. S., and Riebe, S. M., 1974, Electron microscopic observations on the developing chondrocytes in vitamin A-treated mouse limb buds, *Anat. Rec.* **178**:394.

Kochhar, D. M., Aydelotte, M. B., and Vest, T. K., 1976, Altered collagen fibrillogenesis in embryonic mouse limb cartilage deficient in matrix granules, *Exp. Cell Res.* **102**:213.

Köhler, E., Barrach, H., and Neubert, D., 1970, Inhibition of NADP dependent oxidoreductase by the 6-aminonicotinamide analogue of NADP, *FEBS Lett.* **6**:225.

Konigsberg, I. R., 1970, The relationship of collagen to the clonal development of embryonic skeletal muscle, *in: Chemistry and Molecular Biology of the Intercellular Matrix* (E. A. Balazs, ed.), pp. 1779–1810, Academic Press, New York.

Laitinen, O., Uitto, J., Iivanainen, M., Hannuksela, M., and Kivirikko, K. I., 1968, Collagen metabolism of the skin in Marfan's syndrome, *Clin. Chim. Acta* **21**:321.

Lamberg, S. I., and Dorfman, A., 1973, Synthesis and degradation of hyaluronic acid in the cultured fibroblasts of Marfan's disease, *J. Clin. Invest.* **52**:2428.

Lamberg, S. I., and Stoolmiller, A. C., 1974, Glycosaminoglycans. A biochemical and clinical review, *J. Invest. Dermatol.* **63**:433.

Landauer, W., 1957, Niacin antagonists and chick development, *J. Exp. Zool.* **136**:509.

Landauer, W., 1965, Nanomelia, a lethal mutation of the fowl, *J. Hered.* **56**:131.

Landauer, W., and Sopher, D., 1970, Succinate, glycerophosphate and ascorbate as sources of cellular energy and as antiteratogens, *J. Embryol. Exp. Morphol.* **24**:187.

Larsson, K. S., 1960, Studies on the closure of the secondary palate. II. Occurrence of sulphomucopolysaccharides in the palatine processes of the normal mouse embryo, *Exp. Cell Res.* **21**:498.

Larsson, K. S., 1962a, Studies on the closure of the secondary palate. III. Autoradiographic and histochemical studies in the normal mouse embryo, *Acta Morphol. Neerl-Scand.* **4**:349.

Larsson, K. S., 1962b, Studies on the closure of the secondary palate. IV. Autoradiographic and histochemical studies of the mouse embryos from cortisone-treated mothers, *Acta Morphol. Neerl-Scand.* **4**:369.

Larsson, K. S., 1962c, Studies on the closure of the secondary palate. V. Attempts to study the teratogenic action of cortisone in mice, *Acta Odontol. Scand.* **20**:1.

Larsson, K. S., 1970, Action of salicylate on prenatal development, *in: Congenital Malformations of Mammalia* (H. Tuchmann-Duplessis, ed.), Masson et Cie., Paris.

Larsson, K. S., 1974, Mechanisms of cleft palate formation, *in: Congenital Defects, New Direction in Research* (D. T. Janerich, R. G. Skalko, and I. H. Porter, eds.), pp. 255–273, Academic Press, New York.

Larsson, K. S., Boström, H., and Carlsöö, S., 1959, Studies on the closure of the secondary palate. I. Autoradiographic study in the normal mouse embryo, *Exp. Cell Res.* **16**:379.

Larsson, K. S., Boström, H., and Jutheden, G., 1968, Influence of *in vitro* salicylate treatment on ^{35}S-sulfate *in vitro* incorporation in mesenchymal tissue, *Ark. Kemi* **29**:389.

Lash, J. W., and Saxén, L., 1972, Human teratogenesis: *In vitro* studies on thalidomide-inhibited chondrogenesis, *Dev. Biol.* **28**:61.

Lash, J. W., Glick, M. C., and Madden, J. W., 1964, Cartilage induction *in vitro* and sulfate-activating enzymes, *in: Metabolic Control Mechanisms in Animal Cells, Natl. Cancer Inst. Monogr. No. 13*, pp. 39–49.

Lasher, R., and Cahn, D., 1969, The effects of 5-bromodeoxyuridine on the differentiation of chondrocytes *in vitro*, *Dev. Biol.* **19**:415.

Lenaers, A., Ansay, M., Nusgens, B. V., and Lapiere, C. M., 1971, Collagen made of extended α chains, procollagen in genetically-defective dermatosparaxic calves, *Eur. J. Biochem.* **23**:533.

Levitt, D., and Dorfman, A., 1972, The irreversible inhibition of differentiation of limb-bud mesenchyme by bromodeoxyuridine, *Proc. Natl. Acad. Sci. U.S.A.* **69**:1253.

Levitt, D., and Dorfman, A., 1973, Control of chondrogenesis in limb-bud cell cultures by bromodeoxyuridine, *Proc. Natl. Acad. Sci. U.S.A.* **70**:2201.

Levitt, D., and Dorfman, A., 1974, Concepts and mechanisms of cartilage differentiation, *Curr. Top. Dev. Biol.* **8**:103.

Linsenmeyer, T. F., Trelstad, R. L., and Gross, J., 1973, The collagen of chick embryonic notochord, *Biochem. Biophys. Res. Commun.* **53**:39.

Lipson, M. J., and Silbert, J. E., 1968, Glycosaminoglycans of adult frog back skin, *Biochim. Biophys. Acta* **158**:344.

Liu, T. Z., and Bhatnagar, R. S., 1973, Inhibition of protocollagen proline hydroxylase by dilantin, *Proc. Soc. Exp. Biol. Med.* **142**:253.

Loewi, G., and Meyer, K., 1958, The acid mucopolysaccharides of embryonic skin, *Biochim. Biophys. Acta* **27**:453.

Lowther, D. A., and Natarajan, J., 1972, The influence of glycoprotein on collagen fibril formation in the presence of chondroitin sulphate proteoglycan, *Biochem. J.* **127**:607.

Lowther, D. A., and Toole, B. P., 1968, The interaction between acid mucopolysaccharide-protein complexes and tropocollagen, *in: Symposium on Fibrous Proteins, Australia* (W. G. Crewther, ed.), pp. 229–232, Butterworths, Sydney.

Macek, M., Huryck, J., Chvapil, M., and Kadlecová, V., 1966, Study on fibroblasts in Marfan's syndrome, *Humangenetik* **3**:87.

Marsh, L., and Fraser, F. C., 1973, Chelating agents and teratogenesis, *Lancet* **1**:876.

Martin, G. R., Byers, P. H., and Piez, K. A., 1975, Procollagen, *Adv. Enzymol.* **42**:167.

Marzullo, G., 1972, Regulation of cartilage enzymes in cultural chondrocytes and the effect of 5-bromodeoxyuridine, *Dev. Biol.* **27**:20.

Marzullo, G., and Lash, J. W., 1970, Control of phenotypic expression in cultured chondrocytes: Investigation on the mechanism. *Dev. Biol.* **22**:638.

Matalon, R., and Dorfman, A., 1968, The accumulation of hyaluronic acid in cultured fibroblasts of the Marfan syndrome, *Biochem. Biophys. Res. Commun.* **32**:150.

Mathews, M. B., 1965, The interaction of collagen and acid mucopolysaccharides. A model for connective tissue, *Biochem. J.* **96**:710.

Mathews, M. B., 1967a, Macromolecular evolution of connective tissue, *Biol. Rev.* **42**:499.

Mathews, M. B., 1967b, Chondroitin sulfate and collagen in inherited skeletal defects of chickens, *Nature* **213**:1255.

Mathews, M. B., 1971, Comparative biochemistry of chondroitin sulphate-protein of cartilage and notochord, *Biochem. J.* **125**:37.

Mathews, M. B., and Lozaityte, I., 1958, Sodium chondroitin sulfate-protein complexes of cartilage. I. Molecular weight and shape, *Arch Biochem.* **74**:158.

Matukas, V. J., Panner, B. J., and Orbison, J. L., 1967, Studies on ultrastructural identification and distribution of protein–polysaccharide in cartilage matrix, *J. Cell Biol.* **32**:365.

Maurice, D. M., 1957, The structure and transparency of the cornea, *J. Physiol.* **136**:263.

McGee, J. O'D., Langners, U., and Udenfriend, S., 1971, Immunological evidence for an inactive precursor of collagen proline hydroxylase in cultured fibroblasts, *Proc. Natl. Acad. Sci. U.S.A.* **68**:1585.

McKusick, V., 1972, *Heritable Disorders of Connective Tissue*, Mosby, St. Louis.

McKusick, V. A., and Martin, G. R., 1975, Molecular defects in collagen, *Ann. Intern. Med.* **82**:585.

McMahon, D., 1974, Chemical messengers in development: A hypothesis, *Science* **185**:1012.

Miller, E. J., and Matukas, V. J., 1974, Biosynthesis of collagen, *Fed. Proc.* **33**:1197.

Murata, K., Yukiyama, Y., and Horiuchi, Y., 1973, Metabolic changes of urinary acidic glycosaminoglycans in Weber Christian disease, *Clin. Chim. Acta* **49**:129.

Nakagawa, H., and Bentley, J. P., 1971, Salicylate-induced inhibition of collagen and mucopolysaccharide biosynthesis by a chick embryo cell-free system, *J. Pharm. Pharmacol.* **23**:399.

Nanda, R., 1970, The role of sulfated mucopolysaccharides in cleft palate production, *Teratology* **3**:237.

Neubert, D., 1970, Protocollagen hydroxlase in mammalian embryos and the influence of thalidomide and some of its metabolites, *in: Metabolic Pathways in Mammalian Embryos during Organogenesis and their Modification by Drugs* (R. Bass, F. Beck, H. J. Merker, D. Neubert, and B. Randhahn, eds.), SFB 29, pp. 505–512, Free University Press, Berlin.

Neufeld, E. F., and Fratantoni, J. C., 1970, Inborn errors of mucopolysaccharide metabolism, *Science* **169**:141.

Neutra, M., and Leblond, C. P., 1966, Radioautographic comparison of the uptake of galactose-^{3}H and glucose-^{3}H in the Golgi region of various cells secreting glycoproteins and mucopolysaccharides, *J. Cell Biol.* **30**:137.

Nevo, Z., and Dorfman, A., 1972, Stimulation of chondromucoprotein synthesis in chondrocytes by extracellular chondromucoprotein, *Proc. Natl. Acad. Sci. U.S.A.* **69**:2069.

Nigra, T. P., Friedland, M., and Martin, G. R., 1972, Controls of connective tissue synthesis: Collagen metabolism, *J. Invest. Dermatol.* **59**:44.

Oegema, T. R., Jr., Laidlaw, J., Hascall, V. C., and Dziewiatkowski, D. D., 1975, The effect of proteoglycans on the formation of fibrils from collagen solutions, *Arch. Biochem. Biophys.* **170**:698.

Ogston, A. G., 1970, The biological functions of the glycosaminoglycans, *in: Chemistry and Molecular Biology of the Intercellular Matrix*, Vol. 3 (E. A. Balazs, ed.), pp. 1231–1240, Academic Press, New York.

O'Hare, M. J., 1973, A histochemical study of sulphated glycosaminoglycans associated with the somites of the chick embryo, *J. Embryol. Exp. Morphol.* **29**:197.

Olsen, B. R., Berg, R. A., Kishida, Y. and Prockop, D. J., 1975, Further characterization of embryonic tendon fibroblasts and the use of immunoferritin techniques to study collagen biosynthesis, *J. Cell Biol.* **64**:340.

Oohira, A., Kimata, K., Suzuki, S., Takata, K., Suzuki, I., and Hoshino, M., 1974, A correlation between synthetic activities for matrix macromolecules and specific stages of cytodifferentiation in developing cartilage, *J. Biol. Chem.* **249**:1637.

Palmoski, M. J., and Goetinck, P. F., 1972, Synthesis of proteochondroitin sulfate by normal, nanomelic, and 5-bromodeoxyuridine-treated chondrocytes in cell culture, *Proc. Natl. Acad. Sci. U.S.A.* **69**:3385.

Partridge, S. M., Davids, H. F., and Adair, G. S., 1961, The chemistry of connective tissues. 6. The constitution of the chondroitin sulphate-protein complex in cartilage, *Biochem. J.* **79**:15.

Piez, K. A., 1968, Cross-linking of collagen and elastin. *Annu. Rev. Biochem.* **37**:547.

Pinnell, S. R., and Martin, G. R., 1968, The crosslinking of collagen and elastin: Enzymatic conversion of lysine in peptide conversion of lysine in peptide linkage to α-aminoadipic-δ-semialdehyde (allysine) by an extract from bone. *Proc. Natl. Acad. Sci. U.S.A.* **61**:708.

Pinnell, S. R., Krane, S. M., Kenzora, J. E., and Glimcher, M. J., 1972, A heritable disorder of connective tissue. Hydroxylysine-deficient collagen disease, *N. Engl. J. Med.* **286**:1013.

Pinsky, L., and Fraser, F. C., 1959, Production of skeletal malformations in the offspring of pregnant mice trated with 6-aminonicotinamide, *Biol. Neonate* **1**:106.

Pratt, R. M., and King, C. T. G., 1971, Collagen synthesis in the secondary palate of the developing rat, *Arch. Oral Biol.* **16**:1181.

Pratt, R. M., and King, C. T. G., 1972, Inhibition of collagen crosslinking associate with β-aminopropionitrile-induced cleft palate in the rat, *Dev. Biol.* **27**:322.

Pratt, R. M., Goggins, J. F., Wilk, A. L., and King, C. T. G., 1973, Acid mucopolysaccharide synthesis in the secondary palate of the developing rat at the time of rotation and fusion, *Dev. Biol.* **32**:230.

Pratt, R. M., Larsen, M. A., and Johnston, M. C., 1975, Migration of cranial neural crest cells in a cell-free hyaluronate-rich matrix, *Dev. Biol.* **44**:298.

Rennert, O. N., 1975, Syndrome of the defective lysosome—the genetic mucopolysaccharidoses, *Ann. Clin. Lab. Sci.* **5:**355.

Revel, J. P., 1970, Role of the Golgi apparatus of cartilage cells in the elaboration of matrix glycosaminoglycans, *in: Chemistry and Molecular Biology of the Intercellular Matrix,* Vol. 3 (E. A. Balazs, ed.), pp. 1485–1502, Academic Press, New York.

Robinson, H. C., and Dorfman, A., 1969, The sulfation of chondroitin sulfate in embryonic chick cartilage epiphyses, *J. Biol. Chem.* **244:**348.

Rodén, L., 1970a, Biosynthesis of acidic glycosaminoglycans (mucopolysaccharides), *in: Metabolic Conjugation and Metabolic Hydrolysis,* Vol. II (W. H. Fishman, ed.), pp. 345–442, Academic Press, New York.

Rodén, L., 1970b, Structure and metabolism of the proteoglycans of chondroitin sulfates and keratan sulfate, *in: Chemistry and Molecular Biology of the Intercellular Matrix,* Vol. 2 (E. A. Balazs, ed.), pp. 797–821, Academic Press, New York.

Rosenberg, L., Hellmann, W., and Kleinschmidt, A. K., 1975a, Electron microscopic studies of proteoglycan aggregates from bovine articular cartilage, *J. Biol. Chem.* **250:**1877.

Rosenberg, L., Margolis, R., Wolfenstein-Todel, C., Pal, S., and Strider, W., 1975b, Organization of extracellular matrix in bovine articular cartilages, *in: Extracellular Matrix Influences on Gene Expression* (H. C. Slavkin and R. C. Greulich, eds.), pp. 415–421, Academic Press, New York.

Schettler, G., and Kahlke, W., 1967, Niemann–Pick disease, *in: Lipids and Lipidoses* (G. Schettler, ed.), pp. 288–309, Springer-Verlag, New York.

Schiller, S., and Dorfman, A., 1957, The metabolism of mucopolysaccharides in animals: The effect of cortisone and hydrocortisone on rat skin, *Endocrinology* **60:**376.

Schiller, S., and Dorfman, A., 1960, Effect of age on the heparin content of rat skin, *Nature* **185:**111.

Schimmelpfennig, K., 1971, Problems connected with *in vivo* labelling of embryonic glycosaminoglycans with $Na_2{}^{35}SO_4$ in teratological studies, *Naunyn-Schmiedebergs Arch. Pharmakol.* **271:**320.

Schimmelpfennig, K., Baumann, I., and Kaufmann, C., 1972, Studies on glycosaminoglycans (GAG) in mammalian embryonic tissue. II. Influence of vitamin A and Na-salicylate on embryonic GAG, *Naunyn-Schmiedebergs Arch. Pharmakol.* **272:**65.

Schulte-Holthausen, H., Chacko, S., Davidson, E. A., and Holtzer, H., 1969, Effect of 5-bromodeoxyuridine on expression of cultured chondrocytes grown in vitro, *Proc. Natl. Acad. Sci. U.S.A.* **63:**864.

Seegmiller, R. E., Overman, D. O., Runner, M. N., 1972, Histological and fine structural changes during chondrogenesis in micromelia induced by 6-aminonicotinamide, *Dev. Biol.* **28:**555.

Serafini-Fracassini, A., and Smith, J. W., 1966, Observations on the morphology of the proteinpolysaccharide complex of bovine nasal cartilage and its relationship to collagen, *Proc. R. Soc. London Ser.* **165:**440.

Serafini-Fracassini, A., and Smith, J. W., 1974, *The Structure and Biochemistry of Cartilage,* Churchill and Livingstone, Edinburgh.

Shapira, Y., and Shoshan, S., 1972, The effect of cortisone on collagen synthesis in the secondary palate of mice, *Arch. Oral Biol.* **17:**1699.

Skalko, R. G., Packard, D. S., Schwendimann, R. N., and Raggio, J. F., 1971, The teratogenic response of mouse embryos to 5-bromodeoxyuridine, *Teratology* **4:**87.

Slavkin, H. C., 1972, Intercellular communication during odontogenesis, *in: Developmental Aspects of Oral Biology* (H. C. Slavkin and L. A. Bavetta, eds.), pp. 165–199, Academic Press, New York.

Smith, J. G., Jr., Davidson, E. A., and Taylor, R. W., 1965, Human cutaneous acid mucopolysaccharides: The effects of age and chronic sun damage, *in: Advances in Biology of Skin,* Vol. 6 (W. Montagna, ed.), pp. 211–218, Pergamon Press, Oxford.

Smith, J. W., 1970, The disposition of proteinpolysaccharide in the epiphysial plate cartilage of the young rabbit, *J. Cell Sci.* **6:**843.

Solursh, M., and Meier, S., 1973, The selective inhibition of mucopolysaccharide synthesis by vitamin A treatment of cultured chick embryo chondrocytes, *Calcif. Tissue Res.* **13:**131.

Solursh, M., Vaerewyck, S. A., and Reiter, R. S., 1974, Depression by hyaluronic acid of glycosaminoglycan synthesis by cultured chick embryo chondrocytes, *Dev. Biol.* **41**:233.

Steffek, A. J., Verrusio, A. C., and Watkins, C. A., 1971, Cleft palate in rodents after maternal treatment with various lathyrogenic agents, *Teratology* **5**:33.

Stoolmiller, A. C., and Dorfman, A., 1969, The metabolism of glycosaminoglycans *in: Comprehensive Biochemistry*, Vol. 17 (M. Florkin and E. H. Slotz, eds.), pp. 241–275, Elsevier, Amsterdam.

Sundaresan, P. R., 1972, Recent advances in the metabolism of vitamin A, *J. Sci. Ind. Res.* **31**:581.

Takacs, E., and Warkany, J., 1968, Experimental production of cardiovascular malformations in rats by salicylate poisoning, *Teratology* **1**:109.

Telser, A., Robinson, H. C., and Dorfman, A., 1965, The biosynthesis of chondroitin-sulfate protein complex, *Proc. Natl. Acad. Sci. U.S.A.* **54**:912.

Thomas, L., 1956, Reversible collapse of rabbit ears after intravenous papain and prevention of recovery by cortisone, *J. Exp. Med.* **104**:245.

Thorp, F. K., and Dorfman, A., 1967, Differentiation of connective tissue, *Curr. Top. Dev. Biol.* **2**:151.

Toole, B. P., 1973, Hyaluronate and hyaluronidase in morphogenesis and differentiation, *Am. Zool.* **13**:1061.

Toole, B. P., and Lowther, D. A., 1968, The effect of chondroitin sulphate-protein on the formation of collagen fibrils *in vitro*, *Biochem. J.* **109**:857.

Toole, B. P., Jackson, G., and Gross, J. 1972, Hyaluronate in morphogenesis: Inhibition of chondrogenesis *in vitro*, *Proc. Natl. Acad. Sci. U.S.A.* **69**:1384.

Trelstad, R. L., 1973, The developmental biology of vertebrate collagens, *J. Histol. Cytol.* **21**:521.

Trelstad, R. L., Kang, A. H., Cohen, A. M., and Hay, E. D., 1973, Collagen synthesis *in vitro* by embryonic spinal cord epithelium, *Science* **179**:295.

Trelstad, R. L., Hayashi, K., and Toole, B. P., 1974, Epithelial collagens and glycosaminoglycans in the embryonic cornea, *J. Cell Biol.* **62**:815.

Walker, B. E., 1961, The association of mucopolysaccharides with morphogenesis of the palate and other structures in mouse embryos, *J. Embryol. Exp. Morphol.* **9**:22.

Walker, B. E., 1971, Induction of cleft palate in rats with anti-inflammatory drugs, *Teratology* **4**:39.

Walker, B. E., and Fraser, F. C., 1956, Closure of the secondary palate in three strains of mice, *J. Embryol. Exp. Morphol.* **4**:176.

Warkany, J., and Takacs, E., 1959, Experimental production of congenital malformations in rats by salicylate poisoning, *Am. J. Pathol.* **35**:315.

Weinstock, M., and Leblond, C. P., 1974, Formation of collagen, *Fed. Proc.* **33**:1205.

Weissmann, B., 1955, The transglycosylative action of testicular hyaluronidase, *J. Biol. Chem.* **216**:783.

Weston, P. D., Barrett, A. J., and Dingle, J. T., 1969, Specific inhibition of cartilage breakdown, *Nature* **222**:285.

Yamagata, T., Saito, H., Habuchi, O., and Suzukis, 1968, Purification and properties of bacterial chondroitinases and chondrosulfates, *J. Biol. Chem.* **243**:1523.

Yamato, K., Handa, S., Yamakawa, T., Saida, T., and Ideda, S., 1974, The urinary glycosaminoglycans in pachydermoperiostosis, *Jpn. J. Exp. Med.* **44**:19.

Time–Position Relationships

WITH PARTICULAR REFERENCE TO CLEFT LIP AND CLEFT PALATE

11

DAPHNE G. TRASLER and
F. CLARKE FRASER

I. INTRODUCTION

Most anatomical birth defects can be classified either as *deformations*, alterations in shape or structure of a normally formed part, such as congenital torticollis and some types of clubfoot; or *malformations*, primary structural defects that result from an error in morphogenesis. (Morphogenesis refers, in this context, to the progressive emergence of organized structures in the embryo.) Normal morphogenesis depends on a system of highly integrated and interacting sets of reactions. The embryonic cells are programmed in such a way that certain genes are turned on in the cells of certain tissues and others remain inactive, according to a progressive set of nuclear–cytoplasmic interactions. At this level of interaction, the production of abnormal molecules (enzymes, organelle components, membrane constituents, intercellular matrix) or the production of normal molecules at altered rates or at the wrong time may change the properties of the cells or their products, leading to errors in development. At another level of interaction, normal development depends on the properties of groups of cells, tissues, or organs and their relations to one another. There may be interaction of cells within a tissue, for

---

DAPHNE G. TRASLER and F. CLARKE FRASER • Department of Biology, McGill University, Montreal, Quebec, Canada.

example, when a group of cells has to reach a certain *critical mass* for its cellular components to proceed with differentiation and organ formation. Or there may be interaction between tissues. One tissue may require an *inductive stimulus* from another before it will proceed further along its developmental pathway. Or two or more tissues may need to meet one another by *merging or fusion* to form a structure. These kinds of interaction often involve a developmental *threshold,* in which the structure is morphologically normal if the threshold is reached and abnormal if it is not; this threshold concept has been dealt with in Chapter 3, Vol. 1.

When a process depends on interactions between groups of cells or tissues, errors in morphogenesis may result from developmental asynchronies, in which a malformation results from failure of some structure to be in the right place at the right time to interact with another structure. Presumably these developmental errors also depend, initially, on the production of abnormal molecules or on altered reaction rates, but the critical factor in determining whether a malformation occurs is whether or not the interacting tissues achieve a critical degree of synchrony that allows them to reach the developmental threshold. To understand the causes of such malformations it is necessary to identify the nature of the asynchrony, so that one knows better where to look for the primary defect. Also, such classes of malformations have certain general properties which are worthy of note (Chapter 3, Vol. 1).

One can classify malformations involving developmental thresholds into those where a critical mass is not reached, those where epithelia that should fuse do not touch (or having touched, will not fuse), those where an inductive stimulus fails because the interacting tissues are not close enough to one another, and those where abnormal growth patterns result in abnormal anatomical relationships of otherwise normal structures.

A classical example of a malformation involving a critical mass is absence of the third molar tooth in the mouse, from which Grüneberg (1952) developed the concept of a quasicontinuous variant, a trait resulting from separation of a continuous distribution of a developmental variable (size of tooth anlagen) into discontinuous parts (presence or absence of tooth) by a developmental threshold (critical mass).

Among those malformations resulting from failure of epithelia to meet or fuse are clefts of the lip and palate (to be discussed at length subsequently), most cases of anencephaly/spina bifida, and hypospadias. Of course the reason for the failure may differ from case to case of the same malformation— what might be referred to as "pathogenetic heterogeneity." For instance, cleft palate may occur because the tissues supposed to fuse do not reach each other or because the epithelia will not fuse when they do meet, and the latter category may be subdivided into those in which the epithelia are intrinsically incapable of fusing and those that have lost their competence to fuse because their contact has been delayed beyond a critical point.

The third group, resulting from disturbed inductive interactions include anophthalmia/microphthalmia resulting from delay in growth of the optic cup toward the overlying ectoderm (Chase and Chase, 1941) so that lens induction does not occur (anophthalmia) or occurs late (microphthalmia), and absent or small kidneys resulting from delayed growth of the ureteral bud toward the renal blastema (see Chapter 2, this volume). Again, the same developmental error may arise for different reasons, that is, there is pathogenetic heterogeneity. Anophthalmia, for instance, may occur because of delay in outgrowth of the optic cup or by a failure of the overlying ectoderm to respond to the stimulus.

Finally, there are malformations resulting simply from certain structures failing to achieve their proper anatomical relationships to one another. An example is atrial septal defect of the heart (secundum type) where the upward growth of the atrial septum primum is delayed so much that its upper border fails to close off the ostium secundum (Fraser and Rosen, 1975). Perhaps congenital dislocation of the hip could be classified here also, since the inadequate shape of the acetabular roof may permit the femoral head to slip out of its socket.

Cleft lip and cleft palate provide good examples of how deviations can and do occur in the timing of events and positioning of each of the several parts involved in normal lip and palate development. We will describe first the normal development of the structures and then the possible alterations leading to malformation and examples of these.

II. PRIMARY PALATE FORMATION

A. Normal Development of the Primary Palate

The development of the primary palate has been described in detail in mice (Trasler, 1968; Pourtois, 1972), rats (Lejour-Jeanty, 1965; Smith and Monie, 1969), and hamsters (Waterman and Meller, 1973a), while in humans a reasonably clear picture is emerging with the increased availability of early human embryos (Warbrick, 1960; Anderson and Matthiessen, 1967, 1968; Vermeij-Keers, 1972; Frederiks, 1973; Iizuka, 1973). Central face development starts with the appearance of the nasal placodes as local ectodermal thickenings on either side of the inferior aspect of the face. The mesoderm between the placodes and prosencephalon contains some mesenchymal cells that have previously migrated there from the neural crest (Johnston and Listgarten, 1972). The mesenchyme soon condenses and proliferates at the borders of the placode, and the placode epithelial cells also actively divide. This results in an apparent sinking inward of the central part of the placode as its lateral and medial rims expand outward thus forming a pit or groove. The olfactory pit becomes deeper as the lateral and medial processes grow, and it is further bounded on its floor, especially in human embryos, by the maxillary process. The definitive formation of the primary palate begins at the bottom of the olfactory groove, in the isthmus between the lateral and medial nasal processes in the mouse, and the isthmus between maxillary and medial processes in man. This isthmus has the deepening nasal pit behind and superior to it, and it is soon consolidated by the lateral (maxillary in humans) and medial processes converging above the isthmus (Figure 1). The medial surface of the lateral process and the lateral surface of the medial process meet, with their apposed epithelia forming an epithelial plate (epithelial plate of Hochstetter = nasal fin). Lejour (1970b) states that necrotic changes in the epithelial areas ready to fuse anticipate their meeting. The epithelial plate is almost immediately obliterated by autolysis, and thus the isthmus between the processes progressively widens as the external nasal pit is becoming smaller. In human embryos (Vermeij-Keers, 1972) the meeting and fusing of maxillary and medial processes is soon followed by fusion of the lateral process, more superiorly, and anteriorly with the medial process.

Deeper within the nasal pit the lateral (or maxillary in human) and medial processes are also meeting and forming part of the epithelial plate or nasal fin. In this area, however, the epithelia of the nasal fin will later pull apart, forming the passage leading from the nasal pit to the internal pit or primary choana. While this has been happening the primary choana has originated from the coalescence of epithelial cells to form a knot on the surface of the mouth posterior to the isthmus (Pourtois, 1972). Autolytic and phagocytic activities within the knot result in the formation of a "pinhole," the internal pit, which burrows inward to meet the passage from the external pit. For a period the junction of the two is separated by a thin epithelium, the bucconasal membrane.

There has been considerable confusion in the literature concerning the fusion between the processes and subsequent mesodermal penetration of the isthmus—in fact Anderson and Matthiessen (1967) deny that this happens—and the more posterosuperior fusion of the processes forming that part of the nasal fin where there is no breakdown of the epithelial seam or mesodermal penetration, but instead a pulling apart of the apposed epithelia.

B. Abnormal Primary Palate Formation—Cleft Lip

It is apparent from the above description of primary palate formation that cleft lip could result from a mistiming in the growth of any one or more of the parts involved in the complex process of midface development. The critical thing is the convergence of the facial processes to permit merging or fusion. This requires that the processes appear in the right place, achieve the correct position and form, and have no obstruction to fusion. Thus a list of possible alterations in development can be drawn up which would lead to cleft

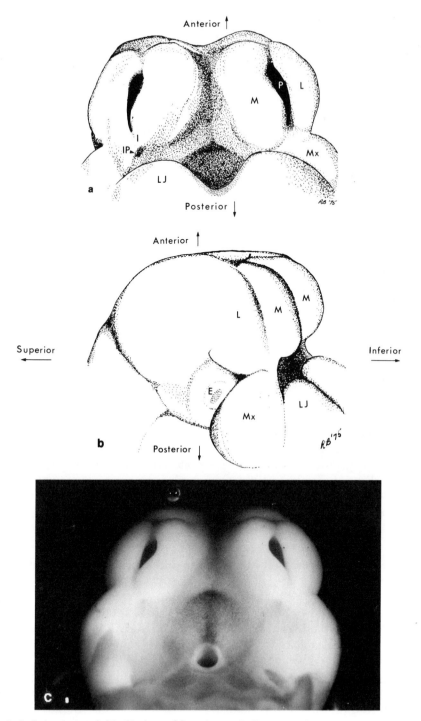

Fig. 1. Inferior (a,c) and side (b) views of face. Arrows indicate anterior, posterior, superior, and inferior directions for orientation. E = eye, I = isthmus, IP = internal pit, L = lateral nasal process, LJ = lower jaw, M = medial nasal process, Mx = maxillary process.

lip; experimental work in the mouse and rat has already provided examples for a number of the possibilities.

1. Position and Time of Placode Induction in Relation to Brain and Other Adjacent Structures

Presumably the nasal placodes in mammals arise by induction from the forebrain, as they do in amphibia (Jacobson, 1966), and their size and position will be determined in part by the size and shape of the forebrain at the time of induction, although this has not been experimentally demonstrated. If the placodes are too far apart, either because they are induced late, when the brain is larger, or on time but more laterally, and if the relative amounts of mesenchyme in lateral and medial processes remain the same, the medial processes may fail to merge, partially or completely, resulting in a median cleft. Midline clefts are rarely found in rodents but treatment with aspirin or 6-aminonicotinamide on gestation day 9½ appears to reduce the size of the medial processes which leads to median cleft in the C5BL/6 mouse strain, where the placodes are already widely spaced in the normal embryo (Trasler, 1965; Trasler and Leong, 1974). Thus in terms of the quasicontinuous model, one may hypothesize that the space between the placodes is the continuous variable and that the threshold is a critical distance between them, beyond which merging fails. Various factors, genetic and environmental, would contribute to the underlying variation.

Median cleft lip in humans, which is also rare (Sharma, 1974), could be explained in this way. Possibly the nasal placodes are more widely spaced in embryos of racial groups with relatively high frequencies of median cleft lip and low frequencies of lateral cleft lip, such as American blacks (Chung and Myrianthopoulos, 1968).

2. Rate of Neural Crest Cell Migration to and Mitosis in the Facial Processes

The mesenchyme of the three processes or prominences surrounding the nasal groove is formed from cells which have migrated from the neural crest and therefore the size and shape of the processes is governed by the number, time of arrival, and rate of division on arrival of the neural crest cells. A treatment applied at this period could cause cleft lip by influencing one or more of these features. Irradiation in mice before the neural crest cells migrate has been shown to cause death of the neural crest cells (Bahkdinaronk et al., 1976) and, since this treatment also results in lateral cleft lip and median cleft lip (as well as brain and eye defects), it has been suggested that the clefts are a result of a reduced neural crest population.

The *dancer* gene appears to act on the neural crest cells to cause an inner ear defect and, with increased expressivity, lateral cleft lip as well (Deol and Lane, 1966; Leong, 1973). The cleft lip in this case appears to result from a reduction in size of the medial nasal process, perhaps due to a lack of migration of neural crest cells since their rate of proliferation does not appear to differ from that in controls (Trasler and Leong, 1974, and unpublished observations). Treatment with 6-aminonicotinamide brings about a reduction in mitotic index and a medial nasal process which is reduced in size (Trasler and Leong, 1974).

3. Relative Sizes and Positions of Facial Processes

As we have said, any treatment or genetic predisposition which tends to lead to a lesser degree of contact between the processes at the critical time when consolidation of the isthmus takes place can lead to malformation. The degree of contact is determined by the size and shape of the processes at this time. These can be influenced by the position of the nasal placodes and the rates of neural crest cell migration as discussed above. In some examples susceptibility to cleft lip can be related to size and shape of the processes, but the relation, if any, to nasal placodes or neural crest has not been illustrated. Thus the clefts produced by hadacidin (Lejour, 1970a; Lejour-Jeanty, 1966) are associated with a delay in growth of the maxillary process. In the A/J and CL/Fr strains, which are genetically predisposed to cleft lip, the position of the medial process is such that it barely meets with the lateral at the critical time (Trasler, unpublished; Leong, 1973).

4. Change in Rate of Autolysis in the Formation of the Primary Choana and in the Dissolution of the Nasal Fin

Hadacidin has been shown to increase the normal rate of autolysis and phagocytosis in the areas of fusion between the processes, raising the question of whether the induced clefts result from alteration of the maxillary processes or from interference with fusion, by increasing the propensity to pull apart in the nasal fin, or perhaps both (Lejour, 1970a).

5. Mechanical Interference with Convergence of Facial Processes

Any of the above examples lead to a lack of convergence of the processes and thus to a failure to meet and fuse. A further possibility is mechanical interference with convergence. In the case of the homozygote for the mutant gene *patch,* for instance, a bleb of fluid collects between the medial processes and prevents their merging, thus leading to median cleft lip (Grünberg and

Truslove, 1960). A further possible example was postulated by Trasler (1966), who suggested that divergence of the medial nasal processes toward the lateral processes is aided by the presence of the embryonic heart. A decrease in heart size could decrease this divergence and thereby decrease the convergence between medial and lateral process. Perhaps this may account for at least some of the association between cleft lip and congenital heart malformations (Fraser and Rosen, 1975).

6. Failure to Coalesce or Fuse

If the processes meet, but subsequent formation and dissolution of the epithelial plate does not occur, there will be a separation and no fusion, followed by breakdown of the isthmus. Incomplete coalescence could lead to partial separation and to cleft lip of varying degrees of severity.

III. SECONDARY PALATE FORMATION

A. Normal Development of the Secondary Palate

Closure of the secondary palate appears to take place in the following manner: The palatal shelves, which are initially in a vertical position on either side of the tongue, reorient themselves to attain a horizontal position overlying the tongue. The means of doing this differs from species to species. In the mouse there appears to be a force within the shelves that promotes movement from the vertical to the horizontal position. According to Walker and Fraser (1956) the shelves appear to flow dorsally and "wedge" themselves between the tongue and nasal septum. This process has been dramatically illustrated by Greene and Kochhar (1973a) using frozen sections, which greatly reduce fixation artifacts, and give a more representative picture of the anatomical relationships. On reaching the horizontal the shelves are immediately in contact and strongly adhere to one another. Ultrastructural changes (described in more detail subsequently) occur in the epithelia at the future point of fusion before the shelves actually meet (Waterman et al., 1973). Recently a sulfated mucopolysaccharide layer has been shown to appear on the surface of the shelves, prior to their reorientation, that may account for the strong adhesion of the shelves to each other when they meet (Greene and Kochhar 1974). The apposed epithelia fuse to form an epithelial seam which soon breaks down allowing complete fusion and mesenchymal continuity within the palate. In the mouse (Figures 2 and 3), shelf movement begins posteriorly and the anterior part becomes horizontal last, although fusion first occurs somewhat anterior to the midpoint of the shelves (Walker and Fraser, 1956). Movements of the tongue may play a role in aiding the shelves to force their way into the space above them.

In the rat (Coleman, 1965), the rostral part of the shelf moves by rotation, while caudally the shelves wedge their way over the tongue by changing shape, bulging medially and retracting the original free edge. Wragg *et al.* (1972b), from observations of frozen sections of rat embryos, noted no spaces in the oral cavity and suggest that a shelf force drives the shelves medially, trapping the tongue well above the shelves. This has also been seen in the baboon (Bollert and Hendrickx, 1971). Subsequently the tongue musculature may contract slightly, lowering the upper surface, and the semifluid shelves push into the depression in the tongue and leave a lateral space into which the tongue expands laterally, very much as in the "wedging" process described in the mouse. The oral cavity walls collapse medially into the same space to meet the widening tongue, and the oral cavity width is decreased, allowing the horizontal shelves to meet.

In the rabbit the shelves first become horizontal rostrally while still being vertical caudally (Walker, 1971b). No evidence could be found by Walker and Ross (1972) for a pronounced intrinsic shelf force in rabbit embryos where, in contrast to mouse embryos, the palatal shelves did not move when the tongue was displaced from between them.

A plausible suggestion for the "internal shelf force" observed in the rat and mouse is that straightening of the cranial base may increase the tension at the base of the shelf, thus causing the shelf to "curl" medially (Verrusio, 1970). It has been observed that before and during palate closure the cranial base angle increases, thus straightening the cranial base (Hart *et al.*, 1969; Harris, 1964, 1967). It is also possible that the vascular plexus that appears within the shelves 24 hr before closure and expands by the time of palate closure may contribute to the shelf force by engorgement (Gregg and Avery, 1971). Similar claims have been made for an increase in tissue turgor (Lazzaro, 1942). Acid mucopolysaccharides are synthesized rapidly in the shelf area shortly before the shelves move (Walker, 1961) and have been implicated as a possible physical basis for the shelf force (Walker and Fraser 1956, 1957; Larsson, 1962). Analysis of the shelves for actin-like and myosin-like proteins shows their presence at palate closure time, thus providing evidence that these contractile proteins might provide the shelf force (Lessard *et al.*, 1974). Collagen increases in an exponential manner in the embryo and in the shelves at palate closure and thus has been implicated in assisting normal closure (Pratt and King, 1971).

The tongue may play an active part in palate closure. Wragg *et al.* (1972a) showed that electrical stimulation of the rat tongue at the time of palate closure would cause the tongue muscle to contract. Human embryos show mouth opening reflexes (Humphrey, 1969), and rabbit and mouse display active body movements at palate closure time (Walker, 1969, 1974). These observations suggest that the tongue can perhaps move from between the

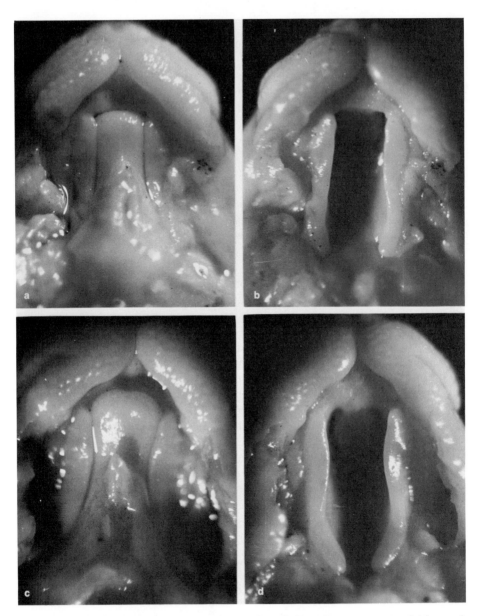

Fig. 2. Palate closure stages. (a) Stage 0, tongue in place; (b) stage 0, tongue removed; (c) stage 1, tongue in place; (d) stage 1, tongue removed.

shelves momentarily or enough to allow the shelves to enter the space thus created above it.

The mandible and tongue grow steadily before and during palate closure, carrying the tongue tip beyond and below the primary palate as observed in the mouse, rat, and human (Shih *et al.*, 1974; Wragg *et al.*, 1970; Hart *et al.*, 1969, 1972).

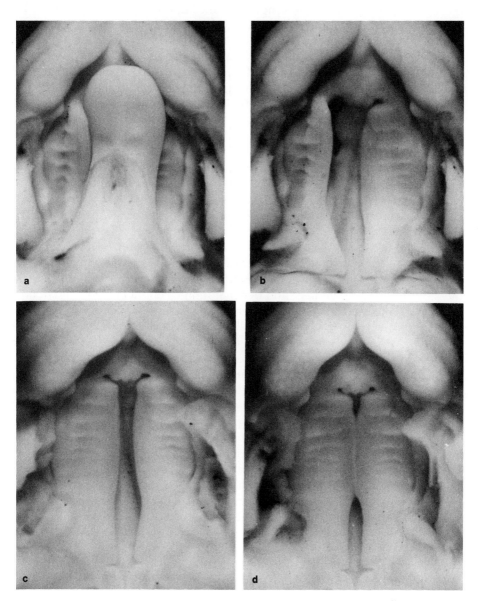

Fig. 3. Palate closure stages. (a) Stage 3, tongue in place; (b) stage 3, tongue removed; (c) stage 4; (d) stage 6.

Fusion of the apposed palatal shelves has been well studied in recent years at the microscopic and ultramicroscopic level. Recently special staining and electron-microscopic observations have shown that the shelf surface is covered with an acid mucopolysaccharide substance (Greene and Kochhar, 1974), and a carbohydrate is found just prior to fusion (Pratt *et al.*, 1973); these are suggested as being the "glue" for shelf adhesion. On initial contact the epithelia put out desmosomes between apposed surface cells and the basement lamina is still intact, according to Morgan's (1969) observations of rat, DeAngelis and Nalbandian's (1968) of mouse and rat, and Chaudhry and Shah's (1973) of hamster palates. Farbman (1968), on the other hand, noted discontinuities of the basement lamina even before contact between the shelves of mouse embryos. All authors agree that soon after initial fusion the basal lamina breaks down and the epithelial seam degenerates by autolysis and phagocytosis (Pourtois, 1970). Accompanying these morphological changes, there is a rise in acid phosphatase activity in the shelf tips before fusion (Pourtois, 1970) and an increase in glycogen granules (Morgan, 1969). Agreement is not complete as to whether degenerative changes occur in the epithelium before contact (Smiley, 1970), whether autophagic vacuoles are present in the epithelium before initial fusion (Mato *et al.*, 1967), and whether fusion is a necessary condition for completion of autolysis in presumptive fusing cells (Vargas *et al.*, 1972).

Morgan's (1969) studies showed that there were a few electron-microscopically dense bodies in prefusion cells, but their number did not change before fusion. A few hours after fusion he found autophagic changes, the basal lamina fragmented, and degeneration of epithelia. For human embryos the same kinds of changes after fusion were described by Matthiesen and Anderson (1972). Vargas *et al.* (1972) used histochemical stains of palates *in vitro* and found only an increase in the lysosomal enzyme glucosaminidase during and after fusion which was localized in the degenerating cells.

Scanning electron-microscopic studies in mouse and human of the surface of the shelves where fusion will take place show changes in these cells. The cells round up, lose cell boundaries, retract, and lose their regular polygonal shape with microvilli outlining their edges (Waterman *et al.*, 1973; Waterman and Meller, 1973b). It would appear then that in some way the cells are prepared for fusion and furthermore are, in fact, programmed to go through the changes that take place at fusion. This was shown by examining shelves that had been prevented from meeting each other, as in the A/J mouse with cleft lip and cleft palate, in shelves cultivated *in vitro* (Angelici and Pourtois, 1968; Smiley and Koch, 1972), and in cleft palate induced with meclozine hydrochloride or mechanically by amniotic fluid withdrawal (Morgan and Harris, 1973). These unfused shelves in the specific area of presumptive fusion showed epithelial cell death, increased acid phosphatase (indicating lysosomal enzyme activity), and accumulation of glycogen and membrane-bounded vacuoles containing heterogenous material. Morgan also noted that the basal lamina beneath this zone was absent, although it was intact on either side. Thus, this specific zone of the palate undergoes degenerative changes, even without tissue contact, as though there must be a programmed cell death taking place. This was also suggested by Hudson and Shapiro (1973), who used radioautography to study mitotic rate in rat shelf epithelia and found that palatal tip epithelia showed a faster decline in mitotic activity than other areas.

B. Cleft Palate

It is obvious from the foregoing that there are a large number of closely interrelated processes taking place at the proper time and position in the embryo to bring about palate closure. Should any one or several of these processes be mistimed or displaced, a cleft palate could result. Thus: (1) A structural abnormality within the shelves would prevent their movement to the horizontal. (2) The shelves could be normal but too narrow and so unable to meet each other in the midline. (3) A reduction in shelf force would bring about a delay in their movement to the horizontal. (4) Anything that interferes with displacement of the tongue at the right time from between the shelves would delay or prevent them from becoming horizontal. (5) Disproportionate growth of the head could make it wider than normal so that the shelves could fail to reach each other. (6) The epithelia of the shelves could fail to fuse (Fraser, 1971).

Many investigators have attempted to find out what particular change in the oral morphological relationships led to cleft palate. Given the complex and closely interrelated nature of the processes involved in palate closure, it has become apparent that although it is easy to cause a cleft palate with a large number of teratogens and procedures, it is very hard to pinpoint the initial deviation. Thus, in the following examples of the possible mechanisms for the production of cleft palate listed above, further investigation may show that the effect of the agent involved is actually being exerted on some other part of the process.

1. Structural Abnormality in Shelves

The palate of the *phocomelic* mutant mouse appears to be cleft because rods of cartilage form where shelf movement normally occurs (Fitch, 1957).

2. Narrow Shelves

A deficient shelf width was observed by Poswillo (1968) and by Masuyama (1959) after doses of X-irradiation to rat embryos sufficient to cause cleft palate in all embryos. These narrow shelves were able to move to the horizontal on time even though there was maxillary and mandibular retardation and the tongue was delayed in its descent. Hypervitaminosis A (Kochhar and Johnson, 1965; Kochhar, 1973) in doses that result in more than 80% cleft palate in rat embryos produced shelves that had a reduced amount of mesenchyme and were either rounded, reduced in size, or even completely absent at their posterior ends. These shelves moved to the horizontal on or before normal time. This teratogen also produced abnormal chondrogenesis in and around the maxillary areas and maxillomandibular ankylosis. The *urogenital* mutant mouse appears to have cleft palate because the shelves are smaller than normal (Fitch, 1957).

3. Reduction in Shelf Force

In cleft palate caused by cortisone treatment (Walker and Fraser, 1957), delay in movement of the shelves from vertical to horizontal was demonstrated both in relation to chronological age, developmental age as measured by various external features (Walker and Fraser, 1957), and embryonic weight (Dostal and Jelinek, 1974). The delay may result from a reduction in shelf force, but this has not been proven; Walker (1974) has suggested that it results, rather, from a myopathy in the tongue muscle, reducing its mobility. Conclusive proof would come from the demonstration of delayed shelf movement in cortisone-treated embryos in the absence of the tongue. If cortisone does act by reducing shelf force, we still do not know how. Does it act by causing an internal change in cell structure or perhaps delay in vascular development, as shown by Gregg and Avery (1971) or an external change such as the cranial base failing to straighten (Harris, 1964) and exert the needed tension on the shelves (Verrusio, 1970)? The cranial base shows selective impairment of cell proliferation in the presphenoid and a delay in straightening of the presphenoid after 6-aminonicotinamide (6-AN) treatment (Long *et al.*, 1973). If the shelf force is indeed due to the cranial base straightening then 6-AN may be reducing it by its effect on the cranial base. The difference between the A/Jax and C57BL/6 strains in the stage at which the shelves move (C57BL/6 earlier than A/Jax, Walker and Fraser, 1956) may reflect a genetically determined difference in the strength of the shelf force, although again, this has not been proven.

4. Tongue Obstruction

a. Cleft Palate Secondary to Cleft Lip.   The cleft palate that is often associated with cleft lip appears to result from mechanical obstruction by the tongue, preventing or delaying the shelves from becoming horizontal. The defect in lip formation results in a large premaxilla or prolabium in the region of the primary palate which appears to stop the tongue from moving forward and downward; rather it remains arched between the shelves and obstructs their movement toward the horizontal (Trasler and Fraser, 1963).

b. Cleft Palate Associated with Oligohydramnios.   Oligohydramnios caused by puncture of the amniotic sac produces cleft palate in mice (Trasler *et al.*, 1956; Walker, 1959) and rats (DeMyer and Baird, 1969; Poswillo, 1966). The reduction in amniotic fluid constricts the embryo so that the flexion of the head increases, the jaw is pressed on the chest, and the movement of the tongue is impeded. Harris (1964) noted that there was a reduction of amniotic fluid, and embryos had an attitude of extreme flexion at palate closure time after cortisone treatment, suggesting that cortisone might produce clefts via tongue obstruction. However, Fraser *et al.* (1967) demonstrated that cortisone-treated embryos had their amniotic fluid reduced equally whether their palates were closed or cleft. Thus the oligohydramnios was probably not the primary factor in determining which embryos in a litter would be cleft, although it could have been a contributing factor by imposing a delay in shelf movement that would cause clefts in embryos that were relatively late in shelf movement (i.e., closer to the threshold of abnormality, see Chapter 3, Vol. 1).

c. Cleft Palate Associated with Micrognathia. Retarded mandible growth has often been found in association with cleft palate, and it has been postulated that a small mandible would prevent the tongue's normal downward and forward movement at palate closure time, thus increasing resistance of the tongue to shelf movement. Abramovich (1972) reported that cleft palate in rats caused by a *lathyrus odoratus* diet was partly associated with short, thick mandibles. However, some fetuses with closed palates had mandibular shortening and some with cleft palates had normal mandibles. Collagen cross-linkage is affected by lathyrogens, and it has been suggested that it may act by reducing shelf force within the shelves (Pratt and King, 1971), rather than by reducing mandibular growth. Although hypervitaminosis A and 6-aminonicotinamide can cause micrognathia and cleft palate in mice, measurements showed no reduction in mandible or tongue length at palate closure time after treatment with either of these drugs or cortisone (Shih *et al.*, 1974). Thus one cannot assume that cleft palate associated with micrognathia at birth (as in the Robin syndrome) was caused by the micrognathia preventing tongue movement. Morgan and Harris (1968) found that meclozine causes cleft palate and that in treated rat embryos Meckel's cartilage was shortened and the lower jaw was short (perhaps by interference with an inductive stimulus of Meckel's cartilage on mandibular growth). They too suggested that the short lower jaw prevented the elevation of the palatal process by retaining the tongue high in the oral cavity.

d. Other Examples. The cleft palate produced by the homozygous mutation *shorthead* in mice apparently results from the disproportions of the head retaining the tongue between the shelves at palate closure time, although the mechanism has not been worked out in detail (Fitch, 1961). Another mutation with cleft palate in mice, *muscular dysgenesis,* has most skeletal muscles affected by day 14½ of gestation, and disorganization of the tongue musculature is seen at palate closure time (Pai, 1965). It is possible that the tongue cannot contract to facilitate shelf closure, as suggested by Wragg *et al.* (1972a). On the other hand, lack of embryonic movements (Walker, 1969, 1974) might prevent head flexion or mouth opening. Since these embryos also have micrognathia this too might prevent the tongue from moving down in the mouth.

5. Wide Head

A disproportionately wide head could prevent the shelves meeting. No proven case of this situation has been reported, although the association of ocular hypertelorism with cleft palate in certain human families might be an example.

6. Failure or Inhibition of Shelf Epithelial Fusion

Contrary to earlier ideas, it has been found recently that, in spite of the delays in shelf movement caused by cortisone (Walker and Fraser, 1957), the shelves do appear in close apposition to each other when examined in frozen sections (Greene and Kochhar, 1973b). Thus the epithelia, although in contact, do not fuse. *In vitro* experiments with excised palates show that hydrocortisone which produces cleft palate *in vivo* causes a retardation of the fusion process but never prevents it (Saxén, 1973; Lahti *et al.*, 1967, 1972). Thus cortisone is probably not interfering with the inductive influence from the mesenchyme which the epithelia appear to depend upon for their competence to fuse (Pourtois, 1972). More likely the programmed epithelial changes have started before the shelves meet, and the shelves fail to fuse because they are no longer competent to do so as a result of the delay in shelf movement. An alternative explanation is that when they do meet the epithelia cannot hold together in a head that is wider than at the normal time of fusion (Dostal and Jelinek, 1974).

Palate shelves from meclozine-treated embryos cultured *in vitro* were able to fuse (Morgan, 1969). Other *in vitro* experiments used β-2-thienylalanine, a phenylalanine inhibitor, which delayed or completely inhibited shelf fusion. It was found that the shelves could meet and form an epithelial band along the full length of the palate; however, there was incomplete autolysis of the epithelial cells, the basement membrane remained intact, and there was no mesenchymal penetration across the epithelial band (Baird and Verrusio, 1973). When hadacidin was applied to palates *in vitro*, the epithelia adhered but as with the previous teratogen no epithelial breakdown or mesenchymal contact took place. If the hadacidin was removed from the culture, the shelves proceeded with the breakdown of the epithelial band between them and there was mesenchymal penetration across it (Fairbanks and Kollar, 1974). Thus direct interference with the process of epithelial fusion seems a likely explanation for the mode of action of the latter two teratogens.

7. Identifying the Mode of Action of Teratogens

It should be clear from the above discussion that it is very difficult to identify in a multifactorial system, the precise means by which a given teratogen acts to produce a given malformation. In very few of the above examples can one say, unequivocally, that the given agent causes a cleft by acting on a particular component of the system in a particular way.

At the biochemical level, a great deal of knowledge has accumulated about changes in the palate during its formation and the alterations in them produced by a variety of teratogens, but there are very few examples where the biochemical changes can be conclusively linked with the morphogenetic process. The nature of the shelf force, for instance, is still not clear. Various authors have related it to mucopolysaccharide synthesis (Walker, 1961; Larsson, 1962), collagen synthesis (Pratt and King, 1971), microfilaments (Lessard *et al.*, 1974), hydration (Jacobs, 1966), engorgement (Gregg and Avery, 1971), tumescence (Lazzaro, 1942), and extension of the cranial base (Harris, 1964, 1967; Verrusio, 1970; Long *et al.*, 1973). Perhaps many or all of them play a role.

Much of the evidence has come from observations on the effects of teratogens known to cause cleft palate, but it is very difficult to prove that the observed change is the one responsible for the error in development (Burdi *et al.*, 1972). Corticoids, for instance, produce many changes. They inhibit mitosis (Bullough, 1962), inhibit messenger RNA synthesis (Zimmerman *et al.*, 1970), inhibit sulfomucopolysaccharide synthesis (Larsson, 1962), increase water retention (Jacobs, 1966), stabilize lysosomal membranes (Weissmann and Thomas, 1963), produce myopathy (Walker, 1971a), and decrease amniotic fluid (Harris, 1964). Nevertheless it is still not clear which of these effects are causally involved in the cleft palate. The reduction in amniotic fluid is equal in those embryos with and without a cleft (Fraser *et al.*, 1967) but it could be a contributing factor, shifting the distribution of liability toward the threshold.

Perhaps some of the lack of progress stems from a tendency to make observations at the wrong time (for example, at the time the shelves are closing rather than right after the teratogen was given, usually several days earlier) or in the wrong place (for example, in the shelves if the primary effect is in the cranial base).

Perhaps a few examples of possible pitfalls would be illuminating. One such is a failure to recognize technical artifacts. It was maintained, for instance, that cortisone causes cleft palate by delaying movement of the shelves (which it does) so that when they do become horizontal, growth of the head has carried them so far apart they are unable to meet (Walker and Fraser, 1957). The conclusion was based on the observation that in fixed A/J embryos from mothers treated with cortisone under circumstances that causes 100% cleft palate, the shelves never do meet. However, use of frozen embryos shows that the shelves do in fact meet (Greene and Kochhar, 1973b), and the space observed in fixed embryos must have been a fixation artifact. Presumably then, the cleft results because the shelves meet at a time when the epithelia are no longer competent to fuse or are not in apposition long enough to achieve fusion. Another difficulty is in distinguishing induced alterations responsible for the developmental error from secondary changes resulting from the developmental error. For example, the changes in sulfur incorporation and hydration observed (Jacobs, 1966) in palatal shelves following cortisone treatment may not be causally related to the failure to close, since they are observed in an abnormal shelf, several days after the beginning of treatment and at a time when the palate should already be closed. The same is true for the changes in sulfation following treatment with vitamin A (Nanda, 1969).

Thus we have very few examples where the mode of action of a teratogen is conclusively demonstrated. Mechanical obstruction by the tongue seems a well-established example, at least in the case of cleft palate associated with cleft lip (Trasler and Fraser, 1963) and following amniotic puncture (Trasler *et al.,* 1956; Poswillo, 1966). Reduction in shelf width is the major cause of cleft palate produced by the homozygous mutant gene *ur* (Fitch, 1957), since the shelves appear to become horizontal at the right time. It is reasonable to assume that lathyrogens act by inhibiting collagen cross-linking, but is the primary effect on the shelf or the cranial base? Obviously, much further work is needed.

In closing, we should mention one further problem that often besets the study of time–position relationships. Every developmental event occurs in relation to many other developmental events, and it often becomes a problem to decide what criterion of embryonic age to use as a measure of when a particular event happens.

Chronological age is not an entirely satisfactory measure of developmental age, as it can be measured only from the time of insemination, not fertilization, and there may be marked variations in developmental stage among embryos in the same uterine horn. Embryonic weight probably correlates better with what is happening developmentally, but the data are scanty. One may also turn to various features indicating the rate of differentiation, such as number of somites (if one is working at appropriate stages), or "morphological" rating as estimated by the stage of development of various external features such as webbing of digits, growth of ear pinna, and appearance of follicles. The relation of palate stage to morphological rating is much clearer than it is to chronological age, since much of the "random" variation is removed (Walker and Fraser, 1956). The problem becomes particularly difficult when one is trying to assess the effect of a teratogen on the timing of a developmental process and the teratogen also affects the timing of other processes. For instance, cortisone delays palate shelf movement with respect to both chronological age and morphological rating, but also speeds up the development of the embryo, as measured by morphological rating, with respect to chronological age. On the other hand, 6-aminonicotinamide delays both palate shelf movement and morphological rating with respect to chronological age. Thus the delaying effect of cortisone is much greater than that of 6-AN if morphological rating is used as the standard rather than chronological age. Cortisone reduced embryonic weight in comparison to morphological rating and/or chronological age, and shelf movement does not appear to be delayed, using weight as the comparison (Dostal and Jelinek, 1974). Which is the "correct" method of comparison?

Probably what is needed is a multidimensional comparison where the effect of the teratogen on palate closure is considered in relation to chronological age, embryonic weight, and morphological rating for various genotypes. Perhaps by this approach one would gain a proper appreciation of the effect of a teratogen on the time–position relationships of a developmental event.

REFERENCES

Abramovich, A., 1972, Cleft palate in the fetuses of lathyric rats and its relation to other structures: Nasal septum, tongue and mandible, *Cleft Palate J.* **9**(1):73.

Angelici, D., and Pourtois, M., 1968, The role of acid phosphatase in the fusion of the secondary palate, *J. Embryol. Exp. Morphol.* **20**:15.

Andersen, H., and Matthiessen, M., 1967, Histochemistry of the early development of the human central face and nasal cavity with a special reference to the movement and fusion of the processes, *Acta Anat.* **68**:473.

Andersen, H., and Matthiessen, M. E., 1968, Single-egg human twin foetuses with harelip and cleft palate, *Acta Anat.* **70**:219.

Bahkdinaronk, A., Eto, K., and Johnston, M. C., 1976, The pathogenesis of malformation following X-irradiation of mouse embryos during neural plate formation (in preparation).

Baird, G., and Verrusio, A. C., 1973, Inhibition of palatal fusion *in vitro* by β-2-thienylalanine, *Teratology* **7**:37.

Bollert, J. A., and Hendrickx, A. G., 1971, Morphogenesis of the palate in the baboon (*Papio cynocephalus*), *Teratology* **4**(3):343.

Bullough, W. S., 1962, Growth control in mammalian skin, *Nature* **193**:520.

Burdi, A., Feingold, M., Larsson, K. S., Leck, I., Zimmerman, E. F., and Fraser, F. C., 1972, Etiology and pathogenesis of congenital cleft lip and cleft palate, an NIDR state of the art report, *Teratology* **6**(3):255.

Chase, H. B., and Chase, E. B., 1941, Studies on an anophthalmic strain of mice, 1. Embryology of the eye-region, *J. Morphol.* **68**:279.

Chaudhry, A. P., and Shah, R. M., 1973, Palatogenesis in hamster. II. Ultrastructural observations on the closure of palate, *J. Morphol.* **139**(3):329.

Chung, C. S., and Myrianthopoulos, N. C., 1968, Racial and prenatal factors in major congenital malformations, *Am. J. Hum. Genet.* **20**:44.

Coleman, R. D., 1965, Development of the rat palate, *Anat. Rec.* **151**:107.

DeAngelis, V., and Nalbandian, J., 1968, Ultrastructure of mouse and rat palatal processes prior to and during secondary palate formation, *Arch. Oral Biol.* **13**:601.

DeMyer, W., and Baird, I., 1969, Mortality and skeletal malformations from amniocentesis and oligohydramnios in rats: Cleft palate, clubfoot, microstomia and adactyly, *Teratology* **2**(1):33.

Deol, M. S., and Lane, P. W., 1966, A new gene affecting the morphogenesis of the vestibular part of the inner ear in the mouse, *J. Embryol. Exp. Morphol.* **16**:543.

Dostal, M., and Jelinek, R., 1974, Morphogenesis of cleft palate induced by exogenous factors VI the question of delayed palatal and process horizontalization, *Teratology* **10**:47.

Fairbanks, M. B., and Kollar, E. J., 1974, Inhibition of palatal fusion in vitro by hadacidin, *Teratology* **9**(2):169.

Farbman, A. I., 1968, Electron microscope study of palate fusion in mouse embryos, *Dev. Biol.* **18**:93.

Fitch, N. S., 1957, An embryological analysis of two mutants in the house mouse, both producing cleft palate, *J. Exp. Zool.* **136**:329.

Fitch, N., 1961, Development of cleft palate in mice homozygous for the shorthead mutation, *J. Morphol.* **109**:151.

Fraser, F. C., 1971, Etiology of cleft lip and palate *in: Cleft Lip and Palate* (W. C. Grabb, S. Rosenstein, and K. R. Bzoch, eds.), Chapter 3, Little Brown, Boston.

Fraser, F. C., and Rosen, J., 1975, Association of cleft lip and atrial septal defect in the mouse, *Teratology* **11**(3):321.

Fraser, F. C., Chew, D., and Verrusio, A. C., 1967, Oligohydramnios and cortisone-induced cleft palate in the mouse, *Nature* **214**:417.

Frederiks, E., 1973, Vascular pattern in embryos with clefts of primary and secondary palate, *Ergeb. Anat. Entwicklungsgesch.* **46**(6):1.

Greene, R. M., and Kochhar, D. M., 1973a, Palatal closure in the mouse as demonstrated in frozen sections, *Am. J. Anat.* **137**(4):477.

Greene, R. M., and Kochhar, D. M., 1973b, Spatial relations in the oral cavity of cortisone-treated mouse fetuses during the time of secondary palate closure, *Teratology* **8**(2):153.

Greene, R. M., and Kochhar, D., 1974, Surface coat on the epithelium of developing palatine shelves in the mouse as revealed by electron microscopy. *J. Embryol. Exp. Morphol.* **31**:683.

Gregg, J. M., and Avery, J. K., 1971, Experimental studies of vascular development in normal and cleft palate embryos, *Cleft Palate J.* **8**:101.

Grüneberg, H., 1952, Genetical studies on the skeleton of the mouse. IV. Quasi-continuous variations, *J. Genet.* **51**:95.

Grüneberg, H., and Truslove, G. M., 1960, Two closely linked genes in the mouse, *Genet. Res. Camb.* **1**:69.

Harris, J. W. S., 1964, Oligohydramnios and cortisone-induced cleft palate, *Nature* **203**:533.

Harris, J. W. S., 1967, Experimental studies on closure and cleft formation in the secondary palate, *Sci. Basis Med. Annu. Rev.*, Chapter XX, p. 356–370.

Hart, J. C., Smiley, G. R., and Dixon, A. D., 1969, Sagittal growth of the craniofacial complex in normal embryonic mice, *Arch. Oral Biol.* **14**:995.

Hart, J. C., Smiley, G. R., and Dixon, A. D., 1972, Sagittal growth trends of the craniofacial complex during formation of the secondary palate in mice. *Teratology* **6**(1):43.

Hudson, C. D., and Shapiro, B. L., 1973, A radioautographic study of deoxyribonucleic acid synthesis in embryonic rat palatal shelf epithelium with reference to the concept of programed cell death, *Arch. Oral Biol.* **18**(1):77.

Humphrey, T., 1969, The relation between human fetal mouth opening reflexes and closure of the palate, *Am. J. Anat.* **125**(3):317.

Iizuka, T., 1973, Stage of the formation of the human upper lip, *Okajimas Folia Anat. Jpn.* **50**(5):307.

Jacobs, R. M., 1966, Effects of cortisone acetate upon hydration of embryonic palate in two inbred strains of mice, *Anat. Rec.* **156**:1.

Jacobson, A. G., 1966, Inductive processes in embryonic development, *Science* **152**:25.

Johnston, M. C., and Listgarten, M. A., 1972, The migration interaction and early differentiation of oro-facial tissues, *in: Developmental Aspects of Oral Biology* (H. S. Slavkin and L. A. Bavetta, eds.), p. 53, Academic Press, New York.

Kochhar, D. M., 1973, Limb development in mouse embryos. I. Analysis of teratogenic effects of retinoic acid, *Teratology* **7**(3):289.

Kochhar, D. M., and Johnson, E. M., 1965, Morphological and autoradiographic studies of cleft palate induced in rat embryos by maternal hypervitaminosis A. *J. Embryol. Exp. Morphol.* **14**:223.

Lahti, A., and Saxen, L., 1967, Effect of hydrocortisone on the closure of palatal shelves *in vivo* and *in vitro*, *Nature* **216**:1217.

Lahti, A., Antilla, E., and Saxén, L., 1972, The effect of hydrocortisone on the closure of the palatal shelves in two inbred strains of mice *in vivo* and *in vitro*, *Teratology* **6**(1):37.

Larsson, K. S., 1962, Closure of the secondary palate and its relation to sulphomucopolysaccharides, *Acta Odontol. Scand.* **20**(Suppl. 31):5.

Lazzaro, C., 1942, Sul meccanismo di chiusura del palato secondario, *Monit. Zool. Ital.* **51**:249.

Lejour, M., 1970a, Cleft lip induced in the rat, *Cleft Palate J.* **7**:169.

Lejour, M., 1970b, Pathogénie des fentes labiomaxillairy provoquées chez le rat par l'hadacidine, *in: Editions de Archives de Biologie,* pp. 1–108, Vaillant-Carmanne, Liège, Belgium.

Lejour-Jeanty, M., 1965, Étude morphologique et cytochemique du développement du palais primaire chez le rat (Complete des observations chez le lapin et l'homme), *Arch. Biol.* **76**:97.

Lejour-Jeanty, M., 1966, Becs de lièvre provoqués chez le rat par un derive de la penicilline, l'hadacidin, *J. Embryol. Exp. Morphol.* **15**:193.

Leong, S., 1973, Face shape and mitotic index in mice with teratogen-induced and inherited cleft lip, M.Sc. thesis, McGill University.

Lessard, J. L., Wee, E. L., and Zimmerman, E. F., 1974, The presence of contractile proteins in mouse fetal palate prior to shelf elevation. *Teratology* **9**(1):113.

Long, S. V., Larsson, K. S., and Lohmander, S., 1973, Cell proliferation in the cranial base of A-J mice with 6-AN-induced cleft palate, *Teratology* **8**(2):127.

Masuyama, Y., 1959, Experimental studies on the genesis of cleft palate due to X-ray irradiation, *Univ. Osaka, J. Dent.* **4**:847.

Mato, M., Aikawa, E., and Katahira, M., 1967, Alteration of fine structure of the epithelium on the lateral palatine shelf during the secondary palate formation, *Gunma. J. Med. Sci.* **16**:79.

Matthiessen, M., and Andersen, H., 1972, Disintegration of the junctional epithelium of human fetal hard palate, *Z. Anat. Entwicklungsgesch.* **137**(2):153.

Morgan, P. R., 1969, Recent studies on the fusion of the secondary palate, *London Hosp. Gaz. Clin. Sci. Suppl.* **62**(3):VI.

Morgan, P. R., and Harris, J. W. S., 1968, Observations on meclozine-induced cleft palate in rat embryos, *J. Dent. Res.* **46**:1271A.

Morgan, P. R., and Harris, J. W. S., 1973, The fate of the palatal epithelium in rat embryos with experimentally induced cleft palate—an ultrastructural study, *Teratology* **8**:230.

Nanda, R., 1969, The normal palate and induced cleft palate in rat embryos, Ph.D. thesis, University of Nijmegen, The Netherlands.

Pai, A. C., 1965, Developmental genetics of a lethal mutation, muscular dysgenesis(mdg), in the mouse. I. Genetic analysis and gross morphology. II. Developmental analysis, *Dev. Biol.* **11**:82; 93.

Poswillo, D., 1966, Observations of fetal posture and causal mechanisms of congenital deformity of palate, mandible and limbs, *J. Dent. Res. Suppl.* 3 **45**:584.

Poswillo, D. E., 1968, Cleft palate in the rat, *Lab. Anim.* **2**:181.

Pourtois, M., 1970, La resorption des murs epitheliaux au cours de la formation du palais primaire et du palais secondaire, *Bull. Group. Int. Rech. Sci. Stomatol.* **13**:465.

Pourtois, M., 1972, Morphogenesis of the primary and secondary palate, *in: Developmental Aspects of Oral Biology* (H. S. Slavkin and L. A. Bavetta, eds.), Chapter 5, Academic Press, New York and London.

Pratt, R. M., Jr., and King, C. T., 1971, Collagen synthesis in the secondary palate of the developing rat, *Arch. Oral Biol.* **16**(10):1181.

Pratt, R. M., Gibson, W. A., and Hassell, J. R., 1973, Concanavalin A binding to the secondary palate of the embryonic rat, *J. Dent. Res.* **52**:111.

Saxén, I., 1973, Effects of hydrocortisone on the development *in vitro* of the secondary palate in two inbred strains of mice, Arch. Oral Biol. **18**(12):1469.

Sharma, L. K., 1974, Median cleft of the upper lip, *Plast. Reconstr. Surg.* **53**(2):155.

Shih, L. Y., Trasler, D. G., and Fraser, F. C., 1974, Relation of mandible growth to palate closure in mice, *Teratology* **9**(2):191.

Smiley, G. R., 1970, Fine structure of mouse embryonic palatal epithelium prior to and after midline fusion, *Arch. Oral Biol.* **15**:287.

Smiley, G. R., and Koch, W. E., 1972, An *in vitro* and *in vivo* study of single palatal processes, *Anat. Rec.* **173**(4):405.

Smith, S. C., and Monie, I. W., 1969, Normal and abnormal nasolabial morphogenesis in the rat, *Teratology* **1**:1.

Trasler, D. G., 1965, Aspirin-induced cleft lip and other malformations in mice, *Lancet* **1**:606.

Trasler, D. G., 1966, Mouse heart size during lip formation in cleft lip embryos, Abstracts of papers of Sixth Annual meeting of the Teratology Society, p. 25, Corpus Christi, Texas.

Trasler, D. G., 1968, Pathogenesis of cleft lip and its relation to embryonic face shape in A/J and C57BL mice, *Teratology* **1**:33.

Trasler, D. G., and Fraser, F. C., 1963, Role of the tongue in producing cleft palate in mice with spontaneous cleft lip, *Dev. Biol.* **6**:45.

Trasler, D. G., and Leong, S., 1974, Face shape and mitotic index in mice with 6-aminonicotinamide-induced and inherited cleft lip, *Teratology* **9**:A39.

Trasler, D. G., Walker, B. E., and Fraser, F. C., 1956, Congenital malformations produced by amniotic-sac puncture, *Science* **124**:439.

Vargas, Idoyaga V., Nasjleti, C. E. D., and Axcurra, J. M., 1972, Cytodifferentiation of the mouse secondary palate *in vitro;* morphological, biochemical and histochemical aspects, *J. Embryol. Exp. Morphol.* **27**(2):413.

Vermeij-Keers, C., 1972, Transformations in the facial region of the human embryo, *Ergeb. Anat. Entwicklungsgesch.* **46**(5):3.

Verrusio, A. C., 1970, A mechanism for closure of the secondary palate, *Teratology* **3**:17.

Walker, B. E., 1959, Effects on palate development of mechanical interference with the fetal environment, *Science* **130**(3381):981.

Walker, B. E., 1961, The association of mucopolysaccharides with morphogenesis of the palate and other structures in mouse embryos, *J. Embryol. Exp. Morphol.* **9**(1):22.

Walker, B. E., 1969, Correlation of embryonic movement with palate closure in mice, *Teratology* **2**:191.

Walker, B. E., 1971a, Induction of cleft palate in rats with antiinflammatory drugs, *Teratology* **4**(1):39.

Walker, B. E., 1971b, Palate morphogenesis in the rabbit, *Arch. Oral Biol.* **16**:275.

Walker, B. E., 1974, Palate closure in strain CD-1 mice, *J. Dent. Res.* **53**:1497.

Walker, B. E., and Fraser, F. C., 1956, Closure of the secondary palate in three strains of mice, *J. Embryol. Exp. Morphol.* **4**:176.

Walker, B. E., and Fraser, F. C., 1957, The embryology of cortisone-induced cleft palate, *J. Embryol. Exp. Morphol.* **5**:201.

Walker, B. E., and Ross, L. M., 1972, Observation of palatine shelves in living rabbit embryos, *Teratology* **5**(1):97.

Warbrick, J. G., 1960, The early development of the nasal cavity, *J. Anat.* **94**:351.

Waterman, R. E., and Meller, S. M., 1973a, Nasal pit formation in the hamster; a transmission and scanning electron microscopic study, *Dev. Biol.* **34**:255.

Waterman, R. E., and Meller, S. M., 1973b, A scanning electron microscope study of secondary palate formation in the human, *Anat. Rec.* **175**(2):464.

Waterman, R. E., Ross, L. M., and Meller, S. M., 1973, Alterations in the epithelial surface of A-Jax mouse palatal shelves prior to and during palatal fusion: A scanning electron microscopic study, *Anat. Rec.* **176**(3):361.

Weissmann, G., and Thomas, L., 1963, Studies on lysosomes II. The effect of cortisone on the release of acid hydrolases from a large granule fraction of rabbit liver induced by an excess of vitamin A, *J. Clin. Invest.* **42**:661.

Wragg, L. E., Klein, M., Steinvorth, G., and Warpeha, R., 1970, Facial growth accommodating secondary palate closure in rat and man, *Arch. Oral Biol.* **15**:705.

Wragg, L. E., Smith, J. A., and Borden, C. S., 1972a, Myoneural maturation and function of the foetal rat tongue at the time of secondary palate closure, *Arch. Oral Biol.* **17**(4):673.

Wragg, L. E., Diewert, V. M., and Klein, M., 1972b, Spatial relations in the oral cavity and the mechanism of secondary palate closure in the rat, *Arch. Oral Biol.* **17**(4):683.

Zimmerman, E. F., Andrew, F., and Kalter, H., 1970, Glucocorticoid inhibition of RNA synthesis-responsible for cleft palate in mice: A model, *Proc. Natl. Acad. Sci. U.S.A.* **67**(2):779.

Teratogenesis and Oncogenesis

ROBERT PAUL BOLANDE

I. INTRODUCTION

In recent years, numerous and varied relationships have been shown to exist between congenital malformations and neoplasms (Di Paolo and Kotin, 1966; Miller, 1969; Foulds, 1969; Warkany, 1971; Bolande, 1973b). The extent and complexity of these relationships is now only beginning to be appreciated as more and more associations are documented. Indeed, the kinship of teratogenesis and oncogenesis appears to be of fundamental importance in human developmental pathobiology. Its appreciation significantly adds to the understanding of the neoplastic process in general. It is the purpose of this chapter to document and examine the existing body of data concerning these relationships and show how in certain instances they may be held accountable for the pathogenesis of neoplastic disorders.

We shall be concerned primarily with structurally demonstrable anomalies of organs and tissues, rather than those teratologic disorders whose primary or sole manifestations are congenital derangements or insufficiency of cell function.

II. TUMORS IN EARLY LIFE

A. General Features

In order to deal with the subject in depth, we must consider the peculiarities and unique features of neoplasms of early life, distinguishing them in

ROBERT PAUL BOLANDE • Director of the Department of Pathology, The Montreal Children's Hospital, and Professor of Pathology and Paediatrics, McGill University, Montreal, Quebec, Canada.

many respects from those occurring in later life. Common adult cancers appear to arise by a regressive mutation of cells within mature tissue, in cells retaining an ability to multiply and regenerate in adulthood. Other adult cancers may arise in developmentally anomalous tissue, or their appearance is enhanced or predetermined by other inborn defects. In contrast, the most common solid tumors of childhood are manifested in very early life, sometimes at birth, and are characterized by unique cellular features indicating an origin in abnormal embryogenesis. When malignant, they are rapidly progressive and highly lethal. The susceptibility of the young host to the malignant process has been demonstrated repeatedly by an enhanced growth of transplanted tumors in young animals. Moreover, oncogenic viruses and chemical carcinogens more readily induce tumors in the young host than in mature ones (Foulds, 1969). Paradoxically, regression and cytodifferentiation occur most often in human tumors of early life (Everson and Cole, 1966; Bolande, 1971).

Human tumors characteristic of early life have been admirably summarized and classified by Willis (1962). They are often characterized by an overgrowth of well-differentiated cells and tissues in either orderly or chaotic arrangements. The cells may be indigenous or alien to the site of involvement. A tumor may also contain persistent embryonal or fetal tissues, indicating a failure of proper maturation or cytodifferentiation in intrauterine or postnatal life. Some tumors that are themselves nonmalignant become the seat of malignant transformation in later life. In other instances, tumors may persist in a latent or cryptic form for long periods after birth, becoming manifest later in childhood or even in adult life.

B. Definitions and Descriptions

The important tumors to be considered here are hamartomas and hamartoses, teratomas, and embryomas (Willis, 1962; Bolande, 1967).

1. Hamartomas

"Hamartoma" is a term used to describe a tumor-like mass of tissue, apparent at or near the time of birth, that is composed of an excess of more-or-less normal tissue indigenous to its site of origin. Its capacity for growth is limited, often paralleling that of the host, and its biologic behavior is benign. The distinction from malformation is often difficult. A hamartoma may be unifocal or multifocal. Multifocal hamartomas are referred to as *hamartoses*. Hamartomas may exhibit a multiplicity of forms at different sites within a given individual, wherein they are referred to as *pleiotropic hamartoses*. The important hamartomas and hamartoses are listed in Table 1. Some of these, as indicated, are genetically determined.

Table 1. Hamartomas and Hamartoses

Tissue of origin	Pathologic examples	Genetics	Neoplastic transformation
Vascular	Congenital hemangiomas and vascular nevi of skin		
	Lymphangiomas and cystic hygromas		
	Milroy disease	Autosomal dominant	
	Multiple glomangiomas	Autosomal dominant	
	Angiomatoses:		
	Skin and Viscera		
	Hereditary mucocutaneous telangiectasia (Rendu-Osler)	Autosomal dominant	
	Facial and intracranial (Sturge-Weber)	?	
	Brain and retina (von Hippel-Lindau)	Autosomal dominant	Hypernephroma, pheochromocytoma
Connective tissue	Congenital fibromatosis	Autosomal dominant ?	
	Familial cervical lipo-dystrophy (symmetrical lipomatosis)	Autosomal dominant	
Skeletal	Multiple exostosis	Autosomal dominant	Chondrosarcoma, rare
	Enchondromatosis (Ollier disease)		Chondrosarcoma, rare
	Fibrous dysplasia of bone	Autosomal dominant?	Osteogenic sarcoma, rare
Skin	Congenital melanotic nevi		Melanoma, rare
	Linear nevus sebaceous (Jadassohn)		Brain tumor
	Basal cell nevus syndrome	Autosomal dominant	Basal cell carcinomas Medulloblastoma
Intestine	Multiple familial polyposis	Autosomal dominant	Carcinoma of colon, common
	Peutz-Jegher syndrome	Autosomal dominant	Carcinoma, rare
Pleiotrophic hamartoses: (multiple sites of origin)	Tuberous sclerosis	Autosomal dominant	Brain glioma
	Maffucci syndrome		Chondrosarcomas of bone, angiosarcomas of skin, rare
	Gardner syndrome	Autosomal dominant	Carcinoma of colon, fibrosarcomas of bone or soft tissue

(cont'd)

Table 1 (cont'd).

Tissue of origin	Pathologic examples	Genetics	Neoplastic transformation
	Cowden syndrome	Autosomal dominant	Carcinoma of breast, thyroid, bowel
Neurocristopathic hamartoses	von Recklinghausen's neurofibromatosis	Autosomal dominant	Neurogenic sarcoma, pheochromocytoma
	Multiple mucosal neuroma syndrome	Autosomal dominant	Medullary carcinoma of thyroid, pheochromocytoma
	Sipple syndrome		Same as above
	Neurocutaneous melanosis		Neurogenic sarcoma, melanoma

2. Teratomas

A teratoma is a tumor formed of a multiplicity of tissues derived from more than one primitive germ layer. Component tissues are often alien to the site of development and tend to be arranged in a haphazard and confused fashion. Besides the diverse and heterotopic constituents of teratomatous tissue, there is also asynchronous maturation of its various parts. Cells of embryonal, fetal, or adult character may be jumbled together. Frequently much of the tissue is of glial and neuroid type, including ganglion cells, ependyma, and choroid plexus. Epithelium may be differentiated into acinar, cystic, and ductlike structures. Abortive attempts at organogenesis are apparent. Teratomatous tissue may be solid, multicystic, or arranged about a single large cyst. Increase in size of a teratoma may result from distension of cystic spaces with the accumulated products of lining cells. It may also result from the proliferative activity of fetal or embryonal constituents. This form of growth may be transient and rapid, only to abate with maturation and cytodifferentiation of the active tissues at some later stage of development. In some instances, teratomas may become the seat of unbridled malignant growth, in which malignant embryonal tissue all but replaces the organoid patterns of the original tumor. Typically, teratomas develop in the testis, ovary, retroperitoneum, mediastinum, and sacrococcygeal regions. Less common sites include the bases of the skull, pineal region, brain, and neck. The sacrococcygeal teratoma is the most common teratoma in infancy.

3. Embryomas

Embryomas are tumors arising in primitive organ and tissue blastema, in cells already committed to a specific type of histogenesis. The tumor is thus restricted to specific organs or tissues and is composed of cells retaining many of the features of a more primitive stage of their development. The tumors in this group are among the most common malignant neoplasms of infancy and childhood. They include neuroblastoma, Wilms' tumor, hepatoblastoma, retinoblastoma, medulloblastoma, embryonal sarcomas, and rhabdomyosarcoma. They also show the highest incidence of regression and cellular maturation.

C. Characteristics of Major Tumors

1. Wilms' Tumor

a. General Features. Wilms' tumor is a malignant embryoma of the kidney derived from metanephric blastema. As such it is composed of an admixture of mesoblastic stroma and primitive nephronoblastic epithelium arranged in sheets and nests containing prominent foci of poorly formed or dysplastic tubules and glomeruli. It has been estimated that about 1 out of 10,000 live births are at risk of developing this tumor (Glenn and Rhame, 1961). The incidence is small: 6–7 per million children under age 15 per year (Young and Miller, 1975). Wilms' tumor is discovered most often between 3 and 4 years of age and accounts for approximately 20% of all malignant conditions in childhood. At this age it is extremely malignant, but cure rates are well in excess of 50% (Martin and Rickham, 1974; Sutow *et al.*, 1970). Cure rates in Wilms' tumor presenting under 1 year of age have always been much better, usually over 80% (Bachmann and Kroll, 1969; Klapproth, 1959).

Wilms' tumors are bilateral in 5–10% of cases. The average age of patients with simultaneously developing bilateral Wilms' tumors is much younger than in those with unilateral Wilms' tumors.

b. Congenital and Infantile Congeners of Wilms' Tumor. Primary renal tumors of infants and children are generally diagnosed as Wilms' tumor. They are thus presumed to have an implacably malignant potentiality and are vigorously treated. In reality, this group of tumors is not monolithic, showing considerable variability in clinical behavior and morphologic characteristics, particularly when detected at birth or within the first few months of life. At this point of development, renal neoplasia is expressed in several clinical and pathologic forms, all of which are significantly less aggressive than conventional Wilms' tumor, if not completely benign. A truly life-threatening, metastasizing Wilms' tumor of conventional morphology occurring within the first few months of life must be exceedingly rare, although a few recent reports seem to have confirmed its existence (Wexler *et al.*, 1975).

Three more or less distinct clinical–pathologic entities are distinguishable from conventional Wilms' tumor in the first months of life (Table 2). The foremost of these is the *congenital mesoblastic nephroma* of infancy (Bolande *et al.*, 1967; Bolande, 1973a), sometimes referred to as fetal renal hamartoma

**Table 2. Congenital and Infantile
Congeners of Wilms' Tumor**

Congenital mesoblastic nephroma of infancy
Well-differentiated epithelial nephroblastomata
   1. Monomorphic epithelial Wilms' tumor
   2. Polycystic nephroblastoma or cystic Wilms'
      tumor
   3. Papillary adenoma
Nephroblastomatosis and nodular renal blastema

(Wigger, 1969) or leiomyomatous hamartoma (Kay *et al.*, 1965; Bogdan *et al.*, 1973). The tumor is typically detected at birth or shortly thereafter by virtue of its huge size. It is composed predominantly of a fibrous or mesenchymal stroma in which bizarre and dysplastic tubules are focally and irregularly scattered. The tumor is essentially benign and curable by nephrectomy alone, although a few cases showing locally invasive and recurrent behavior have been encountered (Fu and Kay, 1973; Bolande, 1974a,b). This tumor is the most common form of congenital renal neoplasia. It has often been misdiagnosed as "congenital Wilms' tumor." Less appreciated is a group of well-differentiated epithelial nephroblastomata. These tumors are essentially monomorphic, composed of well-differentiated nephronoblastic epithelium forming closely apposed tubules or macrocysts, or papillary forms (Bolande, 1974b). These are also relatively benign.

    These tumors are considered to be cytodifferentiated kindred of conventional Wilms' tumor evolving from metanephric blastema during fetal life. As such they are probably initiated earlier in development than conventional Wilms' tumors.

    In recent years, still another type of congenital renal disease has been recognized which is referred to as *diffuse nephroblastomatosis or nodular renal blastema* (Hou and Holman, 1961; Bove *et al.*, 1969). It is generally viewed as neoplastic, although its true nature and malignant potential is uncertain. This disease is characterized by the presence of discrete subcapsular nodules of primitive metanephric epithelium, resembling Wilms' tumorlets. While the lesion may affect one kidney, it is more often bilateral. The nodules are often microscopic in size, and they are identified only as incidental findings at autopsy. They are found in from 1:200 to 1:400 pediatric autopsies under 4 months of age. After this time, they generally disappear from the general pediatric autopsy population, suggesting that a high proportion of the lesions regress. Rarely they may be massive and confluent, replacing, in the extreme situation, the entire outer portion of the renal cortex. A number of cases fitting this description have been reported as diffuse bilateral Wilms' tumor and prolonged survival has been observed following treatment (Bolande, 1974b). In general, the smaller, more localized lesions have been referred to as *nodular renal blastema,* and the more diffuse forms have been called *nephroblastomatosis.*

The ultimate fate and progression of nephroblastomatosis is unclear and controversial, particularly in relationship to the development of frank Wilms' tumor. A number of investigators regard these lesions as nephroblastoma *"in situ,"* emphasizing that a certain proportion of these congenital lesions may give rise to Wilms' tumor later in life (Shanklin and Sotelo-Avila, 1969; Potter, 1972). The situation is somewhat analogous to neuroblastoma *in situ* (see below). Bove *et al.* (1969) found foci of nodular renal blastema in 8 out of 46 kidneys removed for true Wilms' tumor; in 5 of these 8 cases, the Wilms' tumor was bilateral. In their Wilms' tumor population, the incidence of nodular renal blastema was about 50 times greater than the general autopsy population.

c. Familial Wilms' Tumor.   Wilms' tumor occurring in siblings has been reported by a number of authors. Brown *et al.* (1972) reported Wilms' tumor affecting 4 individuals in 3 successive generations in one family. Knudson and Strong (1972) have reviewed and analyzed these familial occurrences of Wilms' tumor. They estimated that about 62% of Wilms' tumor are nonhereditary, while in the remainder genetic factors are involved. In heredofamilial Wilms' tumor, the incidence of bilaterality is much greater and the median age at diagnosis of the tumor(s) is significantly younger. These authors hypothesize that the development of familial Wilms' tumor is determined in part by an autosomal dominant gene. It is apparent that some of these familial cases of bilateral Wilms' tumor are in reality either diffuse nephroblastomatosis or frank Wilms' tumor arising in relationship to nephroblastomatosis.

2. Neuroblastoma

a. General Features.   Neuroblastoma is a clinically important and common malignant embryoma of early life occurring in approximately 1 in 10,000 live births (Beckwith and Perrin, 1963). It is formed of primitive neuroblasts derived from the neural crest. The tumor arises in the adrenal medulla or from some part of the abdominal, thoracic, pelvic, or cervical chains of autonomic ganglia. The organ of Zuckerkandl is an important extra-adrenal site of origin. Although neuroblastoma may be an extremely lethal malignant tumor, particularly when it appears after one year of age, it is characterized by a remarkable incidence of spontaneous regression when present at birth or within the first few months of life (Bachmann and Kroll, 1962), mainly accounting for excellent survival figures at that time (Table 3). Such regressive behavior seems to be the rule in *congenital neuroblastoma,* even in the presence of widespread visceral and skeletal metastases (Becker *et al.,* 1970; Reilly *et al.,* 1968).

Maturation to ganglioneuroma through cytodifferentiation is the best-documented form of regression. Cytodifferentiation is characterized by a transformation of primitive neuroblasts into mature ganglion cells. These

**Table 3. Survival and Age
in Neuroblastoma**[a]

Age of onset	2-year survival (%)
Neonatal or congenital	62–70
Before 1 year of age	35
2nd year	19
After 2nd year	5

[a]Statistics pooled from survey of Surgical Fellows of
American Academy of Pediatrics (1968), Schneider
et al. (1965), and Bachmann and Krolle (1962).

cells become embedded in a dense, proliferating stroma of neuroid connective tissue having the appearance of a neurofibroma or Schwannoma. With progressive degeneration and loss of these ganglion cells in later life, the lesion becomes virtually indistinguishable from neurofibroma or Schwannoma. In those cases where cytodifferentiation of this sort occurs in multiple foci, particularly in the skin, as characteristically occurs in congenital neuroblastoma, the condition produced closely resembles von Recklinghausen's disease (Griffin and Bolande, 1969; Bolande and Towler, 1970).

b. Familial Neuroblastoma. There is little to suggest that neuroblastoma has a heredofamilial basis. Familial recurrence has been described in some 29 cases (Chatten and Voorhees, 1967; Griffin and Bolande, 1969; Knudson *et al.*, 1973). Despite the rarity of familial neuroblastoma, it remains possible that genetic factors exist in its pathogenesis but may be obscured in several ways. Knudson and Amromin (1966) have hypothesized that neuroblastoma may be due to a dominant mutant gene obscured by the early lethality of the disease in the past. In addition to lethality, spontaneous regression or transformation to forms no longer recognizable as neuroblastomatous in origin (ganglioneuroma, neurofibroma, pheochromocytoma) would have the same effect. (Bolande and Towler, 1970; Bolande, 1974b).

c. Neuroblastoma "*in situ.*" The condition designated neuroblastoma "*in situ*" merits serious consideration in any discussion of neuroblastoma. Neuroblastoma *in situ* refers to prominent nodular aggregates of primitive neuroblasts in the central portion of the adrenal glands incidentally found in perinatal autopsies (Beckwith and Perrin, 1963). Depending on the intensity of search and the minimal size of the lesion acceptable to the investigators, neuroblastoma *in situ* has been estimated to occur at anywhere from 1 in 39 to 1 in 500 pediatric autopsies (Beckwith and Perrin, 1963; Shanklin and Sotelo-Avila, 1969: Guin *et al.*, 1969). It is generally agreed that the nodules are no longer identifiable after 3 months of age, either having regressed or cytodifferentiated into patterns indistinguishable from the normal constituents of the adrenal medulla. A recent study suggests that such lesions are almost universally present in fetuses of 10–30 weeks gestation, but tend to regress following 20 weeks gestation (Turkel and Itabashi, 1974).

Neuroblastoma *in situ* over 2 mm in size probably represents a precursive form of overt neuroblastoma. If this is true, its prevalence at birth, as compared to the much lower incidence of clinically overt neuroblastoma, suggests frequent regression or cytodifferentiation of a large number of incipient neuroblastomas. The possibility of this pathogenetic relationship should be appreciated when attempting to delineate heredofamilial patterns and teratogenic relationships in neuroblastoma.

d. Associated Malformations. A wide variety of malformations have been described in isolated case reports of children with neuroblastoma, but no special pattern or relationships are apparent from reviews of the literature and the large series of pediatric cancer deaths (Sy and Edmonson, 1968; Miller *et al.*, 1968). No statistically significant increases in concurrence of anomalies and the tumor are apparent.

e. Neuroblastoma and the Concept of Neurocristopathy. The neural crest origin of neuroblastoma, the similarity of its regressed and differentiated forms to von Recklinghausen's disease, and its occasional concurrence with that disease, as well as pheochromocytoma, nonchromaffin paraganglioma, and Hirschsprung's disease, places it prominently as one of the *neurocristopathies* (Bolande, 1974c). This term has been used to designate a group of dysgenetic, hamartomatous, neoplastic conditions, occurring as individual lesions or as consortia of multiple, variegated lesions, forming syndromes or complexes. These are all pathogenetically united by a shared origin in the maldevelopment of the neural crest and its derivatives.

3. Retinoblastoma

Retinoblastoma is an embryoma formed of the precursors of rod and cone cells of the retina. These tumors arise in the posterior portions of the inner and outer nuclear layers of the retina; they may be multifocal and bilateral. Mortality is associated with direct extension into the cranial cavity to involve the brain and leptomeninges. Hematogenous metastases to bone, lymph nodes, liver, spleen, and kidney may occur. The mortality of this tumor has been so sharply reduced by prompt enucleation of the affected globe(s), coupled with irradiation and chemotherapy, that many affected individuals have lived to procreate.

The tumor, being of neuroectodermal origin, shares many similarities with neuroblastoma, both in histologic appearance and behavior. Spontaneous regression occurs in retinoblastoma as in neuroblastoma and appears to be heralded by hemorrhagic necrosis of the tumor leading to fibrotic and calcified residua (Hiatt *et al.*, 1961).

Retinoblastoma is most notable among the embryomas of early life for its striking hereditary pattern, best shown by its occurrence in 50% of the offspring of parents cured of bilateral disease, in a manner consistent with the inheritance of a dominant gene. About 60% of all retinoblastomas are nonhereditary and unilateral, 15% are unilateral and hereditary, and 25% are hereditary and bilateral. In hereditary cases, the tumors tend to appear a year earlier than in the nonhereditary cases (Knudson, 1971; Knudson *et al.*, 1973). Sporadic cases may be associated with a spectrum of malformations (Jensen and Miller, 1971). This matter will be discussed later.

4. Sacrococcygeal Teratoma

Sacrococcygeal teratoma is the most important teratoma clinically manifested in the newborn infant. It occurs in 1 out of 20,000–40,000 live births and is four times more frequent in females than males. If present under 4 months of age, less than 10% are malignant, while after this time over 70% are malignant (Dillard *et al.*, 1970; Donnellan and Swenson, 1968; Ghazali, 1973; Vaez-Zadeh *et al.*, 1972), most often an embryonal or yolk-sac carcinoma.

5. Benignity of Neonatal and Infantile Tumors

Tumors manifested at birth or shortly thereafter display surprisingly benign behavior despite malignant cellular features, whereas a tumor of identical morphologic features in later life would be highly lethal. In addition to the examples cited above, hepatoblastoma occurring in infancy has a better prognosis than in later childhood (Ishak and Glunz, 1967). Similarly, yolk-sac carcinoma of the infantile testis is followed by excellent survival following orchiectomy. This favorable prognosis is surprising considering the tumor's bizarre and anaplastic histologic features. The same tumor after 2 years of age is very malignant (Pierce *et al.*, 1970; Young *et al.*, 1970). Congenital leukemia is characterized by spontaneous remissions. The prognosis of Burkitt's lymphoma under a year of age is much better than after this time (Lowry, 1974) (Table 4).

**Table 4. Tumors of Infancy Showing Benign or
Regressive Tendencies**

Infantile congeners of Wilms' tumor
Neuroblastoma
    1. Congenital and infantile neuroblastomas
    2. Neuroblastoma *"in situ"*
Sacrococcygeal teratoma before 4 months of age
Yolk-sac carcinoma of infantile testis before 2 years of age
Hepatoblastoma (?) before 1 year of age
Congenital leukemia
Burkitt's lymphoma under 1 year of age
Retinoblastoma (?)

The realization of these facts about tumors in early life has led us to postulate that an oncogenic *period of grace* exists, beginning *in utero* and extending through the first months of life. During this time the host is resistant to the full expression or progression of malignant disease (Bolande, 1971). Neoplasms seem repressed during this period, tending toward benignity through arrested growth, regression, or cytodifferentiation. The mechanism(s) responsible for this are unknown.

III. ORIGIN OF NEOPLASMS IN CONGENITALLY ANOMALOUS OR DYSPLASTIC TISSUE

An important factor predisposing to cancer in later life is the presence of developmentally anomalous tissue. Thus cancer may develop in heterotopias (fetal rests and choristomas), developmental vestiges, hamartomas, and dysgenetic gonads (Willis, 1962; Bolande, 1967, 1973b). Some oncologists have suggested that most tumors of early life arise in embryonal rests (Foulds, 1969).

A. Tumors Arising in Developmental Vestiges and Heterotopias

Developmental vestiges are the residue of transient structures of normal embryogenesis (Willis, 1962). Examples of tumors arising in developmental vestiges are well-established histogenetically (Warkany, 1971; Willis, 1962; Bolande, 1973b). The craniopharyngioma develops from parapituitary vestiges of Rathke's pouch. Chordoma is derived from notochordal remnants persisting within the nucleus pulposus. Residua of the mesonephric duct or Wolffian ducts in the female give rise to a peculiar papillary adenocarcinoma of the cervix and uterus or ovary and broad ligament in young girls. A number of rare carcinomas have been described as developing in various vestiges such as branchial cleft cysts, thyroglossal duct cysts, and urachal remnants.

B. Tumors Arising in Undescended Testes

One of the most common forms of heterotopia is cryptorchidism, affecting 0.28% of the male population. The incidence of tumor is 15–40 times more common in undescended testes than it is in intrascrotal testes. Between 3 and 11% of all testicular neoplasms develop in cryptorchid testes (Kissane and Smith, 1967). By far the most common tumor that develops is seminoma. It has been suggested that the extrascrotal testis is a minimal manifestation of disordered sexual differentiation, since undescended testes are the rule in male pseudohermaphroditism. It has been further suggested that foci of testicular dysgenesis are present in undescended testes, and it is from these foci that tumors arise in later life, typically after puberty (Sohval, 1956).

C. Gonadal Tumors in Somatosexual Disturbances

Aberrant sexual development favors the development of gonadal neoplasms in later life (Table 5). The undescended testes of male pseudohermaphrodites share the susceptibility to seminoma seen in cryptorchid normal males. Of tumors occurring in undescended testes of male pseudohermaphrodites, 60% are seminomas (Gilbert, 1942). By contrast, Sertoli cell tumors are more common in the feminizing, estrogenic testes of the testicular feminization syndrome (Morris, 1953), occurring in about 20% of these patients.

A unique form of neoplasm occurring in dysgenetic gonads is the gonadoblastoma or gonocytoma (Teter and Boczkowski, 1967; Scully, 1970). The gonadoblastoma is defined by Scully as a neoplasm containing an intimate mixture of germ cells and elements resembling immature granulosa or Sertoli cells. Nests and masses resembling huge Call–Exner bodies are formed. These bodies are subject to marked calcification. The tumor may progress into or be associated with a dysgerminoma–seminoma type of pattern. This neoplastic process is often bilateral. Less frequently, embryonal teratoma, embryonal carcinoma, endodermal sinus tumor, or choriocarcinoma may supervene.

It has been estimated that gonadoblastomas and/or dysgerminomas develop in 30–40% of dysgenetic gonads (Hamerton, 1971; Talerman, 1971). Most of the tumors arise in a gonad of indeterminate nature, but 22% develop in a primitive atretic ovary and 18% in a cryptorchid, dysgenetic testis (Scully, 1970). Eighty percent of gonadoblastomas develop in phenotypic females, often showing some evidence of virilization or primary amenorrhea. The remainder develop in phenotypical males with cryptorchidism, hypospadias, and female internal secondary sex characteristics.

Buccal smears are chromatin-negative in most gonadoblastoma patients, the majority exhibiting a normal male karyotype. A lesser number show XO/XY mosaicism. Although some of the individuals may have one or more features of Turner syndrome, gonadoblastomas have not been shown to occur commonly in XO individuals. The presence of a Y chromosome in the karyotype seems to be necessary to increase the expectation of gonadoblastoma.

D. Malignant Transformation in Hamartomas

Hamartomas are often the seat of malignant transformation. This usually occurs later in life. It is difficult to determine the actual increase in susceptibil-

Table 5. Tumors in Dysgenetic Gonads

Condition	Clinical features	Karyotype	Gonad	Tumor
Male, normal	Normal	XY	Undescended testis	Seminoma
Male, pseudohermaphrodite	Undescended testes, hypospadias, variable internal genitalia	XY or mosaics with Y chromosome	Undescended testes	Seminoma, gonadoblastoma
Testicular feminization syndrome	Normal female habitus, vagina, no uterus or tubes, undescended testes	XY	Dysgenetic testes	Sertoli cell or tubular adenoma
Pure gonadal dysgenesis	Eunuchoid female habitus, uterus, and fallopian tubes	XY or mosaics	Dysgenetic testes	Gonadoblastoma–dysgerminoma–seminoma
Female pseudohermaphrodite, nonadrenal	Female habitus with variable masculinization	XY or XO/XY and other mosaics with Y chromosome	Streak ovaries or indeterminate gonads	Gonadoblastoma–dysgerminoma
Mixed gonadal dysgenesis	Asymmetric gonads and internal genitalia	XO/XY and other mosaics with Y chromosome	Streak or indeterminate gonad on one side	Gonadoblastoma
	Ambiguous external genitalia		Contralateral testis	Seminoma, Sertoli cell adenoma
True hermaphrodite	Variable external and internal genitalia and secondary characteristics	XX or mosaicism	Bilateral ovo-testes; asymmetrical testis and ovary	Dysgerminoma
	Usually male habitus		Ovo-testis and contralateral ovary or testis	

ity to cancer over nonhamartomatous tissue, and in some hamartomas the incidence of cancer is very high. Thirty percent of cases of von Reckling-hausen's disease may develop sarcomatous transformation of neurofibroma after age 50 (Brasfield and Das Gupta, 1972). Virtually all patients with multi-ple familial polyposis of the colon eventually develop carcinoma of the colon.

The salient features of malignant transformation in hamartomas and hamartoses are documented in Table 1. Malignancy seems to be documented most frequently in genetically determined hamartoses, rather than in the sporadic isolated lesions. The basis for this is unknown.

IV. INCREASED EXPECTANCY OF NEOPLASMS IN TERATOLOGIC DISORDERS

A. General Considerations

It is now well established that congenital malformations and neoplasms occur together with unusual frequency, both as a general phenomenon and in specific teratologic conditions. The elucidation of many of these important relationships is largely due to the pioneering epidemiologic studies of R. W. Miller and co-workers at the National Cancer Institute (Miller, 1966, 1969).

In an analysis of 371 carefully studied cases of childhood malignant dis-ease at the University of Tokyo, 41% of these children showed evidence of congenital malformation, in contrast to a 13% incidence in children without malignant neoplasms. Wilms' tumor showed the highest incidence (58%), fol-lowed by lymphoreticular malignancies (48%), hepatoblastoma (45%), leukemia (44%), neuroblastoma and retinoblastoma (35% each), brain tumors (28%), and testicular–ovarian tumor (17%) (Kobayashi *et al.*, 1968). Berry *et al.* (1970) could find no significant increase in anomalies in neuroblastoma, Wilms' tumor, and hepatoblastoma, but the search for malformation was not as detailed as in the Japanese study.

B. Specific Teratologic Disorders

1. Aniridia–Wilms' Tumor

Aniridia is a dominantly inherited condition affecting no more than 1:50,000 of the general population (Miller, 1966, 1969; Fraumeni and Glass, 1968). About 30% of cases are sporadic and presumed to represent new germinal mutations. The presence of this congenital anomaly somehow ren-ders the affected child prone to the development of Wilms' tumor. Aniridia is present in 1 out of 80 Wilms' tumor cases. The sporadic aniridias seem more at risk, as about one third of these develop Wilms' tumor.

Additional but less common features of the aniridia–Wilms' tumor syndrome are microcephaly, physical and mental retardation, anomalies of the genitourinary tract, and recurved aural pinnae, and a variety of developing Wilms' tumor is highest when sporadic aniridia is accompanied by genitourinary malformations and mental retardation.

2. Hemihypertrophy

Congenital hemihypertrophy is a condition characterized by gross asymmetry of the body. While detectable at birth, it usually becomes more apparent with increasing age. It seems to occur more often on the right side. These children are at unusually high risk for the development of one of three types of malignancy: Wilms' tumor (Fraumeni *et al.*, 1967), hepatoblastoma (Fraumeni *et al.*, 1968), and adrenocortical carcinoma (Fraumeni and Miller, 1967; Haicken *et al.*, 1973). While the majority of the tumors develop on the hypertrophied side of the body, 30% develop contralaterally. It has been shown that an excessive incidence of pigmented nevi and vascular hamartomas also occur with hemihypertrophy.

Some 16 cases of adrenocortical carcinoma occurring in hemihypertrophy have been described. These cases are clinically striking as they often present as an endocrinopathy—Cushing's syndrome or precocious puberty. Those patients who have their tumors diagnosed and removed under 1 year of age have a better prognosis than those diagnosed and treated later in life (Haicken *et al.*, 1973). One of the surviving patients subsequently developed a fatal Wilm's tumor (Riedel, 1952). These observations underline the necessity of the early diagnosis of hemihypertrophy and close surveillance of affected individuals.

Meadows *et al.* (1974) recently reported that a mother with hemihypertrophy gave birth to three children who developed Wilms' tumor. The fourth child had shown no evidence of Wilms' tumor but had a double collecting system in one kidney. None of the offspring showed hemihypertrophy. These workers also document a family with hemihypertrophy in which another child developed Wilms' tumor.

3. Omphalocele–Macroglossia Syndrome (Beckwith–Wiedemann Syndrome)

Beckwith *et al.* (1964) and Wiedemann (1964) independently described a complex constellation of anomalies including omphalocele, macroglossia, bilateral cytomegaly of the fetal adrenal cortex, hyperplasia of gonadal interstitial cells, renal medullary dysplasia, and hyperplastic visceromegaly in several other organs, including the kidneys and pancreas. The pancreas exhibits islet hyperplasia. In some cases, neonatal hypoglycemia is present which may prove fatal. In some instances, postnatal development is associated with the development of hemihypertrophy. As in isolated hemihypertrophy, Beckwith's syndrome has been found to be excessively associated with adrenocortical carcinoma and Wilms' tumor (Beckwith, 1969) and nephroblastomatosis. Cutaneous vascular nevi may also occur.

4. Basal Cell Nevus Syndrome

The basal cell nevus or nevoid basal cell carcinoma syndrome is another clear example of a teratologic–neoplastic syndrome. The principal features of the syndrome consist of multiple basal cell hamartomas of the skin, multiple epithelial-lined jaw cysts, and multiple skeletal anomalies including scoliosis, spina bifida, bifid ribs, fused and hemivertebrae, and dolichocephaly with prominent supraorbital ridges and frontal eminences, a broad nasal root and briding of the sella turcica. There may be agenesis of the corpus callosum and low intelligence. Ocular abnormalities may include congenital cataracts, coloboma, and glaucoma. In addition, there may be dyskeratosis of the palms and soles (Walike and Karas, 1969; Schwartz *et al.*, 1970; Howell and Anderson, 1972).

Most important is the neoplastic predisposition. The skin lesions tend to transfer into frank basal cell carcinomas before 15 years of age, where ordinary basal cell carcinoma typically does not appear until after 50 years. In addition, approximately 20% of children with the basal cell nevus syndrome develop medulloblastoma at an early age (Knudson *et al.*, 1973; Neblett *et al.*, 1971). Lipomas and fibromas may also be present.

The basal cell nevus syndrome is inherited as an autosomal dominant with a high degree of penetrance. The syndrome is estimated to occur in 1–2% of patients diagnosed as having multiple basal cell carcinomas, cystic epitheliomas, and/or odontogenic cysts (Knudson *et al.*, 1973).

5. Genitourinary Tract Malformations and Wilms' Tumor

There is an excess concurrence of Wilms' tumor with genitourinary tract malformations. Approximately 6% of patients with Wilms' tumor exhibit upper urinary tract anomalies including horseshoe kidneys, ectopic or solitary kidneys, hypoplastic kidney, and duplication of the upper urinary tract, i.e., double kidneys, pelves, and ureters (Miller *et al.*, 1964; Jagasia and Thurman, 1965).

Wilms' tumor has been described in association with male pseudohermaphroditism manifested by cryptorchidism and hypospadias (Stump and Garrett, 1954; Di George and Harley, 1966; Le Marec *et al.*, 1971). Barakat *et al.* (1974) have suggested that in addition to Wilms' tumor, an unusual concurrence of congenital "nephron disorders" may occur in this teratologic–neoplastic complex. By nephron disorders they refer to nephropathies manifested as the congenital nephrotic syndrome and infantile glomerulonephritis. They point out that various combinations of pseudohermaphroditism, congenital nephron disorders, and Wilms' tumor have been reported. They suggest that these concurrences are linked by common teratogenic factors, all affecting renal embryogenesis.

6. Poland Syndrome and Leukemia

It has recently been appreciated that an unusual concurrence of the Poland syndrome and leukemia may exist (Lanzkowsky, 1975). The two main components of the Poland syndrome are symbrachydactyly and a pectoralis muscle defect. The muscle defect is a unilateral aplasia of the sternal portion of the pectoralis major muscle. The symbrachydactyly is ipsilateral and is characterized by syndactyly and short digits of the hand. Thus far, at least 6 instances of acute leukemia have been described with this teratologic disorder.

7. Nephroblastomatosis, Nodular Renal Blastema, and Related Anomalies

There is a distinct association of these lesions with teratologic disorders. Bove *et al.* (1969) showed that 5 of 8 infants with nodular renal blastema had trisomy 18 and its characteristic constellation of anomalies, including horseshoe and multicystic kidneys. Bilateral nephroblastomatosis has been described in association with splenic agenesis and malformation of the liver (Vlachos and Tsakraklides, 1968) and the Beckwith–Wiedemann syndrome (Sotelo-Avila and Gooch, 1976).

Liban and Kozenitzky (1970) and Perlman *et al.* (1973) have described familial nephroblastomatosis occurring in an Israeli sibship displaying a unique teratologic syndrome characterized by fetal gigantism, visceromegaly, hypoglycemia, hyperplasia of the islets of Langerhans, and bizarre facies. This syndrome bears a superficial resemblance to Beckwith syndrome. More recently Mankad *et al.* (1974) reported a case of diffuse bilateral nephroblastomatosis occurring in the Klippel–Trenaunay syndrome, i.e., cutaneous hemangiomas with bony and soft tissue hypertrophy of extremities.

The nephroblastomatosis complex must be distinguished from Wilms' tumor, particularly bilateral Wilms' tumor. It is our contention that nephroblastomatosis, or Wilms' tumor arising in it, shows the most striking teratologic associations and hereditary patterns.

8. Congenital Heart Disease and Neoplasia

Isolated case reports have appeared in the literature suggesting an association of neuroblastoma *in situ* with a variety of anomalies including congenital heart disease (Reisman *et al.*, 1966), adrenal cyst (Tubergen and Heyn, 1970), and trisomy D (Nevin *et al.*, 1972). Chatten and Voorhees (1967), in their review of the rare instances of familial neuroblastoma, reported one kinship in which three sisters died of neuroblastoma. One of these girls was shown to have a concurrent patent ductus arteriosus and hypertrophic pyloric stenosis. An unaffected brother died in infancy of long-segment aganglionosis of the colon and ileum and showed a neuroblastoma *in situ* in one adrenal at autopsy. In the original study of Beckwith and Perrin (1963) defining neuroblastoma *in situ*, of the 13 infants they described 9 had severe malformations; 7 of these 9 had congenital heart disease. Their observations have tended to focus attention on the possibility of an association of neuroblastoma and/or neuroblastoma *in situ* with congenital heart disease. At present, there is no good evidence to substantiate any unusual concurrence of congenital heart disease with neuroblastoma.

While about 6 cases of Wilms' tumor associated with congenital heart malformations have been reported (Gaulin, 1951; Lynch and Green, 1968), excessive concurrence cannot be documented. Long-term follow-up of 779 patients who had undergone corrective surgery for tetralogy of Fallot revealed no evidence of excess cancer development (Mulvihill *et al.*, 1973).

The case reported by Lynch and Green (1968) merits some special consideration. This was a 6-year-old child with corrected transposition of the great vessels and Ebstein anomaly, who developed fatal bilateral Wilms' tumor. The remarkable feature is that the mother had Sipple syndrome (pheochromocytoma and medullary thyroid carcinoma), one of the genetically determined neurocristopathies. This was probably a unique event.

9. Sacrococcygeal Teratoma and Anomalies

Hickey and Layton (1954) reviewed the English literature on sacrococcygeal teratoma from 1938 to 1954, and in an analysis of 112 patients found 11% associated with congenital anomalies. The anomalies were largely along the longitudinal axis of the body (anencephaly, spina bifida, patent urachus, cleft palate, rectovaginal fistulas, meningoceles, spinal anomalies, and undescended testicles). A strong familial history of twinning was present in 15% of the cases. Berry *et al.* (1970), reviewing the concurrence of anomalies in a series of 350 pediatric tumors, found the most significant increase in incidence in sacrococcygeal teratoma as compared to the Wilms' tumor, neuroblastoma, and hepatoblastoma groups. The defects observed included imperforate anus, rectovaginal fistula, ectopia vesicae with epispadias, tracheoesophageal fistulae, talipes equinovarus, and hydrocephalus. These anomalies occurred in 6 of 63 sacrococcygeal teratomas. Three of the five major malformations seen here were defects of the hindgut and cloacal region and as such could be attributed to local growth disturbances secondary to the presence of the teratoma during intrauterine development. Fraumeni *et al.* (1973) confirmed the associations with pelvic anomalies described earlier, but he also noted an increased incidence of hindgut and genitourinary tract duplications.

C. Cytogenetic Abnormalities

Specific chromosomal abnormalities are directly related to teratologic syndromes, and in selected instances they are associated with an increased incidence of neoplasms. In addition, chromosome fragility and breakage without specific karyotype abnormality occurs in a group of heredofamilial disorders, known as the chromosomal breakage syndromes. These syndromes, which are recessively inherited, are also associated with a high incidence of neoplasms.

Table 6. Specific Teratologic Disorders Associated
with Neoplasm

Anomaly	Tumor
Aniridia	Wilms' tumor
Hemihypertrophy	Wilms' tumor, hepatoblastoma, adrenocortical carcinoma
Beckwith's syndrome	Adrenocortical carcinoma, Wilms' tumor
Genitourinary tract malformations	Wilms' tumor, nephroblastomatosis
Basal cell nevus syndrome	Basal cell carcinoma, medulloblastoma, rhabdomyosarcoma
Poland syndrome	Leukemia
Mongolism (21 trisomy)	Leukemia
13q−syndrome	Retinoblastoma
Fanconi's anemia	Leukemia, squamous cell carcinoma, hepatoma (androgen-induced)
Bloom's syndrome	Leukemia, gastrointestinal carcinoma
Ataxia–telangiectasia	Lymphoma, leukemia, others

1. Specific Chromosomal Abnormalities

a. Down Syndrome. The incidence of acute leukemia is increased 10–20 times in mongolism (Miller, 1970). The increased risk is encountered in translocation and mosaic forms, as well as the usual 21 trisomy form. Also the peak mortality from leukemia in Down syndrome is younger by 2–3 years than in the general pediatric population (Knudson *et al.*, 1973). The leukemia developing in mongolism is characteristically of the acute lymphoblastic type. Congenital leukemia, which is typically of the acute myelogenous type, occurs even more frequently in mongolism. In this form, spontaneous remission is common (Bolande, 1971). Although a clear-cut relationship of mongolism with solid tumors is not apparent, few cases of concurrent retinoblastoma have been described (Bentley, 1975).

b. 13q– and Retinoblastoma. Approximately 40 individuals have been reported with a partial deletion of a long arm of a D-group chromosome (Wilson *et al.*, 1973; Ladda *et al.*, 1973; Orye *et al.*, 1974; O'Grady *et al.*, 1974; Pruett and Atkins, 1969). Banding techniques have shown this to be chromosome 13 (13q–). In about 20% of these patients, cases of retinoblastoma (unilateral and bilateral) and asplastic or hypoplastic thumbs occur. More often there is microcephaly, trigonencephaly, protruding upper teeth, low simplified ears, cleft or high-arched palate, micrognathia, and retarded growth and mental development. Mental retardation has been the most consistent finding (Wilson *et al.*, 1973; Knudson *et al.*, 1973). The 13q– effect is highly variable, and some individuals are only minimally affected.

The incidence of 13q– in a large series of retinoblastomas is unknown. Approximately 50 cases of retinoblastoma, both sporadic and familial, have been cytogenetically studied and no karyotypic abnormality found (Ladda *et al.*, 1973; Pruett and Atkins, 1969). The group associated with 13q– may be pathogenetically different from both the sporadic and the familial cases.

c. Trisomy 18 and Nephroblastomatosis. The association of nephroblastomatosis and trisomy 18 shown by Bove *et al.* (1969) and Shanklin and Sotelo-Avila (1969) has been discussed above.

2. Chromosomal Instability Syndromes

a. Fanconi's Anemia. Fanconi's anemia is a progressive pancytopenia associated with multiple congenital anomalies including cryptorchidism, renal anomalies, growth retardation, microcephaly, hypogenitalism, microphthalmia, and skeletal abnormalities, particularly deficiencies on the radial aspects of the forearms and hands. The condition is inherited as an autosomal recessive. Cytogenetic studies reveal frequent chromosome breaks, chromatid exchanges, and sometimes endoreduplications resulting in polyploidy. This occurs at all ages and during all clinical stages of disease at a constant rate (Dosik *et al.*, 1970; Schroeder and Kurth, 1971).

Individuals with Fanconi's anemia are predisposed to develop acute leukemia, often myelomonocytic, and a few cases of hepatoma and squamous cell carcinoma in young adults have been described (Swift *et al.*, 1971). Androgenic steroid treatment of the anemia seems to enhance the development of liver tumors (Mulvihill *et al.*, 1975).

An important *in vitro* correlate of this increased susceptibility to neoplasia is the observation that cultured cells of the heterozygotes for Fanconi's anemia are far more susceptible to oncogenic transformation by SV40 virus than are cultured normal cells (Todaro *et al.*, 1966). A similar susceptibility has been shown in Down syndrome and E-trisomy fibroblasts (Todaro and Martin, 1967).

b. Bloom's Syndrome.   Bloom's syndrome is a condition associated with low birth weight, retarded growth, and photosensitivity causing telangiectatic erythema of the skin. Additional abnormalities of the skin, skeleton, and genital region occur (Bloom, 1966). There is often hypogammaglobulinemia (Schoen and Shearn, 1967). The disease is inherited as an autosomal recessive trait. The gene is rare, occurring mainly in Ashkenazaic Jews. Leukemia frequently develops in these individuals in early life. After age 30 a variety of gastrointestinal cancers may develop. The pattern of chromosome abnormality in Bloom's syndrome differs somewhat from that in Fanconi anemia in that symmetrical quadriradial configurations, in addition to breaks, are produced (Schroeder and Kurth, 1971; German, 1972, 1974). These unusual patterns are the result of sister chromatid exchanges. There is also retarded DNA growth (German, 1974).

c. Ataxia–Telangiectasia.   Ataxia–telangiectasia is a disease characterized by prominent oculocutaneous telangiectasia, cerebellar ataxia, and immunologic deficiency predisposing to chronic infection. The ataxia is due to neuronal degeneration in the cerebellum. The immune defect is characterized by reduced cellular immunity and various abnormalities of serum immunoglobulins, the most common being reduced amounts of IgA and IgE (Peterson and Good, 1968). The thymus is generally small, markedly deficient in lymphocytes, and devoid of Hassall's corpuscles. The disease is inherited as an autosomal recessive trait.

Ataxia–telangiectasia is associated with an increased occurrence of leukemia and other lymphoreticular malignancies; medulloblastoma, frontal lobe glioma, and ovarian dysgerminoma have also been reported. The fragility and spontaneous breakage of chromosomes found in Fanconi anemia and Bloom syndrome also occur in about 50% of cases of ataxia–telangiectasia (Schroeder and Kurth, 1971).

d. Other Conditions.   It is clear that this group of heritable diseases shares the presence of chromosomal breakage, immune deficiency, and a tendency to develop leukemia and other malignant neoplasms. To a variable extent anemia, hypogammaglobulinemia, or dysgammaglobulinemia, and skin disorders may occur. In addition to the conditions discussed above, glutathione-reductase deficiency, Kostmann agranulocytosis, and pernicious anemia fit this pattern. The presence of chromosomal instability and imbalance thus appears to be associated with a tendency to develop cancer. Karyotype abnormalities also predispose cell cultures to transformation by oncogenic virus. Similarly cells damaged by radiation are transformed more readily by oncogenic virus (Stoker, 1963). The basis of the chromosomal instability and resultant tendency to develop malignant lesions cannot as yet be adequately explained.

Cytogenetic studies on cancer cells and noncancerous cells of cancer-bearing or cancer-predisposed individuals has intensified through the use of banding techniques. Such strides seem to suggest that specific chromosomal banding defects predispose to particular types of cancer (German, 1974).

V. TERATOGENICITY OF CARCINOGENIC AGENTS

A. General Considerations

A striking observation has emerged from the enormous body of data on experimental carcinogenesis: many agents known to be carcinogenic postnatally are teratogenic to the fetus or embryo. DiPaolo and Kotin (1966) listed 26 chemical agents that had been tested for both carcinogenic and teratogenic activity in animals. Of the 20 listed as carcinogenic, 19 had been shown to be teratogenic. This is not surprising, as carcinogenicity implies interference with some important cell function and such interference would be likely to alter developmental processes. Conversely, of 25 agents shown to be teratogenic, 21 were also carcinogenic. Some of the important agents showing combined oncogenic and teratogenic effects are listed in Table 7. Direct evidence for the carcinogenicity of teratogenic agents given or experienced during intrauterine life is nonetheless very limited. Stewart *et al.* (1958) in the Oxford Survey of Childhood Malignancies showed that the frequency of three prenatal events—direct fetal radiation, maternal virus infection, and threatened abortion—was slightly but significantly higher in children dying of cancer than for a group of healthy children of comparable age.

B. Irradiation

Agents exhibiting both teratogenic and oncogenic effects following intrauterine administration are rare. The best example of the latter is irradiation. An enormous body of information has been accumulated concerning both its postnatal oncogenicity and prenatal teratogenic effects. The data concerning the true oncogenicity of prenatal radiation are equivocal. A later increase in tumor incidence has been observed in animals and man. MacMa-

Table 7. Some Agents That Are Both Oncogenic and Teratogenic

	Oncogenic in adult	Teratogenic	Oncogenic in fetus
Physical agents			
X-ray	+	+	+
			−
Alkylating agents			
Nitrogen mustard	+	+	−
Triethylenemelamine	+	+	−
Alkylnitrosoureas	+	+	++
Polycyclic hydrocarbons			
Benzypyrene	+	+	+
			−
Methylcholanthrene	+	+	?
Dibenzanthracene	+	+	+
2-Acetylaminofluorine	+	+	?
N-2-Fluorenylacetamide	+	+	?
Miscellaneous			
Estrogens and stilbesterol	+	+	+
Trypan blue	+	+	?
Actinomycin D	+	+	−
Ethyl carbamate (urethane)	+	++	+
Aflatoxins	+	+	?

hon (1962) and Stewart and Kneale (1970) have shown that a slight excess of childhood cancer appears in the offspring of mothers receiving prenatal diagnostic obstetric X-rays. Studies of survivors of the atomic bomb explosions at Hiroshima and Nagasaki showed that those heavily exposed to radioactive fallout in intrauterine life have not experienced an excess incidence of tumors thus far. Individuals exposed during childhood or as adults experience an excess of leukemia and other cancers (Miller, 1956).

C. Chemical Agents

The animal experiments aimed at showing carcinogenicity in the fetus, in addition to or separate from teratogenicity, are hampered by a high incidence of fetal death, abortions, and resorption, particularly when the agent is administered early in gestation or intra-amniotically. Often in this type of experiment, the surviving offspring are not observed long enough for evidence of tumor development. The administration of dibenzanthracene, urethane, benzpyrene, and alkylnitrosoamines have been followed by increased incidence of tumors in offspring (Rice, 1973). Those agents showing teratogenic and oncogenic effects are summarized in Table 7.

1. Urethane

Urethane readily traverses the placenta and persists briefly in the fetus. Experiments where a single dose is administered to gestating mice on different days of gestation make it possible to study the varying spectrum of teratogenic and oncogenic effects. Administration of urethane between 7 and 11 days tends to produce offspring with profound brain, tail, and lung anomalies. The critical period for carcinogenesis comes later, between days 13 and 17 of gestation, where an increasing incidence of offspring were produced which developed multiple hepatomas and papillary adenomas of the lung. The number and incidence of these tumors of the lung increased with advancing gestation until days 15–17, after which the susceptibility to the oncogenic effects of urethane diminishes until birth (Nomura and Okamoto, 1972; Di Paolo, 1962; Vessilinovitch *et al.*, 1967).

2. Alkylnitrosoureas

Of great importance are the investigations carried out by Druckrey and co-workers (1966) with alkylnitrosoureas, especially ethylnitrosourea. These alkylating agents are highly effective fetal carcinogens when given even as a single injection to gestating rats and hamsters. In 90–100% of offspring, multiple neurogenic tumors develop leading to death of the animal at 150–600 days after birth. The tumors involve the peripheral and central nervous systems and consist of multiple neurinomas, gliomas, and ependymomas of varying degrees of differentiation (Wechsler *et al.*, 1969; Koestner *et al.*, 1971; Cravioto *et al.*, 1973; Druckrey, 1973). There is a certain resemblance to von Recklinghausen's disease in the human. The fetus appears to be 50 times more sensitive to the effect of these carcinogens than does the adult animal.

In order to test the susceptibility to oncogenesis at different stages of development, a series of pregnant rats was given a single dose of ethylnitrosourea on each day of pregnancy. It was found that tumors could be induced only after the 12th day of gestation. After this time, there was an abrupt increase in tumors with the greatest number being induced between 18 and 22 days of gestation. Beginning shortly before and accelerating after birth, there is a rapid decrease in susceptibility to the carcinogenic action of the agent (Druckrey *et al.*, 1966). With high intravenous doses given on the 15th day of gestation, a high incidence of malformations of the paws (oligodactyly and syndactyly) occurred in the newborn (Druckrey *et al.*, 1966). The characteristic neurogenic tumors did not become apparent until 5–6 months after birth. On the other hand, when ethylnitrosourea was given to pregnant hamsters on the 8th day of gestation, severe malformations of the entire cephalic region, eye, and thorax developed, along with a high incidence of fetal death and resorption (Givelber and Di Paolo, 1969); carcinogenesis was not apparent.

More recent experiments have shown that malformations of the brain can be produced by alkylnitrosoureas in rats prior to 11½ days gestation after which time no teratogenic effects were noted (Koyama *et al.*, 1970). These anomalies included microcephaly, exencephaly, and encephalocele. Diwan and Meier (1974) showed that transplacental treatment of mice with alkylnitrosoureas induces tumors in several strains of mice. The most commonly produced tumors were pulmonary adenomas and leukemia. The most sensitive period for the inducibility of neoplasm was during days 16–18 of gestation. As the injections were begun on the 12th day of gestation, teratogenicity was not observed.

The experiments cited, which are only representative of the intensive investigations going on with these agents, are of great importance in delineating the temporal relationships between oncogenic and teratogenic reactions to transplacental injury. It seems clear that carcinogenesis tends to be maximal in the latter stages of gestation, occurring in rodents only after organogenesis is complete. Prior to this time, teratogenesis is the prevalent response (Rice, 1973).

3. Stilbesterol and Estrogens

The first clear example of transplacental carcinogenesis in man was elucidated by Herbst and Scully (1970) and Herbst *et al.* (1971). They showed that a rare and peculiar clear-cell cervicovaginal adenocarcinoma developed in adolescent girls whose mothers had been given stilbesterol during their gestation for the prevention of threatened abortion. Since their original report, numerous articles have appeared confirming this association. The tumor is a papillary adenocarcinoma with clear cells. It is thought to arise by malignant transformation of Müllerian or mesonephric duct vestiges in the vagina, known as vaginal adenosis (Herbst and Scully, 1970). Vaginal adenosis is characterized by the presence of foci of endocervical and endometrial-like glands immediately beneath the vaginal squamous mucosa usually along the anterior wall of the vagina. Their anterior, superficial location favors Müllerian origin. The presence of vaginal adenosis is the most common pathologic effect of maternal estrogens and occurs in about 30% of exposed offspring (Carrington, 1974). That these teratologic vestiges are precursors of the clear-cell carcinoma is based on the frequent presence of vaginal adenosis in frank carcinoma (Herbst and Scully, 1970) and histologic evidences of dysplastic changes developing in adenosis and leading to carcinoma (Barber *et al.*, 1974).

The actual risk of exposed girls developing adenocarcinoma is not certain. It has been estimated as lower than 0.1% (Ulfelder, 1973). Lanier *et al.* (1973) suggests that it is no greater than 4 in 1000 for all girls whose mothers received estrogens during pregnancy; for those whose mothers received estrogens in the first trimester, the incidence may reach 9 in 1000.

The oncogenicity of estrogens has been previously shown in mice, rats, guinea pigs, and hamsters (Hueper and Conway, 1964). Gardner (1959) reported epidermoid carcinomas of the cervix and vagina in mice after intravaginal installation of estrogen. Meissner *et al.* (1957) reported endometrial hyperplasia and carcinoma in young rabbits given diethystilbesterol intramuscularly. Dunn and Greene (1963) showed that injections of estrogens into newborn mice produced carcinomas of the cervix and vagina.

The teratogenicity of estrogens is also appreciated. They typically produce abnormalities in somatosexual differentiation. Paradoxical masculinization of female offspring has been described in humans and rodents (Bongiovanni *et al.*, 1959). This may include retention of Wolffian duct remnants. It thus seems likely that stilbesterol given to pregnant women significantly elicits anomalies in primitive sex duct development and involution, and thus accounts for the presence of vaginal adenosis.

VI. PATHOGENETIC CONCEPTS

A. Possible Mechanisms

It is clear that links between oncogenesis and teratogenesis are numerous and varied. An explanation for all these relationships is not now available, yet certain generalizations are possible.

Rapidly dividing cells are a prerequisite for teratogenesis and oncogenesis. During mitosis, the cell is most vulnerable to structural changes in DNA and to disruption of transcription to RNA and subsequent translation into protein–enzyme synthesis. The resting cell is more capable of a certain degree of repair and restoration of this system than the dividing cell. Moreover, mitosis is required to produce a population of altered or injured cells in order for tumor or malformation to become manifest. Mitotic activity is generally present to a greater extent in intrauterine life than at any later stage in the life cycle.

It seems clear from the experimental studies cited that the timing of the initiating event(s) may be critical in determining the form of the outcome. The degree of cytodifferentiation and the metabolic or immunologic state of the organism may determine whether the effect is teratogenic, oncogenic, or both. Additional variables might be the changing placental permeability to the agent and its persistence in both the mother and fetus.

We tend to view teratogenesis and oncogenesis as different developmental stages in the intrauterine reaction to a special type of injury, teratogenesis being the more primitive reaction. The effect of an inciting agent in early gestation would be teratogenic, with a gradual shifting toward combined oncogenic–teratogenic forms in later gestation and, ultimately, pure oncogenic expression following late gestational or postnatal exposure. In man, the excessive coexistence of tumors and anomalies could be explained on this basis.

On the other hand, a primary teratologic event in the fetus may in some fashion predispose the organism to a secondary oncogenic induction in later life. This might explain the neoplastic transformations occurring in hamartomas and hamartoses, developmental vestiges, heterotopias, and dysgenetic tissues. It is also possible that these structures harbor a latent oncogenic agent, which may even have been originally responsible for the anomaly in intrauterine life. Environmental insults in later life, e.g., trauma, infection, hormonal or metabolic change, might cause a de-repression or activation of this cryptic genome and result in the production of cancer.

For example, the heightened hormonal stimulation accompanying puberty might be largely responsible for the development of tumors in dysgenetic gonads. These anomalous tissues are probably unable to respond appropriately to increasing levels or cyclical fluctuations in gonadotropins, and the result might be the peculiar neoplasms encountered. A similar mechanism may be involved in the development of the adolescent carcinoma or the vagina induced by stilbesterol therapy during gestation. Here vestiges of anomalous sex duct differentiation (vaginal adenosis) may also respond to gonadotropic or other hormonal stimulation of puberty by tumor formation.

In those teratologic conditions associated with chromosomal instability, fragility, or imbalance, the unusual tendency to develop cancer is paralleled by a predisposition of cell cultures from these patients to be transformed by oncogenic virus. It seems that certain cytogenetic derangements render cells unusually sensitive to many carcinogenic agents (Schroeder and Kurth, 1971). When this is associated with an immune-deficiency state, as is typical of many of the conditions in this group, a failure in the host's surveillance system against cancer might be anticipated. This would allow for the establishment and growth of a clone of cancer cells, culminating in a frank neoplasm.

B. Applicability of the "Two-Hit" Theory of Carcinogenesis

Any discussion of oncogenesis must perforce deal with the two-hit hypothesis forwarded by Ashley (1969), Knudson (1971), and Knudson and Strong (1972). This theory assumes that carcinogenesis is related to discrete mutational events occurring at a random and constant average rate. For a tumor to develop, two such mutational events must occur sequentially, the first may be prezygotic or postzygotic and the second always postzygotic. When the first mutation is prezygotic, thus affecting all somatic cells, the tumor is hereditary, usually bilateral or multiple, and of very early appearance in life. When the first mutation is postzygotic affecting only certain somatic cells, the tumor is nonhereditary and single. Knudson has gathered impressive evidence in support of this hypothesis, particularly in comparing sporadic and familial embryomas of early life—retinoblastoma, Wilms' tumor, and possibly neuroblastoma.

It is not difficult to integrate this theory into the explanation of the relationships between teratogenesis and oncogenesis. It has relevance if the first hit were expressed as an heritable anomaly, or constellation of anomalies, and the subsequent appearance of tumor, expressive of the second hit. The increased incidence of tumor in these conditions depends on a heightened susceptibility of the defective genome to the oncogenic transformation of the second hit. The theory may be further applicable in nonhereditary combinations of anomaly and tumor as well. Here one must assume a postzygotic initiation of the sequence of events.

REFERENCES

Ashley, D. J. B., 1969, The two "hit" and multiple "hit" theories of carcinogenesis, *Br. J. Cancer* **23:**313.

Asman, R. B., and Pierce, E. R., 1970, Familial multiple polyposis, *Cancer* **25:**972.

Bachmann, K. D., and Kroll, W., 1962, Über das pränatal Enstandene. Neuroblastoma sympatheticum, *Kinderheilk* **103:**61.

Bachmann, K. D., and Kroll, W., 1969, Der Wilms-Tumor in ersten Lebensjahr insbesondere über 62 Nephroblastome des Neugeborenen, *Dtsch Med. Wochenschr.* **94:**2598.

Barakat, A. Y., Papadopoulou, Z. L., Chandra, R. S., Hollerman, C. E., and Calcagno, P. L., 1974, Pseudohermaphroditism, nephron disorder and Wilms' tumor: A unifying concept, *Pediatrics* **54:**366.

Barber, H. R. K., and Sommers, S. C., 1974, Vaginal adenosis, dysplasia, and clear-cell adenocarcinoma after diethylstilbestrol treatment in pregnancy, *Obstet. Gynecol.* **43:**645.

Becker, J. M., Schneider, K. M., and Krasna, I. H., 1970, Neonatal neuroblastoma, *Progr. Clin. Cancer* **4:**382.

Beckwith, J. B., 1969, Macroglossia, omphalocele, adrenal cytomegaly, gigantism and hyperplastic visceromegaly, *Natl. Found-March of Times Orig. Art. Ser.* **2:**188.

Beckwith, J. B., and Perrin, E. V., 1963, *In situ* neuroblastomas: A contribution to the natural history of neural crest tumors, *Am. J. Pathol.* **43:**1089.

Beckwith, J. B., Wang, C. I., Donnell, G. N., and Gwinn, J. L., 1964, Hyperplastic fetal visceromegaly with macroglossia, omphalocele, cytomegaly of fetal adrenal cortex and other abnormalities, *J. Pediatr.* **65:**1053.

Bentley, D., 1975, A case of Down's syndrome complicated by retinoblastoma and celiac disease, *Pediatrs.* **56:**131.

Berry, C. L., Keeling, J., and Hilton, C., 1970, Coincidence of congenital malformations and embryonic tumors of childhood, *Arch. Dis. Child.* **45:**229.

Bloom, D., 1966, The syndrome of congenital telangiectatic erythema and stunted growth, *J. Pediatr.* **68:**103.

Bogdan, R., Taylor, D. E. M., Mostofi, F. K., 1973, Leiomyomatous hamartoma of the kidney, *Cancer* **31:**462.

Bolande, R. P., 1967, *Cellular Aspects of Developmental Pathology,* Lea and Febiger, Philadelphia.

Bolande, R. P., 1971, Benignity of neonatal tumors and the concept of cancer repression in early life, *Am. J. Dis. Child.* **122:**12.

Bolande, R. P., 1973a, Congenital mesoblastic nephroma of infancy, *Perspect. Pediatr. Pathol.* **1:**227.

Bolande, R. P., 1973b, Relationships between teratogenesis and oncogenesis, *in: Pathobiology of Development* (E. Perrin and M. Finegold, eds.), pp. 114–134, Williams & Wilkins, Baltimore.

Bolande, R. P., 1974a, Congenital mesoblastic nephroma. Letter to the Editor, *Arch. Pathol.* **98:**357.

Bolande, R. P., 1974b, Congenital and infantile neoplasia of the kidney, *Lancet,* **2:**1497.

Bolande, R. P., 1974c, The neurocristopathies: A unifying concept of disease arising in neural crest maldevelopment, *Hum. Pathol.* **5:**409.

Bolande, R. P., and Towler, W. F., 1970, A possible relationship of neuroblastoma to von Recklinghausen's disease, *Cancer* **26:**162.

Bolande, R. P., Brough, A. J., and Izant, R. J., Jr., 1967, Congenital mesoblastic nephroma of infancy, *Pediatrics* **40:**272.

Bongiovanni, A. M., Di George, A. M., and Grumbach, M. M., 1959, Masculination of the female infant associated with estrogenic therapy alone during gestation: Four cases, *J. Clin. Endocrinol. Metab.* **19:**1004.

Bove, K., Koffler, H., and McAdams, A. J., 1969. Nodular renal blastema. Definition and possible significance, *Cancer* **24:**323.

Brasfield, R. D., and Das Gupta, T. K., 1972, Von Recklinghausen's disease: A clinico-pathological entity, *Ann. Surg.* **175:**86.

Brown, W. T., Puranik, S. R., Altman, D. H., and Hardin, H. C., Jr., 1972, Wilms' tumor in three successive generations, *Surgery* **72:**756.

Carrington, E. R., 1974, Relationship of stilbesterol exposure *in utero* to vaginal lesions in adolescence, *J. Pediatr.* **85:**295.

Chatten, J., and Voorhess, M. L., 1967, Familial neuroblastoma, *N. Engl. J. Med.* **277:**1230.

Chretien, P. B., Milam, J. D., Foote, F. W., and Miller, T. R., 1970, Embryonal adenocarcinomas (a type of saccrococcygeal teratoma) of the sacrococcygeal region, clinical and pathologic associations of 21 cases, *Cancer* **26:**522.

Cravioto, J. F., Weiss, J. F., Weiss, K. de C., Goebel, H. H., and Ransokoff, J., 1973, Biological characteristics of peripheral nerve tumors induced with ethylnitrosourea, *Acta Neuropathol.* **23:**265.

Di George, A. M., and Harley, R. D., 1966, The association of aniridia, Wilms' tumor and genital abnormalities, *Arch. Ophthalmol.* **75:**796.

Dillard, B. M., Mayer, J. H., McAlister, W. H., McGavrin, M., and Strominger, D. B., 1970, Sacrococcygeal teratoma in children, *J. Pediatr. Surg.* **5:**53.

Di Paolo, J. A., 1962, Effects of oxygen concentration on carcinogenesis induced by transplacental urethan, *Cancer Res.* **22:**299.

Di Paolo, J. A., and Kotin, P., 1966, Teratogenesis–oncogenesis: A study of possible relationships, *Arch. Pathol.* **81:**3.

Diwan, B. A., and Meier, H., 1974, Strain and age dependent transplacental carcinogenesis by 1-ethyl-1-nitrosourea in inbred strains of mice, *Cancer Res.* **34:**764.

Donnellan, W. A., and Swenson, O., 1968, Benign and malignant sacrococcygeal teratomas, *Surgery* **64:**834.

Dosik, H., Hsu, L. V., Todaro, G. J., Lee, S. L., Hirschhorn, K., Selirio, E. S., and Alder, A. A., 1970, Leukemia in Fanconi's anemia, cytogenetic and tumor virus susceptibility studies, *Blood* **36:**341.

Druckrey, H., 1973, Specific carcinogenic and teratogenic effects of indirect alkylating methyl and ethyl components, and their dependency on stages of oncogenic developments, *Xenobiotica* **3:**271.

Druckrey, H., Ivankovic, S., and Preussmann, R., 1966, Teratogenic and carcinogenic effects in the offspring after a single injection of ethylnitrosourea to pregnant rats, *Nature* **210:**1378.

Dunn, T. B., and Greene, A. W., 1963, Cysts of epididymis, cancer of the cervix, granular cell myoblastoma, and other lesions after estrogen injection in newborn mice, *J. Natl. Cancer Inst.* **31:**425.

Everson, T. C., and Cole, W. H., 1966, *The Spontaneous Regression of Cancer,* W. B. Saunders, Philadelphia.

Foulds, L., 1969, *Neoplastic Development Vol I,* Academic Press, London and New York.

Fraumeni, J. F., and Glass, A. G., 1968, Wilms' tumor and congenital aniridia, *J.A.M.A.* **206**:825.

Fraumeni, J. F., Jr., and Miller, R. W., 1967, Adrenocortical neoplasms with hemihypertrophy, brain tumors and others, *J. Pediatr.* **70**:129.

Fraumeni, J. F., Jr., Geiser, C. F., and Manning, M. D., 1967, Wilms' tumor and congenital hemihypertrophy: A report of five new cases and review of literature, *Pediatrics* **40**:886.

Fraumeni, J. F., Jr., Miller, R. W., and Hill, J. A., 1968, Primary carcinoma of the liver in childhood: An epidemiological study, *J. Natl. Cancer Inst.* **40**:1087.

Fraumeni, J. F., Li, F. P., Dalager, N., 1973, Teratomas in children: Epidemiologic features, *J. Natl. Cancer Inst.* **5**:1425.

Fu, Y.-S., and Kay, S., 1973, Congenital mesoblastic nephroma and its recurrence, *Arch. Pathol.* **96**:66.

Gardner, W. U., 1959, Carcinoma of the uterine cervix and upper vagina: Induction under experimental conditions, *Ann. N.Y. Acad. Sci.* **75**:543.

Gatti, R. A., and Good, R. A., 1971, Occurrence of malignancy in immunodeficiency diseases. A literature review. *Cancer* **28**:29.

Gaulin, E., 1951, Simultaneous Wilms' tumors in identical twins, *J. Urol.* **66**:547.

German, J., 1972, Genes which increase chromosomal instability in somatic cells and predispose to cancer, *Prog. Med. Genet.* **8**:61.

German, J., 1974, *Chromosomes and Cancer,* Wiley, New York.

Ghazali, S., 1973, Presacral teratomas in children, *J. Pediatr. Surg.* **8**:915.

Gilbert, J. B., 1942, Studies in malignant testis tumors. VIII. Tumors in pseudohermaphrodites: A review of sixty cases and a case report, *J. Urol.* **48**:665.

Givelber, H., and Di Paolo, J. A., 1969, Teratogenic effects of *N*-ethyl-*N*-nitrosourea in the Syrian hamster, *Cancer Res.* **29**:1151.

Glenn, J. F., and Rhame, R. C., 1961, Wilms' tumor: Epidemological experience, *J. Urol.* **85**:911.

Griffin, M., and Bolande, R. P., 1969, Familial neuroblastoma with regression and maturation to ganglioneurofibroma, *Pediatrics* **43**:377.

Guin, G. H., Gilbert, E. E., and Jones, B., 1969, Incidental neuroblastoma in infants, *Am. J. Clin. Pathol.* **51**:126.

Haicken, B. N., Schulman, N. H., and Schneider, K. M., 1973, Adrenocortical carcinoma and congenital hemihypertrophy, *J. Pediatr.* **83**:284.

Hamerton, J. L., 1971, Abnormal sex chromosome complements in the female, *Human Cytogenetics,* Vol. II, p. 65, Academic Press, New York.

Herbst, A. L., and Scully, R. E., 1970, Adenocarcinoma of the vagina in adolescence. A report of 7 cases including 6 clear-cell carcinomas, *Cancer* **25**:745.

Herbst, A. L., Ulfelder, H., and Poskanzer, C. D., 1971, Registry of clear-cell carcinoma of genital tract of young women, *N. Engl. J. Med.* **285**:407.

Hiatt, R., Kendrick, D. L., and Guerry, D., 1961, Retinoblastoma, regression and progression, *Am. J. Ophthalmol.* **52**:717.

Hickey, R. C., and Layton, J. M., 1954, Sacrococcygeal teratoma, *Cancer* **7**:1031.

Hou, L. T., and Holman, R. L., 1961, Bilateral nephroblastomatosis in a premature infant, *J. Pathol. Bacteriol.* **82**:249.

Howell, J. B., and Anderson, D. E., 1972, The nevoid basal cell carcinoma syndrome, *in: Cancer of the Skin* (A. Andrade, S. L. Gumport, G. L. Popkin, and T. D. Rees, eds.), W. B. Saunders, Philadelphia.

Hueper, W. C., and Conway, W. D., 1964, *Chemical Carcinogenesis and Cancers,* Charles C Thomas, Springfield, Ill.

Ishak, I. G., and Glunz, P. R., 1967, Hepatoblastoma and hepatocarcinoma in infancy and childhood, *Cancer* **20**:396.

Jagasia, K. H., and Thurman, W. G., 1965, Congenital anomalies of the kidney in association with Wilms' tumor, *Pediatrics* **35**:338.

Jensen, R. D., and Miller, R. W., 1971, Retinoblastoma: Epidemological characteristics, *N. Engl. J. Med.* **285**:307.

Kay, S., Pratt, C. B., and Salzberg, A. M., 1965, Hamartoma (leiomyomatous type) of the kidney, *Cancer* **19**:1825.

Kissane, J. M., and Smith, M., 1967, *Pathology of Infancy and Childhood*, C. V. Mosby, St. Louis.

Klapproth, H. J., 1959, Wilms' tumor: A report of 45 cases and an analysis of 1351 cases reported in the world literature from 1940 to 1958, *J. Urol.* **81**:633.

Knudson, A. G., 1971, Mutation and cancer: Statistical study of retinoblastoma, *Proc. Natl. Acad. Sci. U.S.A.* **68**:820.

Knudson, A. G., Jr., and Amromin, G. D., 1966, Neuroblastomas and ganglioneuroma in a child with multiple neurofibromatosis—implication for the mutational origin of neuroblastomas, *Cancer* **19**:1032.

Knudson, A. G., and Strong, L. C., 1972, Mutation and cancer: A model for Wilms' tumor of the kidney, *J. Natl. Canc. Inst.* **48**:313.

Knudson, A. G., Strong, L., and Anderson, D. E., 1973, Heredity and cancer in man, *Prog. Med. Genet.* **9**:113.

Kobayashi, N., Furukawa, T., and Takatsu, T., 1968, Congenital anomalies in children with malignancy, *Paediatr. Univ. Tokyo* **16**:31.

Koestner, A., Swanberg, J. A., and Wechsler, W., 1971, Transplacental production with ethylnitrosourea of neoplasms of the nervous system in Sprague-Dawley rats, *Am. J. Pathol.* **63**:37.

Koyama, T. J., Hanada, H., and Matsumoto, S., 1970, Methylnitrosourea-induced malformations of brain in SD-JCL rat, *Arch. Neurol.* **22**:342.

Ladda, R., Atkins, L., Littlefield, J., and Pruett, R., 1973, Retinoblastoma: Chromosome banding in patients with heritable tumor, *Lancet* **1**:506.

Lanzkowsky, P., 1975, Absence of pectoralis major muscle in association with acute leukemia. *J. Pediatr.* **86**:817.

Lanier, A. P., Noller, K. L., Decker, D. G., Elveback, L. R., and Kurland, L. T., 1973, Cancer and stilbesterol: A follow-up of 1719 persons exposed to estrogens *in utero* and born 1943–1959, *Mayo Clin. Proc.* **48**:793.

Le Marec, B., Lautredou, A., Urvoy, M., Renault, A., Fonlupt, J., Dary, J., Ardouin, M., and Coutel, Y., 1971, Un cas d'association de nephroblastome avec aniridie et malformations génitales, *Arch. Fr. Pediatr.* **28**:457.

Liban, E., and Kozenitzky, I. L., 1970, Metanephric hamartomas and nephroblastomatosis in siblings, *Cancer* **25**:885.

Lowry, W. S., 1974, Passive immunity against cancer, *Lancet* **1**:602.

Lynch, H. T., and Green, G. S., 1968, Wilms' tumor and congenital heart disease, *Am. J. Dis. Child.* **115**:723.

MacMahon, B., 1962, Prenatal x-ray exposure and childhood cancer, *J. Natl. Cancer Inst.* **28**:1173.

Mankad, V. N., Gray, G. F., and Miller, D. R., 1974, Bilateral nephroblastomatosis and Klippel-Trenaunay syndrome, *Cancer* **33**:1462.

Martin, J., and Rickham, P. P., 1974, Wilms' tumor—an improved prognosis, *Arch. Dis. Child.* **49**:459.

Meadows, A. T., Leichtenfeld, J. L., and Koop, C. E., 1974, Wilms' tumor in three children of a woman with congenital hemihypertrophy, *N. Engl. J. Med.* **291**:23.

Meissner, W. A., Sommers, S. C., and Sherman, G., 1957, Endometrial hyperplasia, endometrial carcinoma, and endometriosis produced experimentally by estrogen, *Cancer* **10**:500.

Miller, R. W., 1956, Delayed effects occurring within the first decades after exposure of young individuals to the Hiroshima atomic bomb, *Pediatrics* **18**:1.

Miller, R. W., 1966, Relation between cancer and congenital defects in man, *N. Engl. J. Med.* **275**:87.

Miller, R. W., 1969, Childhood cancer and congenital defects. A study of U. S. death certificates during the period 1960–66, *Pediatr. Res.* **3**:389.

Miller, R. W., 1970, Neoplasia and Down's syndrome, *Ann. N.Y. Acad. Sci.* **171**:637.

Miller, R. W., Fraumeni, J. F., Jr., and Manning, M. D., 1964, Association of Wilms' tumor with aniridia, hemihypertrophy and other congenital malformations, *N. Engl. J. Med.* **270:**922.

Miller, R. W., Fraumeni, J. F., and Hill, J. A., 1968, Neuroblastoma: Epidemiologic approach to its origin, *Am. J. Dis. Child.* **115:**253.

Morris, J. M., 1953, The syndrome of testicular feminization in male pseudohermaphrodites, *Am. J. Obstet. Gynecol.* **65:**1192.

Mulvihill, J. J., Miller, R. W., and Taussig, H. B., 1973, Long-time observations on the Blalock-Taussig operation. V. Neoplasms in teratology of Fallot. *Johns Hopkins Med. J.* **133:**16.

Mulvihill, J. J., Ridolfi, R. L., Schultz, F. R., Borzy, M. S., and Naughton, P. B. T., 1975, Hepatic adenoma in Fanconi's anemia treated with oxymetholone, *J. Pediatr.* **87:**122.

Neblett, C. R., Waltz, T. A., and Anderson, D. E., 1971, Neurological involvement in the nevoid basal cell carcinoma syndrome, *J. Neurosurg.* **35:**577.

Nevin, N. C., Dodge, J. A., and Allen, I. V., 1972, Two cases of trisomy D associated with adrenal tumors, *J. Med. Genet.* **9:**919.

Nomura, T., and Okamoto, E., 1972, Transplacental carcinogenesis by urethan in mice: Teratogenesis and carcinogenesis in relation to organogenesis, *Gann* **63:**731.

O'Grady, R. B., Rothstein, T. B., and Romano, P. E., 1974, D-group relation syndromes and retinoblastoma, *Am. J. Ophthalmol.* **77:**40.

Orye, E., Delbeke, M. J., and van den Beele, 1974, Retinoblastoma and long-arm deletion of chromosome, *Clin. Genet.* **5:**457.

Perlman, M., Goldberg, G. M., Bar-Ziv, J., and Danovitch, G., 1973, Renal hamartomas and nephroblastomatosis with fetal gigantism: A familial syndrome, *J. Pediatr.* **83:**414.

Peterson, R. D. A., and Good, R. A., 1968, Ataxiatelangiectasia, *Natl. Found.-March Dimes, Birth Defects Orig. Art. Ser.* **4:**370.

Pierce, G. B., Bullock, W. K., and Huntington, R. W., Jr., 1970, Yolk-sac tumors of the testis, *Cancer* **25:**644.

Potter, E. L., 1972, *Normal and Abnormal Development of the Kidney,* p. 271ff, Year Book Medical Publishers, Chicago.

Pruett, R. C., and Atkins, L., 1969, Chromosome studies in patients with retinoblastoma, *Arch. Ophthalmol.* **82:**177–181.

Reilly, D., Mesbit, W. D., and Krivit, W., 1968, Cure of 3 patients who had skeletal metastases in disseminated neuroblastoma, *Pediatrics* **41:**47.

Reisman, M., Goldenberg, E. D., and Gordon, J., 1966, Congenital heart diseases and neuroblastoma, *Am. J. Dis. Child.* **111:**308.

Rice, J. M., 1973, An overview of transplacental carcinogenesis, *Teratology* **8:**113.

Riedel, H. A., 1952, Adrenogenital syndrome in a child due to adrenocortical tumor, *Pediatrics* **10:**19.

Schneider, K. M., Becker, J. M., and Krasna, I. H., 1965, Neonatal neuroblastoma, *Pediatrics* **36:**359.

Schoen, E. J., and Shearn, M. A., 1967, Immunoglobulin deficiency in Bloom's syndrome, *Am. J. Dis. Child.* **113:**594.

Schroeder, T. M., and Kurth, M., 1971, Spontaneous chromosomal breakage and high incidence of leukemia in inherited diseases, *Blood* **37:**96.

Schwartz, S. H., Blankenship, B. J., and Stout, R. A., 1970, Multiple basal cell nevus syndrome, *J. Oral Surg.* **28:**523.

Scully, R. E., 1970, Gonadoblastoma: A review of 74 cases, *Cancer* **25:**1340.

Shanklin, D. R., and Sotelo-Avila, C., 1969, *In situ* tumors in fetuses, newborns and young infants. *Biol. Neonate* **14:**286.

Sohval, A. R., 1956, Testicular dysgenesis in relation to neoplasm of the testicle, *J. Urol.* **75:**285.

Sotelo-Avila, C., and Gooch, M., 1976, Neoplasia in Beckwith's syndrome, *Perspect. Pediatr. Pathol.* **3:**255.

Stewart, A. M., and Kneale, G. W., 1970, Age distribution of cancers caused by obstetrical x-rays and their relevance to cancer latent periods, *Lancet* **2:**4.

Stewart, A. M., Webb, J., and Hewitt, D., 1958, A survey of childhood malignancies, *Br. Med. J.* **1:**1495.

Stoker, M., 1963, Effect of x-irradiation on susceptibility of cells to transformation by polyoma virus, *Nature* **200:**756.

Stump, T. A., and Garrett, R. H., 1954, Bilateral Wilms' tumor in a male pseudohermaphrodite, *J. Urol.* **72:**1146.

Survey on neuroblastoma among the Surgical Fellows of the American Academy of Pediatrics, 1968, *J. Pediatr. Surg.* **3:**191.

Sutow, W., Gehan, E. A., Heyn, R. M., Kung, F. H., Miller, R. W., Murphy, M. L., and Traggis, D. G., 1970, Comparison of survival curves, 1956 versus 1962 in children with Wilms' tumor and neuroblastoma, *Pediatrics* **32:**880.

Swift, M., Zimmerman, D., and McDonough, E. R., 1971, Squamous cell carcinomas in Fanconi's anemia, *J.A.M.A.* **216:**325.

Sy, W. M., and Edmonson, J. H., 1968, The developmental defects associated with neuroblastoma—etiologic implications, *Cancer* **22:**234.

Talerman, A., 1971, Gonadoblastoma and dysgerminoma in two siblings with dysgenetic gonads, *Obstet. Gynecol.* **38:**416.

Teter, J., and Boczkowski, K., 1967, Occurrence of tumors in dysgenetic gonads, *Cancer* **20:**1301.

Todaro, G. J., and Martin, G. M., 1967, Increased susceptibility of Down's syndrome fibroblasts to transformation by SV40, *Proc. Soc. Exp. Biol. Med.* **124:**1232.

Todaro, G. J., Green, H., Swift, M. R., 1966, Susceptibility of human diploid fibroblast strains to transformation by SV40 virus, *Science* **153:**1252.

Tubergen, D. G., and Heyn, R. M., 1970, *In situ* neuroblastoma associated with an adrenal cyst, *J. Pediatr.* **76:**451.

Turkel, S. B., and Itabashi, H. H., 1974, The natural history of neuroblastic cells in the fetal adrenal gland, *Am. J. Pathol.* **76:**225.

Ulfelder, H., 1973, Stilbesterol, adenosis, and adenocarcinoma, *Am. J. Obstet. Gynecol.* **17:**796.

Vaez-Zadeh, K., Sieber, W. K., Sherman, F. E., and Kieseweller, W. B., 1972, Sacrococcygeal teratomas in children, *J. Pediatr. Surg.* **7:**152.

Vessilinovitch, S. D., Mihailovich, N., and Pietra, G., 1967, The prenatal exposure of mice to urethan and the consequent development of tumors in various tissues, *Cancer Res.* **27:**2333.

Vlachos, J., and Tsakraklides, V., 1968, A case of renal dysplasia and its relation to bilateral nephroblastomatosis, *J. Pathol. Bacteriol.* **95:**560.

Walike, J. W., and Karas, R. P., 1969, Nevoid basal cell carcinoma syndrome, *Laryngoscope* **79:**478.

Warkany, J., 1971, *Congenital Malformations. Notes and Comments,* Year Book Medical Publishers, Chicago.

Wechsler, W., Kleinhues, P., Matsumoto, S., and Zulch, K. J., 1969, Pathology of experimental neurogenic tumors induced during prenatal and postnatal life. *Ann. N.Y. Acad. Sci.* **159:**360.

Wexler, H. A., Poole, C. A., and Fujacco, R. M., 1975, Metastic neonatal Wilms' tumor, *Pediatr. Radiol.* **3:**179.

Wiedemann, H. R., 1964, Complexe malformatif familial avec hernie ombilicale et macroglossie, *J. Genet. Hum.* **13:**223.

Wigger, H. J., 1969, Fetal hamartoma of the kidney, *Am. J. Clin. Pathol.* **51:**323.

Willis, R. A., 1962, *The Borderland of Pathology and Embryology,* 2nd ed., Butterworths, Washington and London.

Wilson, M. G., Towner, J. W., and Fujimoto, A., 1973, Retinoblastoma and D-deletions, *Am. J. Hum. Genet.* **25:**57.

Young, J. L., and Miller, R. W., 1975, Incidence of malignant tumors in U.S. children, *J. Pediatr.* **86:**254.

Young, P. G., Mount, B. M., Foote, F. W., and Whitmore, W. E., Jr., 1970, Embryonal adenocarcinoma of the prepubertal testis, *Cancer* **26:**1065.

Examples of Abnormal Organogenesis IV

Abnormal Organogenesis in the Eye

13

ALFRED J. COULOMBRE and
JANE L. COULOMBRE

I. INTRODUCTION

A. The Roles of Tissue Interactions in Ocular Development

The vertebrate eye focuses images of the environment in the plane of the retinal photoreceptors. Optimal visual function requires not only transparent dioptric media, but also that the relative sizes, shapes, and orientations of ocular tissues fall within the geometric tolerances imposed by the laws of optics. These functional requirements are met during embryonic development by the differentiation of highly specialized populations of cells from undifferentiated precursors and by the coordinated morphogenesis of the resulting tissues. Both processes are controlled to a remarkable extent by interactions which occur among the emerging tissues. Such interactions occur in an orderly, temporal sequence and help to mediate the developmental expression of one of the set of possible ocular phenotypic combinations defined by the individual genotype.

B. Genetic Constraints

The gene pool of any species defines its total set of possible phenotypic combinations. The combinations of phenotypic characteristics that can be

ALFRED J. COULOMBRE and JANE L. COULOMBRE • Laboratory of Vision Research, National Eye Institute, National Institutes of Health, Bethesda, Maryland 20014. J. L. C. is deceased.

developed by an individual of the species rests upon the constraints inherent in a unique subset of the gene pool, the individual genotype. The expression of one or another of the phenotypes inherent in such a subset is a function of environmental inputs at each stage during the developmental expression of the individual genotype. Each phenotype falls somewhere along a continuum from unfit to highly adaptive, whether evaluated by human judgments of functional efficiency, measured as a departure from the statistical norm for the population, or judged on the basis of reproductive success in the sterner court of evolution.

C. Parameters of Abnormal Ocular Development

Defective development of the eye results from maladapting genetic or environmental inputs, or from both. The genome may be so constituted that, with normally encountered environmental inputs, its developmental expression results in eyes which are structurally or functionally unfit. Alternatively, a genotype which, given a normal developmental environment, would generate a normal ocular phenotype, may generate an unfit phenotype if untoward environmental inputs channel its expression toward one of the unfit possibilities within its potential range.

Environmental influence may be exerted at a number of levels of organization: DNA replication; transcription; translation; self-assembly; the elaboration or alteration of specific cell products by enzymes; spatial compartmentalization of cell products, including their excretion or secretion; mitosis; cell death; cell migration; tissue interactions, including inductions and trophic influences; etc. At whatever level an adverse environmental influence initially operates, the chain of events it sets in motion often, but not always, results in interference with essential tissue interactions. Many of the phenotypic defects which result can, therefore, be understood by analyses of development conducted at the level of tissue interactions.

The nature of an insult during embryonic development is frequently less important to the nature and extent of the outcome than is the developmental stage at which it occurs. This is so, because many steps in the causally linked sequence of tissue interactions are maximally at hazard for a restricted period of time and can be disrupted directly or indirectly during that susceptible period by diverse influences, each operating initially at one or more of the organizational levels mentioned above. For this reason, the identification and analysis of each step in the series of normal developmental events which leads to phenotypic expression not only increases our understanding of teratogenetic processes but also increases our power to predict and, more rarely, to rationally control the consequences of genetic or environmental influences upon such processes.

D. Suitability of the Eye for Teratogenetic Study

The vertebrate eye affords an excellent opportunity for such an analysis. It contains an unusually large number of differentiated tissues. The numerous developmental interactions among these tissues have been intensively studied. Furthermore, the eye is subject to long lists of genetically determined or environmentally initiated teratological defects.

Each tissue is brought to its final state of differentiation, its cell-population size and its definitive geometry (size, shape, and position and orientation relative to other tissues), not only by intrinsic processes, but also by extrinsic influences exerted by neighboring tissues. The developmental changes which result, occur in precise sequences which are spatially ordered and temporally coordinated. The mechanisms of the tissue interactions underlying this orderly coordination are largely unknown. However, especially for the eye, many tissue pairs have been identified, one of which exerts influences to which the other responds. For many of these pairs of tissues we know the developmental durations of the interactions and the nature of the elicited responses. Further, we know enough such steps in ocular development to permit their sequential ordering in flow sheets (Figure 1) that, at least at the morphogenetic level of analysis, partially explain the development of the eye. More importantly, these flow sheets permit us to predict in testable terms the teratological consequences of interfering at one or another step in the morphogenetic sequence.

Several publications have reviewed the vast literature bearing on the several facets of this topic. Descriptive summaries of normal development of the eye are provided by Mann (1950), Duke-Elder (1963), Keeney (1951), and O'Rahilly (1966). The dynamics of ocular development, as revealed by experimental embryological studies, is covered by Coulombre (1965). Abnormal ocular development is treated by Mann (1957) and by Duke-Elder (1963). Genetic and metabolic eye disease is surveyed by Sorsby (1951) and in a volume edited by Goldberg (1974). This chapter in no way substitutes for the extensive coverage provided in those publications. Rather, it stresses how tissue interactions coordinate vertebrate ocular morphogenesis and gives a few examples of the spectrum of quite different defects which results from genetic or environmental interference with one or another of these interactions.

II. EARLY STAGES IN OCULAR DEVELOPMENT

A. Anophthalmos

The most severe defect of the eye is its absence, *anophthalmos*. This condition may be: *primary,* when it results from failure of the presumptive retina to

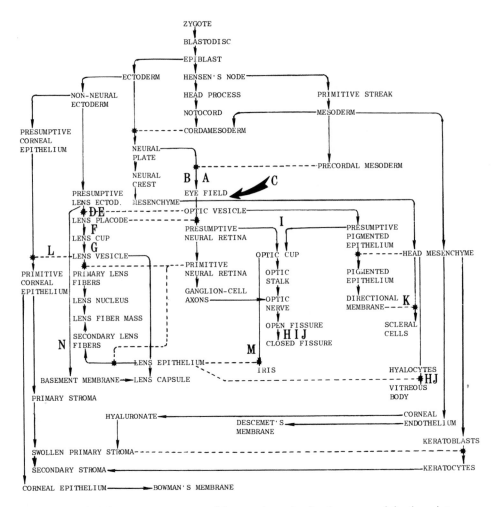

Fig. 1. This flow sheet represents some of the steps in ocular development and the tissue interactions which effect them. In general, earlier stages are represented above and later stages below. Solid lines (arrows) indicate the direction of each step. Asterisks indicate that the step is mediated by a tissue interaction and dotted lines connect with the interacting tissue or tissue product. Alphabetic letters represent those defects treated in the text which result from disruption of one or another step: A, anophthalmos; B, cyclopia or synophthalmos; C, congenitally cystic eye; D, primary aphakia; E, nanophthalmos; F, lens at hazard to viral infection; G, persistent corneal–lenticular stalk; H, coloboma of the embryonic fissure; I, metaplastic formation of neural retina from pigmented epithelium; J, cornea plana and microcornea; K, scleral defects; L, corneal agenesis; M, iris agenesis; N, abnormal development of the lens capsule.

appear early in development; *secondary,* when it accompanies more general and more severe anomalies such as anancephaly; or *degenerative,* when it results from the breakdown of eye tissues later in development. Primary anophthalmos is easily distinguished from the secondary form, but is easily confused diagnostically with degenerative anophthalmos or with cases of severe underdevelopment of the eye. Any evaluations of reports of anophthalmos in man must take this difficulty into account.

Genetically inherited anophthalmos has been reported in man (Roberts, 1937; Sjøgren and Larsson, 1949) and in other animals (Landauer, 1932). Suitably timed, adverse environmental influence also results in anophthalmos in many vertebrate animals. The long list of implicated agents includes hypoxia (chicken: Gallera, 1950–1951; Rübsaamen, 1948; rat: Werthemann and Reiniger, 1950), X-irradiation (Wilson and Karr, 1951), and deficiencies or excesses of many chemical substances.

The embryo is at greatest hazard to anophthalmos early in development, concurrent with the first tissue interaction known to be directly related to ocular development. This is the induction of the presumptive retinal field in the anterior neural plate by precordal mesenchyme (Figures 1A and 2A). It remains to be determined whether the eye field is initially a single midline region which is later subdivided into the two lateral eye fields, or whether the lateral eye fields are separately induced at the outset (reviewed by Adelmann, 1936). Settlement of this basic point would be helpful diagnostically, since, if the eye field is initially single, primary anophthalmos would probably be bilateral, whereas if the lateral eye fields are induced separately, unilateral primary anophthalmos would be a possibility. In either event, the neuroepithelium of the eye field initiates a temporal sequence of tissue interactions which is essential for ocular development. Thus, when this inductive step is not completed, no eye develops (primary anophthalmos). Failure of induction of the eye field is simply a manifestation of the action of an adverse or defective genetic or environmental input. We have no knowledge currently of the initial sites, or the mechanisms, of action of any of the numerous agents which precipitate primary anophthalmos. By whatever diverse pathways they act, these strikingly different teratogens appear to have in common that they finally prevent or disrupt the induction of the eye fields.

B. Cyclopia

Cyclopia, the presence of a single median eye in a single median orbit, stands at one extreme of a spectrum of disorders. Between this condition and the development of two normally lateralized eyes, any degree of partial separation or fusion of the eyes (*synophthalmos*) may be encountered. Examples are: two retinal cups and lenses in a common scleral envelope, with separate or with fused corneas and with a common optic nerve; two completely separated eyes with separate optic nerves, resident in a common median orbit; two incompletely lateralized orbits; and many gradations among these configurations.

While several instances of familial occurrence of cyclopia have been recorded for man, the condition is lethal and genetic inheritance has not been demonstrated conclusively. Inherited cyclopia has been reported in the

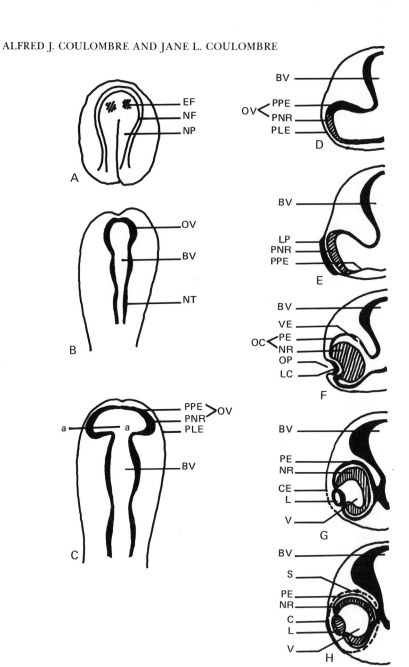

Fig. 2. These diagrammatic representations of stages in ocular development depict dorsal views of early embryos in A, B, and C, and transverse sections (at the level a–a, as depicted in C) in D, F, F, G, and H. A: At the open-neural-plate stage the eye is represented by two precordal retinal fields in the neural ectoderm. B: Following closure of the neural folds, the optic vesicles (anlagen of the retina) evaginate from the forebrain. C and D: The tip of each optic vesicle (presumptive neural retina) contacts the surface ectoderm, undergoes induction, and becomes the primitive neural retina. E: Reciprocally, the presumptive neural retina induces, in the ectoderm, a patch of lens cells which thickens to form the lens placode. F: The optic vesicle invaginates to form the lens cup.

domestic fowl (Landauer, 1956) and in the guinea pig (Wright and Wagner, 1934). Synophthalmos has been produced in animals (teleosts, amphibia, birds, mammals) by a number of different treatments and agents (destruction of anterior midline structures by surgery or by localized X-irradiation, temperature shock, hypoxia, and excesses of a number of chemicals), provided that these act prior to the time at which the optic vesicles begin to evaginate from the prosencephalic wall (Adelmann, 1936). It is clear, from such studies, that synophthalmos can result from properly timed destruction of tissues in the anterior midline, and therefore from secondary fusion of already lateralized eye fields (Figure 1B). It remains unclear whether true cyclopia arises in this manner and whether cyclopia or synophthalmos can arise also by total or partial failure of an initially single, median eye field to split into two lateral fields.

Once again, the nature and the primary mode of action of the teratogenic agent seems of less moment than the fact that it acts at, or shortly before, the time at which an important morphogenetic event is in progress and vulnerable.

C. Congenitally Cystic Eye

By the time developing eyes are represented by evaginating optic vesicles, they are beyond the hazards of primary anophthalmos, cyclopia, or synophthalmos. Disruption of ocular development at this stage leads to a different spectrum of deformities. The most characteristic of these is congenitally cystic eye. In its extreme form the variably pigmented, neuroepithelial wall of the optic vesicle is greatly distended, is frequently invested partially or completely by choroidal and scleral tissues, but is not associated with a lens, cornea, vitreous body, ciliary body, or iris. Alternatively, a small pigmented epithelium, sometimes associated with scleral tissue, may be found in the depths of the orbit. In slightly less severe forms of congenitally cystic eye, patches of abortive or degenerate neural retina occur in the distal wall of the cyst and small amounts of lens tissue may be present. At its least severe, this series of disorders is represented by eyes that have a normal complement of

---

G: The lumen of the optic stalk disappears, sealing off the eye ventricle from the brain ventricular space. The ventricle of the eye is eliminated by apposition of the outer and inner walls of the optic cup. The lens cup pinches off from the surface to form the lens vesicle, leaving behind the presumptive corneal epithelium at the surface. The embryonic fissure closes and the vitreous body enlarges. H: The sclera is induced in the head mesenchyme by the pigmented epithelium. The primary lens fibers elongate to eliminate the lumen of the lens vesicle. Labels: BV, brain ventricle; C, cornea; CE, corneal epithelium; EF, eye field; L, lens vesicle or lens; LC, lens cup; LP, lens placode; NF, neural fold; NP, neural plate; NR, neural retina; NT, neural tube; OC, optic cup; OP, lens pore; OV, optic vesicle; PE, pigmented epithelium; PLE, presumptive lens epithelium; PNR, presumptive neural retina; PPE, presumptive pigmented epithelium; S, sclerotic coat; VE, ventricle of the eye; V, vitreous body.

tissues, but in which the neural retina remains more or less extensively separated from the pigmented epithelium.

Heritability of this series of anomalies has not been demonstrated in man, but the so-called "anophthalmos" inherited by some mice (Chase and Chase, 1941) appears not to be a primary anophthalmos, but an aborted development of the eye at the optic-vesicle stage, which resembles congenitally cystic eye in man.

The time at which the developing eye is at hazard to this condition (just prior to the 7-mm crown–rump length in man) corresponds to the time when each optic vesicle expands laterally, bringing the basement membrane at its tip into contact with that of the lateral head ectoderm (Figures 1C; 2B). In the area of contact, neural retina is induced in the vesicle and lens ectoderm is induced in the surface ectoderm. When the vesicle fails to contact the surface ectoderm, these reciprocal interactions, and all tissue interactions subsequently flowing from them, do not occur. Such a vesicle typically does not invaginate to form an optic cup and tends to become cystic, presumably after closure of the optic stalk severs confluence with the ventricles of the brain. Pigmented epithelium develops from the neuroepithelium of the wall of the abnormal vesicle. This may explain why sclera is so frequently the only other ocular tissue associated with congenitally cystic eyes, since pigmented epithelium is known to induce scleral tissue in the periocular mesenchyme during normal development (Figures 1K; 2H). It is not known at present what factors assure proper expansion of the optic vesicle, effect confrontation of its tip with the surface ectoderm, and regulate its timely invagination to form the optic cup. However, the conclusion seems inescapable that untoward genetic or environmental influences give rise to the spectrum of anomalies which characterize this period in ocular development because they ultimately abort or interfere with these morphogenetic events.

D. Primary Aphakia

A characteristically different spectrum of eye disorders is associated with events attending and following the expansion of the optic vesicles. Primary aphakia, the failure of the lens to form, is a concomitant of the types of severe deformity discussed above, which result from very early disruption of ocular development. It has been produced experimentally by treatment of animals with phlorizin, dinitrophenol, vitamin deficiency, anoxia, temperature shock, or X-irradiation. Indeed, primary aphakia will result from the properly timed action of any agent or event which interferes with lens induction. Extensive experimental analyses of lens induction (reviewed by Jacobson, 1966) suggest that this disorder may be precipitated by: destruction of, or failure to develop, either lens-inducing potency in the presumptive neural retina or ectodermal competence to respond to this influence; shifts in the developmental windows during which potency and competence are present; or, as is probably the usual case, interference with contact between the two interacting tissues (Figures 1D; 2C–E). Since the lens inductively initiates the development of the cornea (Figure 1L) and the iris (Figure 1M), these structures are absent as a secondary consequence of true primary aphakia.

E. Nanophthalmos

Eyes that fail to achieve normal size, for whatever reason, are referred to as microphthalmic. In pure microphthalmos (nanophthalmos), all parts of the eye are present, are normal in shape and position, but are proportionately reduced in size. This condition is frequently familial and both dominant and recessive transmission have been documented in man. Roberts (1948) also reported hereditary transmission in pigs. Unpublished experiments with embryos of domestic fowl (Jajszczak and Coulombre) indicate that mechanical interference with closure of the neural folds, just prior to evagination of the optic vesicles, sometimes results in unilateral or bilateral nanophthalmos. In these cases, the optic vesicle was small in diameter and made contact with a subnormal area of surface ectoderm. The small lens and eye cup that resulted in such cases gave rise to a small but well-formed eye. This finding suggests that a reduced area of contact between the optic vesicle and the surface ectoderm at the time when lens placode and neural retina are reciprocally induced may be one of the pathways to nanophthalmos (Figure 1E)

III. LATE STAGES IN OCULAR DEVELOPMENT

A. Defects Related to Lens and Optic-Vesicle Formation

Successful induction of the major ocular tissues puts the developing eye safely beyond the time at which it is subject to the derangements discussed above, but it makes possible other forms of defective development. One set of deformities is related to the invagination of the lens placode.

1. Virus-Induced Cataracts

An example of ocular teratology flowing from disturbance at the time of lens invagination, is the impact of virus infections, which can be powerfully teratogenic. The lens of the eye is particularly vulnerable. Cataracts, produced by extensive disorganization and degeneration of lens fibers, accompany some viral infections (rubella: Cotlier *et al.*, 1968; mumps: Robertson *et al.*, 1964; Karkinen-Jääskeläinen, 1973). However, these cataracts characteristically develop when infection has occurred early in embryonic development and not when exposure occurs at later stages. In the case of mumps-related cataract, one possible explanation of this phenomenon can be excluded. Lens fibers do not lose their susceptibility to damage by the viruses as development proceeds, since viruses experimentally introduced within the capsules of older lenses destroy lens fibers (Karkinen-Jääskeläinen, 1973). The period when the lens is at hazard to viral infection appears to be terminated by the developmental completion of the primitive lens capsule which seals off the enclosed lens cells from access by virus particles (Robertson *et al.,* 1964; Karkinen-Jääskeläinen, 1973; Figures 1F; 2F and G). When chick embryos are experimentally infected with mumps virus prior to the closure of the lens pore, virus particles are seen in the pit of the open lens cup. When the lens cup pinches off from the surface ectoderm, it forms a lens vesicle which is, from the outset, completely enveloped by a basement membrane, the primitive lens capsule (O'Rahilly and Meyer, 1960; Figure 1N). Any virus particles present in the lens cup are trapped within the lens as the vesicle forms; such lenses subsequently become, and remain, cataractous. By contrast, viruses inoculated into the egg following formation of the lens vesicle and capsule do not enter the lens, nor do cataracts develop in such protected lenses. This example illustrates how a morphogenetic event can terminate the period when a tissue is at hazard by setting up a physical barrier to a teratogenic agent.

2. Persistent Corneal–Lenticular Stalk

This anomaly is characterized by a persistent continuity of the corneal epithelium with the lens epithelium through an abnormal pore at the center of the cornea. Most probably, it results from failure of the lens cup to pinch off from its parental surface ectoderm (Figures 1G; 2F and G). Presumably, closure of the optic vesicle is possible only during a restricted developmental time, and this condition results from any interference with completion of closure of the lens vesicle during this interval.

B. Defects Related to Closure of the Embryonic Fissure

Proper closure of the embryonic fissure, an event which occurs shortly after formation of the lens vesicle and optic cup, seals off the eye cup and permits the accumulation of vitreous substance between the lens and the neural retina (Figure 1H). Failure at this pivotal step can seriously derange the subsequent morphogenesis of the eye, or result in less serious but visible defects of the iris and/or choroid such as coloboma.

1. Microphthalmos

A second category of microphthalmos, with an etiology distinct from that of nanophthalmos, is associated with cystic extensions of the vitreous space, usually through a colobomatous opening in the eye wall at the location of the embryonic (choroid) fissure. Frequently, these small eyes have extensively folded neural retinas but unfolded pigmented epithelia. This condition occurs as a part of quite different syndromes, some of them genetically inherited, and can be assumed to have quite diverse etiologies. It also results from a variety of properly timed environmental insults. What all of these agents may have in common is that, directly or indirectly, they prevent complete closure of the embryonic fissure and allow the escape of vitreous substance as it forms. Experiments with otherwise normal embryos of domestic fowl have demonstrated: (1) the accumulation of vitreous substance following closure of the embryonic fissure is essential for the normal enlargement of the eye (Coulombre, 1956), and (2) the presence of both the lens and the neural retina is required for the accumulation of the vitreous substance (Coulombre and Coulombre, 1964, 1965). When, in suitably controlled experiments, vitreous substance is drained from the growing eye or its accumulation is stopped by surgical removal of either the embryonic lens or the neural retina, a profoundly microphthalmic eye develops.

2. Metaplasia of the Pigmented Epithelium

Metaplastic transformation of patches of pigmented epithelium into neural retina, especially near the embryonic fissure, is often a concomitant of this type of microphthalmos. This aspect of the anomaly is secondary to the subnormal growth of the vitreous body and is the result of interference with a complex tissue interaction between the neural retina and the pigmented epithelium that occurs at this stage in development (Figures 1I; 2F–H). Experiments, including those cited above, demonstrated: (1) The pigmented epithelium and neural retina increase in area by different mechanisms and are kept in apposition during development by the expanding vitreous body (Coulombre et al., 1963). (2) When vitreous substance accumulates at a sufficiently subnormal rate, the neural retina becomes progressively more folded while the pigmented epithelium remains small in area and smooth (Coulombre, 1956). (3) When and where such folding separates the neural retina a discrete distance from the pigmented epithelium the unapposed patch of pigmented epithelium often forms neural retina metaplastically (Coulombre and Coulombre, 1965).

3. Cornea Plana and Microcornea

Cornea plana, the absence of a corneal curvature, and microcornea, subnormal corneal size, can arise in conjunction with the type of microphthalmos which results from subnormal growth of the vitreous body (Figure 1J). These anomalies invariably accompany the microphthalmos produced experimentally by draining the vitreous substance in early avian embryos (Coulombre, 1957). The experimental results identify two of the factors involved in the development of corneal curvature: (1) the mechanical tensions produced by the burgeoning vitreous body; and (2) the development of relatively higher resistance to these tensions in tissues at the edge of the developing cornea. When, as in loss of vitreous substance through a colobomatous defect, the intraocular pressure is not maintained, the cornea fails to expand normally in area or to develop its characteristic curvature. It must be stressed that cornea plana and microcornea probably can arise in other ways.

IV. CONCLUSIONS

In these examples, by no means exhaustive, an attempt has been made to relate the spectrum of congenital eye defects, arising from maladaptive genetic or environmental inputs during development, to the chain of tissue interactions which governs the orderly emergence of ocular tissues and their coordinated growth to form a living organ with optical functions. Each interaction puts the eye at hazard to a different, specific type of derangement. Specific anomalies result from any one of a large number of environmental insults or genetic defects, and the nature of the developmental disruption during each of these periods seems to be more a function of which ocular tissues are interacting at the moment, than it is of the nature of the insult. These agents, remarkably diverse in nature, probably enter into very different initial interactions with the embryo. The apparent lack of agent specificity in this class of teratologies seems to have two bases. First, the tissue interactions, and therefore their corresponding sensitive periods, occur in an orderly sequence during development, and each interaction is a necessary antecedent of a group of subsequent steps. In general, those tissue interactions which occur early in development have more succeeding steps dependent upon them than do steps later in development. Thus, disruption of an earlier step tends to result in a more severe defect. Second, tissue interactions are at a highly complex level of organization and depend upon multiple events at many lower levels of organization. Thus, agents, initially acting disruptively at any one of a large number of levels of organization, can have in common that they ultimately disrupt a specific tissue interaction when they are applied during its sensitive period.

REFERENCES

Adelmann, H., 1936, The problem of cyclopia, *Q. Rev. Biol.* **11**:161; 284.
Chase, H., and Chase, E., 1941, Studies on an anophthalmic strain of mice. I. Embryology of the eye region, *J. Morphol.* **68**:279.
Cotlier, E., Fox, J., Bohigian, G., Beaty, C., and Leure DuPree, A., 1968, Pathogenic effects of rubella virus on embryos and newborn rats, *Nature (London)* **217**:38.

Coulombre, A., 1956, The role of intraocular pressure in the development of the chick eye. I. Control of eye size, *J. Exp. Zool.* **133**:211.

Coulombre, A., 1957, The role of intraocular pressure in the development of the chick eye. II. Control of corneal size, *A.M.A. Arch. Ophthalmol.* **57**:250.

Coulombre, A., 1965, The eye, *in: Organogenesis* (R. DeHaan and H. Ursprung, eds.), pp. 219–251, Holt, Rinehart and Winston, New York

Coulombre, A., and Coulombre, J., 1964, Lens development. I. Role of the lens in eye growth, *J. Exp. Zool.* **156**:39.

Coulombre, A., Steinberg, S., and Coulombre, J., 1963, The role of intraocular pressure in the development of the chick eye. V. Pigmented epithelium, *Invest. Ophthalmol.* **2**:83.

Coulombre, J., and Coulombre, A., 1965, Regeneration of neural retina from the pigmented epithelium in the chick embryo, *Dev. Biol.* **12**:79.

Duke-Elder, S., 1963, *System of Ophthalmology. Volume III, Normal and Abnormal development. Part 1. Embryology; Part 2. Congenital deformities,* C. V. Mosby, London.

Gallera, J., 1950–1951, Influence de l'atmosphère artificiellement modifiée sur le développement embryonnaire du poulet, *Acta Anat. (Basel)* **11**:549.

Goldberg, M., ed., 1974, *Genetics and Metabolic Eye Disease,* Little, Brown, Boston.

Jacobson, A., 1966, Inductive processes in embryonic development, *Science* **152**:25.

Karkinen-Jääskeläinen, M., 1973, Spatial and temporal restriction of mumps virus induced lesions in the developing chick lens, *Acta Pathol. Microbiol. Scand. Sect. A., Suppl. No. 234,* 1–52.

Keeney, A., 1951, *Chronology of Ophthalmic Development,* pp. 1–32, Publication 99, American Lecture Series, Charles C Thomas, Springfield, Ill.

Landauer, W., 1932 Über die Entwicklungsmechanischen und genetischen Ursachen des Coloboms und anderen embryonaler Augenmissbildungen, *Albrecht von Graefes Arch. Ophthalmol.* **129**:268.

Landauer, W., 1956, Cyclopia and related defects as a lethal mutation of fowl, *J. Genet.* **54**:219.

Mann, I., 1950, *The Development of the Human Eye,* Grune and Stratton, New York.

Mann, I., 1957, *Developmental Abnormalities of the Eye,* 2nd ed., J. B. Lippincott, Philadelphia.

O'Rahilly, R., 1966, Early development of the eye in staged human embryos, *Contrib. Embryol.* **38**:1.

O'Rahilly, R., and Meyer, D., 1960, The periodic acid-Schiff reaction in the cornea of the developing chick, *Z. Anat. Entwicklungsgesch.* **121**:351.

Roberts, J., 1937, Sex-linked microphthalmia sometimes linked with mental deficiency, *Br. Med. J.* **2**:1213.

Roberts, L., 1948, Microphthalmia in swine, *J. Hered.* **39**:146.

Robertson, G., Williamson, A., and Blattner, R., 1964, Origin and development of lens cataracts in mumps-infected chick embryos, *Am. J. Anat.* **115**:473.

Rübsaamen, H., 1948, Missbildungen am Zentralnervensystem von Tritonen durch algemeinen Sauerstoffmangel bei Normaldruck, *Wilhelm Roux' Arch. Entwicklungsmech. Org.* **143**:615.

Sjøgren, T., and Larsson, T., 1949, Microthalmos and anophthalmos with or without coincident oligophrenia, *Acta Psychiatr. Neurol., Suppl. 56,* pp. 1–103.

Sorsby, A., 1951, *Genetics in Ophthalmology,* C. V. Mosby, St. Louis.

Werthemann, A., and Reiniger, M., 1950, Über Augenentwicklungsstörungen bei Rattenembryonen durch Sauerstoffmangel in der Frühschwangerschaft, *Acta Anat. (Basel)* **11**:329.

Wilson, J., and Karr, J., 1951, Effects of irradiation on embryonic development. I. X-rays on the tenth day of gestation in the rat, *Am. J. Anat.* **88**:1.

Wright, S., and Wagener, K., 1934, Types of subnormal development of the head from inbred strains of guinea pigs and their bearing on the classification and interpretation of vertebrate monsters, *Am. J. Anat.* **54**:383.

Abnormal Organogenesis in the Cardiovascular System

14

OSCAR C. JAFFEE

I. INTRODUCTION

The relative importance of genetic and environmental factors in cardiovascular development has been long debated (cf. Goerttler, 1958, 1970). In a review of the experimental production of cardiovascular defects, Wilson (1960) stressed the importance of nongenetic hemodynamic factors. Analyses of the etiology of congenital heart disease (Nora, 1968; Nora *et al.*, 1970; Neill, 1972) have shown that a clearly defined genetic etiology is demonstrable in only a small proportion of the victims (<5%); these analyses were interpreted in terms of a multifactorial theory, which assumes that environmental stresses produced congenital heart disease to the greatest extent in those individuals most predisposed genetically.

Experimental modifications of the embryonic circulations produced cardiovascular malformations (Jaffee, 1962; Rychter, 1962) and demonstrated the importance of vascular forces in cardiogenesis. The concept that temporary, reversible changes in the hemodynamic equilibrium produced cardiovascular malformations (Wilson, 1960; Rychter, 1962) has received support with the finding that such changes, produced by hypoxia, were indeed followed by the appearance of malformations (Jaffee, 1974a,b).

OSCAR C. JAFFEE • Department of Biology, University of Dayton, Dayton, Ohio 45469.

II. NORMAL CARDIOVASCULAR DEVELOPMENT

A concise review of the normal development of the mammalian cardiovascular system was provided by the late Bradley M. Patten (1960), and a review of avian cardiovascular development is contained in Romanoff's (1960) monograph. A timetable of cardiovascular development in a number of vertebrate forms was constructed by Sissman (1970). References to the normal development of specific regions are included with each region.

A. Formation of the Heart Loop

Cellular movements from bilateral cardiac anlagen to form the heart tube have been described by Rosenquist and DeHaan (1966). Although the early loop forms before the onset of circulation, its formation is influenced by surrounding tissues, as shown by transplantation and radiation studies (Orts Llorca, 1970; Le Douarin and Le Douarin, 1970). While cardiac looping may continue *in vitro,* it is never normal; loop formation is altered in the changed hemodynamic situations (Orts Llorca, 1970; Jaffee, 1962, 1974a). The role of cell death in loop formation has been described (Ojeda and Hurle, 1975).

The importance of the positioning of the original cardiac loop on further development was pointed out by Van Praagh (1967) and Neill (1972). For example, the form of the developing arterial outflow tract (conotruncus) is of special importance in cardiogenesis since all major cyanotic malformations of the heart originate in this region (Neill, 1972).

B. Valvular Development

Normal valvular development was described by Patten (1960). Valvular action is first provided by acellular mounds, referred to as "cardiac jelly" (Patten, 1949), which not only contribute to future valve formation but also take part in the initial interatrial septum (Jaffee, 1963). Cellular proliferation from the endocardium into the mounds transforms them into endocardial cushions (Patten, 1960). Both cardiac jelly mounds and the endocardial cushions have been found to be subject to the molding influence of blood flow (Patten, 1949; Jaffee, 1965).

C. Septation

Cardiac partitioning begins only after the onset of circulation. Atrial septation was reviewed by Fox and Goss (1957); ventricular septal formation has been analyzed by Odgers (1938), Jaffee (1967), and Harh (1975). The aorticopulmonary (conotruncal) septum is initiated by two cellular outgrowths which appear after two clearly separated bloodstreams are found at the juncture of the truncus arteriosus and the aortic arches (Jaffee, 1967). Although the spiral form of the septum was recognized in the last century, its relationship to the flow of blood in this region as described below (Section III) was not appreciated until well into the present century.

D. Blood Vessel Development

The blood vessels most involved in cardiac malformations are the aortic arches. Normal development of the aortic arches has been described for chick embryos by Hughes (1943) and for the mammalian embryos by Moffat (1959). General reviews of blood vessel development were provided by Patten (1960) and Romanoff (1960).

III. DYNAMICS OF CARDIOVASCULAR DEVELOPMENT

Of the physiological parameters that affect cardiovascular development, the best defined is blood flow, as postulated by Spitzer (1923) and discussed in the context of embryonic physiology by Patten (1949). Model experiments demonstrating blood-flow patterns in normal and abnormal embryo hearts were carried out by Goerttler (1958, 1970) and de Vries and Saunders (1962).

A. Rheology of Blood Flow

Rheology of flow in blood vessels was analyzed by Copley and Staple (1962) and the development of two bloodstreams with laminar flow in the embryo heart described by Jaffee (1966). A minimal velocity of flow was found necessary for the maintenance of two separate bloodstreams in the heart; two streams could be fused into a single mass of flowing blood and then made to reappear simply by varying the heart rate (Jaffee, 1966). The varying configurations of the blood streams is dependent upon a laminar flow of blood in the heart, as will be described below (also cf. Goerttler, 1970). Copley and Staple pointed out that in a blood vessel with laminar flow the molecules closest to the vessel wall are relatively immobile. Assuming that a similar situation prevailed between the bloodstreams and the wall of the heart, a minimal resistance to the proliferation of cells would exist in regions between the two streams, as seen in the initial development of the aorticopulmonary septum (Section II.C). In other situations the molding force of the bloodstreams in the developing heart is apparently caused by the collision force of a stream upon the heart wall, as well as by lateral pressure (VI.B.2).

B. Bloodstream Flow Patterns

1. Spiraling

A spiraling of the bloodstreams, making up the venous inflow of the sinus venosus (Jaffee, 1963, 1965), occurs as blood from several sources becomes confluent in this large chamber (Figure 1). This pattern first arises following the fusion of the omphalomesenteric veins in the liver (Patten, 1960; Romanoff, 1960). These fused veins unite with the right Cuvierian duct, giving rise to a large right bloodstream with the smaller left stream, composed of the left Cuvierian duct, spiraling around it (Jaffee, 1963, 1965). The left stream then describes a wide arc which aids in determining the left atrium to be larger than the right atrium. These are but two instances in which flow patterns rather than volume determine the size of a heart chamber.

The inflow streams described above flow approximately parallel to each other through the single atrium and ventricle, but the right stream then spirals around the left in the bulbus arteriosus (Figure 2), the reverse of the

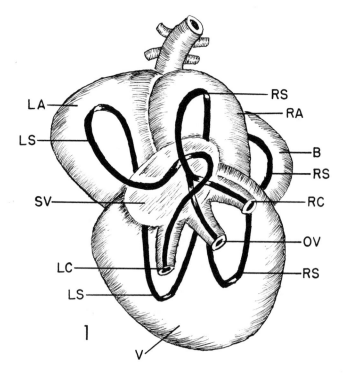

Fig. 1. Dorsal view of 3-day chick heart illustrating bloodstream flow pattern. (From Jaffee, 1965.) Abbreviations: B, bulbus; LA, left atrium; LC, left duct of Cuvier; LS, left stream; OV, fused omphalomesenteric veins; RA, right atrium; RS, right stream; V, ventricle.

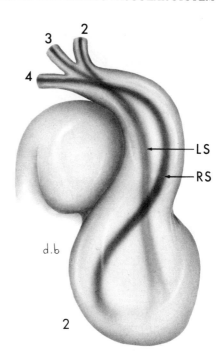

Fig. 2. Ventral view of 3-day chick heart illustrating bloodstream flow patterns of the arterial outflow tract. (From Jaffee, 1967.)

inflowing streams (Figure 1). The direction of bulbar curvature was determined in precirculatory stages (Section II) and the streams initially follow this curvature. The degree of spiraling indicated in Figure 2 becomes exaggerated in subsequent development so that the right stream becomes directed further dorsad and posteriorly (cf. Jaffee, 1967), thus directing it toward the developing sixth arch. This exaggerated spiraling is followed by a rearrangement of the direction of blood flow through the aortic arches (cf. Figures 2 and 3) (for details see Jaffee, 1967), and ends with two clearly defined bloodstreams leading out from the fourth and sixth aortic arches, respectively (Figure 3).

2. Unspiraling

If the spiraling pattern noted on the third day of chick development (Figure 2) were to persist, the aorta would remain associated with the right ventricle. The period between 3.5 and 5 days of chick embryology (Figures 2 and 3) is one of rapid growth (the embryonic weight increases from 31 to 160 mg), heart rate increases by 50% (cf. Jaffee, 1972), and blood pressure increases markedly, from 14 to 32 mm H_2O (Paff *et al.*, 1965). Unspiraling or bulbar rotation requires the aorta and pulmonary artery to rotate around each other with the aorta moving dorsally and to the left. Dynamic factors involved here begin in the sinus venosus with the intricate spiraling pattern

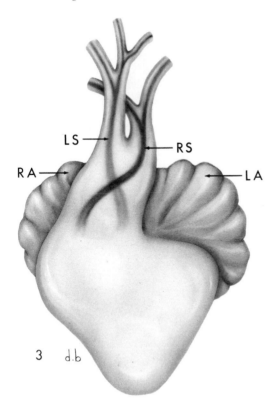

Fig. 3. Arterial outflow tract in heart of 5-day embryo. The aorta (LS) and pulmonary artery (RS) are discrete rheological entities even though septation is as yet incomplete (From Jaffee, 1967.)

breaking down at the higher velocities and pressures of this stage (cf. Jaffee, 1965) so that all the systemic venous return flows directly into the right atrium, with flow into the left atrium secondarily branching from the main flow and reaching the left atrium by way of the incomplete atrial septum. This change in inflow minimizes the exaggerated spiraling into the bulbus arteriosus, allowing the aorta and pulmonary arteries to move into their definitive positions as noted above. The unspiraling process thus provides a pattern of flow which determines the positioning of both the aorta and pulmonary arteries.

IV. FETAL HEART AND CIRCULATORY CHANGES AT BIRTH

A. Fetal Heart

With the exception of the principal bypasses (foramen ovale and ductus arteriosus), the fetal heart is essentially complete (Patten, 1960). While most accounts of fetal physiology have stated or implied that the outputs of both ventricles are equal, recent studies have provided evidence for a higher pressure and greater output in the right ventricle (Assali *et al.,* 1968). This would appear to follow from events arising early in embryonic development (Section III.B). A similar concept of right ventricular dominance may exist in chick "fetuses" (14 days + incubation) according to Hughes (1943).

B. Circulatory Changes at Birth

Circulatory changes at birth are described by Assali *et al.* (1968); Patten (1960), and Edwards (1960). Most defects of atrial septation arise from events during the period of cardiogenesis (Patten 1960; Edwards, 1960; Sections II.C. and V.A.1. Failure of the arterial bypass (ductus arteriosus) to close, unless predisposed by another defect such as aortic stenosis, has not been clearly explained; but the finding that the incidence of this defect increases with altitude, from 12% at sea level (Toronto) to 25% at moderately high altitude (Mexico City) and to even higher frequency at higher altitudes (Zamora *et al.,* 1971; Penaloza *et al.,* 1964), implicates failure of pulmonary resistance to drop as a major factor. Failure of respiratory maturation at high altitudes, with pulmonary resistance remaining high in the immature lungs, has been related to the hypoxia of altitude (McClung, 1969). Since lung immaturity and failure of pulmonary resistance to drop at birth are correlated, the elevated incidence of patent ductus arteriosus at high altitudes (Penaloza *et al.,* 1964) provides an example of cardiovascular defect being secondary to incomplete development in the respiratory system.

V. EXPERIMENTAL PRODUCTION OF CARDIOVASCULAR MALFORMATIONS

Practically all forms of cardiovascular malformations found in man occur spontaneously in other vertebrates (Siller, 1967) and, in fact, have been produced experimentally, which has given rise to a vast literature. Previous reviews include those of Wilson (1960) and Jaffee (1970). The present review is not exhaustive but rather emphasizes studies which have contributed to the elucidation of the development of cardiovascular defects.

A. Septal Defects

1. Atrial Septal Defects

Atrial septal defects were a principal finding following the maternal injection of trypan blue into rats before the onset of embryonic circulation (Fox and Goss, 1957). Abnormal looping of the cardiac tube with consequent lateral displacement and rotation of the atrial chambers, following by misalignment of the atrial septum with other parts of the heart, were considered to be the bases for these findings. Similar cardiac abnormalities were found in association with the high incidences of genetically caused cleft palate in mice (Fraser and Rosen, 1975); the facial defect may have been caused by the development of a small heart which presumably was a part of the syndrome. Ishikawa *et al.* (1975) found atrial septal defects following the administration of acetylcholine to chick embryos.

2. Ventricular Septal Defects

A high correlation of ventricular septal defects and aortic arch anomalies was found in a review of experimental studies by Wilson (1960). Rychter (1962) produced a high incidence of ventricular septal defects by stenosing aortic arches in varying combinations, and he pointed out that these defects were the result of increased vascular resistance. In some cases the silver clips used to produce the stenoses were dislodged and circulation resumed in the involved aortic arches. In such cases septal defects still appeared, demonstrating that even temporary disturbances of the hemodynamic equilibrium led to defects. When temporary, reversible disturbances of hemodynamic equilibrium were produced following moderate hypoxia in chick embryos, ventricular septal defects were a principal finding (Jaffee, 1974a). Ingalls *et al.* (1952) had previously produced ventricular septal defects in mice by stimulating the hypoxic state of an altitude of 25,000 ft.

Trypan blue administration to chick embryos at 44 hr of incubation led to ventricular septal defects which were predictable in blood flow studies in 5-day embryos from the observation that a fusion of the bloodstreams (Figure 4) occurred at the site of the forming septum (Jaffee, 1965). Sympathomimetic amines were found to produce both aortic arch anomalies and ventricular septal defects (Hodach *et al.*, 1974, 1975). These investigators found that the beta-adrenergic activity of the compounds used correlated directly with their ability to produce malformations and that this ability was blocked by propranolol, a beta-adrenergic blocking agent. They felt that these experiments provided circumstantial evidence for a role of hemodynamic factors in the production of the defects. Ishikawa *et al.* (1975) found that acetylcholine, which slowed the chick embryo hearts, led to the formation of ventricular septal defects as well as to the atrial septal defects noted above.

B. Arterial Outflow Tract (Conotruncus)

The form and positioning of the original bulbus arteriosus and its derivatives the conus and truncus arteriosus are dependent upon the form of the

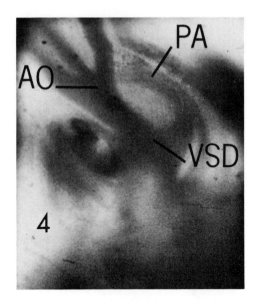

Fig. 4. Ventral view of arterial outflow tract of 5-day chick embryo after trypan blue administration at 44 hr incubation. The site of fusion of the bloodstreams (VSD) represents a ventricular septal defect of the membranous type. (From Jaffee, 1965.) Abbreviations: AO, aorta; PA, pulmonary artery; VSD, ventricular septal defect.

original heart loop (Section II.A). Studies of the dynamics of the development of this area (Section III.B.1 and 2) have revealed its dependence upon the development of other regions of the heart, and a detailed study by Shaner (1949) has established a definite relationship between endocardial cushion tissue and conotruncal defects.

1. Aortic Stenosis

The administration of trypan blue to chick embryos at 24 hr incubation produced chiefly aortic stenosis (Figure 5), compared to the principal finding of ventricular septal defects when this dye was administered at 44 hr (Jaffee, 1970). Underdevelopment of the third and fourth aortic arches was found to be associated with stenosis. Rychter (1962) had previously shown that clipping of the same arches led to the appearance of aortic stenosis; a similar combination of events was found after irradiation of chick embryos (Le Douarin and Le Douarin, 1970; Gilani and Jaffee, 1971), with the Le Douarins also noting a depression of cardiac function following radiation. Study of intermediate stages after the administration of trypan blue and before the anatomical appearance of a stenotic aorta revealed a narrowing of the left bloodstream (Jaffee, 1970).

2. Pulmonary Stenosis

Narrowing of the pulmonary artery, often associated with ventricular septal defects as in the tetralogy of Fallot, is a frequent type of congenital

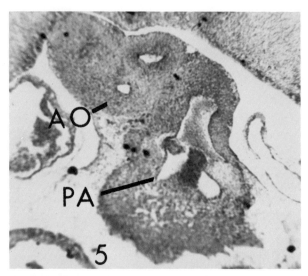

Fig. 5. Aortic stenosis (AO) in heart of 7-day chick embryo following trypan blue administration at 24 hr (From Jaffee, 1970.)

heart disease, yet few significant experimental analyses have been made. Rychter (1962) showed that suppression of the pulmonary arches was followed by frequent appearance of ventricular septal defects, emphasizing that pulmonary stenosis is the basic feature of tetralogy. A detailed study of the effects of thalidomide upon rabbit embryos by Vickers (1968) revealed ventricular septal defect and underdevelopment of the pulmonary trunk as the principal defects. These findings are of special interest since right-sided lesions of the heart, including tetralogy of Fallot, were seen frequently among congenital heart disease found in moderately high incidence (18.4%) in thalidomide victims (Keck *et al.*, 1972).

3. Complete Transposition of the Great Arteries

In complete transposition, the aorta and pulmonary trunk arise from the right and left ventricles respectively and run parallel to each other (Edwards, 1960; Monie *et al.*, 1966) rather than spiraling after leaving the heart. This malformation was a common finding after maternal administration of trypan blue at early (primitive streak) stages of development (Wilson, 1960; Monie *et al.*, 1966). In a detailed analysis by the latter group, shortening of the great arteries, abnormal looping of the heart tube, and a minimal degree of spiraling of the shortened truncus were found. The authors noted that the lack of spiraling may have been a result of altered blood flow. Model experiments by Goerttler (1958, 1970) and de Vries and Saunders (1962) demonstrated a hemodynamic basis for complete transposition. A recent study (Rowland *et al.*, 1973) has shown that complete transposition was the most frequent form of congenital heart disease found among the elevated incidence of cardiac defects in diabetic mothers.

4. Transposition of the Aorta into the Right Ventricle

A condition in which both aorta and pulmonary arteries emerge from the right ventricle was produced by Rychter (1962) by a temporary clamping of the bulbus of the 3-day chick embryo, the clip being removed 48 hr later. The effect of this procedure was said to be the prevention of the morphogenetic movement of the heart loop to the left. When the same procedure was carried out on 4-day embryos, overriding aortas were found. These results were confirmed (Jaffee, 1970) (Figure 6), and blood flow patterns studied at intermediate stages revealed that a inhibition of the unspiraling process (Section III.B.2) was a major cause in the development of this defect.

Double-outlet right ventricle (Figure 7) was a frequent finding after the exposure of chick embryos to hypoxia (Haring *et al.*, 1970; Jaffee, 1974a). The latter study demonstrated that the hypoxic state (cf. V.C.2) led to a swelling and distortion of the heart tube (Figure 10); the normal form of the heart tube at this stage is illustrated in figure 9. When the physiological state of the embryo returned to normal after restoration of atmospheric oxygen, distortion of the heart tube was still evident (Figure 11) in a large percentage of treated embryos. When embryos similarly treated (at 60 hr incubation) were

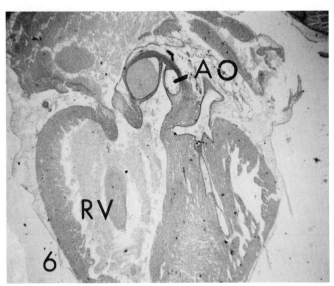

Fig. 6. Overriding aorta in chick embryo heart after temporary compression of the bulbus arteriosus on the fourth day of incubation.

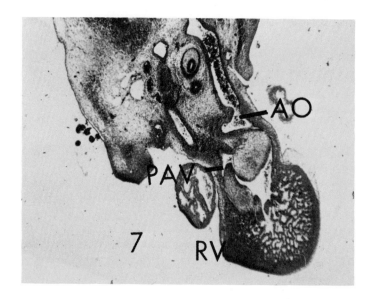

Fig. 7. Double-outlet right ventricle in 7-day chick embryo after exposure to moderate hypoxia at 60 hr of incubation. (From Jaffee, 1974a.) Abbreviations: AO, aorta; PAV, pulmonary artery valve; RV, right ventricle.

studied at five days of incubation, flow pattern studies revealed the major features of double-outlet right ventricle (see Edwards, 1960), with the aorta and pulmonary artery emerging from the right ventricle lateral to each other and the entire outflow tract still displaced to the right as in earlier stages (Figure 12). Double-outlet right ventricle was found after irradiation of chick embryos by Le Douarin and Le Douarin (1970), with a depression of cardiac function noted directly following the radiation.

5. Common Truncus Arteriosus

Complete failure of septation of the arterial outflow tract was a frequent sequel when an antimetabolite (cytosine arabinoside) was administered to chick embryos at 48 hr (Jaffee, 1970); a severe inhibition of growth was evidenced by the finding of smaller-than-normal hearts. The examination of 5-day embryos revealed a single stream of blood flow through the bulbus, representing a fusion of the two streams normally found. Common truncus has been found after irradiation of chick embryos (Gilani and Jaffee, 1971), occasionally after hypoxia (Figure 8) in chicks (Jaffee, 1974a), and after trypan blue in rat embryos (Monie et al., 1966).

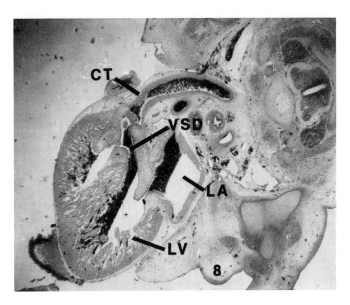

Fig. 8. Common truncus arteriosus following hypoxic exposure as in Figure 7. The common trunk (CT) emerges from the right ventricle. Abbreviations: LA, left atrium; LV, left ventricle; VSD, ventricular septal defect.

C. Left Heart Hypoplasia

Underdevelopment of the left side of the heart was produced by a clamping of the left atrium (Rychter, 1962) and also by the placement of a nylon device in the left atrioventricular canal (Harh *et al.*, 1973). Both authors emphasized hemodynamic aspects in the interpretation of their results, with Harh *et al.* noting that a flow-volume hypoplasia appeared to be the mechanism involved. The mechanism whereby a narrowed stream of flow becomes progressively narrower is discussed below (Section VI).

D. Aortic Arch Anomalies

The relationship between aortic arch anomalies and cardiac defects is well established (Section V.A.2 and V.B.1). Depression of cardiac function was found followed by failure of aortic arch vascularization in studies with hypoxia (Grabowski and Schroeder, 1968) and irradiation (Le Douarin and Le Douarin, 1970). Probable hemodynamic factors in the production of aortic arch anomalies following catecholamine administration was noted above (Section V.A.2).

E. Valve Anomalies

The influence of blood flow upon early stages of valvular development was noted above (Section II.B), and distortion of cardiac jelly mounds following hypoxia was also found (Jaffee, 1974b). Le Douarin and Le Douarin (1970) noted that endocardial cushion tissue was especially sensitive to radiation, and valvular malformations were a frequent finding following radiation by Gilani and Jaffee (1971). The observation that the normally muscular right atrioventricular valve of the chick heart was transformed into a connective tissue valve by modifications of blood flow (Rychter, 1962) has demonstrated a hemodynamic factor in valvulogenesis.

VI. MECHANISMS OF CARDIOVASCULAR TERATOGENESIS

A. Primary Lesion

The best-documented primary lesions known to influence later cardiovascular development are abnormalities in the form of the primary heart tube and/or interference with formation of the great vessels, especially the aortic arches. Careful analyses of atrial septal defects (Section V.A.1) and of complete transposition of the great arteries (Section V.B.3) revealed that the anatomical lesions were of primary importance. These experiments were carried out in precirculatory stages using trypan blue which acted mainly either as a general inhibitor of growth (Monie *et al.,* 1966) or as a specific mesodermal inhibitor (Fox, 1972). When hypoxia was used as a teratogenic agent after the onset of circulation (Section V.B.4 and Figures 10 and 11), distortion of the form of the heart tube and failure of vascularization of aortic arches were found to be the primary lesions. Similar results were observed when embryos were examined immediately following radiation (Le Douarin and Le Douarin, 1970).

B. Development of the Lesion into a Malformation

1. Precirculatory

The influence of an abnormal heart tube upon further development has been outlined above (Section V.A.1 and V.B.3). The direction of original heart looping and further development has been analyzed by Van Praagh (1967).

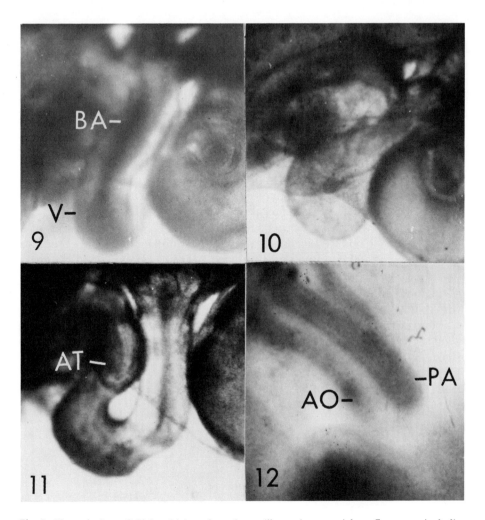

Fig. 9. Ventral view of 66-hr chick embryo heart illustrating arterial outflow tract including ventricle (V) and bulbus arteriosus (BA).

Fig. 10. Sixty-six hr heart shown immediately after 6 hr exposure to moderate hypoxia. View as in Figure 9. Note swelling of ventricle and distortion of bulbus.

Fig. 11. Three day (72-hr) embryo heart after hypoxic exposure as in Figure 10. Normal function has returned, but the form of the heart loop remains anomalous and the atrium (AT) is displaced ventrally (cf. Figure 9).

Fig. 12. Arterial outflow tract of 5-day chick embryo after exposure to moderate hypoxia at 60 hr. Blood flow patterns indicate the development of double-outflow right ventricle; for normal heart (cf. Figure 3). Further details in text. Abbreviations: AO, aorta; PA, pulmonary artery.

2. Hemodynamic

Failure of septation following fusion of the normally separate bloodstreams has been demonstrated experimentally (Section V.A.2 and Figure 4), as well as following the narrowing of a great artery, e.g., pulmonary stenosis (Section V.B.2). The development of an abnormal flow pattern, foreshadowing a malformation, was shown in experimental partial transposition (Section V.B.4 and Figure 12) and in model experiments illustrating complete transposition (Section V.B.3).

The importance of lateral pressure in determining the diameter of a great artery was noted in a model experiment by Rodbard (1959). The narrowing of a tube with an elastic lining, as produced by a drop in head pressure, would be followed by an increased velocity at the site of the narrowing concomitant with a decreased lateral pressure, in accordance with hydraulic principles. This was demonstrated in experiments in which narrowing of a vessel at the site of the reduced lateral pressure stimulated distal proliferative activity by the vessel endothelium (Rodbard, 1959). This explained the presence of well-developed blood vessels downstream to stenoses.

Hemodynamic influence on the structure of the heart wall is illustrated in normal development when the wall of the truncus arteriosus, originally muscular and contractile, becomes transformed into two elastic-walled arteries as valvular efficiency changes the hemodynamics of flow within the arteries (Van Mierop, 1970).

VII. GENERAL TERATOLOGICAL PRINCIPLES

A. Genotype

Crosses between strains of rats with varying spontaneous incidences of cardiac malformations have shown that genotype rather than intrauterine environment determined the incidence of these malformations in the offspring (Fox, 1967). High rates of ventricular septal defects were found in specific inbred strains of chickens by Siller (1958). Several syndromes with some hereditary component, e.g., Down syndrome, include a high rate of congenital heart disease. On the other hand, surveys of familial aggregates and twin studies have been interpreted as not indicating Mendelian inheritance of cardiovascular defect, but rather as supporting the multifactorial concept (see Section I). Analysis of the thalidomide victims led Lenz to rule out genetic factors and to come to the general conclusion: "Heredity is of minor importance in cardiac abnormalities, and if heredity is involved, it is specific." (Lenz, 1966, p. 10).

B. Stage of Development

1. Vulnerable Period

Cardiovascular malformations may arise at any time during the major developmental period of this system, 6–7 weeks in the human (Dankmeijer, 1964), and some, such as aortic stenosis and atresia, may occur even later. Patent ductus arteriosus may develop either in association with structural defects, with immaturity of the respiratory system (Section IV.B), or with unknown functional disorders.

2. Critical Period

The classical concept of critical period as related to the heart was discussed by Rychter (1962). Periods of increased sensitivity of the cardiovascular system to teratogenic agents depend upon the mechanism of action of the agent, its mode of transport, and probably specific features of the metabolism of the embryo at any given stage. Trypan blue was found most effective at early stages in mammals (Section VI.A); radiation appeared to be most effective on the third day of incubation of chick embryos, presumably because of higher mitotic rates in the cardiovascular system at this time (Le Douarin and Le Douarin, 1970; Gilani and Jaffee, 1971), and sympathetomimetic compounds were found most effective on the fifth day of chick incubation (Sections V.A.2 and V.D).

C. Nature of the Teratogenic Agent

Similar defects may be produced by different teratogenic agents, and different agents may have the same effect (Wilson, 1960), emphasizing the nonspecific action of most of these agents. The mechanisms of action of well-documented teratogens such as rubella and thalidomide remain obscure, but both affect more than one organ system. In order to evaluate the importance of the hypoxia of altitude, which is known to increase the incidence of patent ductus arteriosus in clinical surveys (Section II.A), reliable statistics on the background incidence of congenital heart disease among populations living at high altitudes are needed. Experimental studies (Ingalls *et al.,* 1952;

Jaffee, 1974b) have indicated that hypoxia, as found at high altitudes, should increase the incidence of congenital cardiovascular malformations. Preliminary evidence that catecholamines (Sections V.A.2 and V.E) and acetylcholine (Sections V.A.1 and 2) produced malformations by interfering directly with cardiac function, if confirmed, will strengthen the thesis advanced in this review that functional factors influence development.

VIII. CONCLUSIONS

The contention that "environmental factors operate to a far greater degree than formerly appreciated; that these can be evaluated on a firm scientific basis; and that intelligently conceived measures to protect the fragile embryo or vulnerable fetus will ultimately be devised" (Ingalls *et al.,* 1952, p. 767) is gathering support. An environmental factor, hypoxia, found to produce reversible physiological changes in the embryo, affected cardiac structure by altering hemodynamics. Hearts thus treated, when examined shortly after the hypoxic exposure, were found to be malformed but functioning normally, a situation commonly observed in congenital heart disease. Thus variations in embryonic physiology must be regarded as potentially teratogenic to the cardiovascular system. Obviously not all changes in the embryological physiologic state will be teratogenic, and further research is needed to establish such relationships. In the future it may become feasible to monitor maternal physiology in early embryonic stages, especially in cases where there is reason to suspect that abnormal development may be taking place.

Changes similar to those resulting from hypoxia follow the action of a number of other teratogenic agents (Grabowski, 1970) and may similarly affect cardiovascular development. The finding that slowing of the embryonic heart with acetylcholine led to septal defects suggests that altered cardiac metabolism per se may affect development. While studies of embryonic physiology are accumulating, in most instances these have not been related to cardiogenesis.

Emphasis upon hemodynamics in this review does not imply disregard of other factors, e.g., intracellular events, synthesis of connective tissues, morphogenetic movements not influenced by circulation, all of which are especially important in precirculatory stages where the effect of such processes upon the developing heart tube are in need of experimental analyses. Once circulation becomes established, all these processes plus myocardial metabolism* become expressed in a final common pathway represented by cardiac output. Elucidation of the quantitative relationships of such vascular forces with the tissues of the developing cardiovascular system presents a promising field of investigation.

---

*Measurements of myocardial oxygen uptake in embryonic hearts following hypoxia provide evidence for this concept (Jaffee and Gornet, 1976).

ACKNOWLEDGMENTS

Research in the author's laboratory was aided by NIH grant HL 11432 and a grant from the Miami Valley Heart Association.

REFERENCES

Assali, N. S., Bekey, G. A., and Morrison, L. W., 1968, Fetal and neonatal circulation, *in: Biology of Gestation,* Vol. II (N. A. Assali, ed.), pp. 52–116, Academic Press, New York.

Copley, A. L., and Staple, P. H., 1962, Haemorheological studies on the plasmatic zone in the microcirculation of the cheek pouch of Chinese and Syrian hamsters, *Biorheology* **1**:3.

Dankmeijer, J., 1964, Cardiac malformations and the stages of their origin during embryonic development, *Arch Biol. (Liege)* **75**(Suppl):1133.

de Vries, P. A., and Saunders, J. B. de C. M., 1962, Development of the ventricles and spiral outflow tract in the human heart, *Carnegie Inst. Wash., Publ. 621, Contribs. Embryol.* **37**:87.

Edwards, J. P., 1960, Congenital malformations of the heart and great vessels, *in: Pathology of the Heart* (S. E. Gould, ed.), pp. 260–496, Charles C Thomas, Springfield, Ill.

Fox, M. H., 1967, Genetic transmission of congenital membranous ventricular septal defects in selectively inbred strains of rats, *Circ. Res.* **20**:422.

Fox, M. H., 1972, The influence of the fetal genome on trypan blue-induced cardiovascular defects, *Teratology* **6**:339.

Fox, M. H., and Goss, G. M., 1957, Experimentally produced cardiac malformations of the heart and great vessels in rat fetuses. Atrial and caval abnormalities, *Anat. Rec.* **129**:309.

Fraser, F. C., and Rosen, J., 1975, Association of cleft lip and atrial septal defect in mice: a preliminary report, *Teratology* **11**:321.

Gilani, S. H., and Jaffee, O. C., 1971, Congenital cardiovascular anomalies in chick embryos exposed to X-irradiation and hypothermia, *Am. J. Obstet. Gynecol.* **111**:47.

Goerttler, K., 1958, *Normale und Pathologische Entwicklung des Menschlichen Herzens,* George Thieme Verlag, Stuttgart.

Goerttler, K., 1970, Embryology, teratology and congenital heart disease: A correlation, *in: Cardiac Development with Special Reference to Congenital Heart Disease* (O. C. Jaffee, ed.), University of Dayton Press, Dayton, Ohio.

Grabowski, C. T., 1970, Embryonic oxygen deficiency—a physiological approach to analysis of teratological mechanisms, *in: Advances in Teratology,* Vol. 4 (D. H. M. Woolam, ed.), Academic Press, New York.

Grabowski, C. T., and Schroeder, R. E., 1968, A time-lapse photographic study of chick embryos exposed to teratogenic doses of hypoxia, *J. Embryol. Exp. Morphol.* **19**:347.

Harh, J. Y., 1975, Experimental cardiac morphogenesis. I. Development of the ventricular septum in the chick, *J. Embryol. Exp. Morphol.* **33**:13.

Harh, J. Y., Paul, M. H., Gallen, W. J., Freidberg, D. Z., and Kaplan, S., 1973, Experimental production of hypoplastic left heart syndrome in the chick embryo, *Am. J. Cardiol.* **31**:51.

Haring, O. M., 1966, Cardiac malformations in the rat induced by maternal hypercapnea with hypoxia, *Circ. Res.* **19**:544.

Haring, O. M., Patterson, J. R., and Sarche, M. A., 1970, Prenatal development of the cardiovascular system in the chicken. Effects of hypercapnea and hypoxia, *Arch. Pathol.* **89**:537.

Hodach, R. J., Gilbert, E. F., and Fallon, J. F., 1974, Aortic arch anomalies associated with administration of epinephrine in chick embryos, *Teratology* **9**:203.

Hodach, R. J., Hodach, A. E., Fallon, J. F., Folts, J. D., Bruyere, H. J., and Gilbert, E. F. 1975, The role of beta adrenergic activity in the production of cardiac and aortic arch anomalies in chick embryos, *Teratology* **12**:33.

Hughes, A. F. W., 1943, The histogenesis of the arteries of the chick embryo, *J. Anat.* **77**:266.

Ingalls, T. H., Curley, F. J., and Prindle, R. A., 1952, Experimental production of congenital anomalies. Timing and degree of anoxia as factors causing fetal deaths and congenital anomalies in the mouse, *N. Engl. J. Med.* **247**:758.

Ishikawa, S., Kawamura, T., Takao, A., Ando, M. Miwa, H., and Okai, O., 1975, Cardiovascular malformations following acetylcholine chloride administration to chick embryos, *Teratology* **12**:198.

Jaffee, O. C., 1962, Hemodynamics and cardiogenesis. The effects of altered vascular patterns on cardiac development, *J. Morphol.* **110**:217.

Jaffee, O. C., 1963, Bloodstreams and the formation of the interatrial septum in the anuran heart, *Anat. Rec.* **147**:355.

Jaffee, O. C., 1965, Hemodynamic factors in the development of the chick embryo heart, *Anat. Rec.* **151**:69.

Jaffee, O. C., 1966, Rheological aspects of the development of blood flow patterns in the chick embryo heart, *Biorheology* **3**:59.

Jaffee, O. C., 1967, The development of the arterial outflow tract in the chick embryo heart, *Anat. Rec.* **158**:35.

Jaffee, O. C., 1970, Comparative and experimental studies of the development of blood flow patterns in embryonic hearts, *in: Cardiac Development with Special Reference to Congenital Heart Disease* (O. C. Jaffee, ed.), Univ. Dayton Press, Dayton, Ohio.

Jaffee, O. C., 1972, Effects of propranolol on the chick embryo heart, *Teratology* **5**:153.

Jaffee, O. C., 1974a, The effects of moderate hypoxia and moderate hypoxia with hypercapnea on cardiac development in chick embryos, *Teratology* **10**:275.

Jaffee, O. C., 1974b, The influence of altitude upon the effects of hypoxia and hypercapnea on cardiovascular development in chick embryos, *I.R.C.S. Med. Sci.* **2**:1422.

Jaffee, O. C., and Gornet, S., 1976, A metabolic factor in cardiogenesis: oxygen uptake by the embryonic myocardium of the chick heart following hypoxia, *I.R.C.S. Med. Sci.* **4**:528.

Keck, E. W., Roloff, D., and Markworth, P., 1972, Cardiovascular findings in children with thalidomide dysmelia syndrome, *Proc. Assoc. Eur. Pediatr. Cardiol.* **8**:66.

Le Douarin, N., and Le Douarin, G., 1970, The effects of radiation on cardiac development, *in: Cardiac Development with Special Reference to Congenital Heart Disease* (O. C. Jaffee, ed.), Univ. Dayton Press, Dayton, Ohio.

Lenz, W., 1966, Malformations caused by drugs in pregnancy, *Am. J. Dis. Child.* **112**:99.

McClung, J., 1969, *Effects of High Altitude on Human Birth*, Harvard Univ. Press, Cambridge, Mass.

Moffatt, D. B., 1959, Developmental changes in the aortic arch system of the rat, *Am. J. Anat.* **105**:1.

Monie, I. W., Takacs, E., and Warkany, J., 1966, Transposition of the great vessels and other cardiovascular abnormalities in rat fetuses induced by trypan blue, *Anat. Rec.* **156**:175.

Neill, C. A., 1972, Etiology of congenital heart diseases, *Cardiovasc. Clin.* **4**:138.

Nora, J. J., 1968, Multifactoral inheritance hypothesis for the etiology of congenital heart disease: The genetic–environmental interaction, *Circulation* **38**:604.

Nora, J. J., Nihill, M. R., and Vargo, T. A., 1970, Etiology of congenital heart disease: Genetic aspects, *in: Cardiac Development with Special Reference to Congenital Heart Disease* (O. C. Jaffee, ed.), Univ. Dayton Press, Dayton, Ohio.

Odgers, P. N. B., 1938, The development of the parts membranacea septi of the human heart, *J. Anat.* **72**:247.

Ojeda, J. L., and Hurle, J. M., 1975, Cell death during the formation of the tubular heart of the chick embryo, *J. Embryol. Exp. Morphol.* **33**:523.

Orts Llorca, F., 1970, Curvature of the heart: Its first appearance and determination, *Acta Anat.* **77**:468.

Paff, G. H., Boucek, R. J., and Gutten, G. S., 1965, Ventricular blood pressures and competency of valves in the early embryonic chick heart, *Anat. Rec.* **151**:119.

Patten, B. M., 1949, Initiation and early changes in the character of the heart beat in chick embryos, *Physiol. Rev.* **29**:31.

Patten, B. M., 1960, The development of the heart, *in: Pathology of the Heart* (S. E. Gould, ed.), Charles C Thomas, Springfield, Ill.

Penaloza, D., Arias-Stella, J., Sime, F., Recarrarren, S., and Marticarena, E., 1964, The heart and pulmonary circulation in children at high altitudes, *Pediatrics* **34**:568.

Rodbard, S., 1959, Physical forces and the vascular lining, *Ann. Intern. Med.* **50**:1339.

Romanoff, A., 1960, *The Avian Embryo*, Macmillan, New York.

Rosenquist, G. C., and DeHaan, R. L., 1966, Migration of precardiac cells in the chick embryo: A radiographic study, *Carnegie Inst. Wash. Publ. 625 Contrib. Embryol.* **38**:111.

Rowland, T. W., Hubbell, J. P., Jr., and Nadas, A. S., 1973, Congenital heart disease in infants of diabetic mothers, *J. Pediatr.* **83**:815.

Rychter, Z., 1962, Experimental morphology of the aortic arches and heart loop in chick embryos, *Adv. Morphol.* **2**:333.

Shaner, R. F., 1949, Malformation of the atrioventricular endocardial cushions of the embryo pig and its relation to defects of the conus and truncus arteriorsus, *Am. J. Anat.* **84**:431.

Siller, W. G., 1958, Ventricular septal defects in the fowl, *J. Pathol. Bacteriol.* **76**:431.

Siller, W. G., 1967, Aortic dextroposition complexes in the fowl: A study in comparative pathology, *J. Pathol. Bacteriol.* **94**:155.

Sissman, N. J., 1970, Developmental landmarks in cardiac morphogenesis: Comparative chronology, *Am. J. Cardiol.* **25**:141.

Spitzer, A., 1923, Uber den Bauplan des normale und missbildeten Herzens (Versuch einer phylogenetischen Theorie), *Virchows Arch Anat.* **243**:81.

Van Mierop, L. H. S., 1970, Morphological and functional development of the chick cardiovascular system during the first week of incubation, *in: Cardiac Development with Special Reference to Congenital Heart Disease* (O. C. Jaffee, ed.), Univ. Dayton Press, Dayton, Ohio.

Van Praagh, R., 1967, Chamber localization, *in: Heart Disease in Infancy and Childhood* (J. D. Keith, R. D. Rowe, and P. Vlad, eds.), Macmillan, New York.

Vickers, T. H., 1968, The cardiovascular malformations in the rabbit thalidomide embryopathy, *Br. J. Exp. Pathol.* **49**:179.

Wilson, J. G., 1960, Experimental production of cardiac defects, *in: Congenital Heart Disease* (A. D. Bass and G. K. Moe, eds.), American Association for the Advancement of Science, Washington, D.C.

Zamora, C., Espino-Vela, J., Padilla, S., Hindojosa, H. R., and de la Garza, M. L., 1971, Frecuencia estadistica de las cardioparias congenitas, *Arch. Inst. Cardiol. Mex.* **41**:373.

Abnormal Organogenesis in the Urinary Tract

15

IAN W. MONIE

I. INTRODUCTION

The following presentation, unless otherwise indicated, is concerned with normal and abnormal development in the Long–Evans rat fetus and gestational ages are based on the day of finding sperm in the vagina being designated day "zero." Such times, of course, show some variation not only between different strains but within the same strain and even within the same litter.

II. GENERAL CONSIDERATIONS

A. Outline of Development of the Urogenital System

The principal events in the development of the urogenital system of the rat are depicted in Figure 1. According to Torrey (1943), in the albino rat about day 10 some vestigial pronephric tubules form in the nephrogenic ridge but soon disappear. Thereafter, within the ridge the mesonephros is established during day 11 by the appearance of numerous tubules some of which become S-shaped and acquire capsular (Bowman's) spaces but no glomeruli. Nearby, about the same time, the mesonephric (Wolffian) duct appears and

IAN W. MONIE • Department of Anatomy, University of California, San Francisco, San Francisco, California 94143.

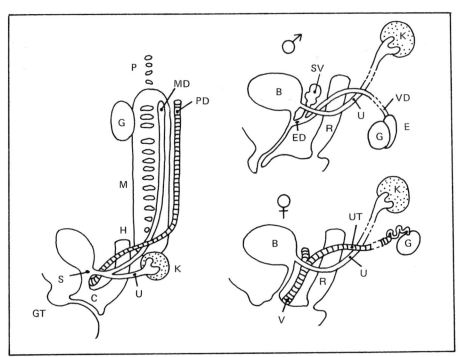

Fig. 1. Left: structure of the developing urogenital system common to both sexes. Right: definitive urogenital structures of the male and female. B, bladder; C, cloaca; E, epididymis; ED, ejaculatory duct; G, gonad; GT, genital tubercle; H, hindgut; K, kidney (metanephros); M, mesonephros; MD, mesonephric (Wolffian) duct; P, vestigial pronephric tubules; PD, paramesonephric (Müllerian) duct; R, rectum; S, urogenital sinus; SV, seminal vesicle; U, ureter; UT, uterus; V, vagina; VD, vas deferens.

begins to grow caudally. The mesonephros eventually disappears except for its duct and a few tubules which unite the latter with the testis.

The mesonephric duct terminates in the cloaca during day 12 and about the same time gives off the ureteric bud around which the metanephros, the definitive kidney, forms. Within the latter many renal (nephric) vesicles now appear developing into nephrons which unite with outgrowths from the ureteric bud. The caudal end of the ureter leaves the mesonephric duct about day 15 and ascends the urogenital sinus (urethral) wall to the bladder.

By day 17 the urorectal septum has divided the cloaca into the urogenital sinus ventrally and the rectum dorsally; the former gives rise to the bladder cranially, and to the urethra caudally.

Paramesonephric (Müllerian) ducts form about day 13 and, migrating caudally close to the mesonephric ducts, ultimately unite the dorsal to the urogenital sinus. The gonads differentiate on the medial aspect of each mesonephros. All the foregoing events are common to both sexes.

In males the paramesonephric ducts degenerate between days 19 and 22. In the mesonephroi some tubules remain as vasa efferentia to connect the testis with the mesonephric duct which now becomes the duct of the epididymidis, the vas deferens, and the ejaculatory duct. A seminal vesicle arises from the vas about day 19.

In females the mesonephric ducts degenerate between days 19 and 21 but the paramesonephric ducts remain to form the uterine tubes, uteri, and part (?) of the vagina. The ovaries take up a position close to the mouth of each uterine tube. Mesonephric tubules largely disappear.

In both sexes the kidneys (metanephroi) form in the pelvis and migrate cranially. Their development as well as that of other components of the urinary system is considered in more detail in later sections.

B. Development of Some Related Structures

The development of the urinary system, besides being closely related to that of the genital system, is also influenced by the development of neighboring structures. Thus, as the caudal end of the mesonephric duct approaches the cloacal wall (days 11 and 12), it passes between the primary and definitive umbilical arteries (Figure 2). Should the former vessel fail to disappear, as normally occurs, further development of the duct, ureteric bud, and metanephros of the same side may be severely disturbed (Monie, 1961; Monie and Khemmani, 1973). Likewise, retarded or abnormal growth of the lumbar,

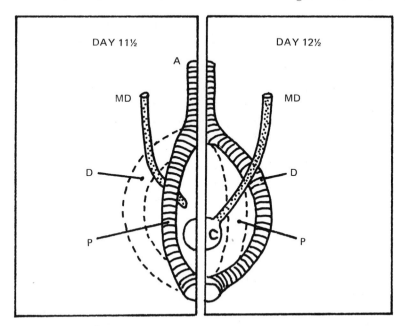

Fig. 2. Relationship of mesonephric duct to developing umbilical arteries in the rat fetus. C, cloaca; D, definitive umbilical artery; MD, mesonephric duct; P, primary umbilical artery.

sacral, and caudal vertebrae, and of the pelvis, can result in ectopic kidneys (Monie *et al.,* 1957).

Hindlimb buds appear during day 11, and their development may be adversely affected by conditions which also cause malformations of the urogenital system, e.g., symmelia. Since the neural tube extends almost the entire length of the early embryo, defects of its caudal end, such as spina bifida, are often accompanied by derangement of the urinary system. Disturbed development of the hindgut, which opens into the cloaca in common with the urogenital sinus, also can have a profound effect on the formation of the caudal portion of the urinary tract.

C. Genetically Determined Defects

Abnormalities of the urinary system due to genetic factors are well recognized in man and many animals. Table 1 lists some strains of mice, rats, and hamsters in which such malformations occur. Often they are accompanied by defects of other systems. The genesis of urinary tract malformations in embryos of the *my* strain of mouse was studied in detail by Brown (1931).

D. Environmentally Determined Defects

Hale (1935) reported that maternal vitamin A deficiency in pigs resulted in young with "misplaced kidneys" as well as eye defects, and since that time many teratogenic agents have been found which adversely affect the development of the mammalian urinary system. Table 2 shows a number of teratogenic procedures and some of the abnormalities of the urinary system produced by them. The nature of the latter depends on the time of administration, dosage, and duration of exposure to the teratogen, and often different types of malformations will be produced by advancing or delaying treatment by a day or less.

In addition to the teratogens presented in Table 2, the following are also credited with producing malformations of the urinary system in rat (r), rabbit (rb), rhesus monkey (rh), mouse (m), or guinea pig (gp). They are: antibodies (r); *l*-asparaginase (rb); 6-azauridine (rh); bradykinin (m); carzinophilin (r); Congo red (r); diphenylamine (r); griseofulvin (r); hyperthermia (gp); metahexamide (r); 4-methyl-5-oxyuracil (r); monomethylformamide (r); nitrous oxide (r); rubidomycine (r); sarkomycin (r); serotonin (r); tetracycline (r); thallium (r); trauma (r); vitamin B_{12} deficiency (r); and zinc deficiency (r). For further information and references consult *Catalog of Teratogenic Agents* by T. H. Shepard (1973). Other teratogens affecting the development of the mammalian urinary system such as alloxan, retinoic acid, and thalidomide are mentioned in Section III.

Table 1. Some Genetic Factors Causing Urinary System Defects[a]

Animal	Strain	Urinary system defects	Other defects	Reference
Mouse	Luxate (lx)	HN, HU, HSK, renal cysts; kinked terminal ureter	Lumbar vertebrae; hind limb	Carter (1953)
	Myelencephalic blebs (my)	Absent or small kidney; HN, HU, renal cysts	Eye, foot, nervous system	Bagg (1925); Brown (1931)
	Shorttail (Sd)	Absent or small kidney; absent or partial ureters	Lumbar, sacral, and caudal vertebrae; intestine	Gluecksohn-Schoenheimer (1943)
	Shortear (Se)	HN with and without obstructed ureters	Ears	Wallace and Spickett (1967)
Rat	ACI (sub)	Absent kidney; absent or partial ureters	Genital tract	Cramer and Gill (1975)
	Gunn (sub)	HN, HU, renal cysts; obstructed ureters	Glucuronyl transferase absent; jaundice	Lozzio et al., (1967)
	Stub	Absent kidney, HN, absent bladder, obstructed ureters, urethral atresia	Tail	Ratcliffe and King (1941)
	Wistar albino (sub)	Absent or small kidney; HN, HU		Hain and Robertson (1936)
	Wistar albino (sub)	HN (mostly males); no obstructed ureters		Astarbaldi and Bell (1962)
	?	Renal cysts		Solomon (1973)
Hamster	Piebald	Absent kidney (females)	Genital tract	Orsini (1952)
	VSY	HN	Diabetes mellitus	Meier and Yerganian (1961)

[a]Abbreviations: HN, hydronephrosis; HSK, horseshoe kidney; HU, hydroureter; sub, substrain.

Table 2. Some Teratogenic Procedures Causing Urinary System Defects[a]

Teratogen	Principal defects[b]	Maternal treatment	Reference
Vitamin A deficiency	HSK, small kidney, ec. ureter, ureteric sten. or atr., hypospadias	Vitamin A-deficient diet since weaning, up to and through gestation.	Wilson and Warkany (1948)
	As above with HN and HU	As in above but single dose of vitamin A (16,000 IU) orally from 10th through 15th day of gestation.	Wilson et al. (1953)
Vitamin A excess	Abs. kidney, fused kidney, crossed renal ectopia, HN, HU, double ureter, ureteric atresia, bladder diverticula	Single dose of vitamin A (10,000 IU) orally from day 7⅓ through 12⅓ of gestation.	Kalter and Warkany (1961)
	HN, HU, ec. ureter, ureteric atr., abs. bladder, pers. ureteric memb.	Vitamin A (50,000 IU) given three consecutive days: days 5–7, 8–10, 11–13, and 14–16.	Giroud et al. (1958)
PGA (folic acid) deficiency	HN, small kidney, HU, ureteric atr., small bladder, urethral sten. or atr.	PGA-deficient diet containing 1% SST and 0.5% x-methyl PGA from day 11 of gestation onward.	Monie et al. (1954)
	As above with ec. kidney, double kidney, ec. ureter	Same regimen as in above maintained for days 10–13 of gestation only.	Monie et al. (1957)
Niacin deficiency	HN, divided kidney, ec. kidney, HU, distended bladder	Single i.p. injection of aqueous solution of 6-AN (8 mg/kg body weight) from days 9 through 20 of gestation.	Chamberlain and Nelson (1963)

Pantothenic acid deficiency	HN and HU	Pantothenic acid-deficient diet for first 9–11 days of gestation, then sodium omega methyl pantothenate (5–10 gm/kg diet) added for 2–3 days. Thereafter vitamin supplemented diet given.	Nelson et al. (1957)
Riboflavin deficiency	HN and HU	Riboflavin-deficient diet containing 5–216 mg/kg diet galactoflavin for duration of gestation.	Nelson et al. (1956)
Chlorambucil	Abs., fused, small, or ec. kidneys; HN, HU, abs. or incomp. ureter, urethral atr., pers. cloaca, enlarged gen. tubercle	Chlorambucil (6 mg/kg body weight) in peanut oil i.p. on day 10 of gestation.	Monie (1961)
Methyl salicylate	Abs., small, double, or ec. kidneys; crossed renal ectopia, HN, HU, ureteric atr., ec. ureter, distended bladder	Methyl salicylate (0.1–0.2 ml) subcutaneously on day 10 or 11 of gestation.	Bertone (1965); Monie (1970)
Trypan Blue	Fused kidneys, absent bladder, urethral atr., diverticula of ugs	1 ml of 1% aqueous solution trypan blue (Grübler) subcutaneously at least once before and once during gestation.	Goldstein (1957)
X-irradiation	Abs. kidney, HSK	Single exposure to 25–200 r, 9th–11th days of gestation.	Wilson (1954)

[a]Rats used in all studies except that of Kalter and Warkany (1961) in which mice were employed.
[b]Abbreviations: Abs, absent; atr, atretic; ec, ectopic; gen, genital; HN, hydronephrosis; HSK, horseshoe kidney; HU, hydroureter; incomp, incomplete; memb, membrane; pers, persistent; sten, stenosis; ugs, urogenital sinus.

III. SPECIFIC CONSIDERATIONS

A. Kidney

The definitive kidney (metanephros) is established on day 12 when the ureteric bud, springing from the caudal end of the mesonephric duct, contacts the metanephric blastema and stimulates it to undergo differentiation. Thereafter the kidney migrates from the pelvis ventral to the umbilical (later common iliac) arteries to reach its definitive position in the lumbar region about day 15 (Fig. 6). This "ascent" of the kidney (Figure 3) has been attributed to reduction of flexion of the vertebral column and to the decrease in size of the mesonephros (Gruenwald, 1939, 1943). Also of importance appear to be the growth of the vertebral column and pelvis, the enlargement of the kidney, the lengthening of the ureter, and the pressure of surrounding structures.

While the kidney is migrating, the cranial end of the ureter dichotomizes to form the renal pelvis, then through repeated divisions, the future collecting duct system. By day 15 it has produced four or five generations of tubules, the terminal portion (ampulla) of each evoking a renal vesicle, future nephron,

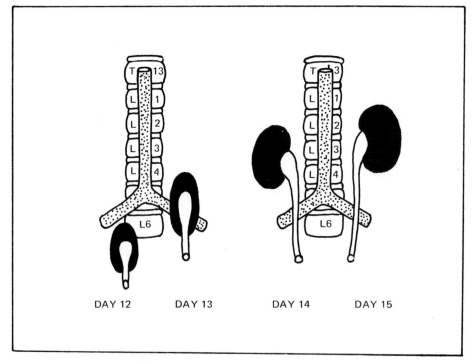

Fig. 3. Renal migration in the rat fetus.

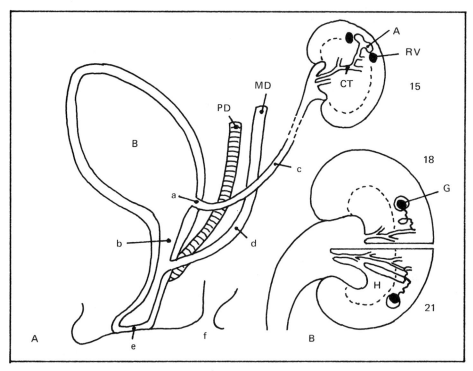

Fig. 4. Development of urinary system of the rat. Numbers indicate day of renal development. A: Ureter (a) opens into bladder between days 19 and 21; urethra (b) rewidens about day 17 or 18; ureters (c) rewiden about day 18; mesonephric ducts (d) patent on day 15; urogenital membrane (e) breaks down about day 20; anus (f) patent about day 16. Kidney development is that seen at day 15. B: Kidney development about days 18 and 21. During this time the renal papilla appears and the loop of Henle gradually extends into the medulla. A, ampulla; B, bladder; CT, collecting tubule; G, glomerulus; H, loop of Henle; RV, renal vesicles.

from the nephrogenic cells of the outer cortex (Figure 4). This process is repeated many times ending about four weeks after birth as judged by cortical structure and nephron function at that time (Baxter and Yoffey, 1948). In man the equivalent process ceases about the 35th week of gestation (Vernier and Smith, 1968).

By day 16 each of the older renal vesicles has evolved into a renal corpuscle and a renal tubule, the former giving rise to a glomerulus and capsular (Bowman's) space, and the latter, acquiring first a C-shaped then an S-shaped form, joins a collecting tubule. From the renal tubule come the proximal and distal convoluted tubules and the loop of Henle, the latter beginning to extend into the medulla during day 20. Capsular spaces in the inner cortex become expanded about day 17, indicating that kidney secretion has become established (Figure 4).

The renal pelvis, initially a narrow crescentic cleft, becomes somewhat triangular on cross-section about day 18, with the gradual advance into it of the renal (medullary) papilla, and acquires its characteristic reniform shape by day 21 (Figure 9). The rate of development of the renal papilla, however, can vary in different strains and within the same strain and may not be fully formed until several days after birth (Woo and Hoar, 1972).

1. Absent Kidneys

When the kidney is missing from a near-term fetus or newborn, it is often difficult or impossible to determine whether it never actually formed (renal agenesis) or whether it developed for a certain time than atrophied (renal aplasia). In view of this, the term "absent kidney," a statement of fact and not of conjecture, is often preferable to either of the foregoing.

Absence of kidneys due to genetic factors operating in the mouse, rat, and hamster occurs in several of the strains listed in Table 1. In all instances abnormalities also appear in other parts of the urinary tract and in other systems. Some of the combinations vary, however: Bagg (1925) found hydronephrosis and cystic kidneys in the *my* strain of mice, but Gluecksohn-Schoenheimer (1943) never encountered these in the *Sd* strain.

Lack of kidneys occurred in fetal rats and mice from hypervitaminosis A, chlorambucil, methyl salicylate, and X-irradiation (Table 2), while sodium arsenate produced the same abnormality in fetal hamsters (Ferm *et al.*, 1971), and in fetal rats (Beaudoin, 1974). Tuchmann-Duplessis *et al.*, (1959) found this defect in a kitten as a consequence of maternal deficiency of pteroylglutamic acid (PGA), and it has been noted in guinea pig fetuses from hyperthermia during pregnancy (Edwards, 1969). Absence of the kidney was again reported in rat young from 4-methyl oxyuracil when accompanied by maternal stress (Kosmachevskaya and Chebotar, 1968), and from the administration of retinoic acid during pregnancy (Dumas, 1964). One kidney was found to be missing in an 18-week human fetus whose mother received chlorambucil for treatment of Hodgkin's disease during pregnancy (Shotton and Monie, 1963).

The cause of absent kidneys was sought in mouse embryos of the *my* strain by Brown (1931), who concluded that marked lateral flexion of the embryo which separated the ureteric bud from the renal blastema, abnormal umbilical arteries, and hemorrhages were important factors in determining this defect.

A study of the offspring of female rats given chlorambucil during pregnancy (Monie, 1961) revealed that while kidneys formed in young embryos, they were degenerated or absent in old ones. Abnormal umbilical arteries were common, but there were no hemorrhages. Failure of the mesonephric ducts to reach the cloacal wall, and discontinuous ureters, occurred frequently. It is possible that abnormal development of the umbilical arteries hindered the caudal advance of the mesonephric duct, setting off a train of events which ended in renal atrophy. An examination of abnormal umbilical arteries in rat embryos from mothers given retinoic acid during pregnancy gives some support to this hypothesis (Monie and Khemmani, 1973). It is possible that a few cases in this study were due to true renal agenesis, but most fell into the category of renal aplasia. The characteristic facies and lung hypoplasia observed by Potter (1946) in infants with bilateral absence of the kidneys were not detectable in these fetuses.

2. Dysplastic Kidneys

Dysplasia, in the sense of local failure of differentiation, is difficult to ascertain in fetal rat kidneys before birth but is recognized more readily in the postnatal period when the kidney is more mature. Bernstein (1971) discusses the complex problem of renal dysplasia and describes its occurrence in the kidneys of rats following ureteric ligation in the newborn. Percy and Albert (1974) also have produced it by the injection of 5-iododeoxyuridine (IUdR) into rats shortly after birth.

3. Small Kidneys

Small kidneys (Figures 5c, 7, and 10) are often encountered in fetuses from the influence of genetic or environmental factors (Tables 1 and 2). Frequently they are described as "hypoplastic kidneys," but such a term, if

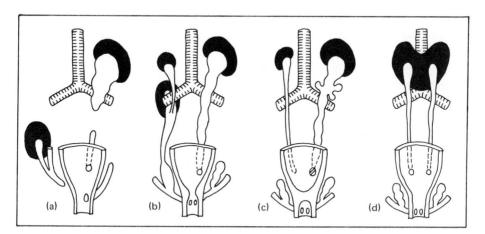

Fig. 5. Various abnormalities of the urinary system. (a) Left—pelvic kidney with ureter entering blind-ending mesonephric duct; right—hydronephrosis and hydroureter; partial ureteric atresia. (b) Left—accessory kidney and ectopic kidney dorsal to umbilical (common iliac) artery; ectopic ureter ends in seminal vesicle; right—hydronephrosis and dilated ureter. Urethral stenosis. (c) Left—small kidney; blind-ending ureter; right—hydronephrosis and dilated ureter with diverticula and persistent ureteric membrane. Urethral atresia. (d) "Horseshoe" kidney.

implying that the kidneys are small but normally differentiated, is probably incorrect in many cases. Usually they are found to be incompletely developed or atrophic. Some reflect an overall retardation of the development of the fetus, others a disparity between the growth of the kidney and other fetal structures.

Microscopic study usually reveals reduction in the number of collecting tubules, nephrons, and a diminution of the nephrogenic cells of the outer cortex. Often the ureter is stenosed or atretic, and the urethra likewise. Occasionally a few cystlike structures are present in the renal cortex, suggesting that some juxtamedullary nephrons, less affected than others, are functioning. Such kidneys have been observed in rat fetuses as a result of maternal PGA deficiency from days 11 to 21 (Monie *et al.*, 1954). It is possible that some of these kidneys, in which nephrons of the inner cortex survive while their formation in the outer cortex ceases, could become oligonephronic kidneys of the type seen in man.

4. Fused Kidneys

Various degrees of renal fusion resulting in "horseshoe" (Figure 5d) or "lump" kidneys have been observed in mice of the *lx* strain (Table 1) and in rat young as a consequence of vitamin A deficiency, chlorambucil, trypan blue, and X-irradiation but not following PGA deficiency (Table 2). Crossed renal ectopia, a combination of fusion and displacement, was observed in fetal mice as a result of maternal hypervitaminosis A and in rat young following the administration of methyl salicylate (Table 2). In fused kidneys, the renal pelves usually face ventrally due to absence of rotation, suggesting that union occurs early in gestation when the renal anlagen lie close together in the pelvis. Warkany (1971) noted that in rat fetuses from vitamin A-deficient mothers, "horseshoe" kidneys were found mostly in younger fetuses suggesting that they were not conductive to survival.

5. Ectopic Kidneys

Arrested "ascent" of the kidney so that it remains in the pelvis (pelvic kidney), or comes to rest in a place other than its definitive position, has been produced in fetal rats by niacin deficiency, chlorambucil, methyl salicylate, and PGA deficiency from days 10 to 13 but not from days 11 to 21 of gestation (Table 2).

Examination of young rat embryos from mothers given a PGA-deficient regimen from days 10 to 13 of pregnancy revealed that retarded growth of the lumbosacral region and of the pelvis was an important factor in hindering renal migration. In some embryos the kidney projected around an umbilical (common iliac) artery, lay dorsal to it, or was split into two portions by it

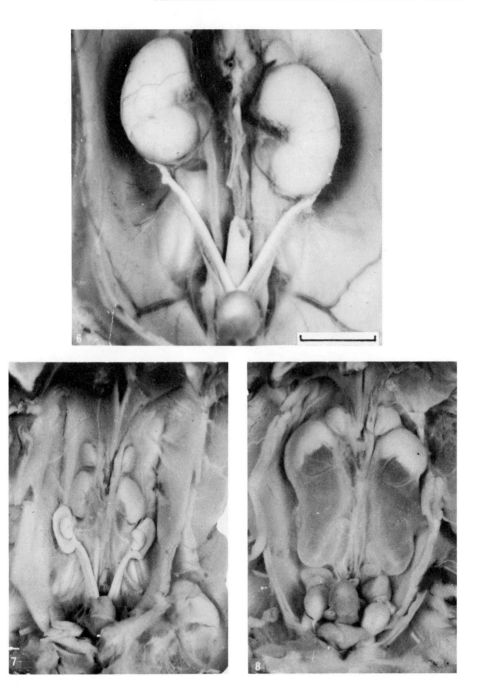

Figs. 6–8. Day 21 rat fetuses. Scale presents 3 mm: Fig. 6. Control fetus showing adrenals, kidneys, ureters, uteri, hindgut, and bladder. Fig. 7. Maternal PGA-deficiency, days 11–21. Small kidneys and narrow ureters. Fig. 8. Maternal PGA-deficiency, days 11–21. Hydronephrosis and hydroureter. (Reproduced by permission from Monie *et al.*, 1954.)

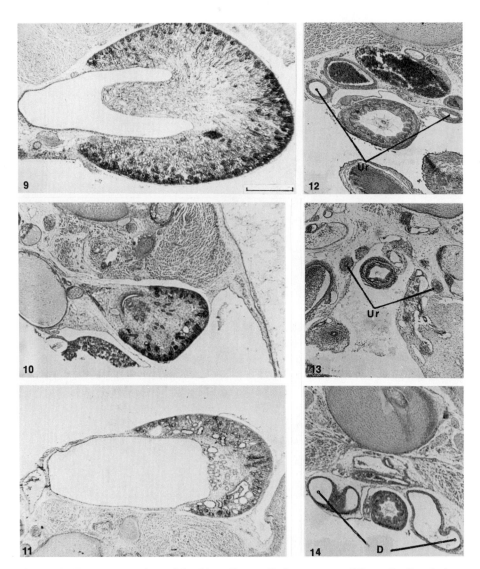

Figs. 9–14. Transverse sections of day 21 rat fetuses. Scale represents 450 μm. D, diverticulum; Ur, ureter: Fig. 9. Control kidney showing cortex, medulla, renal papilla, and pelvis. Fig. 10. Maternal PGA deficiency, days 11–21; small, poorly developed kidney. Fig. 11. Maternal PGA deficiency, days 11–21; hydronephrosis. Fig. 12. Control; ureters shown near their midpoints. Fig. 13. Maternal PGA deficiency, days 11–21; ureteric atresia. Fig. 14. Maternal PGA deficiency, days 11–21; dilated ureters and diverticula. (Reproduced by permission from Monie *et al.*, 1954.)

(Figure 5b, 15). Migration of a kidney dorsal to the umbilical artery into the psoas muscle was also seen. Day 13 is a critical time for renal "ascent" in the rat since then it should escape from the pelvis into the abdomen (Figure 3).

6. Double and Supernumerary Kidneys

Double kidneys, where two distinct kidneys on the same side are united, and supernumerary kidneys, where they remain separate, have been observed in rat fetuses following methyl salicylate administration, or maternal PGA deficiency (day 10–13) during pregnancy (Table 2). They are attributable to early bifurcation of the ureteric bud, or, in some instances, to splitting of the kidney by impingement on the umbilical artery during its "ascent" (Figure 15).

7. Cystic Kidneys

Cysts have been encountered in the kidneys of *lx* and *my* strains of mice, in the Gunn rat, and in a rat strain described by Solomon (1973) (Table 1). They are not commonly seen with the teratogens listed in Table 2 although sometimes a few are found in small poorly formed kidneys, or in hydronephrotic kidneys, following maternal PGA deficiency from days 11 to 21 in rats. On the other hand, Crocker *et al.* (1972) noted cysts of the proximal tubules of rat young following the administration of commercial diphenylamine (DPA) to females during the last week of gestation. These authors also found that the collecting tubules were unaffected in contrast to the cystic changes which occur in these structures when DPA is given to adult rats over several months.

Crocker and Vernier (1970) have cultured kidneys from fetal mice and observed that cysts of the collecting tubules can be reduced or prevented by increasing the potassium content of the culture medium. Crocker (1973) has obtained similar results on culturing human fetal kidneys. The effect of potassium in such situations, however, has been questioned by Resnick *et al.* (1973).

When day-15 fetal rat kidneys were cultured, cysts of the collecting tubules formed after three days and cysts of the nephric tubules after six days, whereas when day-18 kidneys were cultured such cysts were both present after three days (Monie and Morgan, 1975). These variances are explained by the different level of nephron maturity on the two days selected (Figures 16 and 17). Perey *et al.* (1967) and Ojeda *et al.* (1972) have produced cysts in the kidneys of rabbits by the injection of adrenal corticosteroids shortly after birth. Ojeda *et al.* (1972) concluded that different types of cystic kidneys probably have the same etiology. How such cysts relate, if at all, to those encountered in human infants with cystic disease of the kidney is not yet clear, but the techniques employed may well yield useful information on teratogenic mechanisms.

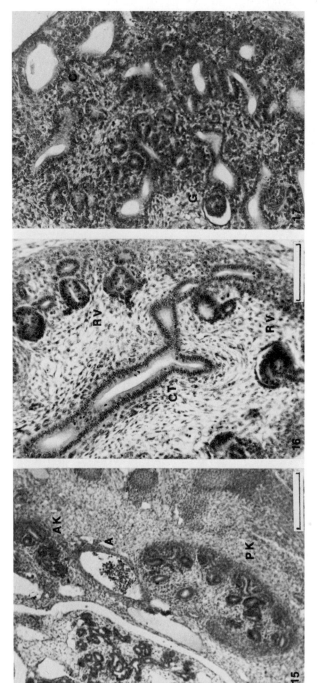

Fig. 15. Saggital section of day-16 PGA-deficient rat fetus showing accessory kidney (AK), left umbilical artery (A), and principal kidney (PK) which is ectopic. Scale represents 150 μm. (Reproduced by permission from Monie et al., 1957.)

Fig. 16. Section of day-15 fetal rat kidney. CT, collecting tubule; RV, renal vesicle. Scale represents 75 μm.

Fig. 17. Section of day-15 fetal rat kidney cultured for 3 days. Cysts (C) are forming from the ampullae of a collecting tubule. A renal vesicle with early glomerular formation (G) is seen (same scale as Figure 16).

8. Hydronephrosis

Hydronephrosis (Figures 8 and 11), expansion of the renal pelvis generally with reduction in the amount of the renal parenchyma, is a frequently encountered abnormality and often is accompanied by a dilated and sometimes tortuous ureter. Many of the strains of animals in Table 1 develop hydronephrosis, and it results from most of the teratogens listed in Table 2. It has also been found in fetuses from female rats with alloxan-induced diabetes (Takano and Nishimura, 1967). Whether hydronephrosis results from a teratogenic procedure seems to depend on how severely the kidney is affected. Thus, Wilson and Warkany (1948) did not encounter it in rat fetuses from mothers maintained on a vitamin A-deficient diet before and during pregnancy but did find it when vitamin A supplementation was provided. For hydronephrosis to arise, the kidney must be capable of secretion.

In hydronephrotic kidneys resulting from teratogens, the cortex is often greatly diminished and the renal papilla small or absent (Figure 11). Glomeruli are reduced in number, renal tubules are short, and there are fewer collecting tubules. Some of the tubules, renal and collecting, may be dilated in all or part of their course while the nephrogenic layer in the outer portion of the cortex is reduced or absent. Dilatation of the ureter, especially in its upper third, is common and there may or may not be obstruction to some part of the urinary tract, such as at the point of entry of the ureters into the bladder or in the urethra.

Cases of hydronephrosis and dilated ureter where there is no obstruction of the urinary tract have been observed in rat young following maternal PGA deficiency from days 10 to 13 of gestation and from the administration of 0.2 ml methyl salicylate on day 11 of gestation (Table 2). In such cases the renal papilla was usually absent or miniscule, and nephrons and collecting tubules few in number; other abnormalities such as distended bladder or rectovesical fistula were also seen. Specimens of this type may arise from a defective ureteric bud which results in a ureter and renal pelvis deficient in muscle, fewer collecting tubules and, consequently, fewer nephrons. Where there are functioning nephrons with short tubules, the reduced absorption will tend to aggravate the situation.

Woo and Hoar (1972) have observed that the rate of growth of the renal papilla in normal Charles River rats is variable and can result in "apparent hydronephrosis" in kidneys in late gestation. They also noted that while methyl salicylate (0.05 or 0.1 ml on day 10 or 11) can accentuate this situation, the damage is not necessarily permanent and normal renal development can be regained soon after birth. These findings make it clear that a diagnosis of hydronephrosis in late fetal life may be erroneous if based only on the size and configuration of the renal pelvis and in the absence of definite abnormality of the urinary tract.

While fluid back-pressure from obstruction can cause hydronephrosis and dilated ureters in postnatal life, and very probably in the fetal period of animals (like man) with a long gestation, its role in the genesis of these abnormalities in the prenatal rat and mouse is less clear. Obstructions of the urinary tract may, at times, simply reflect the existence of a spectrum of concomitant but independent malformations. However, in a study involving rabbits, where gestation is longer than in the rat and mouse, Springman *et al.* (1973) found that removal of young from the mothers at 29 days and incubating them for 24 hr resulted in distended bladders in 70% and dilated ureters and renal pelves in 9%, compared to the nonincubated fetuses; in some there was also flattening of the renal papilla.

Delayed breakdown of the ureteric membrane as a possible cause of some cases of hydronephrosis and dilatation of the ureter in man was suggested by Vermooten (1939), and evidence seemingly in support of this has been found in rat fetuses (Figures 5c and 20) as a result of maternal PGA deficiency (Monie *et al.*, 1954) and hypervitaminosis A (Giroud *et al.*, 1958). The fact that such a membrane could later resorb without trace perhaps affords an explanation for those cases of hydronephrosis in which there is no visible obstruction to the flow of urine. On the other hand, it is possible that the delay in the disruption of these membranes reflects the greater storage capacity of poorly developed ureters and renal pelves with readily distensible walls. Neuromuscular imbalance affecting the walls of the renal pelves and ureters may be of significance in the genesis of hydronephrosis and dilatation of the ureter, but its role is difficult to determine.

B. Ureter

The ureteric bud arises from the mesonephric duct early on day 12 (the development of its cranial end has been considered in the previous section). Caudally, it separates from the mesonephric duct on day 15, and its opening, covered by epithelium, migrates along the urogenital sinus (urethral) wall to the bladder. The mouth of each ureter remains closed by the ureteric membrane until its disruption between days 19 and 21 (Figure 20). A similar structure occurs in man (Chwalla, 1927) and in the mouse (Szarski, 1937).

As the ureter lengthens the lumen of its intermediate portion narrows or closes, but this is transitory and it is again patent by day 17 or 18. Towards the end of gestation the cranial end of each ureter may show moderate tortuosity while caudally it narrows and follows an S-shaped course as it traverses the bladder wall (Figures 12 and 18).

1. Hydroureter and Megaureter

Dilated ureters are usually termed hydroureters when there is obstruction to the flow of urine, and megaureters when there is none (Figures 5a,b and 14). Both forms occur as a result of genetic or teratogenic factors (Tables 1 and 2) and are often accompanied by hydronephrosis.

Delahunt and Lassen (1964) found dilated ureters as the only visceral abnormality in the young of a cynamologus monkey given thalidomide in early pregnancy. In man this teratogen has produced a variety of abnormalities of the urinary system. "Kinked ureters" have been reported in the *lx* strain of mice and may be responsible for the hydronephrosis and hydroureters often found in these animals. In hydroureter the obstruction may be in the terminal portions of the ureters, in the urethra, or in both of these structures. Some other factors which may cause dilatation of the ureters are considered in Section III.A.8.

2. Ureteric Atresia, Stenosis, and Diverticula

Ureteric atresia or stenosis occurs in the *Sd* and *Se* strains of mice and in the Gunn and Stub strains of rats (Table 1). They have also been found as a result of maternal deficiency or excess of vitamin A, maternal PGA deficiency (Figure 13), or subsequent to the administration of chlorambucil to the mother (Table 2). The site of closure or narrowing is often that part of the ureter which normally undergoes transitory narrowing during development. When chlorambucil was used as a teratogen, the ureters were often incomplete or discontinuous, perhaps because of atrophy of an atretic segment.

In fetal mice of the *my* strain, Brown (1931) observed "retardation points" which later became constrictions with dilatation of the ureter cranially. She also reported a "stricture" at the vesicourethral junction and believed it to be the primary cause of hydronephrosis, which frequently occurred in this strain, in addition to absence of the kidney.

Numerous small diverticula have been noted arising from the intermediate portions of the ureters of some rat fetuses following maternal PGA deficiency from days 11 to 21 (Monie *et al.*, 1954). These were directed cranially and caudally, and the entire ureteric wall was involved in their formation (Figures 5c and 14).

3. Double Ureter

Double ureter, common in man, is not often reported as a result of teratogens given to laboratory animals. It has been seen, however, in fetal mice as a result of maternal hypervitaminosis A and in rats from methyl salicylate given to mothers during gestation (Table 2).

Incomplete double ureter (Y-ureter) is probably due to the early bifurcation of the ureteric bud as it passes cranially, whereas complete double ureter is the result of an extra ureteric bud from the mesonephric duct. In the latter case, both ureters may terminate in the bladder or one of them, usually that

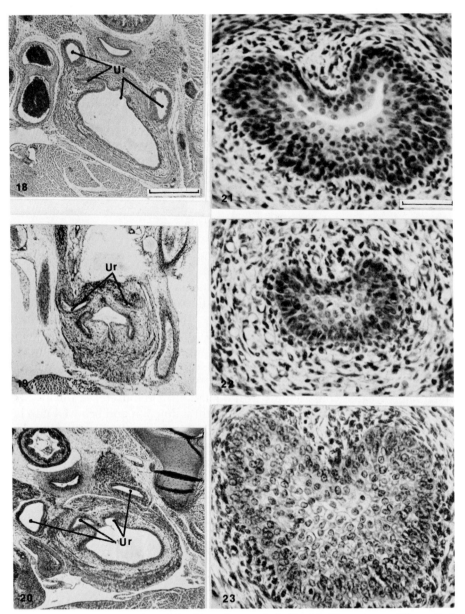

Figs. 18–23. Transverse sections of day-21 rat fetuses. Scale (Fig. 18–20) represents 450 μm. Scale (Figs. 21–23) represents 45 μm. Ur, ureter: Fig. 18. Control; ureters terminating in bladder. Fig. 19. Maternal PGA deficiency; days 11–21. Ureters fail to open into bladder, which is small. Fig. 20. Maternal PGA deficiency, days 11–21. Dilated ureters open into bladder of moderate size. A fragment of the ureteric membrane is visible at the mouth of the left ureter. Fig. 21. Control; transverse section of narrowest point of the urethra. Fig. 22. Maternal PGA deficiency, days 11–21; urethral atresia. Fig. 23. Maternal PGA deficiency, days 11–21; urethral atresia associated with cellular proliferation. (Reproduced by permission from Monie et al., 1954.).

from the more cranial kidney, may end ectopically. Factors involved in the ultimate location of the openings of complete double ureters have been discussed by Monie (1949) and Wharton (1949).

4. Ectopic Ureters

Ectopic ureters terminate in a location other than the bladder (Figure 5a,b). If the ureteric bud fails to separate from the mesonephric duct, in males it will end in one of the derivatives of the latter such as the ejaculatory duct, vas deferens, or seminal vesicle. In females it may terminate in the vagina through the caudal portion of the mesonephric duct which remains open instead of obliterating; however, where vaginal development is incomplete, it will end in the urogenital sinus (urethra). Termination of the ureter in the urethra may occur in both sexes and indicates its tardy migration following separation from the mesonephric duct.

Ectopic ureters can be produced in rats and mice by several of the teratogens listed in Table 2. Wilson and Warkany (1948) found that maternal vitamin A deficiency in rats resulted in ectopic ureters in 37 out of 42 fetuses and of these 36% retained connection with the mesonephric duct, 34% opened into the sinus horns (terminal portions of the mesonephric ducts), and the remaining 30% joined the urethra. Maternal PGA deficiency (days 10–13) resulted in ectopic ureters in rat fetuses opening into the vas deferens, seminal vesicle, or urethra (Monie *et al.*, 1957). Giroud *et al.* (1958) reported ectopic ureters associated with hydronephrosis and hydroureter in rat young following maternal hypervitaminosis A during gestation. In man, ectopic ureters are often dilated and have wide orifices (Warkany, 1971).

C. Bladder

The bladder forms from the cranial portion of the urogenital sinus and the ureters open into it about day 20 when the ureteric membranes break down (Figures 1, 4, and 18). The muscle coat of the bladder is well developed at this time. The allantoic duct, which in human fetuses forms a cord of cells extending from the apex of the bladder into the umbilical cord, is not seen in rats in later fetal life.

1. Absent Bladder

Absence of the bladder is usually associated with severe defects of the urogenital system and has been found in rat fetuses following exposure of the mother to trypan blue (Goldstein, 1957), hypervitaminosis A (Giroud *et al.*, 1958), or to hypervitaminosis A and cortisone (Rao, 1960) during gestation.

Its absence has also been observed in rat fetuses with symmelia induced by giving retinoic acid to the mother during pregnancy (Khemmani, 1974); this is similar to the situation in human symmelia.

2. Small Bladder

Small or "hypoplastic" bladders are common where there is retarded development of some portion of the urinary tract. They are seen in *Sd* mice (Table 1) and in mice with the urorectocaudal (*ur*) syndrome (Grüneberg, 1952). They have also been observed in rat young (Figure 19) following maternal PGA deficiency from days 11 to 21 of gestation (Table 2).

3. Distended Bladder, Diverticula of the Bladder

Distended bladders have been reported in rat fetuses following maternal niacin deficiency, injection of methyl salicylate (Table 2), and from the administration of retinoic acid to pregnant rats (Dumas, 1964). Diverticula of the bladder were observed in mouse fetuses by Kalter and Warkany (1961) after maternal hypervitaminosis A during gestation.

4. Rectovesical Fistula, Exstrophy of the Bladder

Rectovesical fistulae have been reported in mouse fetuses following maternal hypervitaminosis A during pregnancy (Kalter and Warkany, 1961), and after the same teratogenic procedure in rat fetuses (Giroud, 1964). The former authors, while employing the term "rectovesical fistula," point out that the communication is actually with the colon and not with the rectum, which is absent in such cases. Exstrophy of the bladder, well known in man, does not seem to have been produced in laboratory animals by teratogens.

D. Urethra

The cloaca, receiving both the urogenital sinus and the hindgut, is divided by the urorectal septum about day 17 into the urethra ventrally, and the rectum dorsally (Figures 1 and 4). The urogenital membrane sealing the caudal end of the former breaks down about day 20 whereas the anal membrane disappears 3–4 days earlier. The fate of the genital ducts, and of the ureters during their association with the urogenital sinus, has been discussed in a previous section. The lumen of the urethra cranial to the point of entry of the mesonephric ducts is narrowed transitorily about day 18 but thereafter widens (Figure 21).

1. Urethral Atresia and Stenosis

Urethral atresia (Figures 5c, 22, and 23) and stenosis have been observed in fetal rats following maternal PGA deficiency, in fetal mice from maternal hypervitaminosis A, and in rat fetuses subsequent to injections of trypan blue or chlorambucil to the mother during pregnancy (Table 2). Failure of the urethra to widen after its period of narrowing seems to explain most of these abnormalities. It is possible that a stenotic region might later become sculptured out to form posterior urethral valves as sometimes found in man.

2. Hypospadias, Enlarged Genital Tubercle, and Persistent Cloaca

Hypospadias was encountered by Wilson and Warkany (1948) in fetal rats following maternal vitamin A deficiency during pregnancy. Enlargement of the genital tubercle and persistence of the cloaca were observed in rat young after the mothers received chlorambucil on day 10 of gestation (Monie, 1961).

REFERENCES

Astarbaldi, T., and Bell, E. T., 1962, Spontaneous hydronephrosis in albino rats, *Nature* **195:**392.

Bagg, H., 1925, Hereditary abnormalities of the viscera. I. A morphological study with special reference to abnormalities of the kidneys in the descendants of x-rayed mice, *Am. J. Anat.* **36:**275.

Baxter, J. S., and Yoffey, J. M., 1948, The postnatal development of the renal tubules in the rat, *J. Anat.* **82:**189.

Beaudoin, A. R., 1974, Teratogenicity of sodium arsenate, *Teratology* **10:**153.

Bernstein, J., 1971, The morphogenesis of renal parenchymal maldevelopment (renal displasia), *Pediatr. Clin. N. Am.* **18:**395.

Bertone, L., 1965, Congenital malformations produced by methyl salicylate in fetal rats, M.A. thesis, Univ. California, San Francisco.

Brown, A. L., 1931, An analysis of the developing metanephros in mouse embryos with abnormal kidneys, *Am. J. Anat.* **47:**117.

Carter, T. C., 1953, The genetics of luxate mice, *J. Genet.* **51:**441.

Chamberlain, J. G., and Nelson, M. M., 1963, Congenital abnormalities in the rat resulting from single injections of 6-AN during pregnancy. *J. Exp. Zool.* **153:**285.

Chwalla, R., 1927, Über die Entwicklung der Harnblase und der primären Harnröhre des Menschen, *Z. Ges. Anat.* **83:**615.

Cramer, D. V., and Gill, R. J., III, 1975, Genetics of urinogenital abnormalities in ACI inbred rats, *Teratology* **12:**27.

Crocker, J. F. S., 1973, Human embryonic kidneys in organ culture: Abnormalities of development induced by decreased potassium, *Science* **181:**1178.

Crocker, J. F. S., and Vernier, R. L., 1970, Fetal kidney in organ culture: Abnormalities of development induced by decreased amounts of potassium, *Science* **169:**485.

Crocker, J. F. S., Brown, D. M., Borch, R. V., and Vernier, R. L., 1972, Renal cystic disease induced in newborn rats by diphenylamine derivatives, *Am. J. Pathol.* **66:**343.

Delahunt, C. S., and Lassen, L. J., 1964, Thalidomide syndrome in monkeys, *Science* **146:**1300.

Dumas, M., 1964, Effect of maternal hypervitaminosis A on fetal rat development, Ph.D. thesis, Univ. California, San Francisco.

Edwards, M. J., 1969, Congenital defects in guinea-pigs: Fetal resorptions, abortions, and malformations following induced hyperthermia during early gestation, *Teratology* **2**:313.

Ferm, V. H., Saxon, A., and Smith, B. M., 1971, The teratogenic properties of sodium arsenate in the golden hamster, *Arch. Environ. Health* **22**:557.

Giroud, A., 1964, Vitamine A et tératogénèse, *Bull. Acad. Suisse Sci. Med.* **20**:440.

Giroud, A., Martinet, M., and Roux C., 1958, Urétéro-hydronéphrose expérimentale chez l'embryon par hypervitaminose A, *Arch. Fr. Pediatr.* **15**:540.

Glueksohn-Schoenheimer, S., 1943, The morphological manifestation of a dominant mutation in mice affecting tail and urogenital system, *Genetics* **28**:341.

Goldstein, D. J., 1957, Trypan blue induced anomalies in the genitourinary system of rats, *S. Afr. J. Med. Sci.* **22**:13.

Gruenwald, P., 1939, The mechanism of kidney development in human embryos as revealed by an early stage in the genesis of the ureteric buds, *Anat. Rec.* **75**:237.

Gruenwald, P., 1943, Normal changes in the position of the embryonic kidney, *Anat. Rec.* **85**:163.

Grüneberg, H., 1952, *The Genetics of the Mouse*, 2nd ed., Martinus Nijhoff, The Hague.

Hain, A. M., and Robertson, E. M., 1936, Congenital anomalies in rats including unilateral renal agenesis, *J. Anat.* **70**:566.

Hale, F., 1935, The relation of vitamin A to anophthalmus in pigs, *Am. J. Ophthalmol.* **18**:1087.

Kalter, H., and Warkany, J., 1961, Experimental production of congenital malformation in strains of inbred mice by maternal treatment with hypervitaminosis A, *Am. J. Pathol.* **38**:1.

Khemmani, M., 1974, Personal observation.

Kosmachevskaya, É. A., and Chebotar, N. A., 1968, Effect of maternal stress on harmful action of 4-methyl-uracil on embryogenesis of rats, *Bull. Exp. Biol.* **66**:1382.

Lozzio, B. B., Chernoff, A. I., Machado, E. R., and Lozzio, C. B., 1967, Hereditary renal disease in a mutant strain of rats, *Science* **156**:1742.

Meier, H., and Yerganian, G., 1961, Spontaneous diabetes mellitus in the Chinese hamster (*Cricetulus griseus*) II. Findings in the offspring of diabetic parents, *Diabetes* **10**:12.

Monie, I. W., 1949, Double ureter in two human embryos, *Anat. Rec.* **103**:195.

Monie, I. W., 1961, Chlorambucil-induced abnormalities of the urogenital system of rat fetuses, *Anat. Rec.* **139**:145.

Monie, I. W., 1970, Hydronephrosis and hydroureter induced by teratogens in rat fetuses, *Teratology* **3**:206.

Monie, I. W., and Khemmani, M., 1973, Absent and abnormal umbilical arteries, *Teratology* **7**:135.

Monie, I. W., and Morgan, J. R., 1975, Cysts in cultured fetal rat kidneys, *Teratology* **11**:143–152.

Monie, I. W., Nelson, M. M., and Evans, H. M., 1954, Abnormalities of the urogenital system of rat embryos resulting from maternal pteroylglutamic acid deficiency, *Anat. Rec.* **120**:119.

Monie, I. W., Nelson, M. M., and Evans, H. M., 1957, Abnormalities of the urinary system of rat embryos resulting from transitory deficiency of pteroylglutamic acid during gestation, *Anat. Rec.* **127**:711.

Nelson, M. M., Baird, C. D. C., Wright, H. V., and Evans, H. M., 1956, Multiple congenital abnormalities resulting from riboflavin deficiency induced by the antimetabolite galactoflavin, *J. Nutr.* **58**:125.

Nelson, M. M., Wright, H. V., Baird, C. D. C., and Evans, H. M., 1957, Teratogenic effect of pantothenic acid deficiency in the rat, *J. Nutr.* **62**:395.

Ojeda, J. L., Barbosa, E., and Gomez-Bosque, P., 1972, Morphological analysis of renal polycystosis induced by corticoids, *J. Anat.* **111**:399.

Orsini, M. W., 1952, The piebald hamster. A mutation showing growth retardation and urogenital abnormalities, *J. Hered.* **43**:37.

Percy, D. H., and Albert, D. M., 1974, Developmental defects in rats treated postnatally with 5-iododeoxyuridine (IUdR), *Teratology* **9**:275.

Perey, D. Y. E., Herdman, R. C., and Good, R. A., 1967, Polycystic renal disease: A new experimental model, *Science* **158**:494.

Potter, E. L., 1946, Bilateral renal agenesis, *J. Pediatr.* **29**:68.

Rao, V. R., 1960, Urogenital abnormalities induced by hypervitaminosis A and cortisone in the rat, *J. Anat. Soc. India* **8**:61.

Ratcliffe, H. L., and King, H. D., 1941, Developmental abnormalities and spontaneous diseases found in rats of the mutant strain, STUB, *Anat. Rec.* **81**:283.

Resnick, J. S., Brown, D. M., and Vernier, R. L., 1973, Oxygen toxicity in fetal organ culture. 1. The developing kidney, *Lab. Invest.* **28**:437.

Shepard, T. H., 1973, *Catalog of Teratogenic Agents,* Johns Hopkins Univ. Press, Baltimore.

Shotton, D., and Monie, I. W., 1963, Possible teratogenic effect of chlorambucil on a human fetus, *J.A.M.A.* **186**:74.

Solomon, S., 1973, Inherited renal cysts in rats, *Science* **181**:451.

Springman, F. R., Roell, M. G., and Doerr, B. I., 1973, Dilatation of the urinary tract in fetal rabbits following postcesarean incubation, *Teratology* **8**:305.

Szarski, C., 1937, Über die Abtrennung der Ureteren von Urnierengängen bei der Maus, *Anat. Anz.* **84**:241.

Takano, K., and Nishimura, H., 1967, Congenital malformations induced by alloxan diabetes in mice and rats, *Anat. Rec.* **158**:303.

Torrey, T. W., 1943, The development of the urinogenital system of the albino rat. I. The kidney and its ducts, *Am. J. Anat.* **72**:113.

Tuchmann-Duplessis, H., Lefebvres-Boisselot, J., Mercier-Parot, L., 1959, L'action tératogène de l'acide x-methyl folique sur diverses éspèces animales, *Arch. Fr. Pediatr.* **15**:1.

Vermooten, J., 1939, A new etiology for certain types of dilated ureters in children, *J. Urol.* **41**:455.

Vernier, R. L., and Smith, F. G., Jr., 1968, Fetal and neonatal kidney, *in: Biology of Gestation,* Vol. 2 (N. S. Assali, ed.), pp. 225–260, Academic Press, New York.

Wallace, W. E., and Spickett, S. G., 1967, Hydronephrosis in mouse, rat, and man, *J. Med. Genet.* **4**:73.

Warkany, J., 1971, *Congenital Malformations, Notes and Comments,* Year Book Medical Publishers, Chicago.

Wharton, L. R., Jr., 1949, Double ureters and associated renal anomalies in early human embryos, *Contrib. Embryol.* **33**:103.

Wilson, J. G., 1954, Differentiation and the reaction of rat embryos to radiation, *J. Cell. Comp. Physiol.* **43**:11.

Wilson, J. G., and Warkany, J., 1948, Malformations in the genito-urinary tract induced by maternal vitamin A deficiency, *Am. J. Anat.* **83**:357.

Wilson, J. G., Roth, C. B., and Warkany, J., 1953, An analysis of the syndrome of malformations induced by maternal vitamin A deficiency. Effects of restoration of vitamin A at various times during gestation, *Am. J. Anat.* **92**:189.

Woo, D. C., and Hoar, R. M., 1972, "Apparent hydronephrosis" as a normal aspect of renal development in late gestation of rats: The effect of methyl salicylate, *Teratology* **6**:191.

Abnormal Organogenesis in the Reproductive System

16

ALLEN S. GOLDMAN

I. INTRODUCTION

A. Teratology of Sexual Differentiation

The fertilized egg of the human utilizes a program of embryonic differentiation estimated to contain 100,000,000,000 bits of information (yes or no decisions) to attain the chemical constitution of the adult (Dancoff and Quastler, 1953). The information content of a single printed page is about 10,000 bits. Thus, the complexity of the program of human differentiation is equivalent to 10,000,000 printed pages, or the content of a library of considerable size. Yet, it is amazing that errors in readout of this program resulting in teratologic organogenesis are so few. The highest incidence of all malformations reported to occur in man is at most 2 percent of live births.

This chapter will be devoted to the investigation in mammals of errors in the sexual differentiation of the gonads, genital tract, and reproductive structures. These studies have unraveled some of the complexities of this particular developmental program with the use of specific natural and experimental fetal hormone-depriving agents. Teratogenic sexual differentiation has been considered in this chapter to result from the abnormal presence, absence, or quantity, at a critical period of development, of a substance or agent which is

ALLEN S. GOLDMAN • The Teratology Center, Children's Hospital of Philadelphia, Philadelphia, Pennsylvania 19104.

required for or can interfere with the readout of the developmental program. Such interference leads not only to obvious defects of morphology but also to biochemical, physiologic, and behavioral disorders. In this sense interference with normal sexual differentiation over several levels of biological organization can result in teratogenic feminization of males or virilization of females.

The course of sexual differentiation seemed a logical first choice to the early endocrinologists as a subject for study. It appeared reasonable that male differentiation may be controlled by androgens, while female differentiation may be directed by estrogen and progesterone as are the responses of the target organs in the two sexes of the adult mammal. Thus, soon after the initial chemical synthesis of the sex hormones in the 1930s, the individual sex hormones were administered to pregnant mammals and the effects observed on the differentiation of the gonads, genital tract, and reproductive structures in each sex of treated offspring (Burrill and Greene, 1939; Green et al., 1940).

B. Errors in Gonadal Sex Differentiation

The differentiation of the gonads in all species of mammals investigated was not influenced by any sex hormone, with the curious exception that administration of estradiol diproprionate to the pouch-young of the opossum produced sex reversal of the development of the presumptive testes of genetic males to that of ovaries with oocytes (Burns, 1955). Thus, the program of gonadal sex was concluded to be directed in most mammals by the presence of the Y chromosome which converts the indifferent gonad to a testis. This conclusion has had to be modified by the recent observations of Kasdan et al. (1973) of a family in which some sibs had an XX genotype, ovotestes, and male differentiation. These findings suggest that there are some autosomal genes directing the differentiation of the testis as well as those on the Y chromosome.

C. Teratogenic Actions of Estrogens and Progestins on Genital Sex

As was expected from their actions on the genital tract structures of the adult, testosterone stimulated the genitalia and Wolffian duct structures of the female, and estradiol-17β repressed these structures in male fetuses (Burrill and Greene, 1939; Green et al., 1940). However, estradiol-17β produced teratogenic effects as well, i.e., virilization of the genital tract in females and adrenal enlargement in both sexes. Thus, no definitive conclusions about the roles of these hormones in the control of the programing of normal sexual differentiation of genital tract structures could be drawn at this time. Moreover, Courrier and Jost (1942) observed teratogenic effects of certain synthetic derivatives of progesterone, i.e., progestins, on the sexual differentiation of the rabbit. Consequently they warned against the use of these agents to prevent threatened abortion in the pregnant woman. They observed that ethisterone, as an example of a progestin, had three effects on the pregnant rabbit: a progestinal action in maintenance of gestation, a masculinizing effect of the female fetus, and a slight feminizing effect on the genital tract of the male fetus. Although the mechanism by which estrogens and progestins interfered with sexual differentiation was not elucidated until later, the observations of virilization of newborn girls by progestins by Wilkins (1960), and by estrogens by Bongiovanni *et al.* (1959), led to the first pre-thalidomide recognition of the use of teratogenic drugs in man.

II. CONTRIBUTIONS FROM EXPERIMENTAL ENDOCRINOLOGY

A. The Inborn Sexual Pattern is Female

1. Hypothalamic Sex

Pfeiffer (1936) applied the classical endocrinologic techniques of castration and hormone replacement in the neonatal rat. This provided the first definitive evidence concerning the nature of sexual differentiation. The hypothalamic centers controlling gonadotropin secretion and sexual behavior differentiated, in the absence of gonads, in each sex to the female type, i.e., a pattern in postpubertal rats of cyclic secretion of gonadotropins and the assumption of the lordotic position in coitus. The presence of a testis or testosterone replacement in the castrated neonate transformed the indifferent hypothalamus into the male type, i.e., a tonic secretion of gonadotropins and the mounting sexual position.

2. Mammalian Anatomic Sex

The experimental embryological investigations of Jost (1972) in the late 1940s introduced surgical castration with testosterone replacement in the fetus. He also showed that the inborn sexual program of anatomic sex is female unless there is a testis secreting androgens. His removal of the fetal gonad in the rabbit led to feminization of the external genitalia, mammary gland development, and the absence of Wolffian duct development in either sex. Testosterone replacement to the castrated fetus corrected the defects in all androgen-dependent masculine development but did not affect Müllerian duct regression. Decapitation of the fetal rabbit (as a form of hypophysectomy) impaired fetal testicular function and arrested masculine genital differentiation, as manifested by persisting hypospadias. This defect results from incomplete fusion of the labial–scrotal folds and posterior displacement of the urethral orifice from the tip to the under side of the shaft of the penis to the perineum. Complete failure of fusion of the folds would result in the phenotypically female genitalia. Similar studies of fetal decapitation in the rat gave inconclusive results.

Müllerian duct regression appears to be controlled by the production of a proteinaceous Müllerian inhibitor secreted by the fetal testicular Sertoli cells (Josso, 1971, 1973).

3. Human Anatomic Sex

a. Feminization of Males—Congenital Adrenal Hyperplasia. The suggestion that errors in sexual differentiation may occur in man due to a genetic defect in a similar fashion to those produced by fetal or neonatal castration in other mammals was first made by Bongiovanni *et al.* (1967). These investigations observed hypospadias in boys with a rare form of congenital adrenal cortical hyperplasia due to a genetic deficiency of the steroidogenic enzyme system, 3β-hydroxy-Δ^5-steroid oxidoreductase, $\Delta^{5,4}$-3-ketosteroid isomerase (Figure 1). The oxidoreductase, is also called dehydrogenase. They postulated that the genetic defect in this enzyme system in the fetal testis produced a genetic castrate at the critical period of human genital differentia-

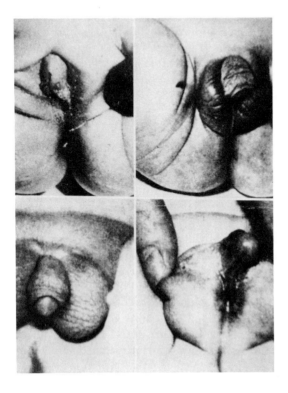

Fig. 1. Genital changes in congenital adrenal hyperplasia. Girl (upper right) with a genetic deficiency in activity of 21-steroid hydroxylase has clitoral hypertrophy and fusion of labia majora. Girl (upper left) with a genetic deficiency of 3β-hydroxy-Δ^5-steroid oxidoreductase has a slight virilization manifested by only minimal clitoral enlargement. Boy with a deficiency of 3β-hydroxy-Δ^5-steroid oxidoreductase (lower left and right) has severe hypospadias with scrotal separation and urethral opening on perineum.

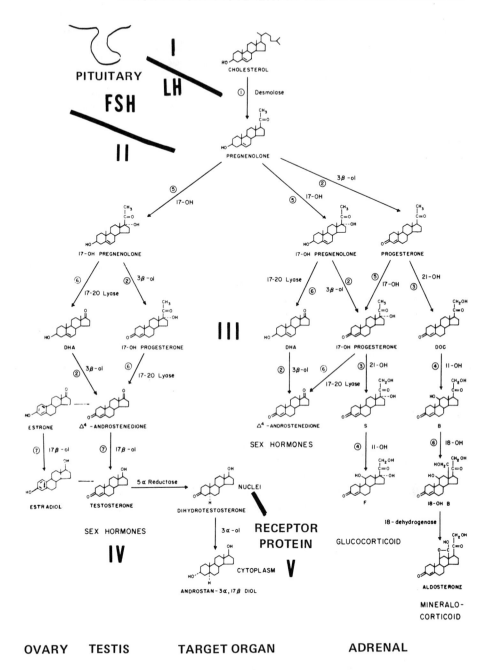

Fig. 2. Scheme of control, synthesis, circulation, uptake, and metabolism of testosterone in androgen-dependent target organs and of steroid biosynthesis in adrenal and ovary. Sites: Gonadotropins: FSH, I; LH, II; enzymes, III; Testosterone, IV; target organ receptor protein, V. Enzymes: (1) cholesterol desmolase, (2) 3β-hydroxy-Δ⁵-steroid oxidoreductase, (3) 21-steroid hydroxylase, (4) 11β-hydroxylase, (5) 17α-steroid hydroxylase, (6) C₁₇₋₂₀ lyase, and (7) 17-ketoreductase.

tion by blocking testosterone biosynthesis at the level of the androgenically weak $\Delta^5 3\beta$-hydroxysteroids, shown at site III—2 in Figure 2. They noted the similarity of the hypospadiac boys with this genetic defect to the hypospadiac fetal decapitated male rabbit and suggested that the etiology of the human disease was due to testosterone deprivation of the male fetus. Testosterone deprivation of the male fetus because of genetic defects in other testosterone biosynthetic enzymes can also explain the occurrence of feminization of the external genitalia in boys born with three other rare forms of congenital adrenal hyperplasia (Figures 2 and 3). These genetic defects include cholesterol desmolase (III–1, Figure 2), 17α-hydroxylase (III–5, Figure 2), and C_{17-20}lyase (III–6, Figure 2), as well as the primarily testicular enzyme defect, 17-ketoreductase (III–7, Figure 2) (Saez *et al.*, 1971).

b. Feminization of Males—Testicular Feminization. One other human genetic disease, testicular feminization, has phenotypically female external genitalia in genetically male boys. This disease is an example of testosterone deprivation in the male fetus resulting from genetic castration related to an inherited defect in androgen responsiveness. The nature of the genetic defect in man (Keenan *et al.*, 1974) and in an animal counterpart, the Stanley–Gumbreck rat male pseudohermaphrodite (Stanley *et al.*, 1973), has been suggested to be a deficiency of androgen receptor proteins in the target organs (Ritzen *et al.*, 1972, Bardin *et al.*, 1973; Milin and Roy, 1973).

c. Virilization of Females—Congenital Adrenal Hyperplasia. Bongiovanni *et al.*, (1967) also suggested that the inborn program of female differentiation may be disrupted in congenital adrenal hyperplasia due to defects in glucocorticoid synthesizing enzymes. Girls born with a defect in 21-or 11β-hydroxylase are markedly virilized as manifested by clitoral hypertrophy to the point of formation of a complete penis in some cases, along with fusion of the labial–scrotal folds (Figures 1 and 2, Table 1). However, girls with the 3β-hydroxy–Δ^5-oxidoreductase defect have a much milder degree of virilization (Figure 1). Girls with a defect of the desmolase, 17α-hydroxylase, and C_{17-20} lyase have normal genitalia presumably because their endocrine glands can produce no androgens or estrogens. High levels of testosterone formed by peripheral conversion of androstenedione, the predominant adrenal androgen in adrenal venous blood, have been demonstrated in patients with a defect in 21-hydroxylase (Camacho and Migeon, 1966; Frasier, *et al.*, 1969). Female fetuses with a defect in 3β-hydroxy-Δ^5-oxidoreductase may be only slightly virilized because of excess adrenal production of the weaker 3β-hydroxy-Δ^5 C_{19} androgens (Goldman, 1970).

III. CONTRIBUTIONS FROM PHYSIOLOGICAL CHEMISTRY

A. Errors of Sex Differentiation Related to Hepatic Enzyme Levels

The next steps in unraveling the teratology of sexual differentiation were made in the 1960s by the work of Kraulis and Clayton (1968) and DeMoor and

Table 1. Natural and Experimental Errors in the Program of Sexual Differentiation

Fetal deficiency	Site[a]	Reversibility		Error demonstrated	
		Testosterone	Corticosterone	Natural	Experimental agents
Feminization of males					
Gonadotropin	I	+		?	Antibodies to LH-FSH
Congenital adrenal hyperplasia					
Cholesterol desmolase	II	+		+	Aminogluthethimide
Isomerase	III-1	+		+	CTM
Oxidoreductase	III-2				Isoxazole
17α-hydroxylase	III-5	+		+	SU 8000; SU 10,603, Δ¹-testosterone-17β-ureide
C_{17-20} Lyase	III-6	+		+	16β-Br-,5α-pregnan-3-one, 3β,17α-diol
17-Ketoreductase	III-7	+		+	
Testosterone	IV	+		?	Antibodies to testosterone
Testicular feminization					
Dihydrotestosterone receptor proteins	V	+		+	Cyproterone acetate
Virilization of females					
Gonadotropins	I, II	−	−	−	Antibodies to LH-FSH
Congenital Adrenal hyperplasia					
Isomerase	I	−	+	+	CTM, progestins
Oxidoreductase	II	−	+	+	isoxazole, estrogens
21-hydroxylase	III-3	−	+	+	
11β-hydroxylase	III-4	−	+	+	Metyrapone

[a] See Figure 2.

Denef (1968) and extended more recently by Gustafsson *et al.* (1975). These investigators have shown that male-specific postpubertal steroid metabolizing enzyme levels in the liver are "imprinted" in the first few days of life in the male rat by the action of testicular androgens. They have demonstrated that neonatal castration interferes with the masculine determination of the biochemical level of organization in the liver so as to allow expression after puberty of the inborn female program, in a fashion similar to the interference with male hypothalamic and behavioral sex differentiation described earlier in the rat.

IV. CONTRIBUTIONS FROM PSYCHOLOGY

A. Sex Behavior

Sex behavior in man is considerably more complex than in other mammals. The development of most of such behavior in man, including gender identity, has been shown by Money and Ehrhardt (1973) to be determined by environmental factors including the home, parents, and peers. However, Money and co-workers, noting that other aspects of sex differentiation in mammals are also directly applicable to man, postulated that some human sexual behavior may also be "imprinted" by the testis and vulnerable to teratogenesis as it is in other mammals. These investigators studied genetic women virilized *in utero* by congenital adrenal hyperplasia due to a defect in the corticoidogenic enzyme, 21-hydroxylase, or by maternal treatment with estrogens or progestins given to prevent threatened abortion (Table 2). Table 2 includes genetic males feminized *in utero* both by testicular feminization, as studied by Money and Ehrhardt (1973), and by treatment of diabetic women for threatened abortions with estrogens and progestins at the critical developmental periods, as studied by Yalom *et al.* (1973). These investigators found significant differences in certain cultural stereotypes of sex behavioral roles in each experimental group, as compared to appropriate controls. It appears that at least some of the program of sexual behavior in man may also be vulnerable as it is in other mammals.

V. ERRORS INDUCED BY REVERSIBLE CHEMICAL CASTRATION OF THE FETUS

The introduction of two types of reversible castration of the fetus in the 1960s provided the opportunity to produce in animals the errors of sexual differentiation which occur naturally in testicular feminization and in congenital adrenal hyperplasia. Animal models of testicular feminization were made in several species of mammals with cyproterone acetate by Neumann *et al.*

Table 2. Differentiation and Teratology of Behavioral Sex in the Human

System gender role	Man	Virilized woman		Woman	Feminized man	
		CAH[a]	PI[a]		TF[b]	PEI[b]
"Tomboyism"		Expressed	Expressed	Suppressed	Suppressed	—
Satisfaction with female role		Suppressed	Suppressed	Expressed	Expressed	—
Aggression and energy in athletics and skills	Expressed	Expressed	Expressed	Suppressed	Suppressed	Suppressed
Playmate preference	Male	Male	Male	Female	Female	Female
Clothing preference	Slacks	Slacks	Slacks	Dresses	Dresses	—
Toy preference	Cars, guns	Cars, guns	Cars, guns	Dolls	Dolls	Dolls
Juvenile interest in infant care-taking	Suppressed	Suppressed		Expressed	Expressed	
Fantasies						
Pregnancy and motherhood	Suppressed	Suppressed		Expressed	Expressed	
Wedding and marriage	Suppressed	Suppressed		Expressed	Expressed	
Adult activity priority	Career	Career		Marriage	Marriage	

[a]Virilism: CAH, congenital adrenal hyperplasia; PI, progestin induced

[b]Feminization: TF, testicular feminization; PEI, progestin estrogen induced, in boys from diabetic mothers given drug to prevent pregnancy complication.

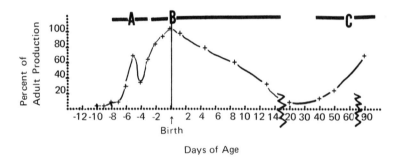

Fig. 3. Schematized plot of testicular production of testosterone during development of the rat.

(1970). This compound is a progestin with potent antiandrogenic action, i.e., it blocks testosterone and dihydrotestosterone uptake in target organs, and prevents androgen-dependent anatomic and hypothalamic differentiation. Animal models of congenital adrenal hyperplasia have been produced by the use of highly specific and potent inhibitors of testosterone biosynthetic enzymes (Goldman, 1970) and antibodies to testosterone (Goldman *et al.*, 1972a). More recently, antibodies to LH-FSH have also been shown to block masculine differentiation of the rat fetus (Goldman *et al.*, 1973c). In short, these teratogens interfere with the readout of male sexual differentiation by producing a reversible chemical castration of the fetus by interfering with testosterone synthesis, circulation, or uptake. These studies will be discussed in depth since they give the fundamental basis for understanding the hormonal control of normal and abnormal sexual differentiation.

VI. CRITICAL PERIODS OF TESTOSTERONE BIOSYNTHESIS AND ANATOMIC SEX ERRORS

There are three critical periods of production of testosterone in the development of the rat when errors in sexual differentiation may occur (Goldman, 1974b). These are depicted schematically as A, B, and C in Figure 3. The first period of testosterone production occurs in the rat on days 14–17 of gestation and in man at 4–6 weeks of pregnancy and is designated period A. The next period, B, occurs around day 17 of gestation to 12–14 days of postnatal age in the rat and from the 4th month of pregnancy to 1–3 months of postnatal age in man. There follows a long period of testicular inactivity in both species—until puberty—when testosterone production reappears, between 40 and 60 days of age in the rat and 12–14 years of age in man, period C. In period A the differentiation of the external genitalia occurs with the formation of the penis and the fusion of the labial scrotal folds to form the scrotum. At this time the differentiation of the mammary glands and ducts also begins. The differentiation of the secondary sex characteristics occurs during C.

VII. CLUES FROM AN EXPERIMENT OF NATURE

A. Congenital Adrenal Hyperplasia

1. Feminization of Males and Inhibition of Testosterone Biosynthetic Enzymes

a. **Experimental Congenital Adrenal Hyperplasia.** The discovery of the highly potent and specific inhibition of the 3β-hydroxysteroid oxidoreductase and $\Delta^{5,4}$-3-ketosteroid isomerase system in mammals by analogs of a C_{19} substrate of this enzyme system (Ferrari and Arnold, 1963; Potts *et al.*, 1963; Goldman, 1972a,b) permitted the experimental validation of the clues derived from the study of congenital adrenal hyperplasia. This provided the first animal models of program readout errors of a genetic defect (Goldman, 1970). These analogs, 2α-cyano-4,4,17α-trimethyl-17β-hydroxy-5-androsten-3-one (CTM) and 17β-hydroxy-4,4,17α-trimethyl-5-androstene-(2,3d)-isoxazole have been shown in several studies to have remarkably tight and specific binding directly to the enzyme-active site in the fetus (Goldman, 1970) (Figure 4). The analogs inactivate the fetal enzyme system at the time of injection. They also produce a long-acting depression of new enzyme synthesis.

b. **Feminization and Inhibition of Oxidoreductase.** Administration of CTM and isoxazole analogs to pregnant rats produces both the anatomic defects (Figures 5 and 6) and deficient activity of the oxidoreductase system characteristic of children with a genetic defect in this enzyme system in this rare form of the adrenal genital syndrome (Goldman, 1970, 1974a; Bloch *et al.*, 1971; Finegold and Basch, 1973). The experimental fetuses have severe adrenal cortical hyperplasia and deficient activity of the oxidoreductase system in the adrenals and testes; at term males have severe hypospadias and females have clitoral hypertrophy. The deficiency in the oxidoreductase is remarkably persistent.

c. **Teratogenic Critical Periods.** The critical developmental period (A) for the production of hypospadias in the experimental male fetuses by a single dose of analog corresponds to the beginning of activity of the oxidoreductase in the testis (Figures 3 and 7). On the other hand, the critical period for production of adrenal cortical hyperplasia and clitoral hypertrophy by the

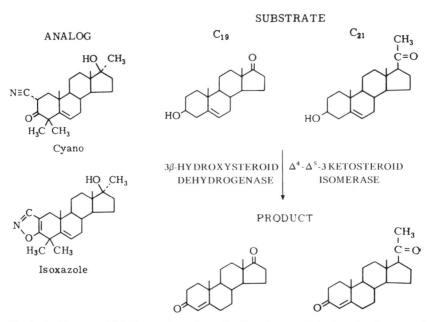

Fig. 4. 2α-Cyano and $2,3d$-isoxazole analogs of a C_{19} substrate (dehydroepiandrosterone) of 3β-hydroxysteroid oxidoreductase and $\Delta^{5,4},3$-ketosteroid isomerase system.

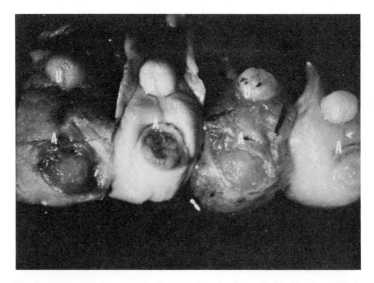

Fig. 5. Phallus and scrotum of normal male fetus (left). Urethra (U) is located in center of tip of penis and anus (A) at base of scrotal raphe. Anourethral distance is 3 mm. Phallus of experimental males of pregnant rats treated with SU 10,603 and inhibition of 17α-hydroxylase (left middle), with aminoglutethimide and inhibition of cholesteral desmolase (right middle), and with CTM and inhibition of $\Delta^5,3\beta$-hydroxysteroid oxidoreductase (right). Note shorter anourethral distances and smaller glans.

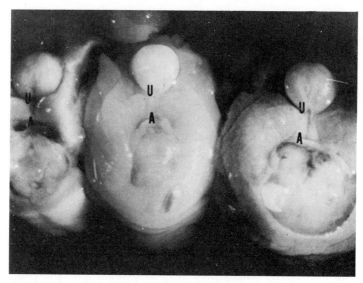

Fig. 6. External genitalia of normal female fetus (left). Genitalia of female fetus of pregnant rat treated with CTM and inhibition of $\Delta^5,3\beta$-hydroxy-steroid oxidoreductase (middle), and with metyrapone and inhibition of 11β-steroid hydroxylase (right). Note enlargement of clitoris in experimental fetuses.

inhibitor has a developmental pattern which is almost identical to that of the enzymatic activity in the normal fetal adrenal, early period B. The cogent implication of the data, that these effects result from the direct inhibition of the fetal enzyme, has been experimentally verified by the production of this defect only in the fetus into which the analog has been directly injected through the uterine wall during the critical period of development and by tracer studies using ^{14}C-labeled isoxazole (Goldman, 1970).

    d. Prevention of Genital Defects.   Testosterone given to pregnant animals completely prevents the production of hypospadias by CTM without affecting the degree of adrenal hyperplasia or inhibition of the dehydrogenase. Testosterone replacement in CTM-inhibited males thereby reestablishes readout of the male program of sexual differentiation. Corticosterone, the main adrenal corticoid of the rat, on the other hand, does not affect the production of hypospadias but completely prevents production of adrenal hyperplasia and clitoral hypertrophy by CTM. Corticosterone suppression of androgen production by the hyperactive adrenals in CTM-treated females allows the expression of the inborn females program of sexual differentiation.

    e. Errors in Mammary Gland Differentiation.   In male fetuses CTM causes development of nipples, which are normally suppressed by fetal testicular function (Figure 8) (Goldman and Neumann, 1969). In light of the feminizing effects of the analog on male external genitalia, this observation

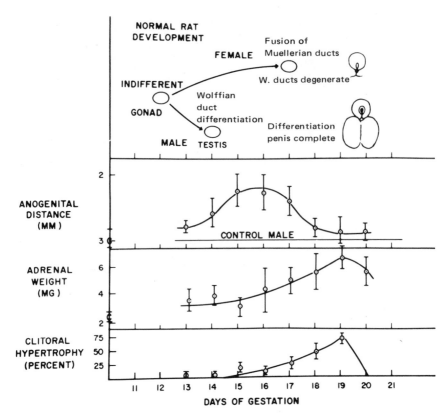

Fig. 7. Critical periods for the production of hypospadias, adrenal hyperplasia, and clitoral hypertrophy by the inhibitor. Vertical bars indicate one standard deviation.

suggests that inactivation of testicular oxidoreductase system leads to insufficient fetal testicular testosterone production, thus allowing development of the nipple in the male. This is completely consistent with the presence of nipples in male offspring of pregnant rats treated with cyproterone acetate (Neumann *et al.*, 1970). In female fetuses CTM inhibits nipple development, probably because of the overproduction of 3β-hydroxy-Δ^5-androgens in response to inhibition of the adrenal oxidoreductase system. Similar to the prevention of the anatomic defects produced by CTM, testosterone prevents the appearance of nipples in male fetuses of analog-treated females, while corticosterone prevents the inhibition of nipple development in experimental females (Neumann and Goldman, 1970).

   f. Early Teratogenic Effectiveness of CTM.   All the manifestations of CTM are produced when it is injected in pregnant rats on days 3–8 of gestation (preimplantation) or to females 6 or 29 days prior to copulation (preovulation) (Goldman, 1970). Thus, CTM is one of the earliest chemical teratogens known. It is effective when injected considerably before the development of demonstrable activity of the oxidoreductase in the endocrine anlagen. This has been demonstrated by the fact that intra-amniotic injection of CTM on day 10 of gestation inhibits activity of the oxidoreductase system present in the rat embryo on day 11, 3–4 days prior to gonadal or adrenal organization (Bloch *et al.*, 1971). During this period of effectiveness inhibitory levels of CTM are only detectable in maternal enzyme-containing tissues. These findings suggest that newly formed fetal enzyme may be either inactivated by CTM made available from maternal enzyme-bound stores during enzyme catabolism or immediately taken up and retained by the rat egg or blastula, or both. Recently, activity of the 3β-hydroxysteroid-Δ^5-oxidoreductase system has been demonstrated in the 5th and 6th day of gestation rat blastula (Dey and Dickman, 1974).

2. Feminization of Males and Inhibition of Other Testosteronogenic Enzymes

   a. Cholesterol Desmolase, 17α-hydroxylase, and C_{17-20} Lyase.   Testosterone-reversible blockade of the readout of male differentiation has also been produced after inhibition of cholesterol desmolase by the nonsteroidal inhibitor, aminoglutethimide, and of 17α-hydroxylase and C_{17-20} lyase both by the nonsteroidal multihydroxylase inhibitors SU 10,603 and SU 8000 (Ciba) (Goldman, 1970) and by the steroidal-specific inhibitors of these enzymes, Δ^1-testos-

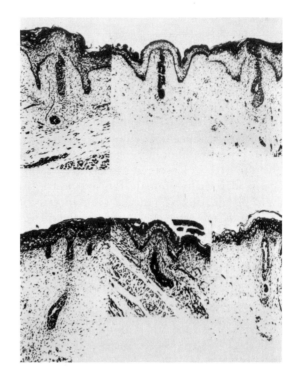

Fig. 8. Top: Mammary gland anlagen of male fetuses. The mother animals were treated with DMSO (left), CTM (middle), and both CTM and testosterone (right). Note nipple induced by CTM is prevented by testosterone (×100). Bottom: Mammary gland anlagen of female fetuses. The mother animals received vehicle (left), CTM (middle), and both CTM and corticosterone (right). Note that the inhibition of nipple development produced in females by CTM is evident in the markedly decreased epidermal thickening at the juncture of the primary glandular bud and the surface and in the circular epidermal invagination. Corticosterone prevents these changes (×100).

terone-17β-ureide and 16α-Br,3β-,17α-dihydroxy-5α-pregnan-20-one (Goldman, 1972b; Goldman *et al.*, 1976) (Figure 5).

3. Feminization of Males by Hormone Antibodies

a. Antibodies to Testosterone and to LH-FSH. The prenatal administration of antibodies to testosterone (Goldman *et al.*, 1972) and of antibodies to LH-FSH (Goldman *et al.*, 1973c) induces the same degree of hypospadias and nipple formation in newborn male rats as that of CTM. These antibodies are γ-globulins but are able to cross the placenta and to inactivate testosterone and gonadotropins in the fetal circulation. These results indicate clearly that gonadotropins of fetal male origin play a role in the errors in the program of masculine organogenesis.

4. Virilization of Females and Inhibition of Glucocorticoid Synthetic Enzyme

Corticosterone-reversible fetal adrenal enlargement and virilization of female genitalia results from maternal administration of metyrapone, a selective nonsteroidal inhibitor of the adrenal 11β-hydroxylase system, during period A and B (Goldman, 1970). As can be predicted, metyrapone is without effect on male genital differentiation.

VIII. ARE PROGESTINS AND ESTRADIOL-17β TERATOGENS LIKE CTM?

When certain progestins or estrogens are given to pregnant women or rats at the time of urethral fold fusion in the male fetus, they produce labial fusion in females (Goldman and Bongiovanni, 1967). When given after this time, they produce only clitoral hypertrophy in female fetuses. In addition a few reports of the occurrence of hypospadias in boys after these agents were given to pregnant women during urethral fold fusion have appeared. These agents also virilize female fetuses, and estrogens produce adrenal enlargement in both sexes. Thus, it seemed possible that the teratologic effects of progestins and estrogens on the sexual differentiation of the male and female fetus are analogous to infants born with congenital adrenal hyperplasia due to a deficiency of the oxidoreductase system. Progestins and estrogens may have a testosterone-depriving action in common with CTM in the experimental model. Moreover, the closer the beginning of treatment was to critical period of penis formation, the more severe was the hypospadias (Goldman and Bongiovanni, 1967).

In vitro estradiol-17β and various progestins inhibited the activity of the 3β-17β-Δ^5-hydroxysteroid oxidoreductase in *Pseudomonas testosteroni* (Goldman and Bongiovanni, 1967). The concentrations of these agents required to inhibit the bacterial oxidoreductase system were in the same relative proportion as were the minimum effective doses required to produce genital malformations in fetal rats.

A. Congenital Adrenal Hyperplasia Produced by Estradiol-17β

Estradiol-17β administered to pregnant rats produces corticosterone-reversible fetal adrenal enlargement and clitoral hypertrophy in female fetuses and testosterone-reversible hypospadias in male fetuses (Goldman, 1970). The activity of the oxidoreductase is reduced in fetal adrenals and testes. These findings supported the hypothesis that estrogens and progestins may interefere with sexual differentiation in a manner similar to that of CTM, thereby providing a possible solution to the paradoxical results of the early endocrinologic experiments on the fetal effects of the sex hormones.

B. Progestins and Human Hypospadias

Aarskog (1970) has reported five cases of hypospadias in the human male fetus with a history of maternal progestin treatment during the critical period of the formation of the penis. When the position of the urethral meatus is related to the week of gestation at which progestin treatment had been started, there seems to be a relationship in that the more proximal openings occurred in the infants of mothers who had been treated in the first month of pregnancy. Although other causes for the defect may not have been excluded, the author has supported the hypothesis that maternal progestin treatment may affect human fetal genital development by inhibiting the activity of the fetal oxidoreductase and thereby mimicking the congenital anomalies observed in both sexes in congenital adrenal hyperplasia associated with deficient 3β-hydroxy-Δ^5-steroid oxidoreductase. This form of hypospadias can be added to the list of preventable human terata.

IX. DELAYED TERATOGENESIS

A. Errors Appearing After Puberty in Congenital Adrenal Hyperplasia

1. Mammary Gland

The first child with a defect in 3β-hydroxy-Δ^5-steroid oxidoreductase to have reached puberty has been reported to have developed pronounced gynecomastia (Parks *et al.,* 1971). Thus, this boy demonstrates that breast development may occur in postpubertal males if the programming for the pubertal sex differentiation of the mammary gland anlagen is disturbed by an enzyme defect which causes a failure of fetal testicular testosterone production. This observation is completely consistent with the findings in the experimental models of defects of mammary gland differentiation (Goldman and Neumann, 1969; Neumann and Goldman, 1970; Neumann *et al.*, 1970).

2. Testicular Tubules and Leydig Cells

A second boy with this rare form of congenital adrenal hyperplasia has been demonstrated to have deficient activity of the oxidoreductase directly in his testes during puberty (Schneider *et al.*, 1975). A biopsy of his testes revealed complete embryonic arrest of sperm cell development and virtually no Leydig cells. These findings are almost identical to those in testes of some adult male rat offspring whose mothers were treated with CTM before differentiation of the testes (Figure 9). These observations suggest that the control of pubertal differentiation of the testis also occurs in the critical periods A or B, since pubertal defects occur due either to a genetic deficiency or experimental inhibition of the oxidoreductase.

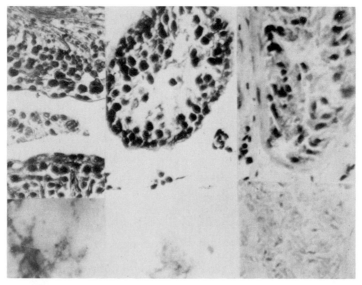

Fig. 9. Section of testis of adult male offspring of pregnant rats treated on day 15 of gestation with vehicle (upper left), with 60 mg/kg CTM (upper middle), and section of testis of pubertal boy with deficiency of 3β-hydroxysteroid oxidoreductase (upper right). Histochemical demonstration of oxidoreductase in Leydig cells of adult offspring treated with vehicle (lower left), with CTM (middle), and of pubertal boy with defect in oxidoreductase (lower right) ($\times 450$). Note complete arrest of spermatogenesis and enzyme activity in Leydig cells of experimental adult male offspring and pubertal boy.

Other biochemical lesions also appear at puberty in Leydig cell differention. The testes of both adult male experimental pseudohermaphrodites produced by treatment with cyproterone acetate during periods A and B (Bottiglioni *et al.*, 1971) and of adult genetic male rat pseudohermaphrodites (Bardin *et al.*, 1973; Coffey *et al.*, 1972; Goldman, 1970) have deficient activity of 17α-hydroxylase and C_{17-20} lyase. Moreover, the testes of adult genetic male rat pseudohermaphrodites fail to have the postpubertal suppression of 5α-reductase and reappearance of 17-ketoreductase characteristic of their normal littermate males (Bardin *et al.*, 1973; Goldman and Klingele 1974a; Goldstein and Wilson, 1972). These observations, taken as a whole, suggest that postpubertal Leydig cell differentiation and sperm cell development require testosterone during periods A and B for their subsequent morphologic and biochemical maturation.

3. Brain—Female Hypothalamic Centers Controlling Puberty

Prenatally administered anti-LH-FSH advances puberty of the female rat, increases the diestrus phase of pubertal estrous cycles, and the estrus phase of adult cycles (Table 5 in Shapiro *et al.*, 1974). Prenatal CTM, by contrast, delays puberty and tends to prolong both phases of pubertal and adult estrous cycles. Each treatment interferes with subsequent pregnancy by reducing fetal and placental weights. An enhancement of the effects of either the antibody or CTM is obtained by additional neonatal treatment during period B. Actinomycin D given to female rat pups in the first 10 days of life also delays onset of puberty and depresses placental weights during the first subsequent pregnancy. Hence, the programming of adult reproductive function of the female hypothalamic–pituitary–ovarian axis probably is also hormonally determined during fetal or neonatal life during period B and is vulnerable to teratogenesis at this time.

These findings also indicate a hitherto unsuspected role of fetal gonadotropins in the female in the onset of puberty and of the function of the postpubertal hypothalamic–pituitary–ovarian axis. Recently, the human female fetus has been shown to have significantly higher levels of FSH during period B than the male fetus (Penny *et al.*, 1973; Winter *et al.*, 1973). Since the ovary is steroidogenically inert during periods A and B (Goldman, 1970), the anti-LH-FSH probably acts on the short feedback loop between LH or FSH and their respective hypothalamic releasing hormones.

Since the rat ovary also does not develop the 3β-hydroxy-Δ^5 steroid oxidoreductase until 10 days of postnatal age, at the end of period B, CTM must produce its effects on the female hypothalamic centers by inhibition of the female adrenal oxidoreductase. The absence of ^{14}C label, when ^{14}C-labeled inhibitor is injected during period B in either the ovary or the hypothalamus, supports this interpretation (Goldman, 1970). The effects of CTM may be due to either the inhibition of the production of a normally required adrenal steroid (e.g., progesterone) or of an abnormal inhibitor-induced Δ^5-3β-hydroxysteroid (e.g., pregnenolone).

Actinomycin D has been shown to block the transcriptional events in the hypothalamus involved in the inhibitory effect of exogenous estradiol-17β on the castration-induced rise in plasma LH in the adult female rat (Schally et al., 1969). This transcriptional block is thought to involve releasing hormone, since the block is reversed by synthetic LH-releasing hormone. Whether actinomycin D acts by blockade of transcription in the hypothalamus of the fetal or neonatal female rat remains to be shown.

X. ANOTHER OF NATURE'S EXPERIMENTS

A. Testicular Feminization

1. Errors in Male Differentiation

Male pseudohermaphroditism due to testicular feminization occurs in the rat, mouse, cow, and man and is an example of an almost total blockade of the readout of masculine differentiation in a genetic and gonadal male (Stanley et al., 1973; Bardin et al., 1973). Each pseudohermaphrodite has an XY karyotype, a chromatin-negative nuclear sex, but an otherwise female phenotype. These errors are due to a sex-linked inheritance of end-organ insensitivity to androgens, which, in turn, is believed by most investigators to be due to deficient end-organ androgen-receptor proteins.

Attention for the remainder of this discussion will be focused on the rat male pseudohermaphrodite. Consistent with their XY karyotype, these rats have inguinal testes with well-developed Leydig cells and no Müllerian duct derivatives. However, the testicular tubules demonstrate germinal arrest, and no male reproductive structures of Wolffian duct origin exist. Consistent with a female phenotype are a short vagina ending in a blind cul de sac, and a feminine number and positioning of nipples.

2. Postpubertal Expression of Errors

a. Female Hepatic Steroid Metabolizing Enzymes. The rat pseudohermaphrodite has a feminine quantitative pattern of urinary corticosteroid excretion and exhibits no sexual behavior. The pseudohermaphrodite also has a feminine qualitative pattern of urinary steroidal metabolites of adrenal and gonadal origin (Gustafsson and Goldman, 1972; Goldman et al., 1973a) due to a phenotypically female pattern of hepatic steroid metabolizing enzymes (Einarsson et al., 1973; Gustafsson et al., 1974a).

b. Brain—Female Preference for Sweet Taste.   Both the female and the genetic male rat pseudohermaphrodite, unlike their male littermates, develop after puberty a preference for sweet solutions, so that if offered a choice of water or 1% saccharin, they will selectively drink 1% saccharin (Shapiro and Goldman, 1973). This observation implies that the development of parts of the brain different from those involved in reproductive function may be controlled by the program of sexual differentiation during periods A and/or B. The observation of a significant elevation of about 10 points in IQ of girls and boys with excess fetal androgens during periods A and B (Money and Ehrhardt, 1973) are also in line with this suggestion.

c. Female Genital Skin Androgen Metabolism.   The genital skin of adult male rat pseudohermaphrodites produced by treatment with cyproterone acetate during periods A and B had female levels of the androgen-metabolizing enzyme, 5α-reductase, which were significantly depressed below those of the male (Bottiglioni *et al.*, 1971). A depression in activity of genital skin 5α-reductase below that of control males was also demonstrated in term rat fetuses after maternal treatment with cyproterone acetate or CTM (Goldman, 1973). This was not a direct inhibition of the enzyme *in vivo* by these two agents but rather a consequence of the blocking of the program of masculine differentiation which they produced in the fetus and which persists to adult life. Since the fetal genital skin included the anlagen of the preputial glands, these observations made two suggestions about the differentiation of genital skin androgen-metabolizing enzymes: (1) The levels of androgen-metabolizing enzymes in the preputial glands may be different in the two sexes. (2) If so, these differences may be "imprinted" by the program of sexual differentiation in periods A and/or B. Since the analysis of the fetal genital skin included the anlagen of the preputial glands, evidence supporting the validity of these suggestions was sought in a study of preputial gland androgen metabolism.

d. Female Preputial Gland Biochemical Sex.   The preputial glands are bilateral apocrine glands at the terminal portion of the urinary tract in both male and female rats. These glands are androgen responsive but not androgen dependent, in that they will enlarge in both sexes under the influence of androgens, but do not require androgens for maintenance of their weight.

The pattern of metabolism of testosterone in the preputial glands of the male rat does differ significantly from that of the female rat primarily by the fact that the female produces less 5α-dihydrotestosterone and more 3α-,17β-dihydroxy-5α-androstane (Goldman and Klingele, 1974c). The difference in pattern of metabolites is due to a significantly higher level of activity of 3α-hydroxysteroid dehydrogenase in the preputial glands of the female than that of male. Furthermore, female elevations in preputial gland 3α-hydroxysteroid dehydrogenase occur in both genetic (Goldman and Klingele, 1974c) and in experimental male pseudohermaphrodites produced by pre- and perinatal treatment with cyproterone acetate (Goldman, 1974c).

Thus, it seems reasonable to conclude that the level of preputial gland 3α-hydroxysteroid dehydrogenase in the male may be depressed by the imprinting action of testosterone during period A and/or B. The elevated levels of this enzyme characteristic of the female will be expressed in male fetuses made unresponsive to testosterone during this period either genetically by an inherited deficiency of androgen-receptor proteins or experimentally by blockade of these receptors with cyproterone acetate. Androgen-receptor proteins are thereby implicated as the next step in the target organs in the imprinting action of testosterone in the program of sexual differentiation.

XI. ANDROGEN-RECEPTOR PROTEINS AND SEXUAL DIFFERENTIATION

It is generally believed that the reversible effects of androgen action in the adult animal are mediated in ventral prostate and other androgen-dependent target organs of the rat by the 5α-reduced metabolite of testosterone, 5α-dihydrotestosterone, by way of the cytosol dihydrotestosterone-receptor proteins. The uptake of these dihydrotestosterone-receptor complexes by nuclear chromatin has been assumed to be the event which initiates nuclear responses to androgen in the adult animal. Recent studies of the dihydrotestosterone receptors in preputial glands of maternal and experimental male pseudohermaphrodites have provided evidence that the same events stimulated by the action of testosterone in the adult animal may occur in the "imprinting" action of testosterone in the fetus or neonate.

A. Defect in Preputial Gland Receptor

1. Genetic Pseudohermaphrodite

There are four dihydrotestosterone-binding proteins [γ_2, isoelectric point (pI) 5.8; δ_3, 6.8; ϵ_3, 7.8; and ζ_1, 8.1) in rat ventral prostate cytosol which are taken up by prostate nuclear chromatin and blocked by cyproterone acetate (Katsumata and Goldman, 1974). After uptake of the γ_2 receptor by prostate nuclear chromatin, DNA and RNA polymerases are activated with ensuing protein synthesis (Mainwaring and Irving, 1973).

The male and female preputial cytosols contain most of the dihydrotestosterone-binding proteins present in prostate cytosol, but only γ_2 is a receptor protein in this gland (Figure 10) (Goldman, 1975). The cytosol of the preputial gland of the genetic male rat pseudohermaphrodite has a selective deficiency of the γ_2 dihydrotestosterone-receptor protein (Figure 10), but the nuclear chromatin of this animal has the capacity to take up the cytosol γ_2 receptor from the glands of both its male and female littermates (Goldman, 1975). Consequently, absence of the cytosol γ_2 dihydrotestosterone receptor can explain the expression of the inborn female program of sexual differentiation of the pre-

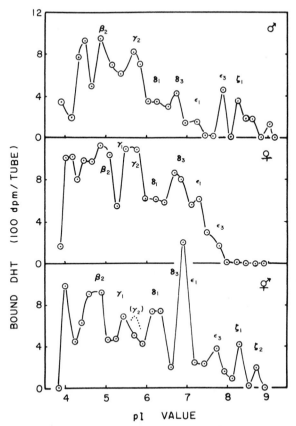

Fig. 10. Isoelectrofocused patterns of dihydrotestosterone binding proteins of preputial gland cytosol of male (upper), female (middle), and genetic pseudohermaphrodite (lower). Pattern of pseudohermaphrodite differs from those of male and female by absence of dihydrotestosterone receptor protein γ_2, pI = 5.8.

putial gland 3α-hydroxysteroid dehydrogenase in the genetic male rat pseudohermaphrodite. If a γ_2 receptor were available, the preputial glands of the genetic pseudohermaphrodite could be expected to exhibit male differentiation.

B. Defect in Preputial Gland Receptor

1. Experimental Pseudohermaphrodite

A role of receptor proteins in the program of sexual differentiation would be supported by the demonstration that the level of the preputial gland cytosol γ_2 receptor is depressed by cyproterone acetate during period A or B, at the critical period when this agent interferes with male differentiation and produces experimental male pseudohermaphroditism. Accordingly, the dihydrotestosterone-binding proteins were studied in the Wolffian ducts of male fetuses of 19 days of gestation after treatment of pregnant rats with cyproterone acetate or vehicle on days 14–19 (Goldman and Katsumata, 1974). The cytosol of the Wolffian ducts during this critical period of the imprinting of masculine differentiation contains the same dihydrotestosterone-binding proteins as the adult preputial glands, including the γ_2 receptor (Figure 11). Cyproterone acetate blocks the binding of this receptor with dihydrotestosterone in the fetus *in vivo* (Figure 11). Thus, in fetuses rendered unresponsive to testosterone, either genetically or with cyproterone acetate, the dihydrotestosterone-receptor protein, γ_2, in the preputial gland cytosol is made either naturally deficient or experimentally blocked with a resulting failure of nucleic acid induction of male-specific protein synthesis.

XII. PROGRAM OF SEXUAL DIFFERENTIATION

These findings taken as a whole suggest a solution to the mystery of sexual differentiation: The irreversible effects produced by testosterone or its

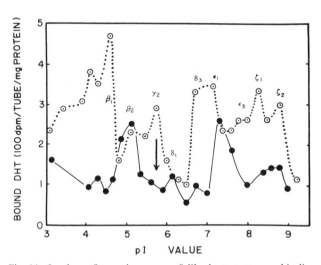

Fig. 11. Isoelectrofocused patterns of dihydrotestosterone binding proteins of the Wolffian ducts of fetuses of 19 days of gestation from pregnant rats treated with vehicle (⊙····⊙) or cyproterone acetate at 30 mg/kg from days 14 to 19 (●———●). Note that γ_2 receptor protein has been experimentally blocked by the antiandrogen.

5α-reduced metabolite in programming male sexual differentiation during periods A and B may have the same chain of molecular events as that of the reversible androgen-dependent effects of androgen action in the adult animal. This molecular chain of events, depicted in Figure 12, would be initiated genetically at different critical periods. The complex of androgen-receptor protein formed after incorporation into nuclei of the various male target-organ anlagen would cause DNA- and RNA-directed synthesis of male specific proteins. Blockade of this program at any of its steps would be expected to allow expression of the innate female program.

XIII. CONCLUSIONS AND THE FUTURE

A. Antiteratogenesis and Fetal Therapy

Summing up would not be complete without some speculation as to the overall meaning of such efforts to unravel a biologic mystery. A valuable consequence of understanding a program of differentiation is that of having knowledge that might help to prevent teratogenesis. The prevention of the anatomic errors in experimental congenital adrenal hyperplasia produced by CTM by appropriate hormonal replacement in the fetus by way of the mother is an example (Goldman, 1970). Presumably, with successful prenatal diagnosis of this and other forms of this disease, this kind of fetal therapy can be performed in man.

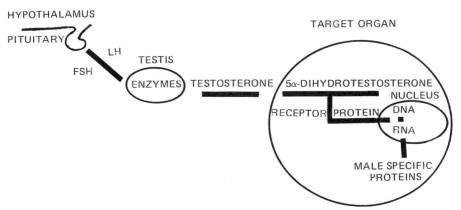

Fig. 12. Scheme of the program of male sexual differentiation.

1. Therapeutic Program Replacement in the Fetus

The nucleus of an adult frog's intestinal cell, after implantation into an enucleated frog's blastula, can direct the development of the blastula into a complete adult frog (Gurdon, 1968). Thus, the complete program of differentiation of the frog is intact but repressed in each adult cell nucleus and is capable of being turned on and read out in the appropriate environment. The possibility of such a program derepression in mammals is suggested by the observation that a slight reversal of the CTM-induced persistent defect in the oxidoreductase is produced in rats by intramuscular injections of phenol–detergent extracts of rat or bovine adrenal RNA or DNA (Goldman, 1970). Moreover, a more complete understanding of the program of sexual differentiation could conceivably permit embryonic reinduction of new organs in the adult to replace diseased ones.

At any rate, although many facets of sexual differentiation remain to be deciphered, it appears that the program of male differentiation is directed by fetal testosterone through the same chain of molecular events by which testosterone produces its hormonal action in the adult. Interference with the readout of this program at any of a variety of sites allows the inborn female program to be expressed.

ACKNOWLEDGEMENTS

The author is a recipient of a Career Development Award (HD 13,682) from the USPHS, and was aided by grants from the USPHS (HD-4683), and the Rockefeller Foundation.

REFERENCES

Aarskog, D., 1970, Clinical and cytogenetic studies in hypospadias, *Acta Paediatr. Scand.* **203** (Suppl).

Bardin, C. W., Bullock, L. P., Sherins, R. J., Movszovicz, I., and Blackburn, W. R., 1973, Part II. Androgen metabolism and mechanism of action in male pseudohermaphroditism: A study of testicular feminization, *Rec. Prog. Horm. Res.* **29:**65.

Bloch, E., Lew, M., and Klein, M., 1971, Studies on the inhibition of testosterone synthesis in rat and rabbit fetal testes with observationn on reproductive tract development, *Endocrinology* **89:**16.

Bongiovanni, A. M., DiGeorge, A. M., and Grumbach, M. M., 1959, Masculinization of the female infant with estrogenic therapy alone during gestation: Four cases, *J. Clin. Endocrinol.* **19:**1004.

Bongiovanni, A. M., Eberlein, W. R., Goldman, A. S., and New, M., 1967, Disorders of adrenal steroid biogenesis. *Rec. Prog. Horm. Res.* **23:**375.

Bottiglioni, F., Collins, W. P., Flamigni, C., Neumann, F., and Sommerville, I. F., 1971, Studies on androgen metabolism in experimentally feminized rats, *Endocrinology* **89:**553.

Burns, R. K., 1955, Experimental reversal of sex in gonads of the opossum *Didelphis Virginiana*, *Proc. Natl. Acad. Sci. U.S.A.* **41:**669.

Burrill, M. W., and Greene, R. R., 1939, Experimental intersexuality: Correlation between treatment and degree of masculinization of genetic female rats, *Am. J. Physiol.* **126:**452.

Camacho, K. M., and Migeon, C. J., 1966, Testosterone excretion and production rate in normal adults and in patients with congenital adrenal hyperplasia, *J. Clin. Endocrinol.* **26:**893.

Coffey, J. C., Aronin, P. A., French, F. S., and Nayfeh, J. N., 1972, Steroid metabolism by testicular homogenates of Stanley Gumbreck pseudohermaphrodite male rat. I. Increased formation of androsterone and androstanediol, *Steroids* **19:**433.

Courrier, R., and Jost, A., 1942, Intersexualité foetale provoquée par la pregneninolone au cours de la grossesse, *C.R. Soc. Biol.* **136:**395.

Dancoff, S. M., and Quastler, 1953, The information content and error rate of living things, *in: Essays on Information Theory in Biology* (H. Quastler, ed.), Univ. of Illinois Press, Urbana, Ill.

DeMoor, P., and Denef, C., 1968, The puberty of the rat liver. I. Feminine pattern of cortisol metabolism in male rats castrated at birth, *Endocrinology* **82:**480.

Dey, S., and Dickmann, Z., 1974, Δ^5-3β-hydroxysteroid dehydrogenase activity in rat embryos on day 1 through 7 of pregnancy, *Endocrinology* **95:**321–323.

Einarsson, K., Gustafsson, J. A., and Stenberg, A., 1973, Neonatal imprinting of liver microsomal hydroxylation and reduction of steroids, *J. Biol. Chem.* **248:**4987.

Ferrari, R. A., and Arnold, A., 1963, Inhibition of β-hydroxysteroid dehydrogenase I and II., *Biochim, Biophys. Acta* **77:**349–364.

Finegold, M. J., and Basch, R. S., 1973, Inhibition of 3β-hydroxysteroid dehydrogenase in the rat adrenal cortex: A metabolic and morphologic study, *Lab. Invest.* **26:**767–777.

Frasier, D. M., Horton, R., and Ulstrom, R. A., 1969, Androgen metabolism in congenital adrenal hyperplasia due to 11β-hydroxylase deficiency, *Pediatrics* **44:**201.

Goldman, A. S., 1970, Animal models of inborn errors of steroidogenesis and steroid action *in: Mammalian Reproduction* (H. Gibian and E. J. Plotz, eds.), p. 389, Springer-Verlag, New York, Heidelberg.

Goldman, A. S., 1972a, Inhibitors of steroidogenesis: Biochemical and Pharmacological implications, *Excerpta Med. Int. Congr. Ser.* **219:**839.

Goldman, A. S., 1972b, Inhibition of 3β-hydroxy-Δ^5-steroid dehydrogenase, hormones and antagonists, *Gynecol. Invest.* **2:**213.

Goldman, A. S., 1973, Rat fetal target organ Δ^5,3β-hydroxysteroid dehydrogenase: Effect of cyanoketone and cyproterone acetate, *Endocrinology* **92:**1043–1050.

Goldman, A. S., 1974a, Influence of hormones on sex differentiation, *Contemp. Obstet. Gynecol.* **3:**69.

Goldman, A. S., 1974b, Sexual programming of the fetus and neonate studied by selective biochemical testosterone depriving agents, *Adv. Biosci.* **13:**17.

Goldman, A. S., 1975, Recent studies on the intersexual programming of the genetic rat male pseudohermaphrodite, *in: Intersexuality in the Animal Kingdom* (R. Reinboth, ed.), p. 422, Springer-Verlag, Berlin.

Goldman, A. S., and Bongiovanni, A. M., 1967, Induced genital anomalies, *Ann. N.Y. Acad. Sci.* **142:**755.

Goldman, A. S., and Katsumata, M., 1974, Separation of androgen binding proteins in rat fetal wolffian duct and genital tissues: Fetal effects of cyproterone acetate administered to the mother (in preparation).

Goldman, A. S., and Klingele, D. A., 1974a, Developmental defects of testicular morphology and steroidogenesis in the male rat pseudohermaphrodite and response to testosterone and dihydrotestosterone, *Endocrinology* **94:**1.

Goldman, A. S., and Klingele, D. A., 1974b, Persistent postpubertal elevation of activity of steroid 5α-reductase in the adrenal of rat pseudohermaphrodites and correction by large doses of testosterone or DHT, *Endocrinology* **94:**1232.

Goldman, A. S., and Klingele, D. A., 1974c, Feminine patterns of androgen metabolism and presence of $\Delta^5,3\beta$-hydroxysteroid oxidoreductase in target organs of genetic rat male pseudohermaphrodites, *Proc. Soc. Exp. Biol. Med.* **146**:539.

Goldman, A. S., and Neumann, F., 1969, Differentiation of the mammary gland in experimental congenital adrenal hyperplasia due to inhibition of $\Delta^5,3\beta$-hydroxysteroid dehydrogenase in rats, *Proc. Soc. Exp. Biol. Med.* **132**:237.

Goldman, A. S., Baker, M. K., Chen, J. C., and Wieland, R. G., 1972, Blockade of masculine differentiation in male rat fetuses by maternal injection of antibodies to testosterone-3-bovine serum albumin, *Endocrinology* **90**:716.

Goldman, A. S., Gustafsson, J. A. And Gustafsson, S., 1973a, Female pattern of metabolism of ^{14}C corticosterone in male pseudohermaphroditic rats, *Proc. Soc. Exp. Biol. Med.* **141**:691.

Goldman, A. S., Gustafsson, J. A., and Gustafsson, S. A., 1973b, New synthetic steroids which inhibit rat gonadal steroid synthesis both *in vitro* and *in vivo*, *Acta Endocrinol.* **73**:146.

Goldman, A. S., Shapiro, B. H., and Root, A. W., 1973c, Inhibition of fetal masculine development in the rat by maternal administration of antibodies to bovine LH, cyanoketone, or antibodies to testosterone-3-bovine albumin, *Proc. Soc. Exp. Biol. Med.* **143**:422.

Goldman, A. S., Eavey, R. D., and Baker, M. K., 1976, Production of male pseudohermaphroditism by two new inhibitors of steroid 17α-hydroxylase and C_{17-20} lyase, *J. Endocrinology* (in press).

Goldstein, J. L., and Wilson, J. D., 1972, Studies on the pathogenesis of pseudohermaphroditism in the mouse with testicular feminization, *J. Clin. Invest.* **51**:1647.

Greene, R. R., Burrill, M. W., and Ivy, A. C., 1940, Experimental intersexuality: The effects of estrogens on the antenatal sexual development in the rat, *Am. J. Anat.* **67**:305.

Gurdon, J. B., 1968, Transplanted nuclei and cell differentiation, *Sci. Am.* **219**(6):24.

Gustafsson, J. A., and Goldman, A. S., 1972, Steroid monosulfates in urine from male pseudohermaphroditic rats, *Endocrinology* **90**:931.

Gustafsson, J. A., Carlstedt-Duke, J. and Goldman, A. S., 1974, On the hepatic sulfurylating activity in male pseudohermaphroditic rats, *Proc Soc. Exp. Biol. Med.* **145**:908.

Gustafsson, J. A., Gustafsson, S. A., Ingelman-Sundberg, M., Pousette, A., Stenberg, A., and Wrange, O., 1975, Sexual differentiation of hepatic steroid metabolism in the rat, *J. Steroid Biochem.* **6**:643.

Josso, N., 1971, Permeability of membranes to the Muellerian-inhibiting substance synthesized by the human fetal testis *in vitro*: A clue to its biochemical nature, *J. Clin. Endocrinol.* **34**:265.

Josso, N., 1973, Activité inhibitrice du testicule de foetus de veau sur le canal de Müller de foetus de rat, en culture organotypique: Rôle des tubes seminifères, *C.R. Acad. Sci. Paris* **274**:3573.

Jost, A., 1972, Embryonic sexual differentiation (morphology, physiology, abnormalities), *in: Hermaphroditism, Genital Anomalies, and Related Disorders*, 2nd ed. (Jones and Scott, eds.), p. 16, Williams & Wilkins, Baltimore.

Kasdan, R., Nankin, H. R., Troen, P., Wald, N., Pan, S., and Yanaihara, T., 1973, Paternal transmission of maleness in XX human beings, *N. Engl. J. Med.* **288**:539.

Katsumata, M., and Goldman, A. S., 1974, Separation of multiple dihydrotestosterone receptors in rat ventral prostate by a novel micromethod of electrofocusing: Blocking action of cyproterone acetate and uptake by nuclear chromatin, *Biochim. Biophys. Acta* **359**:112.

Keenan, B. S., Meyer, W. J., III, Hadjian, A. J., Jones, H. W., and Migeon, C. J., 1974, Syndrome of androgen insensitivity in man: Absence of 5α-dihydrotestosterone binding protein in skin fibroblasts, *J. Clin. Endocrinol. Metab.* **38**:1143.

Kraulis, I., and Clayton, R. B., 1968, Sexual differentiation of testosterone metabolism exemplified by the accumulation of 3β,17α-dihydroxy-5α-androstane-3-sulfate as a metabolite of testosterone in the castrated rat, *J. Biol. Chem.* **243**:3546.

Mainwaring, W. I. P., and Irving, R., 1973, The use of deoxyribonucleic acid cellulose chromatography and isoelectric focusing for the characterization and partial purification of steroid receptor complexes, *Biochem. J.* **134**:113.

Milin, B., and Roy, A. K., 1973, Androgen receptor in rat liver cytosol, "deficiency" in pseudohermaphrodite male rats, *Nature (London), New Biol.* **242**:248.

Money, J., and Ehrhardt, A. A., 1973, *Man & Woman, Boy & Girl: Differentiation and Dimorphism of Gender Identity from Conception to Maturity*, Johns Hopkins Univ. Press, Baltimore.

Neumann, F., and Goldman, A. S., 1970, Prevention of mammary gland defects in experimental congenital adrenal hyperplasia due to inhibition of 3β-hydroxysteroid dehydrogenase, *Endocrinology* **86:**1169.

Neumann, F., von Berswordt-Wallrabe, R., Elger, W., Steinbeck, H., Hahn, J. D., and Kramer, M., 1970, Aspects of androgen-dependent events as studied by antiandrogens, *Rec. Prog. Horm. Res.* **26:**337.

Parks, G. A., Bermudez, J. A., Anast, C. S., Bongiovanni, A. M., and New, M. I., 1971, Prepubertal boy with the 3β-hydroxysteroid dehydrogenase defect, *J. Clin. Endocrinol.* **33:**269.

Penny, R., Olatunji, N., Olambovonnu, and Frasier, S. D., 1973, Follicle stimulating hormone (FSH) and luteinizing-human chorionin gonadotropin (LH-HCG) concentrations in paired maternal and cord sera, *Pediatr. Res.* **70:**331.

Pfeiffer, C. A., 1936, Sexual differences of the hypophyses and their determination by the gonads, *Am. J. Anat.,* **58:**195.

Potts, G. O. Burnham, D. F., and Beyler, A. L., 1963, Inhibitory action of two new steroids in the adrenal cortex of rats, Abstract 35, *Fed. Proc.* **22:**166.

Ritzen, E. M., Nayfeh, S. N., French, F. S., and Aronin, P. A., 1972, Deficient nuclear uptake of testosterone in the androgen-insensitive (Stanley-Gumbreck) pseudohermaphroditic male rat, *Endocrinology* **91:**116.

Saez, J. M., de Peretti, E., Moreau, A. M., David, M., and Bertrand, J., 1971, Familial male pseudohermaphroditism with gynecomastia due to a testicular 17-ketosteroid reductase defect. I. Studies *in vivo, J. Clin. Endocrinol.* **32:**604.

Schally, A. V., Bowers, C. Y., Carter, W. H., Arimura, A., Redding, T. W., and Saito, M., 1969, Effect of actinomycin-D on the inhibitory response of estrogen on LH release, *Endocrinology* **85:**290.

Schneider, G., Genel, M., Goldman, A. S., Rosenfield, R. C., and Bongiovanni, A. M., 1975, Persistent testicular Δ⁵-4 isomerase, 3β-ol dehydrogenase deficiency in the 3β-hydroxysteroid dehydrogenase form of congenital adrenal hyperplasia, *J. Clin. Invest.* **55:**681.

Shapiro, B. H., and Goldman, A. S., 1973, Feminine saccharin preference in the male rat pseudohermaphrodite, *Horm. Behav.* **4:**371.

Shapiro, B. H., Goldman, A. S., and Root, A. W., 1974, Prenatal interference with the onset of puberty, vaginal cyclicity and subsequent pregnancy in the female rat, *Proc. Soc. Exp. Biol. Med.* **145:**334.

Stanley, A. J., Gumbreck, L. G., and Allison, J. E., 1973, Part I. Male pseudohermaphroditism in the laboratory Norway rat, *Rec. Prog. Horm. Res.* **29:**43.

Wilkins, L., 1960, Masculinization of the female fetus due to the use of orally given progestins. J.A.M.A. **172:**1029.

Winter, J. S. D., Boroditsky, R., Faiman, C., and Reyes, F. I., 1973, Serum gonadotropin and sex steroid concentrations in the human, *Pediatr. Res.* **70:**306.

Yalom, I. D., Green, R., and N. Fisk, 1973, Prenatal exposure to female hormones: Effect on psychosexual development in boys, *Arch. Gen. Psychiatr.* **28:**554.

Abnormal Organogenesis of Facial Structures   17

M. C. JOHNSTON, G. M. MORRISS, D. C. KUSHNER, and G. J. BINGLE

I. INTRODUCTION

Much of the early research concerning embryonic facial development has been conducted on amphibian embryos. More recently the emphasis has turned to avian and mammalian embryos, and the development and utilization of the new techniques required have led to considerable advances towards the understanding of developmental mechanisms. The description of facial embryogenesis which is now emerging reflects a sequence of events that is similar in the various types of vertebrate embryos studied.

Facial morphogenesis can be considered as consisting of three major developmental phases. During the first phase, the epithelial framework of the head (brain, gut endoderm, and surface ectoderm) is formed, and within this framework distinct populations of cells undergo extensive migrations. During the second phase these populations, together with adjacent epithelia, develop into the facial processes, which are outgrowths of tissue that fuse with another to form structures such as the primary palate. During the final phase, further differentiation gives rise to specialized tissues and organs. The following pages will give an overview of these three phases of embryonic facial morphogenesis.

M. C. JOHNSTON, G. M. MORRISS, D. C. KUSHNER, and G. J. BINGLE • Laboratory of Developmental Biology and Anomalies, National Institute of Dental Research, National Institutes of Health, Bethesda, Maryland.

A. Origins and Early Migrations of Facial Tissues

As background a brief review will be made of the developmental events that lead up to the stages at which specific tissues contributing to facial development are formed. Since these are well accepted, little documentation is provided. For further description of these early events, the reader is referred to more general texts (e.g., Balinsky, 1975; Langman, 1975).

Early in development the embryo consists of two layers, each of which is composed of a single layer of cells arranged as an epithelium. The layer in contact with the yolk in avian and reptilian embryos, and in an equivalent position in mammalian embryos, is the hypoblast; the other layer is the epiblast. A third germ layer, mesoderm, is formed by a mass movement of the epiblast cell sheet from both sides of the embryo to form a midline thickening, the primitive streak, from which the cells move downwards into the space or potential space between epiblast and hypoblast (Figure 1A) to initiate the formation of mesoderm. During this process (gastrulation) the cells lose their epithelial organization and form a loose meshwork; this type of cellular organization is referred to as mesenchyme.

Following gastrulation the mesoderm cells (primary mesenchyme) (Hay, 1968) migrate laterally and anteriorly. At the same time a more compactly organized axial rod of cells, the notochord or head process, grows anteriorly from the anterior end of the primitive streak (Hensen's node). Mesodermal cells form a continuous sheet between epiblast and hypoblast (Figure 1B), except in the position of the notochord, and the buccopharyngeal and cloacal membranes (these mark the future anterior and posterior extremities of the endodermal gut tube). Those which come to lie adjacent to the notochord are called paraxial mesoderm; in the head these cells form most of the skeletal and connective tissue of the brain and skull. Intermediate mesoderm (Figure 1B) does not differentiate in the head; further posteriorly it forms the urogenital system. Furthest laterally is the lateral plate mesoderm which will be represented in the developing face as mesodermal core cells of the pharyngeal arches.

Paraxial mesoderm and notochord together exert an inductive influence which leads to further differentiation of the cells of the overlying ectoderm (e.g., Toivonen, 1967). This phenomenon is called primary embryonic induction; it is one of many inductive interactions in embryogenesis. The differentiation involves a change in cell shape which can be seen first as a thickened area of ectoderm, the neural plate. The plate then undergoes a folding process to form the neural tube (Figure 1B,C). Neural tube formation (neurulation) begins anteriorly and spreads posteriorly.

Having exercised their inductive capacity, the paraxial mesodermal cells of the posterior head (postotic), trunk, and tail regions organize themselves, in an anterior-to-posterior time sequence, into segmented epithelial blocks. These are the somites, which later give rise to segmental skeletal and connective tissue structures and muscle blocks, forming a pattern which is also im-

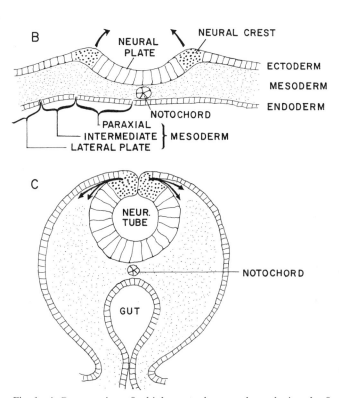

Fig. 1. A:Cross-section of a higher vertebrate embryo during the formation of the middle germ layer (mesoderm). The mesodermal cells move (arrows) from the upper layer (epiblast) into space or potential space between the epiblast and a lower layer (hypoblast). B: After completion of the above cell movement, the embryo is said to consist of three germ layers: ectoderm, mesoderm, and endoderm. An inductive influence emanating from a median strip of (paraxial) mesoderm leads to formation of the neural plate. Neural crest cells form at the junction between the neural plate and surrounding ectoderm. C: Folding processes in the neural plate and lateral body wall lead to formation of the neural tube and gut. Both neural tube and gut endoderm will separate from the surface after subsequent fusions. Crest cells migrate from the neural folds along two paths: under the surface ectoderm and beside the neural tube (arrows).

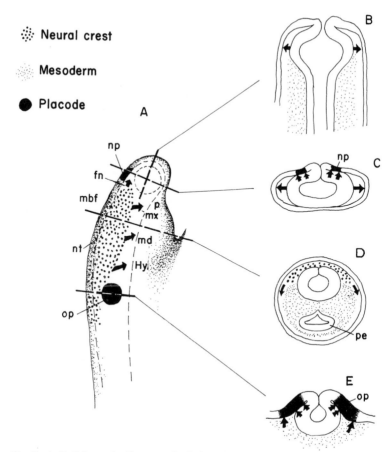

Fig. 2. A–E: Schematic diagrams depicting the interrelationships of migrating neural crest cells and other components observed in early facial development. Broken lines on A represent the section places of B–E.

posed on the developing peripheral nervous system. Paraxial mesoderm anterior to the otocysts contributes to facial development, forming the extrinsic eye muscles and portions of the cranial skeleton (Hazelton *et al.*, 1976); it is sometimes regarded as somitic (Figure 3), although there is little evidence of segmental organization.

As the neural groove becomes more pronounced, its lateral borders (neural folds) rise and then curve toward each other in the midline, supported by the underlying mesoderm (Figure 1C). The end point of this morphogenetic movement is the fusion of the neural folds to form the neural tube. The undifferentiated ectoderm lateral to the neural folds also fuses in the midline, forming surface ectoderm which will later differentiate into epidermis.

Cells along the edge of each neural fold are not involved in this fusion (although they may temporarily fuse with each other in some species); these

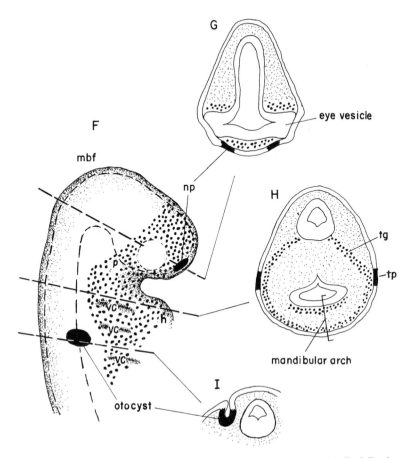

Fig. 2. F–I: The distributions of neural crest and mesodermal cells following completion of the initial migrations. Various structures are designated as follows: frontonasal (fn), maxillary (mx), mandibular (md), and hyoid (hy) neural crest; nasal (np), otic (op), and trigeminal (tp) placodes; and the locations of the prospective pituitary gland (p), neural tube (nt), midbrain flexure (mbf), pharyngeal endoderm (pe), and developing heart (h), trigeminal ganglion (tg), and visceral clefts (vc). (For detailed descriptions, see text).

are the neural crest cells (Figure 1), and they usually begin their migrations at about the time of neural fold fusion. Following migratory movements, which may be extensive, they differentiate into a wide variety of cell types. Crest cells give rise to the neurons of the somatic sensory and autonomic nervous systems, various types of supporting cells of the peripheral nervous system, and pigment cells. In the head and neck the contribution of the neural crest cells is far more extensive and complex. Their role in facial development is of major importance and will be considered in some detail later in this section.

Rapid growth of the anterior neural plate carries this structure up and over the heart-forming region which was initially located anterior to the neural plate. The heart, which forms from lateral plate mesoderm, consequently comes to lie ventral to the pharynx (Figure 2). Between the heart and the structures dorsal to the pharynx, the lateral pharyngeal wall becomes segmented into the visceral arches by inpocketings of ectoderm and outpocketings of endoderm (Figures 2F and 4A). Initially, each of these arches contains only a core of mesoderm which connects the mesoderm of the developing heart with mesoderm dorsal to the pharynx.

As the cephalic part of the neural tube is closing, the brain is already becoming regionalized (a capacity it apparently achieves at the time of primary embryonic induction, Toivonen, 1967). Evaginations of large portions of the lateral wall of the forebrain form the optic vesicles in such a way that their outer surface becomes directly applied to the surface ectoderm of all aspects except caudally, where they are in contact with mesoderm (Figure 2B,C). The neural component (neurohypophysis) of the pituitary primordium forms as a midline evagination from the floor of the caudal region of the forebrain (Figure 2A). Its position relative to other structures marks a point of some significance to subsequent facial development: It is just caudal to the level of the optic vesicles, just anterior to the termination of the notochord, and level with the most anterior extension of the paraxial mesoderm. The ectodermal component (adenohypophysis) forms from dorsal midline oral ectoderm immediately anterior to the buccopharyngeal membrane, i.e., immediately anterior to the pharyngeal endoderm.

Inductive influences from the anterior region of the forebrain, the optic vesicles, and the hindbrain emanate to the overlying ectoderm (Figure 2C, E,G), leading to the differentiation of the olfactory, otic, and lens placodes (Jacobson, 1966; Yntema, 1955). These placodes invaginate inward to form pits, and in the case of the lens and otic pits, finally form vesicles closed off from the surface ectoderm. The otic vesicle (otocyst) subsequently differentiates to form the epithelium of the inner ear.

At about the time of neural tube closure, the neural crest cells begin to migrate (Figure 1C and 2A). They move between the mesoderm and surface ectoderm and between the mesoderm and neural tube. Only the subectodermal population will be considered here: These cells contribute extensively to facial development, eventually forming almost all of the skeletal and connective tissues of the face.

Studies of the migration pattern of cranial neural crest cells have been conducted on avian embryos (Johnston, 1966; Noden, 1975). During migration, they form a multicellular layer immediately beneath the surface as they move ventrally into the facial regions (Figure 2A,D). Crest cells moving dorsal to the eye will eventually form the frontonasal mesenchyme. Crest cells that will form the mesenchyme of the maxillary processes migrate caudal to the eye and, eventually, move in a medial direction to reach the developing pituitary primordium. Together with the frontonasal mesenchyme they form a sheath surrounding the eye (Figure 2F,G). The crest cell layer splits into separate groups of cells as they migrate into each visceral arch (Noden, 1975), and the mesodermal core within each arch is rapidly surrounded by crest cells (Figures 2H and 4A). The trailing edge of the crest cell mass maintains contact with the neural tube at points where the crest cells will initiate formation of the cranial sensory ganglia (Figure 2H).

Studies on avian crest cell migrations and distributions have confirmed observations on amphibian embryos (e.g., Hörstadius, 1950); similar events have recently been shown to occur in mammals. The evidence in mammalian embryos has been derived from histochemical studies of the mouse (Milaire, 1959) and from tritiated thymidine marking of the neural fold cells of rat embryos developing *in vitro* (Johnston and Krames, unpublished). Milaire (1959) concluded that migrating crest cells can be distinguished by high RNA levels, and the labeled crest cells in the study utilizing the thymidine marker were found to migrate as a subectodermal layer of cells coincident with those showing evidence of high RNA levels. We have found (unpublished) that staining for RNA is not sufficiently intense to readily distinguish crest cells from those of the visceral arch mesodermal cores, which also show evidence of high RNA levels. The cell-marking study, however, showed that crest cells surround the mesodermal cores as they do in avian and amphibian embryos.

At about the time crest cells arrive in the visceral arches, blood island formation can be observed in the mesodermal cores (Johnston and Listgarten, 1972). Capillary endothelial buds invade the crest cell mesenchyme, presumably from the mesodermal cores and possibly from other mesoderm. In the facial region, the only vascular components derived from mesoderm are the endothelial cells; all other vessel wall components (e.g., pericytes and smooth muscle cells) are derived from crest mesenchyme (Johnston *et al.*, 1974; LeLievre and LeDouarin, 1975).

Extensive migrations of other cell populations also occur during facial morphogenesis. The migration and derivatives of head paraxial mesoderm cells have been followed by cell-marking procedures (Hazelton, 1970; Hazelton *et al.*, 1977). Following early ganglion formation, a population of paraxial mesodermal cells moves around the ganglia and over the dorsal surfaces of the midbrain and hindbrain (Figure 2G–I). These mesodermal cells will eventually form the meninges, skull base, and calvaria. Others condense around the developing otocyst to form the periotic capsule, which later becomes chondrified and then ossified as the petrous temporal bone. Prospective myoblasts are also derived from paraxial mesoderm; these cells migrate into areas already occupied by crest cell mesenchyme to form the extrinsic ocular and hypoglossal muscle primordia (Figure 3).

Finally, epithelial tissues of both ectodermal and endodermal origin also contribute to facial development. Eye cup and lens development have been described in detail elsewhere in this text (Chapter 14, this volume). Epithelial invaginations from the ectoderm of the oral cavity form the enamel organs which, in addition to participating in the overall organization of the tooth, are responsible for enamel formation (see Slavkin, 1974). Both exocrine (e.g.,)

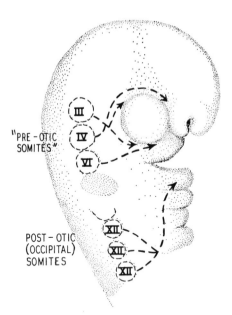

"PRE-OTIC SOMITES"

POST-OTIC (OCCIPITAL) SOMITES

Fig. 3. Major migrations of the extrinsic ocular and hypoglossal muscle primordia. Their origins from paraxial mesoderm are indicated by the number of the cranial nerve by which they are innervated. (From Ross and Johnson, 1972.)

salivary) and endocrine (e.g., anterior pituitary) glands are formed from oral ectoderm. A number of endocrine glands and the thymus form from the endodermal lining of the pharynx.

B. Formation and Coalescence of the Facial Processes

Following completion of crest cell migration into the visceral arches, and the vascularization of the crest cell mesenchyme by endothelia of mesodermal origins, a series of symmetrical outgrowths, termed facial processes, initiates the next phase of facial development. The upper facial processes are termed the medial nasal, lateral nasal, and maxillary processes, while those of the lower face develop from the first two visceral arches and are termed the mandibular and hyoid processes (Figure 4B). The mechanisms of facial process development are not understood, although epithelial–mesenchymal interactions similar to those occurring in limb development may be involved.

As indicated earlier, the nasal placodes are induced in overlying ectoderm by part of the forebrain. The placodal ectoderm then becomes separated from the forebrain by intervening frontonasal mesenchyme (Figure 2). Mesenchymal cells condense on the undersurface of the placode, and an invagination of the placode leads to nasal pit formation. Further outgrowth of tissue associated with an accumulation of the underlying mesenchyme on either side of the nasal pit leads to formation of the medial and lateral nasal processes (Figure 4B). Fusion of these processes below the nasal pit (Figure 4C) constitutes the initial stage of primary palate formation (Chapter 11, this volume).

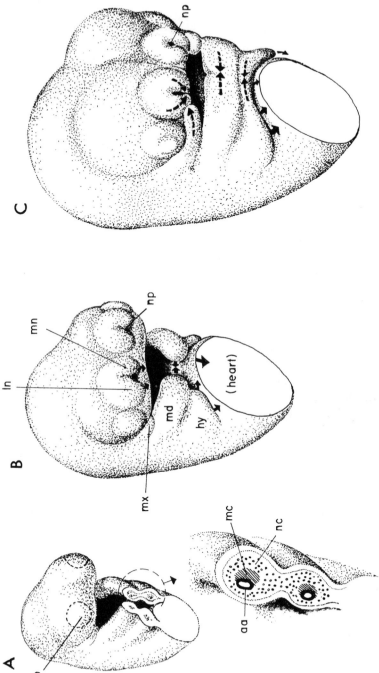

Fig. 4. Schematic representation illustrating development of the facial processes. (From Johnston, 1976.) After completion of crest cell migration (A), facial process development begins (B), and is completed after the fusion of the processes with each other or with other structures (C). Details are given in the text. The heart and adjacent portions of the visceral arches have been removed in A and most of the heart has been removed in B and C. Directions of growth are indicated by arrows. Relevant structures are indicated as follows: aa, mc, and nc, the aortic arch vessel, mesodermal core, and neural crest in first visceral arch; np, nasal placode or nasal pit; mn, ln, and mx, the median nasal, lateral nasal and maxillary processes; md and hy, the mandibular and hyoid processes.

Maxillary process mesenchyme is continuous with other crest-derived mesenchyme, which extends medially to the pituitary gland and eventually forms the connective tissue portions of this structure (Johnston *et al.,* 1974) (Figure 2F). The maxillary process mesenchyme appears to "break through" under the optic primordium, at which time the process becomes considerably more prominent (cf. Figure 4B,C). Coalescence of this mesenchyme with that of the median and lateral nasal processes occurs at about the time of primary palate development (Figure 4C). Medial extensions directed downward from the inner margins of the maxillary processes constitute the palatal shelves which eventually form the secondary palate (Chapter 11, this volume).

Development of the mandibular and hyoid arch processes (Figure 4) is, in some respects, different from the development of the upper facial processes. As noted earlier, the mesodermal cores of the primitive arches are concerned with the formation of blood–vascular elements. Following vascularization of the neural crest-derived mesenchyme, a large amount of cell death can be observed in the mesoderm (Johnston and Listgarten, 1972), and there is some evidence (Hazelton *et al.,* 1977) that the dying cells are partially or completely replaced by other mesodermal cells migrating into the visceral arches from locations close to the neural tube. The latter cells eventually form myoblasts. Thus it appears that the original mesodermal core cells degenerate once their blood-vascular function is completed and that they are replaced by prospective myoblasts originating from regions similar to the sites of origin of other myoblasts forming voluntary skeletal muscle cells (cf. Figure 3).

When the aortic arch vessels of the first two visceral arches develop, the heart is situated almost immediately ventral to the vessels. With further movement of the heart in a caudal direction, and regression of the first two aortic arch vessels, the mandibular and hyoid arches are left as paired "processes" (Figure 4A,B). They grow ventrally and medially as the heart recedes, eventually fusing with their opposite members in the ventral midline. At their dorsal (proximal) ends, three swellings appear on each process, through the accumulation of the underlying mesenchyme. These are the auricular hillocks, which later fuse together to form the pinna (Wood-Jones and Wen, 1934). The external auditory meatus forms from the first cleft. Further growth from the caudal edge of the hyoid arch leads to formation of a skirtlike structure (Figure 4C) which eventually covers most of the neck. This structure fuses with underlying tissues, and its contained mesoderm differentiates into the contractile components of superficial musculature such as the platysma.

C. Terminal Differentiation of Facial Tissues

In the preceding sections, the cell migrations and other morphogenetic movements (folding, invagination, evagination) of the five principal cell populations contributing to facial development have been described. These five populations are the neural crest, mesodermally derived endothelia and skeletal myoblasts, and epithelia of both ectodermal and endodermal origin. As a result of these movements, new associations are established between groups of cells, leading to further inductive interaction and differentiation.

A unique feature of embryonic facial development is that one of the functions performed by mesoderm in other parts of the body, the formation of skeletal and connective tissues, is taken over by neural crest cells (see Hörstadius, 1950; Weston, 1970 for review). Experimental combinations of tissues have shown that in some instances inductive influences which bring about differentiation of mesoderm are without effect on crest cells, and vice versa. For instance, the induction of visceral arch cartilage from crest cells specifically requires pharyngeal endoderm (e.g., Holtfreter, 1968). Other inductive interactions unique to crest cells appear to be required for tooth (Hörstadius, 1950) formation. Some inductive interactions involving crest cells appear to be less specific in their requirements. This appears to be the case, for instance, in the formation of nonendothelial vessel wall components including pericytes and smooth muscle, the connective tissue components of skeletal muscles, dermis, and the meninges of the forebrain (Johnston *et al.,* 1974). Whether or not specific inductive interactions are required, the interface set up between crest and mesoderm cells at an early stage of development (see previous sections) largely determines the origins of cephalic skeletal and connective tissues.

Further discussions of the terminal differentiation of facial tissues are available in a number of recent reviews (Weston, 1970; Sperber, 1973; Johnston *et al.,* 1974).

II. DEVELOPMENTAL ALTERATIONS LEADING TO FACIAL MALFORMATIONS

The first part of this chapter has been concerned with a description of the principal events of normal facial development. The remainder will be primarily concerned with studies in which development has been experimentally altered. Some consideration will also be given to the pathogenesis of malformations of spontaneous origin and to the possible pathogenesis of malformations where no direct information is available.

A. Alterations in Cephalic Neural Plate Development

As noted earlier, the cephalic neural plate (including neural crest) gives rise to a number of facial structures, including the optic cup and crest-derived components of the eyes, and most of the skeletal and connective tissues of the face. Experimental alteration of different components of cephalic neural plate development consequently results in a considerable variety of facial malformations.

1. Generalized Neural Plate Damage

When applied during the early period of neural plate development, X-irradiation results in facial malformations (usually primary palate clefts), eye defects (usually anophthalmia or microphthalmia), and brain malformations (usually microcephaly or encephaloceles) in rats (Wilson *et al.,* 1953) and mice (Murikami *et al.,* 1961; Ritter and Degenhardt, 1963; Bhakdinaronk *et al.,* 1977). The pathogenic mechanisms of X-irradiation appear to be uniform cell death throughout the neural plate (see Figure 5D,E) followed by reduced neural plate growth and a reduction in the numbers of migrating crest cells, as

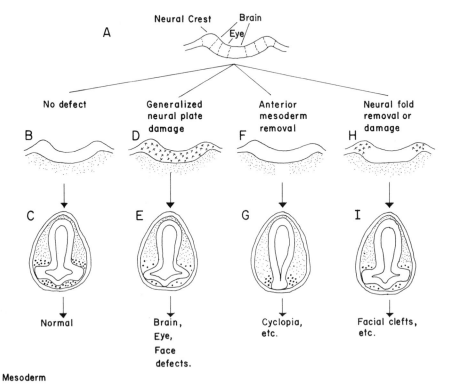

Mesoderm

Neural crest

Fig. 5. Schematic presentation of normal (A–C) and abnormal (D–I) development of the anterior neural plate and its derivatives. Neural crest cells are represented by heavy stippling, mesoderm by fine stippling, and cell death by x's. Sections in the upper rows are comparable to Figure 1B while those in the lower row are comparable to Figure 2G. (For further details, see text).

well as extensive further death of these cells at the time of migration (Bhakdinaronk *et al.,* 1977).

Malformations resulting from administration of a mitotic inhibitor, vincristine, to the rat at similar developmental stages (DeMeyer, 1964a) closely resembled those following X-irradiation. The effects of vinblastine were somewhat different, producing more lower face malformations (e.g., agnathia) in the rat (DeMeyer, 1964b). Only exencephaly resulted from vinblastine and colchicine administered to the hamster (Ferm, 1963a,b). The exact timing of administration, species employed, or other unknown differences may partially account for the variable results. Crest cells appear to be less severely affected by mitotic inhibitors than by X-irradiation.

2. Neural Plate (or Neural Tube) Defects Related to Cyclopia

Experimental reduction of anterior midline mesoderm at the time of neural plate formation results in cyclopia (Adelmann, 1936). Adelmann (1936) also found that removal of neural plate from between the prospective eye vesicles did not result in cyclopia. Similar results have been obtained by numerous other investigators, and the usual interpretation has been that incomplete induction of midline neural plate rudiments allows the optic rudiments to coalesce (Figure 5F,G). However, Källen (1956) has shown that experimental association with mesoderm enhances neural plate growth so that experimental mesoderm removal might, in fact, operate by reducing anterior midplate growth.

It should also be pointed out that, at later stages of development and with different species (Rogers, 1964b; Schowing, 1965), cyclopia may also be produced by somewhat different operative procedures. The latter investigators removed neural tube material from between the optic primordia after tube closure in the chick.

The defects resulting from Roger's (1964b) experiments resembled the more common forms of human cyclopia more closely than most of the defects resulting from Adelmann's procedures. As Rogers (1964b) has noted, pituitary formation failed in many of Adelmann's cases, possibly due to abnormal massing of the extrinsic eye musculature below the eye which may have inhibited normal brain–oral ectoderm interactions (Adelmann, 1936). Adelmann (1936) further suggested that this abnormally located muscle mass may also have resulted in failure of oral cavity development by preventing normal interactions between pharyngeal endoderm and oral ectoderm. In contrast to Adelmann's cyclopia, both pituitary and oral cavity development are usually more normal in human cyclopia (see below), and this also appears to be true for spontaneous cyclopia in the chick (Landauer, 1956) and in many forms of experimentally induced cyclopia. Roger's (1964b) cases, which more closely resembled human cyclopia, resulted from removal of interocular neural fold material anterior to the crest-forming region (cf. Figure 2A). Rogers (1964a) also noted that portions of the neural tube appeared to be at least as much affected as anterior mesoderm in lithium-induced cyclopia in the chick.

The fact that both cyclopia and arhinencephaly (see below) occur in the same families (Cohen et al., 1971) also suggests that human cyclopia may more commonly result from neural fold defects. Adelmann-type cyclopia may be more similar to the cyclopia sometimes associated with otocephaly (see below). Unfortunately, the pathogenesis of spontaneous cyclopia has not been studied and only minimal attention has been paid to the pathogenesis of cyclopia induced by teratogens as described in the material to follow.

A number of teratogic influences, e.g., anoxia (Badtke et al., 1959), alkaloids derived from *Veratrum californicum* (Binns et al., 1968), various ions (e.g., Stockard, 1909), and heat shock (Ingalls, 1966), will produce cyclopia when administered over periods covering the time of neural plate and/or neural tube formation. When the alkaloids were ingested by cattle and sheep at the time of neural plate formation, a series of malformations resulted which varied from slight midfacial underdevelopment through cebocephaly (small "monkey" nose, etc.) to typical cyclopia (Evans et al., 1966). Similar malformations occur in man (e.g., DeMeyer et al., 1964; Cohen et al., 1971) (Figure 6), and the studies on ruminants suggest that they may result from varying degrees of the same primary alterations.

Many features of the cyclopia malformations are theoretically consistent with present information about cephalic development derived mostly from subhuman embryos. The frontonasal mesenchyme, and the median and lateral nasal processes to which this mesenchyme contributes, normally "grow" down between the eyes. Fusion of the optic rudiments would obstruct this downgrowth and prevent the coalescence of the mesenchyme of the maxillary processes with that of the median and lateral nasal processes. Normally, the mesenchymal components of the maxillary incisor tooth buds are derived from the median nasal processes. As would be expected, maxillary incisor formation fails in cyclopia, with only an atypical midline tooth sometimes being present (e.g., Sedano and Gorlin, 1963; Latham, 1971).

The extent of fusion between the optic rudiments seems to be correlated with proboscis formation (Figure 6). A single proboscis (Figure 6A) is usually associated with severe ocular fusion and indicates that the forebrain regions responsible for olfactory placode induction (see earlier sections) have fused together (as have the optic fields) so that a single placode forms. Where ocular fusion is less complete, but the olfactory processes are still prevented (by their position) from interacting with the maxillary processes, two probosci are sometimes observed (e.g., Gorlin et al., 1976). Interaction between probosci and maxillary processes leads to more normal nose formation (Figure 6B).

3. Deficient Crest Cell Formation

Experimental extirpation of chick embryo crest cells prior to migration resulted in specific facial malformations (Johnston, 1964, 1965) (Figure 5H,I).

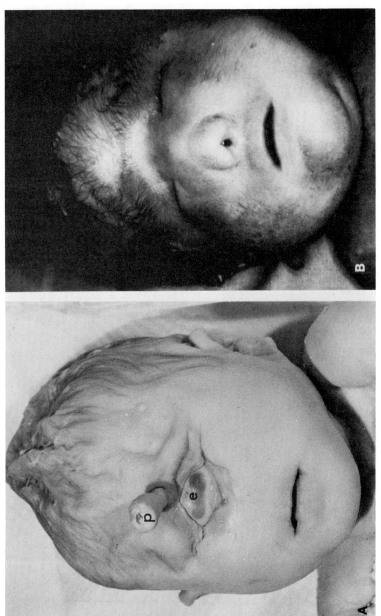

Fig. 6. A: Cyclopia in man. (From Sedano and Gorlin, 1963.) The single midline eye (e) is associated with a single midline proboscis (p). As in the arhinencephaly infant (Figure 7), it appears that the derivatives of the maxillary processes have fused in the midline below the eye. B: Cebocephaly in man. (From Gorlin *et al.*, 1976.) Studies on experimental animals indicate that cyclopia and cebocephaly are part of the same malformation series and that it is possible to put these malformations in a severity sequence. (For details, see text).

Clefts of the primary palate were most frequently observed, while defects of the derivatives of the maxillary processes and visceral arches were less common. Associated malformations of the brain (usually exencephaly or microcephaly) and eye (microphthalmia or anophthalmia) were occasionally found. The brain and eye malformations could be due to excessive extirpation of neural fold material or to the fact that the eye and forebrain depend upon adequate amounts of supporting crest mesenchyme for normal development. In a few embryos, fairly normal eyes, complete with scleral coverings but without surrounding orbit (most of which consists of bones of crest cell origin), were observed; these cases suggest that the eye is capable of essentially normal development with only small amounts of crest mesenchyme.

Cadmium treatment results in an almost identical spectrum of malformations when administered in single doses by injection before crest migration in the hamster (Ferm and Carpenter, 1968; Mulvihill *et al.*, 1970) or the rat (Kushner *et al.*, 1977). Malformations of the brain and eye occur only in the most affected cases of either species and are presumably secondary to effects on the crest cell population.

In human arhinencephaly it appears that deficiencies in frontonasal mesenchyme secondary to deficient neural crest formation result in a range of malformations varying from fairly typical clefts of the lip to almost complete absence of all frontonasal derivatives (Figure 7) (Marin-Padilla *et al.*, 1964).

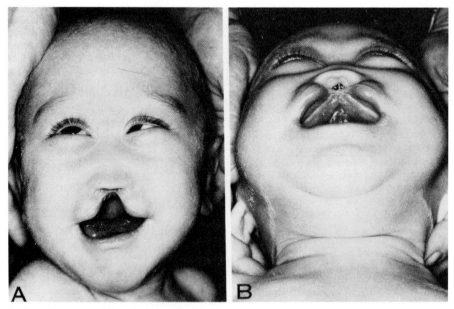

Fig. 7. Infant with arhinencephaly associated with D trisomy. (From Ross and Johnston, 1972.) Derivatives of frontonasal processes (e.g., the nasal septum and anterior portion of the maxilla) appear to be entirely absent. Derivatives of the maxillary process have fused with another in the midline (B).

Affected individuals are consistently holoprosencephalic (single cerebral hemisphere) and usually without olfactory tracts (DeMeyer *et al.*, 1964).

DeMeyer (1964a) examined the brains of rats after treatment with vincristine and found that they were both holoprosencephalic and without olfactory tracts. Murikami *et al.* (1961) also noted holoprosencephalic brains associated with other manifestations of arhinencephaly in mice following X-irradiation. However, while the eyes are more medially positioned and possibly slightly reduced in size in human arhinencepahly, they are not nearly as severely affected as they appear to be after the vincristine or X-irradiation treatments noted above. Current evidence suggests that the arhinencephaly malformation results from a dorsal neural tube deficiency, particularly in the most anterior neural fold region including that portion giving rise to frontonasal mesenchyme (see Figure 5H,I).

B. Interference with Migrating Cells

Evidence for teratogenic effects on migrating cells is accumulating rapidly. Developmental alterations following maternal subcutaneous injection of vitamin A at the time of gastrulation in rat embryos have been studied in some detail (Morriss, 1972, 1973). Cytological alterations (surface blebbing, etc.) were observed almost immediately in all types of cells. Some cell death also occurred. Long-term effects were not evident in ectoderm and endoderm, whereas large spaces between mesodermal cells persisted at 48 hr, suggesting decreased numbers of cells and, possibly, edema or increased matrix production.

Embryos photographed by transmitted light 72 hr after administration of vitamin A (Figure 10 in Morriss, 1972) indicate that many of the crest cells destined to migrate into the maxillary processes and visceral arches did not complete their migration, remaining beside the neural tube. This resulted in differentiation of first arch cartilages which were abnormal in both form and position (Morriss, unpublished). The otocysts were displaced anteriorly relative to the visceral arches, apparently due to the differentiation of a shorter-than-normal cephalic neural plate, thus shifting the otocyst-inducing hindbrain region anteriorly. The maxillary processes were reduced.

In all vitamin A-treated embryos, the resulting malformation complex consisted of face and ear abnormalities, including micrognathia, open-eyelid, and cleft palate, rather similar to the Treacher Collins syndrome (Figure 8) (Poswillo, 1975). In addition, 30% of the embryos had major abnormalities of the central nervous system: exencephaly, meningoencephalocoele, and meningocoele. Vitamin A administered to hamsters by gastric lavage just prior to crest cell migration (Marin-Padilla and Ferm, 1965; Marin-Padilla, 1966) resulted in exencephaly but not cleft palate. Examination of published illustrations (Marin-Padilla, 1966), as well as examination of histological preparations of specimens fixed 10 or 24 hr following vitamin administration

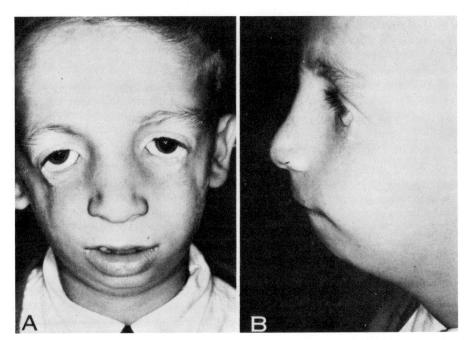

Fig. 8. Treacher Collins syndrome (mandibulofacial dysostosis) (From Ross and Johnston, 1972.) This boy shows typical features of the syndrome including underdevelopment of the zygomatic and mandibular bones and notching of the lower eyelid. Abnormalities of the middle and external ear and isolated cleft palate are also present.

(courtesy, Marin-Padilla), indicated that while there were decreased numbers of migrating crest cells (10-hr specimens) only the ganglia showed appreciable alterations (massive cell death) at 24 hr. Whether the ganglionic cell death is causally related to the exencephaly is not clear (see Johnston, 1975).

It is uncertain whether vitamin A affects migrating cells through a direct effect on the cells or by altering the environment in which the cells migrate. Vitamin A is known to cause increased membrane activity both *in vivo* (Morriss, 1973) and *in vitro* (Daniel *et al.*, 1966; Morriss, 1976). A direct effect is suggested by time-lapse cinemicrographic studies in which migrating primary mesenchyme cells were shown to undergo extensive cell surface blebbing upon exposure to very low doses of vitamin A (Morriss, 1976). In another system vitamin A is known to alter glycoprotein synthesis (Hassell *et al.*, 1975) which might affect migrating cells either directly or indirectly by altering matrix production. Effects on carbohydrate components of extracellular matrices have also been observed (Fell *et al.*, 1962; Solursh and Meier, 1973), although there are no known effects on the production of matrix components usually associated with cell migration, i.e., hyaluronic acid (Toole, 1973; Pratt *et al.*, 1975) and, possibly, collagen (Hay, 1968; Steffek *et al.*, 1975).

A number of spontaneous defects in both animals and man closely resemble the malformations induced by vitamin A. Facial malformations similar to Treacher Collins syndrome are found in mice homozygous for the recessive *oel* gene and exencephaly also occurs in the same strain (Brown and Harne, 1974). The *oel* strain was derived from phocomelic mice (Gluecksohn-Waelch, 1961) which have, in addition to Treacher Collins-like malformations, abnormalities in cartilage development (Fitch, 1957) closely resembling those produced by certain regimens of vitamin A treatment (Kochhar, 1973; also Chapter 18).

Otocephaly ("ear-head," Figure 9A) occurs in man (e.g., Duhamel, 1966) and in experimental animals (Wright and Wagner, 1934). As in the case of the Treacher Collins syndrome, only lower facial structures are affected in the less severe cases of otocephaly (Figure 9B). In the more severe cases, all the crest derivatives and structures they support, such as the eye and forebrain, are absent and the resulting malformation is, literally, an "ear-head" (Figure 9A). Unlike the Treacher Collins syndrome and vitamin A-induced malformations, ear involvement is minimal.

Since the least manifestation of the otocephaly syndrome (defects of the anterior mandibular region) involves structures derived from crest cells with long migration paths, it seems reasonable that these cells would be most affected if interference with migration were the pathogenic mechanism. The more severely affected cases may be cyclopic in both guinea pigs and man, suggesting that there may also be defects (migratory?) involving the more anterior head mesoderm. Cyclopia is also found in mouse embryos treated

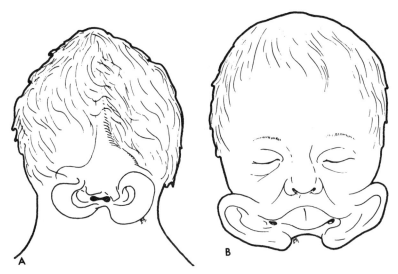

Fig. 9. Gradation of otocephaly in man. (From Duhamel, 1966.) In the more complete form (A), it appears that the crest derivatives and structures they support, such as the eye and portions of the forebrain, are absent.

with vitamin A (Giroud and Martinet, 1956). Progressive facial narrowing with increasing severity of the otocephalic defect (Wright and Wagner, 1934) may compensate for reduced facial mesenchyme and account for the absence of cleft palate.

C. Defects Related to Facial Process and Visceral Arch Development

It is possible to interfere directly with facial process development by a number of teratological procedures. In some cases, such as procedures which cause cell death or retard proliferation, the pathogenic mechanisms are relatively clear. In other cases, such as late interference by vitamin A or procedures which result in edema or vascular hemorrhage, the mechanisms are not so clear. It has been postulated (Hassell *et al.*, 1977) that the principle action of vitamin A administered at these stages is again on cell migration. As indicated earlier, facial process formation is very poorly understood, and a much better understanding of the underlying mechanisms will be necessary before peculiar associations of malformations (e.g., Treacher Collins-like facies with limb defects; Cohen, 1976) can be elucidated.

1. Cell Death and Interference with Cell Proliferation

Antifolates and 6-aminonicotinamide are particularly potent teratogens when applied during the period of facial process development (Chapter 11, this volume). In the rat, intraperitoneal injection of X-methyl pteroyl glutamic acid (a folic acid analog, Nelson *et al.*, 1955) or hadacidin (thought to have antifolate activity, Chaube and Murphy, 1963) results in clefts of the lip or palate or more severe deficiencies of frontonasal process derivatives. Using a regimen similar to that of Chaube and Murphy, Lejour (1970) conducted a histological study of rat embryos shortly after administration of hadicidin and observed cell death in the nasal placodes and adjacent mesenchyme as well as in forebrain structures (see also, Lejour-Jeanty, 1966). 6-Aminonicotinamide (an analog of nicotinic acid) produced either median cleft lip or lateral (typical) cleft lip when injected intraperitoneally in C57B1 mice two days or one day before primary palate formation, respectively (Trasler and Leong, 1974). Decreased cell proliferation, but not cell death, was observed (Trasler, personal communication).

2. Hemorrhage

Defects of both the primary palate and visceral arches can be produced experimentally in which the pathogenic mechanism appears to involve vascular edema and/or hemorrhage. Both aspirin (Larsson *et al.*, 1963) and dilantin (Hetzel and Brown, 1975) cause hemorrhages in the lip region. Both agents are known to cause clefts of the lip when injected prior to primary palate formation (Trasler, 1965; Harbison and Becker, 1969; Hetzel and Brown, 1975). Morphological changes observed in infants born to mothers on dilantin therapy (Jones and Smith, 1975) suggest embryonic edema. Thalidomide also causes vascular hemorrhage (see below) and is known to be capable to producing cleft lip in the ferret (Steffek *et al.*, 1971).

Poswillo (1973) has observed that menorrhages occur at the site of fusion between the external carotid and stapedial arteries (Figure 10) after injection of triazene (an antifolate) in mice or administration of thalidomide by gastric lavage in primates. The subsequent developmental alterations lead to malformations in the mouse closely resembling those of hemifacial microsomia in man (Ross, 1975) (Figure 11). Although less completely studied, the initial alterations following thalidomide administration in the primate appear very similar. Defects in the ear region that sometimes form a component of the human thalidomide syndrome resemble those of hemifacial microsomia (Kleinsaesser and Schlothans, 1965; Livingstone, 1965).

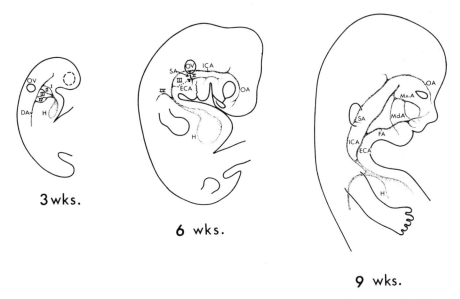

3 wks.

6 wks.

9 wks.

Fig. 10. Three major stages of vascular development in the human embryo. (Modified from Ross and Johnston, 1972.) Various structures are designated as follows: otic vesicle (OV), aortic arch vessels and their derivatives (I–IV), dorsal aorta (DA), heart (H), external carotid (ECA), stapedial (SA), internal carotid (ICA), ophthalmic (OA), maxillary (Mx.A), mandibular (Md.A), and facial (FA) arteries. The external carotid (ECA) and stapedial (SA) arteries fuse (see arrow in 6-week specimen), and it is at this point that hemorrhage can be induced in experimental animals (see text).

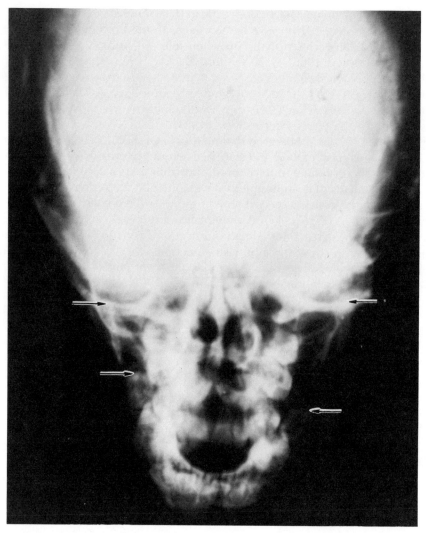

Fig. 11. Hemifacial microsomia (lateral facial dysplasia) is a regional defect involving primarily the ear and temporomandibular joint region. In this particular patient unilateral deficient development of the mandibular ramus (between arrows) has resulted in considerable facial asymmetry. (Photograph, courtesy of R. B. Ross.)

3. Rare Facial Clefts and Amniotic Adhesions

Failure of fusion between facial processes (other than that resulting in the common forms of lip and palate clefts) results in a wide variety of facial clefts, although taken together they are much less common than the usual clefts of the lip and/or palate. Oblique facial clefts between maxillary processes and medial and lateral nasal processes can be produced experimentally by extirpating crest cells in the chick (Johnston, 1965). Presumably, failure of the medial nasal processes to fuse in the midline (Figure 4) leads to the formation of so-called median facial clefts ("severe orbital hypertelorism" or "frontonasal dysplasia," Figure 12). Frequently, there is complete absence of skeletal structures in the midline area and the meninges are located directly under the subcutaneous tissues (Sedano *et al.,* 1970). This abnormality is not associated with any tissue deficiency.

There is evidence (e.g., Jones *et al.,* 1974) that amniotic band adhesions may be involved in the genesis of facial clefts that do not coincide with fusion lines. We have observed amniotic adhesions to neural fold grafts in chick embryos which suggest that tissue damage or breakdown could also be the initiating factor in such adhesions.

D. Other Abnormalities of Facial Development

Defective differentiation of specific cell populations may be the cause of certain facial anomalies. For instance abnormal neural crest cell differentiation may result in defective pigmentation and craniofacial abnormalities in both man and experimental animals. In the Waardenberg (1951) syndrome there is often a patchy or complete absence of pigmentation of the iris, hair, and skin; a variable (in different racial groups, etc.) frequency of deafness and; occasionally, cleft lip and palate (see also, Brown and Podosin, 1966; Brown *et al.,* 1971; Tsafir, 1974). In one mouse model, dancer (Lane, 1955),

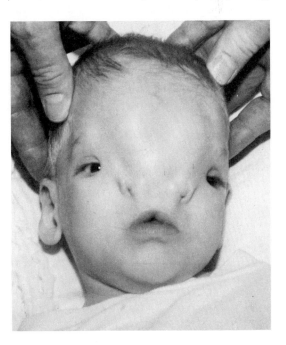

Fig. 12.   Child demonstrating failure of upper facial structures to unite in the midline. (From Sedano *et al.,* 1970.) This malformation has been given several different names including "median facial cleft," "severe orbital hypertelorism," and "frontonasal dysplasia." In this particular case, forebrain structures with their meningeal coverings are located immediately under the skin of the forehead region.

the heterozygote has defects of the vestibular system and a white spot on the forehead while the abnormal homozygote has cleft lip and palate. Deol and Lane (1966) have observed that the abnormal homozygote can be histologically distinguished in early embryonic development by unusually small vestibular ganglia. Crest cells form the supporting (satellite and glial) cells of these ganglia and, presumably, pigment cells associated with the stria vascularis and epithelia adjacent to the vestibular sensory epithelia. These pigment cells are frequently absent in dancer and other mutants (Deol, 1970). Both the functions of the pigment cells and the significance of their absence in relation to other aspects of inner ear pathology are unknown.

In another mouse mutant, pallid (Erway et al., 1971), there appears to be a defect in melanosomes, resulting in dilute coat color. There is also a defect in the mineralization of otoliths in these mice which can be overcome by dietary manganese supplementation (Erway et al., 1971). Pigment cells containing defective melanosomes are closely apposed to cells with secretory characteristics (Limm, 1973), and Erway (personal communication) has suggested that they are in some way responsible for the otolith mineralization defect. Defective crest cell differentiation is also a major factor in the pathogenesis of the recessive familial facial osteodysplasia (Anderson et al., 1972) in which neural crest-derived bone is severely deficient.

Finally, premature closure of the sutures joining facial and cranial bones provides an example of a very late alteration affecting facial morphology. While the "force" necessary for facial growth may be resident in the expansive growth of cartilage and soft tissues, there is now much evidence that open craniofacial interosseous sutures are necessary to permit growth adjustments between bones. For example, scar tissues, secondary to cleft palate surgery, across sutures playing a role in midfacial growth may result in severe growth retardation (e.g., Ross, 1976). Premature suture closure can be produced experimentally in rabbits by hypervitaminosis D (Friedman and Mills, 1969).

In two human conditions, the Crouzon syndrome (craniofacial dysostosis, Figure 13) and the Apert syndrome (acrocephalosyndactyly), premature suture closure is at least partially responsible for severe retardation of subsequent cranial and facial development. More normal growth of the cranial vault can be achieved by artifically creating "sutures" through removal of strips of bone. In the Crouzon syndrome all the malformations appear to be secondary to the premature closure of sutures, whereas in the Apert syndrome the presence of malformations of the hand and clefts of the palate suggest a more complex underlying mechanism.

The material presented in this chapter largely reflects the interests and experience of the authors. Material omitted includes studies conducted on tastebud formation (Farbman, 1972) and regeneration (Zalewski, 1972), as well as studies on normal (e.g., Spooner and Wessels, 1972) and abnormal (Dawe et al., 1968) salivary gland development. Also, of course, many studies have been conducted on tooth formation, both normal (Kock, 1972; Slavkin, 1974) and abnormal (Sofaer et al., 1971; Berry and Germain, 1972; Gorlin et

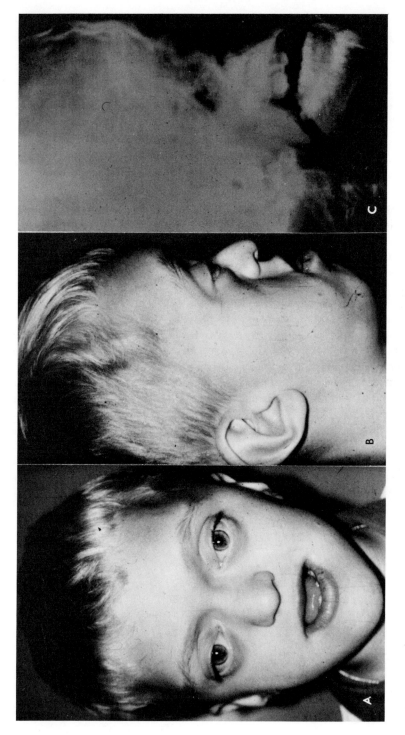

Fig. 13. Crouzon's syndrome. (Courtesy, R. B. Ross.) Premature closure of craniofacial sutures results in retarded growth of cranial and facial sutures (see text). Increased intracranial pressure is thought to contribute to the exophthalmia.

al., 1976). The above references provide leads into the sometimes extensive literature on these subjects.

III. SUMMARY AND CONCLUSIONS

This chapter has been concerned with current concepts of craniofacial development. Early events establishing the basic embryonic architecture preceding facial development were described briefly. These events result in the formation of the several cell populations involved in facial morphogenesis. Facial development itself involves complex patterns of cell migration and interactions between different populations of cells and between cells and their cellular matrix. Unique features of facial development, including the extensive role of neural crest cells and the development of facial processes, were given considerable emphasis.

The very complexity of facial morphogenesis makes this region particularly prone to abnormalities. The fact that many such abnormalities are nonlethal may also be an important factor in their high incidence.

Studies on abnormal development must proceed alongside studies on normal developmental mechanisms. The approaches described in this chapter include experimental interference, through surgical manipulation and the use of teratogenic agents, and the application of our knowledge of normal development in assessing the possible origin of malformations where little or no direct evidence of their pathogenesis is available. The latter approach has proved valuable, for instance, with cyclopia. It is also relevant in those situations where a single end point may be arrived at through more than one pathway, as is the case for cleft lip and palate, exencephaly, and many other malformations.

Finally, it was shown that the study of both induced and spontaneous abnormalities, including those occurring in man, may provide insights into normal developmental mechanisms.

REFERENCES

Adelmann, H. B., 1936, The problem of cyclopia I & II, *Q. Rev. Biol.* **11:**161, 284.
Anderson, L. G., Cook, A. J., Coccaro, P. J., Coro, C. J., and Bosma, J. F., 1972, Familial osteodysplasia, *J.A.M.A.* **220:**1687.
Badtke, G., Degenhardt, K. H., and Lund, O. E., 1959, Tierexperimentelle Beitrag fur Ätiology und Phänogenese kraniofacialer Displasien, *Z. Anat. Entwicklungs-gesch.* **121:**71–102.
Balinsky, B. I., 1975, *An Introduction to Embryology,* 4th ed., W. B. Saunders, Philadelphia.
Bartelmez, G. W., 1960, Neural crest in the forebrain of mammals, *Anat. Rec.* **138:**269.
Berry, C. L., and Germain, J., 1972, Absence of the third molar in the mouse: Its relevance ot human malformation, *J. Pathol.* **108:**35.
Bhakdinaronk, A., Eto, D., and Johnston, M. C., 1977, Cleft lip in mice: Interaction of genetic predisposition and X-irradiation (in preparation).

Binns, W., James, L. F., Keeler, R. F., and Balls, L. D., 1968, Effects of teratogenic agents in range plants, *Cancer Res.* **28**:2323.

Brown, K., and Harne, L. C., 1974, The genetics of two traits: Cleft palate and anencephalus in the oel strain of mice, *Teratology* **9**:14A.

Brown, J. W., and Podosin, R., 1966, A syndrome of the neural crest, *Arch. Neurol.* **15**:294.

Brown, K., Bergsma, D. R., and Barrow, M.V., 1971, Animal models of pigment and hearing abnormalities in man, *Birth Defects: Orig. Art. Ser.* **7**:102.

Chaube, S., and Murphy, M. L., 1963, Teratogenic effect of hadacidin (a new growth inhibitory chemical) on the rat fetus, *J. Exp. Zool.* **152**:67.

Cohen, M. M., 1975, An etiologic and nosologic overview of craniosynostosis syndromes, *Birth Defects Orig. Art. Ser.* **2**:182–189.

Cohen, M. M., 1976, Dysmorphic syndromes, *in: Craniofacial Genetics* (R. E. Steward and G. Prescott, eds.), C. V. Mosby, St. Louis.

Cohen, M. M., and Hohl, T. H., 1976, Etiologic heterogeneity in holoprosencephaly and facial dysmorphia with comments on the facial bones and cranial base, *in: 'Development of the Basicranium* (J. F. Bosma, ed.), U.S. Govt. Printing Office, Washington, D.C.

Cohen, M. M., Jirasek, J. E., Guzman, R. T., Gorlen, R. J., Patterson, M. Q., 1971, Holoprosencephaly and facial dysmorphia: Nosology, etiology and pathogenesis, *Birth Defects Orig. Art. Ser.* **7**:136–142.

Daniel, M. R., Dingle, J. T., Glauert, A. M. and Lucy, J. A., 1966, The action of excess vitamin A alcohol on the fine structure of rat dermal fibroblasts, *J. Cell Biol.* **30**:465.

Dawe, C. J., Morgan, W. D., and Slatick, H. S., 1968, Salivary gland neoplasms in the role of normal mesenchyme during salivary gland morphogenesis, *in: Epithelial–Mesenchymal Interactions* (R. Fleischmajer and R. E. Billingham, eds.), pp. 295–312, Williams and Wilkins, Baltimore.

DeMeyer, W., 1964a, Cleft lip and jaw induced in fetal rats by vincristine, *Arch. Anat.* **48**:179.

DeMeyer, W., 1964b, Vinblastine-induced malformation of face and nervous system in two rat strains, *Neurology* **14**: 806.

DeMeyer, W., Zeman, W., and Palmer, C. G., 1964, The face predicts the brain: Diagnostic significance of median facial anomalies for holoprosencephaly (arhinencephaly), *Pediatries* **34**:256.

Deol, M. S., 1970, The relationship between abnormalities of pigmentation and the inner ear, *Proc. R. Soc. London Ser. A* **175**:201.

Deol, M. S., and Lane, P. W., 1966, A new gene affecting the morphogenesis of the vestibular part of the inner ear in the mouse, *J. Embryol Exp. Morphol.* **16**:543.

Duhamel, B., 1966, *Morphogénèse Pathologique*, Masson et Cie, Paris.

Erway, L. C., Fraser, A. S., and Hurley, L. S., 1971, Prevention of congenital otolith defect in pallid mutant mice by manganese supplementation, *Genetics* **67**:97.

Evans, H. E., Ingalls, T. H., and Binns, W., 1966, Teratogenesis of craniofacial malformations in animals, III, Natural and experimental cephalic deformities in sheep, *Arch. Environ. Health* **13**:706.

Farbman, A. I., 1972, The taste bud: A model system for developmental studies, *in: Developmental Aspects of Oral Biology* (H. S. Slavkin and L. A. Bavetta, eds.), pp. 109–123, Academic Press, New York.

Fell, H. B., Dingle, J. T., and Webb, M., 1962, Studies on the mode of action of excess of vitamin A. 4. The specificity of the effect on embryonic chick-limb cartilage in culture and on isolated rat liver lysosomes, *Biochem. J.* **83**:63.

Ferm, V. F., 1963a, Congenital malformations in hamster embryos after treatment with vinblastine and vincristine, *Science* **141**:426.

Ferm, V. F., 1963b, Colchicine teratogenesis in the hamster embryo, *Proc. Soc. Exp. Biol. Med.* **112**:775.

Ferm, V. F., and Carpenter, S. J., 1968, The relationship between cadmium and zinc in mammalian teratogenesis, *Lab. Invest.* **18**:429.

Fitch, N., 1957, An embryological analysis of two mutants in the house mouse, both producing cleft palate, *J. Exp. Zool.* **136:**329.

Friedman, W. F., and Mills, L. F., 1969, The relationship between vitamin D and the craniofacial dental anomalies of the supravalvular aortic stenosis syndrome, *Pediatrics* **43:**12.

Giroud, A., Delmas, A., and Martinet, M., 1963, Cyclocephalie: Morphogénèse et mécanisme de la production, *Arch. Anat. Histol. Embryol.* **47:**293–311.

Giroud, A., and Martinet, M., 1956, Malformations de la face et hypervitaminoses A., *Rev. Stomotol.* **57:**454.

Glucksohn-Waelsch, S., 1961, *Mouse News Lett.* **25:**12.

Gorlin, R. J., 1970, Developmental anomalities of the face and oral structures, *in: Thomas' Oral Pathology* 6th ed. (R. J. Gorlin and H. M. Goldman, eds.), Vol. 1, pp. 21–95, C. V. Mosby, St. Louis.

Gorlin, R. J., Pindborg, J., and Cohen, M. M., 1976, *Syndromes of the Head and Neck,* McGraw-Hill, New York.

Hanson, J. W., and Smith, D. W., 1975, The fetal hydantoin syndrome, *Teratology* **11:**20A.

Harbison, R. B., and Becker, B. A., 1969, Relation of dosage and time of administration of diphenylhydantoin to its teratogenic effect in mice, *Teratology* **2:**305.

Hassell, J. R., Jones, C. S., and DeLuca, L. M., 1975, Stimulation of rat liver glycopetide synthesis by vitamin A, *J. Cell Biol.* **67**(2, Pt. 2):159a.

Hassell, J. R., Pratt, R. M., and King, C. T. G., 1977, Growth of the secondary palate during normal and abnormal development (in preparation).

Hay, E. D., 1968, Organization and fine structure of epithelium and mesenchyme in the developing chick embryo, *in: Epithelial–Mesenchymal Interactions* (R. Fleischmajer and R. E. Billingham, eds.), pp. 31–55, Williams and Wilkins, Baltimore.

Hazelton, R. D., 1970, A radioautographic analysis of the migration and fates of the occipital somites of the chick embryo with specific reference to the hypoglossal musculature, *J. Embryol. Exp. Morphol.* **24:**455.

Hazelton, R. D., Johnston, M. C., and Larsen, M. A., 1977, Migration and fate of cephalic paraxial mesoderm in the avian embryo (in preparation).

Hetzel, S., and Brown, K. S., 1975, Facial clefts and lip hematoma in mouse fetuses given diphenyl-hydantoin, *J. Dent. Res.* **54A:**83.

Holtfreter, J., 1968, Mesenchyme and epithelia in inductive and morphogenetic processes, *in: Epithelial–Mesenchymal Interactions* (R. Fleischmajer and R. E. Billingham, eds.), pp. 1–30, Williams and Wilkins, Baltimore.

Hörstadius, S., 1950, *The Neural Crest,* Oxford Univ. Press, Oxford, England.

Ingalls, T. H., 1966, Teratogenesis of craniofacial malformations in animals, V. Cyclopian malformations in fish, mice and sheep: A study in comparative epidemiology, *Arch. Environ. Health* **13:**719–725.

Jacobson, A. G., April, 1966, Inductive processes in development, *Science* **152:**25.

Johnston, M. C., 1964, Facial malformations in chick embryos resulting from removal of neural crest, *J. Dent. Res.* **43:**822 (abstract).

Johnston, M. C., 1965, The neural crest in vertebrate cephalogenesis: A study of the migrations and derivatives of cranial neural crest and related cells in the embryos of amphibians, birds and mammals, Ph.D. thesis, Univ. of Rochester, New York.

Johnston, M. C., 1966, A radioautographic study of the migration and fate of the cranial neural crest cells in the chick embryo, *Anat. Rec.* **156:**143.

Johnston, M. C., 1975, The neural crest in abnormalities of the face and brain, *in: Morphogenesis and Malformations of the Face and Brain* (D. Bergsma, ed.), pp. 1–18, Alan R. Lis, New York.

Johnston, M. C., 1976, Development of the face and oral cavity, *in: Orban's Oral Histology and Embryology,* 9th ed. (S. N. Bhaskar, ed.) Mosby, St. Louis.

Johnston, M. C., and Listgarten, M. A., 1972, The migrations, interactions, and early differentiation of oro-facial tissues, *in: Developmental Aspects of Oral Biology* (H. S. Slavkin and L. A. Bavetta, eds.), pp. 53–80, Academic Press, New York.

Johnston, M. C., Bhakdinaronk, A., and Reid, Y. C., 1974, An expanded role for the neural crest in oral and pharyngeal development, *in: Fourth Symposium on Oral Sensation and Perception* (J. F. Bosma, ed.), pp. 37–52, U.S. Govt. Printing Office, Washington, D.C.

Jones, K. L., and Smith, D. W., 1975, The Williams elfin facies syndrome, *J. Pediatr.* **86:**718.

Jones, K. L., Smith, D. W., Hall, B. D., Hall, J. G., Ibbin, A. J., Massoud, H., and Globus, M. S., 1974, A pattern of craniofacial and limb defects secondary to aberrant tissue bands, *J. Pediatr.* **84:**90.

Källen, B., 1956, Contribution to the knowledge of the regulation of the proliferation processes in the vertebrate brain during ontogenesis, *Acta Anat.* **27:**351.

Kleinsasser, Von, O., and Schlothane, R., 1964, Die Ohrmissbildung im Rahmen der Thalidomid-Embryopathie, *Z. Laryngol., Rhinol., Otol.* **43:**344.

Koch, W. E., 1972, Tissue interaction during *in vitro* odontogenesis *in: Developmental Aspects of Oral Biology* (H. S. Slavkin and L. A. Bavetta, eds.), pp. 151–164, Academic Press, New York.

Kochhar, D. M., 1973, Limb development in mouse embryos I. Analysis of teratogenic effects of retinoic acid, *Teratology* **7:**289.

Kushner, D. C., King, C. T. G., and Johnston, M. C., 1977, Cadmium teratogenesis in the rat (in preparation).

Lane, P. W., 1955, *Mouse News Lett.* **19:**25.

Landauer, W., 1956, Cyclopia and related defects as a lethal mutation of fowl, *J. Genet.* **54:**219.

Langman, J., 1975, *An Introduction to Medical Embryology,* 3rd ed., Williams and Wilkins, Baltimore.

Larsson, K. S., Böstrom, H., and Ericson, B., 1963, Salycilate induced malformations in mouse embryos, *Acta Paediatr. Scand.* **52:**36.

Latham, R. A., 1971, Mechanism of maxillary growth in the human cyclops, *J. Dent. Res.* **50:**929.

LeDouarin, N., 1973, A biological cell labeling technique and its use in experimental embryology, *Dev. Biol.* **30:**217.

LeLievre, C. S., and LeDouarin, N. M., 1975, Mesenchymal derivatives of the neural crest: Analysis of chimaeric quail and chick embryos, *J. Embryol. Exp. Morphol.* **34:**125.

Lejour, M., 1970, Cleft lip induced in the rat, *Cleft Palate J.* **7:**169.

Lejour-Jeanty, M., 1966, Bécs de lièvre provoqués chez le rat par un derivé de la penicilline, l'Hadacidin, *J. Embryol. Exp. Morphol.* **15:**193.

Limm, D. J., 1973, Formation and fate of the otoconia: Scanning and transmission elecgron microscopy, *Ann. Otol. Rhinol. Laryngol.* **83:**23.

Livingstone, G., 1965, Congenital ear abnormalities due to thalidomide, *Proc. R. Soc. Med.* **58:**493.

Marin-Padilla, M., 1966, Mesodermal alterations induced by hypervitaminosis A., *J. Embryol. Exp. Morphol.* **15:**261.

Marin-Padilla, M., and Ferm, V. H., 1965, Somite necrosis and developmental malformations induced by vitamin A in the golden hamster, *J. Embryol. Exp. Morphol.* **13:**1.

Marin-Padilla, M., Hoefnagel, D., and Benirschke, K., 1964, Anatomic and histopathologic study of two cases of D_1 (13–15) trisomy, *Cytogenetics* **3:**258.

Milaire, J., 1959, Prédifferenciation cytochemique des divers èbauches céphaliques chez l'embryon des souris, *Arch. Biol.* **70:**588.

Morriss, G. M., 1972, Morphogenesis of the malformations induced in rat embryos by maternal hypervitaminosis A., *J. Anat. (London)* **113:**241.

Morriss, G. M., 1973, The ultrastructural effects of excess maternal vitamin A on the primitive streak stage rat embryo, *J. Embryol. Exp. Morphol.* **24:**219.

Morriss, G. M., 1976, Abnormal cell migration as a possible factor in the genesis of vitamin A-induced craniofacial anomalies, *in: New Approaches to the Evaluation of Abnormal Mammalian Embryonic Development* (D. Neubert, ed.), Second Symposium on Mammalian Prenatal Development, Georg Thieme Verlag, Berlin.

Mulvihill, J. E., Gamin, S. H., and Ferm, V. H., 1970, Facial formation in normal and cadmium-treated golden hamsters, *J. Embryol. Exp. Morphol.* **24:**393.

Murakami, V., Kameyama, Y., Majima, A., and Sakurai, T., 1961, Patterns of radiation malformations of the mouse fetus and subjected stage of development, *Ann. Rep. Res. Inst. Environ. Med.* **9:**71.

Nelson, M. M., Wright, H. V., Asling, C. W., and Evans, H. M., 1955, Multiple congenital abnormalities resulting from transitory deficiency of pteroylglutamic acid during gestation in the rat, *J. Nutr.* **56**:349.

Noden, D. W., 1975, An analysis of migratory behaviour of avian cephalic neural crest cells, *Dev. Biol.* **42**:106.

Poswillo, D., 1973, The pathogenesis of the first and second branchial arch syndrome, *Oral Surg.* **35**:302.

Poswillo, D., 1975, The pathogenesis of the Treacher-Collins syndrome (mandibulo-facial dysostosis), *Brt. J. Oral Surg.* **13**:1.

Pratt, R. M., Johnston, M. C., and Larsen, M. A., 1975, Migration of cranial neural crest cells in cell-free hyaluronate-rich matrix, *Dev. Biol.* **44**:298.

Raven, C. P., and Kloos, J., 1945, Induction by medial and lateral pieces of the archenteron roof with special reference to the determination of the neural crest, *Acta. Neurol. Morphol.* **4**:348.

Ritter, W., and Degenhardt, K. H., 1963, Clefts of the lips, jaws and palate induced in mice by means of x-rays, *Int. Dent. J.* **13**:489.

Rogers, B. O., 1964, Rare craniofacial malformations, in: *Reconstructive Plastic Surgery* (J. M. Converse, ed.), Vol. III, p. 1213, W. B. Saunders, Philadelphia.

Rogers, K. T., 1964a, Radioautographic analysis of the incorporation of protein and nucleic acid precursors into various tissues of early chick embryos cultured *in toto* on medium containing LiCl, *Dev. Biol.* **9**:176.

Rogers, K. T., 1964b, Experimental production of perfect cyclopia by removal of the telencephalon and reversal of bilateralization in somite-stage chicks, *Am. J. Anat.* **115**:487.

Ross, R. B., 1975, Lateral facial dysplasia, in: *Morphogenesis and Malformations of the Face and Brain* (D. Bergsma, ed.), pp. 51–59, Alan R. Lis, New York.

Ross, R. B., 1976, Facial growth in cleft lip and palate, in: *Reconstructive Plastic Surgery* (J. M. Converse, ed.), W. B. Saunders, Philadelphia (in press).

Ross, R. B., and Johnston, M. C., 1972, *Cleft Lip and Palate*, Williams & Wilkins, Baltimore.

Schowing, J., 1965, Obtention expérimentale de cyclopes parfaits par microchirrurgie chez l'embryon de poulet, étude histologique, *J. Embryol Exp. Morphol.* **14**:255.

Sedano, H. O., and Gorlin, R. J., 1963, The oral manifestation of cyclopia, *Oral Surg., Oral Med., Oral Pathol.* **16**:823.

Sedano, H. O., Cohen, M. C., Jirasek, J., and Gorlin, R. J., 1970, Frontonasal dysplasias, *J. Pediatr.* **76**:906.

Slavkin, H. C., 1974, Embryonic tooth formation, *Oral Sci. Rev.* **4**:1.

Smith, D. W., 1970, *Recognizable Patterns of Human Malformation*, W. B. Saunders, Philadelhpia.

Sofaer, J. A., Chung, C. S., Niswander, J. D., and Runck, D. W., 1971, Developmental interaction, size and agenesis among permanent maxillary incisors, *Hum. Biol.* **43**:36–45.

Solursh, M., and Meier, S., 1973, The selective inhibition of mucopolysaccharide synthesis by vitamin A treatment of cultured chick embryo chondrocytes, *Calcif. Tissue Res.* **13**:131.

Sperber, G. H., 1973, *Craniofacial Embryology*, John Wright and Sons, Bristol, U.K.

Spooner, B. S., and Wessells, N. K., 1972, An analysis of salivary gland morphogenesis: Role of cytoplasmic microfilaments and microtubules, *Dev. Biol.* **27**:38–54.

Steffek, A. J., Verrusio, A. C., and Johnston, M. C., 1971, Normal and abnormal oral-facial development in ferrets, *J. Dent. Res.* **50**(Suppl.):97.

Steffek, A. J., Mujwid, D. K., and Johnston, M. C., 1975, Scanning electron microscopy (SEM) of cranial neural crest migration, *J. Dent. Res.* **54**(Suppl. A):165.

Stockard, C. R., 1909, The development of artifically produced cyclopian fish—"the magnesium embryo," *J. Exp. Zool.* **6**:285–339.

Toivonen, S., 1967, Mechanism of primary embryonic induction, in: *Primary Embryonic Induction*, Vol. I, (E. Hagen, W. Wechsler, and F. Zilliken, eds.), p. 1, S. Karger, New York.

Toole, B. P., 1973, Hyaluronate and hyaluronidase in morphogenesis and differentiation, *Am. Zool.* **13**:1961.

Trasler, D. G., 1965, Aspirin-induced cleft lip and other malformations in mice, *Lancet* **1**:606.

Trasler, D. G., and Leong, S., 1974, Face shape and mitotic index in mice with 6-aminonicotinomide-induced and inherited cleft lip, *Teratology* **9**:39A.

Tsafir, J. S., 1974, *Light-eyed Negroes and the Klein-Waardenberg Syndrome,* Macmillan, London.

Waardenberg, P. J., 1951, A new syndrome combining developmental anomalies of the eyelids, eyebrows and nose root, with pigmentary deficiencies of the iris and head hair and with congenital deafness, *Am. J. Hum. Genet.* **3**:195.

Weston, J. A., 1970, The migration and differentiation of neural crest cells, *Adv. Morphog.* **8**:41.

Wilson, J. G., Jordan, H. C., and Brent, R. L., 1953, Effects of irradiation on embryonic develop-, ment II. X-rays on the ninth day of gestation in the rat, *Am. J. Anat.* **92**:153.

Wood-Jones, F., and Wen, I. C., 1934, The development of the external ear, *J. Anat. (London)* **68**:525.

Wright, B., and Wagner, K., 1934, Types of subnormal development of the head from inbred strains of guinea pigs and their bearing on the classification and interpretation of vertebrate monsters, *Am. J. Anat.* **54**:383.

Yntema, C. L., 1955, Ear and nose, *in: Analysis of Development* (B. H. Willier, P. A. Weiss, and V. Hamburger, eds.), pp. 415–428, W. B. Saunders, Philadelphia.

Zalewski, A. A., 1972, Regeneration of taste buds after transplantation of tongue and ganglia grafts to the anterior chamber of the eye, *Exp. Neurol.* **35**:519–528.

Abnormal Organogenesis in the Limbs 18

D. M. KOCHHAR

I. INTRODUCTION

The embryonic limb is a favorite tool of developmental biologists and teratologists. Not only does it serve as a model for highly interdependent developmental events in the embryo as a whole, it provides an excellent example wherein slight deviations in ontogeny are relatively easy to observe and quantify. In this chapter some recent advances in the knowledge of normal limb development in vertebrate embryos will be summarized. The discussion is mostly confined to developmental stages ranging from the initial appearance of a hillock of cells as an outgrowth from the embryonic flank through its elongation and formation of a pentadactyl appendage approximating the adult spatial pattern of cartilage and muscle. Since normal developmental events of the limb are better known in the chick embryo, these are correlated with embryonic stages of mammalian species at which corresponding events are thought to occur. A brief discussion of abnormal organogenesis caused by genetic and environmental factors follows and is illustrated by those few examples for which some information about the initial pathogenesis of the defect is available.

II. NORMAL DEVELOPMENT

The future limbs arise as outgrowths from the embryonic flank and consist of a core of somatopleuric mesoderm covered by ectoderm. The limb bud

D. M. KOCHHAR • Department of Anatomy, School of Medicine, University of Virginia, Charlottesville, Virginia 22901. Present address: Department of Anatomy, Jefferson Medical College, Philadelphia, Pennsylvania 19107.

outgrowth then increases its mass through cell multiplication in such a manner that its dimensions increase only in the proximal–distal direction, resulting in elongation rather than expansion of the organ. When the bud has achieved a certain length, its mesodermal cells, which were structurally and functionally alike, begin differentiation toward a pentadactyl pattern of cartilages and bones: a single long bone proximally, two long bones side by side in the distal part, and various elements of five digits most distally (Figure 1) (Ede, 1972). This general pattern of limb ontogeny is so similar in all land vertebrates that a brief description of developmental events in one species, combined with some knowledge of timing and sequence of these events in other species, will suffice as background for a discussion of abnormal organogenesis (Figure 2).

In humans the upper limb buds make their first appearance when embryos are 26 days old (estimated postovulatory age), at the end of stage 12 (O'Rahilly and Gardner, 1975). At this time the embryos have roughly 28 pairs of somites and measure about 4–5 mm in crown–rump length (CRL). The upper limb bud appears opposite somites 8–12 and is already present when the lower limb buds make their appearance approximately two days later, at stage 13, opposite somites 24–29 (Figure 3).

In the chick embryo the limb buds are first observed as swellings at stage 17, after 2 1/2 days of incubation (Hamburger and Hamilton, 1951), while in ICR/DUB mouse embryos they make their appearance during the 10th day of gestation (plug day = 1st day of gestation). Figure 4A shows a forelimb bud of a 30-somite mouse embryo, 11th day of gestation, which is identical in general anatomic features to the stage 19 chick limb bud (Figure 5) and stage 13 human limb bud (Figure 6). The following account of mouse limb develop-

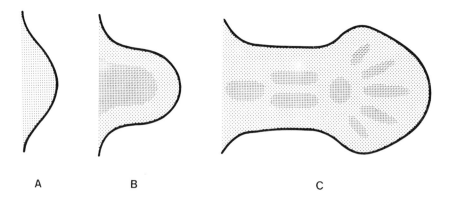

A   B   C

Fig. 1. Development of a generalized vertebrate limb bud. A: Early stage without precartilage condensation; B: elongation stage with central precartilage condensation; C: Later stage showing pentadactyl pattern of precartilage condensations. (Reproduced by permission from Ede, 1972.)

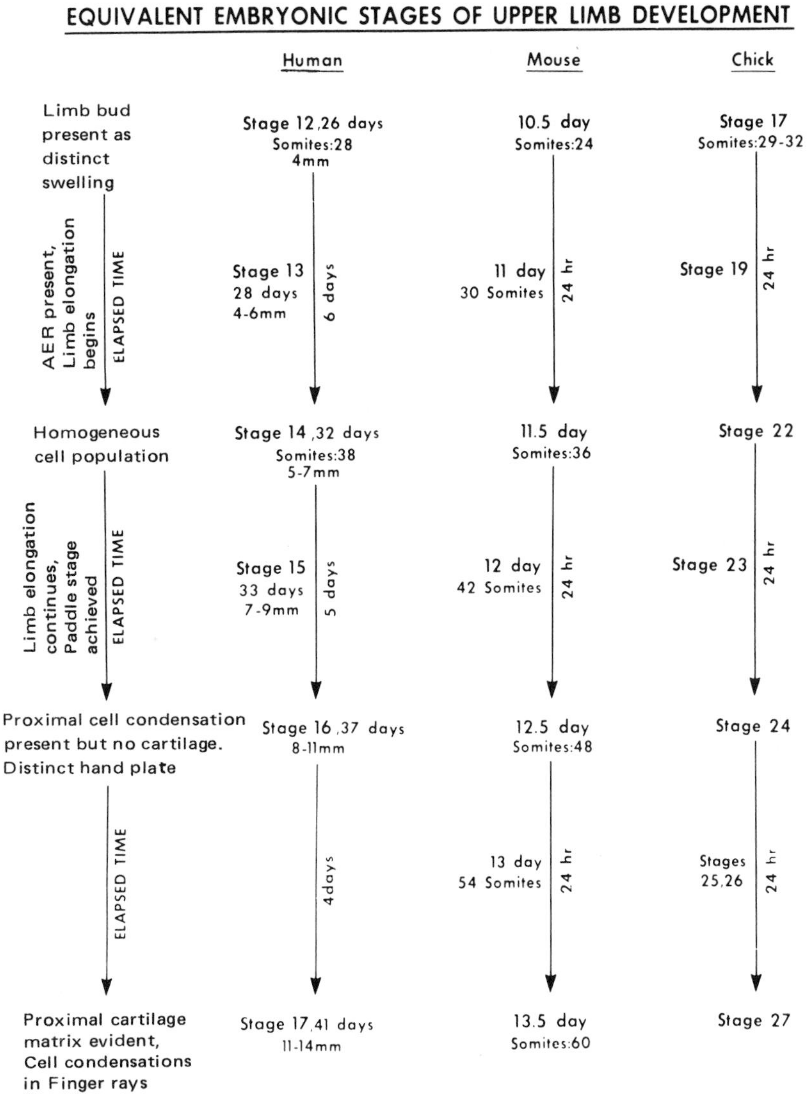

EQUIVALENT EMBRYONIC STAGES OF UPPER LIMB DEVELOPMENT

	Human	Mouse	Chick
Limb bud present as distinct swelling	Stage 12,26 days Somites:28 4mm	10.5 day Somites:24	Stage 17 Somites:29-32
AER present, Limb elongation begins / ELAPSED TIME	Stage 13 28 days 4-6mm / 6 days	11 day 30 Somites / 24 hr	Stage 19 / 24 hr
Homogeneous cell population	Stage 14,32 days Somites:38 5-7mm	11.5 day Somites:36	Stage 22
Limb elongation continues, Paddle stage achieved / ELAPSED TIME	Stage 15 33 days 7-9mm / 5 days	12 day 42 Somites / 24 hr	Stage 23 / 24 hr
Proximal cell condensation present but no cartilage. Distinct hand plate	Stage 16,37 days 8-11mm	12.5 day Somites:48	Stage 24
ELAPSED TIME	/ 4 days	13 day 54 Somites / 24 hr	Stages 25,26 / 24 hr
Proximal cartilage matrix evident, Cell condensations in Finger rays	Stage 17,41 days 11-14mm	13.5 day Somites:60	Stage 27

Fig. 2. Comparative embryonic stages at which early developmental events are known to occur in human, mouse, and chick forelimbs. Based on descriptions by O'Rahilly and Gardner (1975), Kochhar (1976a), and Searls (1973).

ment from this stage onward is identical in essential details to that in the other species except for differences in the developmental rates.

The limb bud doubles its length during a period of 12 hr between 11 and 11.5 days, i.e., from 30 to 36 somite stage (1 somite is added every two hours in the mouse, Kochhar and Aydelotte, 1974). During this time the mesenchyme of the bud is cytologically homogeneous, but the overlying ectoderm

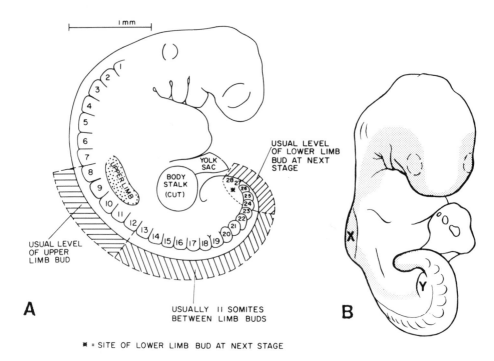

Fig. 3. A: Right lateral view of a stage 12 human embryo (No. 5923, 4 mm, 28 pairs of somites) to show the relationship of the beginning limb buds to specific somites. B: Drawing of a stage 13 (No. 10307, 4.2 mm, 30 pairs of somites) embryo from right ventrolateral aspect. X and Y indicate the right upper and lower limb buds, respectively. (Reproduced by permission from O'Rahilly and Gardner, 1975.).

from the beginning shows a thickened area at the apex of the bud called the apical ectodermal ridge (AER). Small blood vessels are present at the base of the bud at the 30-somite stage, and a marginal vein is seen under the AER at 33-somite stage. The mesenchyme remains uniformly distributed until the 36-somite stage when cells in the distal part of the bud begin to show a slightly less dense distribution than in other parts of the limb (Figure 4A–C).

The mesenchymal cell condensation first becomes evident at 39-somite stage in the proximal part of the bud (Figure 4D). Nerves are observed entering the peripheral area of this condensation at the 42-somite stage (beginning of 12th day of gestation), which implies that premyogenic cells are present peripheral to the area which will become cartilage, although as yet chondrogenic and myogenic cells can not be distinguished cytologically from each other (Figure 4E). Blood vessels course at the boundary between chondrogenic and myogenic zones. As elongation continues, the distal third of the bud acquires a paddle shape. Formation of the elbow or knee joint is foreshadowed by a bend in the bud, and indentations appear at the apex of the paddle on the 13th day, demarcating the digits (Figure 7).

The above-mentioned cell condensations in the proximal regions are initiated slightly before the appearance of the autopod (paddle—hand or foot). Further zones of condensed mesenchyme are observed in more distal regions of the limb on day 13 of development. Cartilage matrix exhibiting typical metachromatic staining with toluidine blue is observed in proximal cartilage late on day 13. It is followed by the appearance of similar metachromatic zones lo-

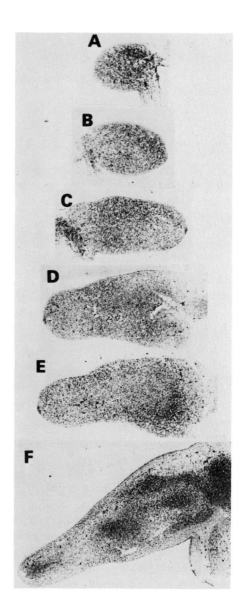

Fig. 4. Forelimbs of mouse embryos on 11th–13th day of gestation. A: 11th day, 30-somite stage; B: 11.25 day, 33-somite stage; C: 11.5 day, 36-somite stage; D: 11.75 day, 39-somite stage; E: 12th day, 42-somite stage; F: 13th day, 54-somite stage. Longitudinal sections through a plane perpendicular to the craniocaudal axis of limb (toluidine blue stained, ×40).

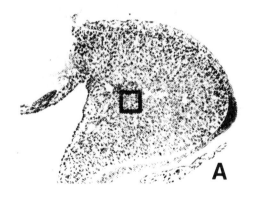

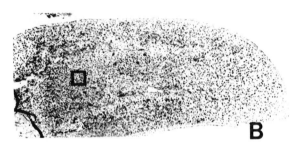

Fig. 5. Wing buds of chick embryos. A: Stage 19 (×90); B: stage 23 (× 60). Dorsal surface of each bud is at top, apical ectodermal ridge to the right. (Reproduced by permission from Searls *et al.*, 1972.)

cated more distally at later stages of development until cell differentiation extends to the apical, subridge region (Figure 4F).

A. Important Developmental Events

1. Role of AER in Limb Outgrowth

Since the work of Saunders (1948) it has been known that AER, the pronounced epithelial thickening around the distal margin of the limb bud (Figure 8), plays a crucial role in development. He demonstrated that surgical removal of the AER from the chick limb bud resulted in cessation of limb outgrowth; the limb developed only terminal deficiencies if AER was removed from older embryos, but more proximal bones failed to develop if the operation was performed on early embryos. These and similar experiments have led to a postulate that AER provides the stimulus for mesenchymal outgrowth (Zwilling, 1961). However, according to other interpretations, a truncated limb may result from destruction of the narrow strip of mesodermal cells that always follows after AER removal. The apical capillary network degenerates after AER removal, and this vascular disturbance may be the reason for necrosis in the apical mesoderm (Barasa, 1964). If mesoderm is re-covered by another AER, vascularization is re-established with little or no necrosis and appropriate limb parts are generated (Gasseling and Saunders, 1961). Stocum (1975), after reviewing the subject, felt that at present conclusive proof for the role of AER was not available, as either an active inductor of mesenchymal outgrowth or as only a protective, biomechanical factor. Some recent experiments by Cairns (1975) re-emphasize that an interaction between AER and the mesoderm is essential to limb morphogenesis. He found that the distal strip of mesoderm subjacent to AER has some distinctive cellular properties which are quite different from those of cells located in more proximal regions of the limb. As mentioned above, cell death occurs in a strip of distal

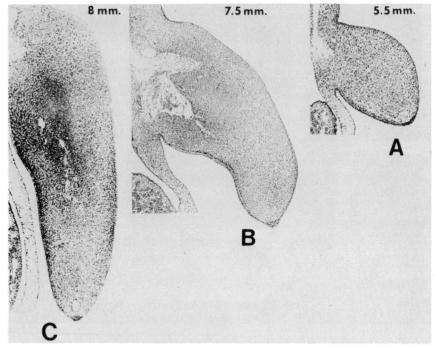

Fig. 6. Arm buds of human embryos. A: Stage 13, 5.5 mm crown–rump length; B: stage 15, 7.5 mm crown–rump length; C: stage 16, 8 mm crown–rump length (×57). (Reproduced by permission from Blechschmidt, 1969.)

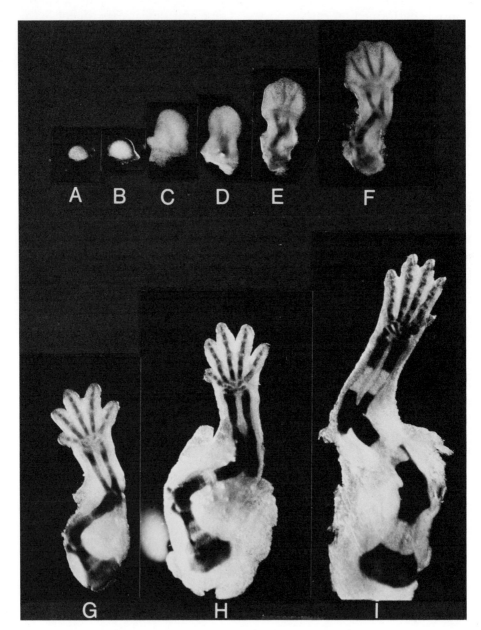

Fig. 7. Development of cartilage pattern in mouse forelimbs on 11th–17th day of gestation. A, B: 11th day; C, D: 12th day; E: 13th day; F: 14th day; G: 15th day; H: 16th day; I: 17th day. Whole mounts of limbs stained with alcian blue (×10).

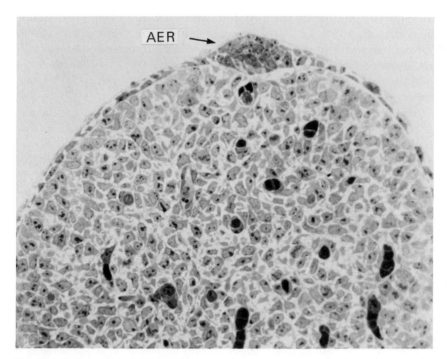

Fig. 8. Apical ectodermal ridge (AER) at the distal margin of the forelimb of a 39-somite (11.75-day) mouse embryo (×780).

mesoderm when the overlying AER is surgically removed. Cairns cultured small blocks of distal mesoderm and found that cells also die under *in vitro* conditions unless AER was included in the explant. The protective action of AER did not require close contact between it and the responding mesoderm; such action was found even if the two tissues were present at a distance of more than 1 mm from each other. This implied that a diffusible factor was involved. Another distinctive property of distal mesoderm is observed when it is explanted in culture as small blocks and cells are allowed to migrate on the plastic substratum. Distal cells differ sharply in morphology and migratory behavior from those derived from proximal mesoderm (Cairns, 1975; Kwasigroch and Kochhar, 1975). It is not yet known if these differences are due to influences of AER.

2. Role of Cell Division

One of the basic questions in developmental biology, how form is achieved in a growing tissue, is exemplified in the limb. Cell division increases the mass of the limb bud, but other factors must operate to attain the definitive shape. During the first 24 hr of limb development, both in chick and mouse embryos (Figure 2), the proliferative index in all regions of mesenchyme is virtually 100%, i.e., all cells are actively dividing (Janners and Searls, 1970; Kochhar and Agnish, unpublished). At the beginning of the next 24-hr period, however, the labeling index (the percentage of cells labeled by tritiated thymidine = the percentage of cells in S-phase of cell cycle) begins to decrease. Analyses of mitotic index (Hornbruch and Wolpert, 1970) and cell cycle (Searls, 1973) reveal that, at least in the chick, this decline in labeling index is due to a decrease in the proliferative index (number of cells in a division cycle at a given time) and not to an increase in the length of cell cycle. Analysis of different regions of limb during early development shows that the labeling index falls drastically in regions of prospective cartilage, only gradually in regions of future muscle, and not at all in the distal subridge mesenchyme. In the 42-somite mouse embryo, early on the 12th day, at a time when the process of cell condensation has not yet extended into the distal mesoderm, the labeling index of distal mesodermal cells remains about 90% (4-hr exposure to [^{3}H]thymidine, Kochhar, 1976a). At this stage, proximal mesoderm shows a labeling index of about 50% in the peripheral (myogenic) areas and only 5–14% in the central chondrogenic area. Later, of course, labeling index of subridge cells also declines to about 70%.

Two points derive from the above discussion. First, at the time when limb outgrowth is initiated, no special zone of the bud can be designated as the growth center. The rate at which the limb bud lengthens closely parallels the rate of cell division; all proximal chick limb bud regions grow with a doubling time of approximately 10 hr, and the length of the cell cycle is also 10 hr (Stark and Searls, 1973; Janners and Searls, 1970). Proximal–distal polarity of outgrowth which results in elongation on an axis perpendicular to the flank must, therefore, be dependent on some other factor. Stark and Searls (1973) have argued that this polarity may be established by the presence of AER. Second, as the cell proliferation rate begins to fall in central–proximal regions preparatory to formation of humeral cartilage, a "progress zone" with high cell proliferation is maintained in a narrow band of mesenchyme directly under the AER. Different experimental manipulations have ascertained that the progress zone measures 100–300 μm, extending inward from AER (Summerbell *et al.*, 1973; Cairns, 1975). During subsequent development, the progress zone maintains its dimensions and continues to provide cells through multiplication. As the progress zone moves distally, cells which are left behind, or become more than a critical distance away from the AER, differentiate into cartilage or muscle according to their positions. Finally, when the limb reaches a certain length, cell differentiation extends to the tip (Stark and Searls, 1973).

3. Cytodifferentiation

At an appropriate stage of development limb mesenchyme begins differentiation into various cell types such as cartilage, muscle, tendon, and connective tissue. A knowledge of temporal and spatial patterns of cytodif-ferentiation in normal limb development forms an important basis for under-standing abnormal organogenesis and has been summarized in chick, mouse, and man (Searls, 1973; Kochhar, 1976a; Sledge, 1966). Cell death which normally occurs in certain limb regions, such as interdigital zones, must also be considered a cell specialization. Chondrogenic and myogenic regions are first identified in histological sections as zones of cell condensation at stage 15 in human, stages 22–23 in chick, and on the 12th day in mouse (Figure 2). Overt differentiation into cartilage gradually becomes apparent in the center of these cell condensations, while peripherally the mesenchyme develops into muscle.

A few comments about the drastic decline in labeling index of presump-tive cartilage are in order. A decline in labeling index is usually due to either an increase in cell generation time or a withdrawal of some of the cells from the proliferative population. Although a careful analysis of cell generation time is still needed in mammalian embryos, if we assume that (just as in the chick) it does not increase in cells of the presumptive cartilage, then the observed fall in labeling index must be due to withdrawal of a large number of cells from the proliferating population. We have argued elsewhere (Kochhar, 1976a) that this withdrawal is only transitory and that actually the presump-tive cartilage cells go through alternative cycles of quiescence and cell division. Further, preliminary observations (unpublished) reveal that, subsequent to the overt differentiation of cartilage in each segment, the labeling index be-gins to increase once again. For example, in the 43-somite mouse embryo, while the labeling index in humerus is still low (5–14%), it has increased to 41% in the scapular cartilage.

What factors determine that some cells become cartilage and others mus-cle? Are the cells determined to become cartilage from the beginning of limb outgrowth or do they acquire this ability only on the basis of the location in the limb where they happen to be positioned at later stages of development? Experimental studies in the chick have shown that prospective cartilage cells do not become stabilized as such until early stage 25 (Zwilling, 1968; Searls and Janners, 1969). Prior to this, these cells can form soft tissue if trans-planted to ectopic sites in the young limb bud. Stabilization as cartilage occurs gradually as the cells begin accumulating visible quantities of the extracellular materials. Caplan and Koutroupas (1973) have postulated that one of the factors influencing the labile mesenchyme cells toward either cartilage forma-tion or muscle formation is differential vascularization of the limb. The vascu-lar pattern of the limb bud prior to the onset of cytodifferentiation essentially consists of a centrally located artery, peripherally located capillaries, and mar-ginally located veins (Figure 9). The consequences of this pattern are that an essentially avascular zone of cells at the core of limb bud is surrounded by vascularized mesenchyme, with the blood vessels coursing at the boundary between the two zones. The avascular zone in the core is the region where cartilage subsequently develops. Low oxygen tension, which could be pre-

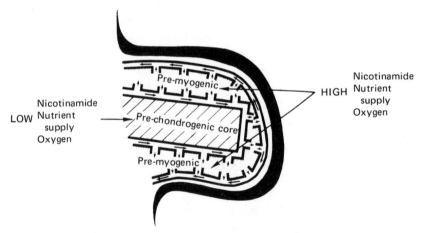

Fig. 9. Schematic illustration of proposed role of the vascular system in establishing gradients of metabolic potential to control mesoderm cell expression. (Reproduced by permission from Caplan and Koutroupas, 1973.)

dicted from the decreased blood supply, has been shown to potentiate chondrogenesis (Pawelek, 1969). Limb bud cells in tissue culture were exposed to various gas mixtures with the result that high oxygen tension permitted myogenesis to progress but prevented chondrogenic expression. Chondrogenesis occurred only when low oxygen tension was applied (Caplan and Koutroupas, 1973). Caplan and Rosenberg (1975) have presented further evidence that shifts in concentrations of essential low-molecular-weight compounds, such as nicotinamide, also play an important role in mesodermal cell expression.

Other views suggest that cell lineage is already determined in the early limb bud, that only after a certain number of cell divisions is the cell able to express itself as a cartilage or a muscle cell (Holtzer *et al.*, 1972; Dienstman *et al.*, 1974). It is probable that resolution of this problem may result from the use of teratogenic agents which inhibit cytodifferentiation.

B. Regulation of Limb Parts

The question of concern here relates to the mechanisms that determine the size and shape of embryonic skeletal elements and the extent of regulation that is possible after extirpation of certain limb parts. As mentioned above the musculoskeletal system is fabricated within the mesenchyme in a proximal–distal sequence. The primordia of proximal parts begin to differentiate first, followed by the appearance at successively later times of more distal parts. Recent studies have established that all prospective regions are potentially present in their normal seriation in the earliest visible wing bud of chick (Figure 10) (Stark and Searls, 1973). Before cell condensations for prospective

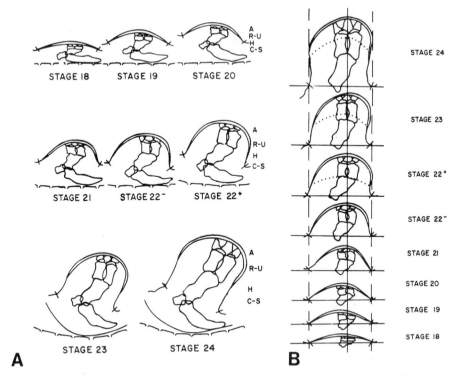

Fig. 10. Maps of prospective bone-forming areas of the developing wing buds of chick embryos. A: (C-S) coracoid-scapula, (H) humerus, (R-U) radius-ulna, (A) autopod or carpals and digits. B: The maps are aligned on an arbitrary axis passing between ulna and radius at the elbow and wrist to illustrate the proximodistal polarity of wing outgrowth. This is also found to align the cranial and caudal junctions of the wing bud and the flank (broken lines). The base was found to be perpendicular to the axis of growth. The dotted line indicates the proximal limit of the subridge region. (Reproduced by permission from Stark and Searls, 1973.)

cartilages appear, each limb segment, which is initially of the order of 300 μm long, grows at a uniform rate (Lewis, 1975). Subsequently the rate of growth of each cartilaginous element is determined by cell division, cell packing, and extracellular matrix secretion (Summerbell, 1976; Wolpert, 1976). Differences in the relative sizes of elements in normal development must, therefore, arise by modulation of these three processes. Gross shortening or defective form of cartilages may be expected if any of these parameters are disturbed either by a defective gene or by a teratogenic agent.

From maps of the location of the cells that form the shoulder, elbow, and wrist in chick wing from stage 18 to 24 (Figure 10), Stark and Searls (1973) have determined the rate of growth of the humerus and the radius–ulna (see Figure 2 for equivalent stages in human and mouse embryos). The prospective humerus region grows with a doubling time of 10 hr from stage 18 to stage 22, equivalent to a cell cycle of 10 hr and a proliferative index of 100%.

From stage 23 to that seen at 10 days of development the humerus lengthens with a doubling time of 38–40 hr, equivalent to a cell cycle duration which is unchanged from earlier stages, i.e., 10 hr, but a proliferative index of only 25%. Searls (1973) maintains that the proliferative index in the limb cartilage region remains constant at about 25% until quite late in development.

Detailed information of the type mentioned above for the chick is not yet available in mammalian embryos. Accurate determination of growth rates of tissues requires not only the proliferative index but also the cell generation time. From preliminary observations on mouse embryos, cell generation time of limb mesenchyme has been estimated to be 12 hr (unpublished). Since relative values of proliferative indices of various limb regions (assessed from labeling indices) are in the same range in the mouse and in the chick, the tentative assumption is that individual limb segments grow at similar rates in the two species.

Alterations produced in adult limb architecture after surgical removal of prospective parts of the limb bud may in some cases be equated with the destruction of cells after administration of some teratogenic agents. The ability of the limb bud to compensate and produce a normal cartilage pattern after surgical intervention has been studied extensively. The precision with which prospective cartilage-forming zones in the chick wing bud can now be excised (Figure 10) has made possible a reinvestigation of this problem by Stark and Searls (1974). They conclude that cartilage-forming regions of the limb mesoderm gradually become stabilized between stage 22 and 24 in the chick (equivalent to 11.5 and 12.5 days in the mouse and stage 14–16 in the human). Stabilization is plausibly due to the advanced state of chondrogenic differentiation and the decreased rate of cell division after stage 22. When 60–70% of the mesenchyme was removed from the center of the wing bud, representing most of the prospective humerus, radius, and ulna, a majority of the experimental limbs from stage 19 or 20 embryos regulated well and a normal limb resulted. Similar intervention in stage 22 embryos resulted in defective wings but the missing bone did not accurately reflect the prospective bone area excised, hence some regulation was still possible. In these cases a delay in the onset of chondrogenesis was observed which was equivalent to the time required for the cell division to replace the missing cells. Removal of individual bone areas from stage 23 and 24 wing buds resulted in loss of only the prospective bones which were excised; other bones developed normally and were of normal size. The results implied that the mesenchymal cells remaining in the experimental limb had already started to differentiate and were unable to regulate the replacement of the excised elements.

It was an interesting observation that, during early development when the regulation of cartilage patterns was still possible, removal of a large number of cells did not greatly reduce the length of cartilages (Barasa, 1964; Summerbell, 1973; Stark and Searls, 1974). Other authors, however, report different results. Wolpert (1976) summarized experiments wherein removal of transverse slices of tissue of varying thickness from early chick limb buds in the region of the humerus and forearm resulted in a proportional reduction of length in these elements. Removal of a 100-μm slice from a stage 21 embryo resulted in a 40% reduction in length, while removal of a 400-μm slice led to an 80% reduction. On the basis of these results Wolpert (1976) concluded that "there does not seem to be any length-regulating mechanism." The types of limb deformities produced by certain teratogens support such a contention. It is probable that the contrasting results on length regulation of chick limb may be reconciled when more attention is focused on the dorsal and ventral ectoderm covering the limb. The ectoderm has been shown to aid in stabilization of the cartilage pattern and thereby to influence the development of the humerus (Stark and Searls, 1974).

III. ABNORMAL DEVELOPMENT

Limb deformities are usually characterized on the basis of bony defects because absence of one or more muscles can easily escape detection if there is no disturbance of function. Moreover, muscle defects are usually associated with anomalies of skin and skeleton in the same region of the body, indicating the presence of a syndrome which involves much more than the muscular system (Warkany, 1971, p. 1025).

A. Nomenclature

The bony defects described in the literature embrace a wide spectrum of deficiencies ranging from mild shortening to a complete absence of one or more long bones, or a reduction or increase in the number of digits. The details of nomenclature of common reduction deformities in man, based on a system devised by Frantz and O'Rahilly (1961), is available elsewhere (Swinyard, 1969, p. 525). For the present purpose, it will be adequate to review only a few descriptive terms. Amelia refers to complete absence of a limb, while meromelia refers to absence of specific limb parts. Partial absence may be of terminal parts, such as the forearm and the hand, and is termed hemimelia, or of only intercalary (intermediate) segments. Terminal and intercalary deficiencies may be either transverse (extending across the width of the limb) or longitudinal (paraxial: pertaining to either the preaxial or postaxial side of the limb). Hence, a broad classification of limb deficiencies into four main groups is made as follows (O'Rahilly, 1969):

Terminal transverse: e.g., hemimelia
Terminal longitudinal: e.g., radial hemimelia
Intercalary transverse: e.g., phocomelia
Intercalary longitudinal: e.g., intercalary radial hemimelia

As is obvious, defects of any part of the free limb including the digits can be adequately classified according to this nomenclature. In descriptive teratological literature, however, digital anomalies are usually referred to as brachydactyly (shortened digits), ectrodactyly (absence of one or more digits), syndactyly (fusion of adjacent digits), and polydactyly (supernumerary digits).

B. Genetic Abnormalities in Chick and Mouse

A large number of inherited abnormalities that affect the limb skeleton are encountered in these species. Although many mutant patterns include systemic disorders in which limbs may also be involved, there are others where deformities are localized in the limb. In contrast to systemic conditions, localized limb defects must be due to disturbances of structures or processes peculiar to the limb bud (Grüneberg, 1963, p. 22). Conceivably, any of the processes in normal limb development enumerated earlier may be altered by the defective gene. Only a few genetic abnormalities have been subjected to experimental analyses with a view to understanding the developmental sequences preceding the final defect. However, four important phenomena have been recognized to date as possible mechanisms of abnormal development due to defective genes: lack or deviation of cellular interactions between AER and mesoderm, deviations of cell death patterns in either normal or ectopic sites, alterations in the behavior of limb mesenchyme, and alterations of molecular events related to chondrogenic differentiation.

Several wingless mutations in chick embryos have been recorded in the literature which have yielded important contributions to the evidence for the Saunders/Zwilling hypothesis of a reciprocal interaction between AER and limb mesenchyme. In one sex-linked wingless (ws) mutation, variable expression of the gene results in a mutant phenotype ranging from embryos with forelimbs absent and hindlimbs normal to embryos in which all limbs are absent (Hinchliffe and Ede, 1973). Most commonly AER is absent in wing buds, and precocious cell death occurs in preaxial mesenchyme much beyond the circumscribed necrotic zone commonly observed in normal chick limb, termed the anterior necrotic zone (ANZ). Limb outgrowth is stopped, and only a rudimentary wing is found in later development. Nothing is known of the mechanism that accentuates cell death, but electron microscopy reveals that neighboring mesenchymal cells become transformed into macrophages which ingest and digest dead cells (Hinchliffe and Ede, 1973). Mutant leg bud mesenchyme suffers less extensive necrosis and often the limb develops normally. In some hindlimbs, where cell death removes a substantial number of preaxial mesenchymal cells, the tibia is reduced or absent and there are fewer digits. The fact that a reduction in the normal amount of mesenchyme results in tibial reduction while the fibula develops normally is interesting in view of the competition for expropriation of mesenchymal cells that exists between the two rudiments in normal development. If a limited amount of mesenchyme is available for the reconstruction of the limb, the tibia always takes precedence over the fibula (Hampé, 1960). In further analyzing limb components in another wingless mutant, Zwilling (1956) used reciprocal interchanges of mutant and normal ectodermal caps and mesenchymal cores. He proposed that there exists a factor in normal mesoderm that maintains the integrity of AER and that this factor, termed apical ectoderm maintenance factor (AEMF), is missing from the mutant mesoderm. The validity of this theory still lacks conclusive proof.

A different set of circumstances is encountered in another chick mutant known as *talpid³* (*ta³*). Genetic suppression of cell death occurs in all major loci of the normal chick limb; specifically cell death is reduced or absent in preaxial and postaxial mesenchyme (anterior necrotic zone, ANZ, and posterior necrotic zone, PNZ), in the "opaque patch" of central limb mesenchyme, and in the interdigital zone (Hinchliffe and Ede, 1967; Hinchliffe and Thorogood, 1974). Associated with this, AER is much more extended than normal, and both forelimb and hindlimb are extremely short and show polydactyly (Figure 11).

The *talpid* mutation is lethal and is characterized as autosomal recessive; mutant embryos show widespread pleiotropic abnormalities such as shortening of vertebral column and fusion of adjacent vertebra, failure of cartilage replacement by bone, subcutaneous edema, abdominal evisceration, and cephalic defects. However, some unique features make it a suitable example for investigations of the cellular basis of limb teratogenesis. The mutant limb

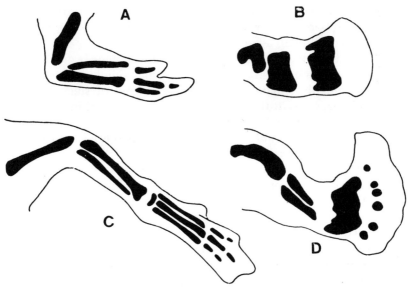

Fig. 11. Cartilage skeleton in wing of (A) normal, (B) *talpid³* mutant, and leg of (C) normal and (D) *talpid³* mutant chick embryos. (Reproduced by permission from Ede, 1972.)

bud is fan-shaped rather than elongated, and this distortion of shape is accompanied by fusion of cartilages and bones (Figure 11) (Ede, 1972). The altered skeletal pattern is foreshadowed by earlier failure of mesenchymal cells to condense properly in precartilage regions. Ede and Flint (1975a,b) studied the behavior of mutant mesenchyme cells cultured either in single-cell suspensions or as explants on plastic substratum. They found that mutant cells were much more adhesive to one another than were normal cells. This abnormally increased adhesiveness could probably explain the production of small spherical aggregates in rotation cultures, in contrast to larger unevenly shaped aggregates of normal cells. Using time-lapse cinephotography, they found that cell movement in cultures of mutant cells was also disturbed, being greater in normal than in mutant cells owing to the fact that mutant cells spent a greater percentage of time at rest than normal cells. Furthermore, mutant cells differed from control cells in being more flattened and having extensively ruffled cell membranes and short spiky microvilli. These alterations in morphology and behavior *in vitro* were interpreted by these authors to be related not only to the malformed shape of the limb but also to the altered pattern of cartilage elements found in the limb.

Of the large number of inherited limb deformities known to date (Grüneberg, 1963; Milaire, 1965; Goetinck, 1966; Lane and Dickie, 1968; Seegmiller *et al.*, 1971; Shepard, 1971), only a few have been subjected to analysis of underlying biochemical causes in defective cartilage and bone. Examples are brachypodism (bp^H) and brachymorphy (*bm*) in the mouse and nanomelia in the chick, all due to autosomal recessive genes and of interest because they result in disproportionately short stature characterized by shorter than normal long bones. Their counterparts exist in man, e.g., brachymorphic mice are similar to human achondroplasiacs.

Recent studies have shown that as normal mesenchyme begins differentiation into cartilage, biosynthesis of some specific macromolecules is either initiated or preferentially enhanced. Limb mesenchyme of stage 18–19 chick embryos synthesizes two distinct species of proteochondroitin sulfate, one of which is selectively increased as chondrogenesis begins at stage 24 (Levitt and Dorfman, 1973; Goetinck *et al.*, 1974). Concurrently there is also a transition from the synthesis of $[\alpha 1(I)]_2 \alpha 2$ collagen to $[\alpha 1(II)]_3$ collagen (Linsenmayer *et al.*, 1973). These macromolecules, once released into the extracellular environment, associate to produce the unique extracellular matrix of cartilage (Chapter 10, this volume). Probably defects in any of these macromolecules or in their interactions would lead to abnormalities in cartilage which could alter the size or shape of developing bone.

Collagen extracted from limb cartilage of 5-day-old brachymorphic mice and from sterna of 14-day nanomelic chick embryos was qualitatively and possibly quantitatively normal (Orkin *et al.*, 1976; Pennypacker and Goetinck, 1976). However, ultrastructure of mutant cartilage matrix was different from normal in that the proteoglycan matrix granules were either absent or greatly reduced in amount and size in both mutants. Chromatography of extracts of nanomelic cartilage showed that the major fraction of proteochondroitin sulfate was much reduced compared with normal cartilage (Pennypacker and Goetinck, 1976). Brachymorphic cartilage, on the other hand, contained normal quantities of glycosaminoglycans; however, the glycosaminoglycans differed from normal in their chromatographic and electrophoretic properties and were found to be undersulfated (Orkin *et al.*, 1976). It is not possible at present to translate the results of these biochemical studies into explanations for altered organogenesis.

It should be emphasized that biosynthesis of macromolecules during organogenesis cannot be viewed in isolation from other developmental events which precede or are concurrent with it. For example, in the brachypod mutant limb, although chondrogenesis and the transition from cartilage to bone are delayed, evidence suggests that the mutant gene acts earlier during blastema formation (Milaire, 1965; Grüneberg and Lee, 1973; Rhodes and Elmer, 1975).

C. Teratogenic Agents

Extensive previous work on teratogen-induced limb deformities has been summarized by Landauer (1969) in the chick and by Warkany (1969) in mammalian embryos. In early attempts to produce limb defects in mammalian embryos, Murphy and DeRenyi (1930) exposed pregnant rats to X-rays and found clubbing of the feet and absence of toes in the offspring. This led to observations that with careful timing and dosing with X-rays several localized defects of the limb could be obtained (Warkany and Schraffenberger, 1947; Wilson, 1954; Russell and Russell, 1954). A large number of chemical agents have been shown to affect limb development, although usually they are not specific to the limbs alone. Some teratogens, however, can selectively produce limb defects if administered to pregnant females over a limited time span.

Before attempting a discussion of sensitive developmental events underlying limb teratogenesis, one must define what constitutes a limb deformity in experimental animals. The term "micromelia" is often used to denote shortening of the limbs regardless of whether reduced size is proportionate to the overall reduction in body or trunk size. If a teratogen produces extensive cytotoxicity, such as nucleic acid antimetabolites, a greater or lesser reduction in general body size is often found. It is assumed that growth retardation due to systemic effects of the teratogen is superimposed on all morphological aspects, making it difficult to recognize strictly localized effects in the limb.

It is now well established that the limb is not susceptible to teratogenesis equally during all stages of morphogenesis. Susceptibility begins as limb bud elongation is initiated and ends by the time limb skeleton is formed as cartilage. During this period individual limb segments—proximal (stylopodium: humerus or femur), middle (zeugopodium: radius–ulna or tibia–fibula), and distal (autopodium: hand or foot parts)—show differential susceptibility in a proximal–distal sequence; a teratogen acting over a limited time span mal-forms only the proximal segments if administered early in the period, and more distal segments if administered later. These facts are illustrated by two classical teratogens which permitted time-limited exposure: X-rays in the chick (Goff, 1962; Summerbell, 1973) and vitamin A acid (retinoic acid) in the mouse (Kochhar, 1973, 1976a).

The proximal–distal progression of sensitivity is overlaid by a cephalocaudal gradient from fore- to hindlimbs of the same embryo. The sequence of developmental events in the forelimbs precedes that in the hindlimbs by 12–24 hr in rodents and by 2–3 days in man (Milaire, 1965; O'Rahilly and Gardner, 1975). A similar lag in susceptibility between the two limbs is seen in teratological studies (Murakami and Kameyama, 1965; Gebhardt, 1970; Kochhar, 1973; Scott et al., 1975).

The timing of effects of thalidomide on human limb, as far as could be determined, places the sensitive period between the 20th and 36th days after conception (Lenz, 1964). This coincides roughly with stages from early limb-bud formation to the formation of the early cartilage blastema (Figure 2). Reduction deformities of the upper limb occurred most frequently if thalidomide was ingested during the 25th and 30th days, postconception and those of the lower limb if it was ingested somewhat later, i.e., 30th–34th days. Digital anomalies, such as triphalangism of the thumbs, were still observed when ingestion took place as late as the 36th day.

Superimposed on the stage-dependent sensitivity of individual limb seg-ments, certain bones are inherently more vulnerable to teratogenesis than others in the chick embryo. Leg bones are more vulnerable than are wing bones and usually suffer more extreme shortening irrespective of the time of treatment. The reverse seems to be true in mammalian embryos; in many cases of micromelia only forelimbs are affected, or the forelimbs are affected more severely than the hindlimbs (Zwilling, 1969; Landauer, 1969). In a review of more than 100 children of thalidomide mothers, Weicker (1965) found that approximately 60% had only upper limb deformities while about 10% had involvement of both upper and lower limbs; legs alone were involved in only a few cases. A definite tendency towards reduction of preaxial bones was noticed; a large number of children had total aplasia of radius and of digits I–III. A greater sensitivity of radius was also revealed experimentally in fetuses of mice exposed to retinoic acid (Kochhar, 1973; 1976a).

One major pathogenetic event linked with the origin of limb defects is the destruction of cells. Cell death has been observed, to a greater or lesser extent, in limbs that have been appropriately examined soon after the application of most teratogenic agents. There is every reason to believe that elimination of cells from the limb through the agency of induced cell death may often ad-versely affect the developing limb; however, it remains a challenge to deter-mine how generalized cell death results in a specific limb deformity. First, one must ascertain whether the limb mesenchymal cells are really being killed indiscriminately or cell death is localized in certain regions of the limb bud.

Only limited information is available, but it suggests that while some teratogens produce widespread cell death others elicit this response in circumscribed areas of the limb bud. A comparison of limb defects produced by teratogens with differing degrees of cytotoxicity is interesting.

Chlorambucil, hydroxyurea, and retinoic acid all produce limb deformities when administered to pregnant mice early on the 12th day of gestation. While all three agents produce micromelia, the manner in which the size and relative proportions of long bones are altered by the first two is very different from the effects of retinoic acid (Figure 12). In the shortened limbs produced by chlorambucil and hydroxyurea all long bones are present in normally articulated sequence, but the micromelia produced by retinoic acid treatment consists of a combination of grossly deformed and almost normal long bones in the same limb. An additional interesting fact is that teratogenic doses of chlorambucil and hydroxyurea produce high embryolethality and reduce general body size of surviving fetuses, while retinoic acid produces no such effects at comparative stages. One explanation for these effects is that cytotoxic agents such as hydroxyurea and chlorambucil produce widespread cell death throughout the embryo including the limb (Kochhar, 1975a; Sadler and Kochhar, 1976; Sadler *et al.*, 1976), for all proliferating cells are particularly sensitive to these agents. It is known, however, that a sizeable population of mesenchymal cells survives in the limb and is responsible subsequently for reconstituting the whole limb, albeit in a smaller version than normal. Retinoic acid, on the other hand, is not generally cytotoxic but attacks only cells of the chondrogenic lineage (Kochhar, 1976a); it eliminates or reduces the cell population undergoing determination in the specific cartilage at the time of treatment, resulting in a specific skeletal defect. The length of time retinoic acid remains available to the embryo determines the severity of deficiency (Kochhar, 1976b).

An example of a teratogen specifically attacking cartilage of developing limbs was reported by Seegmiller *et al.* (1972) in the case of micromelia induced in chick embryos by 6-aminonicotinamide. Degenerative changes in the chondrocytes of cartilage rudiments were accompanied by cytological changes in other cells, suggestive of a lack of synthesis of cartilage matrix. The biochemical mode of action of nicotinamide antagonists is still unresolved but is of considerable interest since the report of Caplan (Caplan and Stoolmiller, 1973) which suggested that 3-acetylpyridine, another nicotinamide antagonist, potentiates chondrogenic expression in cultured limb mesenchymal cells but checks the normal development of muscle.

The importance of AER–mesodermal interactions in normal limb development has been emphasized earlier and has received much attention as regards hereditary limb anomalies, but it has not yet been carefully analyzed in induced defects. Discovery of a specific effect by a teratogen on AER or mesoderm, resulting in cell death or other alterations in one and not in the other, would support but not prove the existence of such interactions. The use of reciprocal combinations between treated vs. control tissues should be ex-

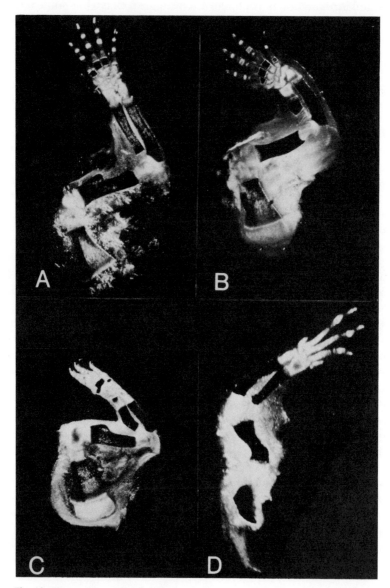

Fig. 12. Comparison of a forelimb of an untreated mouse fetus with those treated with either retinoic acid, hydroxyurea, or chlorambucil on the 12th day of gestation. A: Untreated control. B: Retinoic acid-treated. Note that radius is extremely reduced in size while other long bones are of nearly normal length. C: Hydroxyurea-treated. All long bones are reduced in size but their lengths show normal proportionate relationship to each other. D: Chlorambucil-treated. Alizarin red-stained limbs of 18th day fetuses (×7.5).

ploited in future studies on limb teratogenesis, since organ culture techniques for mammalian embryos are now well established (Kochhar and Aydelotte, 1974; Kochhar, 1975b; Berg *et al.*, 1975).

Abnormalities in AER–mesodermal interactions have been implicated in a recent study on the induction of polydactyly in rat embryos by cytosine arabinoside (Chaube *et al.*, 1968; Scott *et al.*, 1975). Consistently, the extra digit always developed preaxially and was not the result of any unusually high proliferative activity of the mesoderm in this area; in fact, mesodermal cells were less densely packed in the teratogen-treated limb (Scott *et al.*, 1975). An unusual finding was that a small area of necrosis in ectoderm which normally limits the extension of AER was absent in polydactylous limbs. Interestingly, while a majority of the treated limbs were polydactylous (90%), a small percentage of embryos possessed a rudimentary first digit. The fact that polydactyly and adactyly affected the same digit in different embryos suggests a common pathogenesis and emphasizes the probability of delicately balanced factors in the control of normal limb development.

IV. CONCLUSIONS

Experimental studies on vertebrate limb organogenesis have indicated the existence of several developmental mechanisms: interactions between ectoderm and mesoderm, polarized outgrowth of limb bud, cell proliferation, cell death, cell locomotion, and biosynthesis of extracellular macromolecules. Although deviations in a number of these mechanisms have been observed in genetic abnormalities, most teratogen-induced defects appear to result from a process involving cell degeneration. It is likely, however, that enhanced cell death is not the only process involved in teratogenesis, although it is the one most often and most conveniently documented by teratologists. With the aid of the experimental manipulations used by developmental biologists to study inductive interactions in organogenesis, the mode of action of teratogenic agents could be further defined. Specific examples of hereditary and induced deformities are now available which have mostly been studied as regards pathogenesis at the cellular level. Selective use of teratogenic agents is expected to extend the understanding of pathogenetic mechanisms to the molecular level. Teratogenic agents provide valuable tools to those who wish to determine the extent of repair and regulation possible in the developing limb skeleton.

ACKNOWLEDGMENTS

Original work mentioned in this chapter was supported by USPHS grant HD-06550 from the National Institute of Child Health and Human Development, N.I.H. I am deeply grateful to Mrs. Dana Hoffman, Miss Pamela Baker, and Mr. John Penner for valuable assistance, and to Dr. Robert M. Greene for critically reading the manuscript.

REFERENCES

Barasa, A., 1964, On the regulative capacity of the chick embryo limb bud, *Experientia* **20**:443.

Berg, P. L., Krowke, R., and Merker, H. J., 1975, Studies on blastema and epithelium of the limb anlage of mouse embryos, *in: New Approaches to the Evaluation of Abnormal Embryonic Development. Second Symposium on Prenatal Development* (D. Neubert and H. J. Merker, eds.), pp. 151–160, Georg Thieme, Stuttgart.

Blechschmidt, 1969, The early stages of human limb development, *in: Limb Development and Deformity: Problems of Evaluation and Rehabilitation* (C. A. Swinyard, ed.), pp. 24–70, Charles C Thomas, Springfield, Ill.

Cairns, J. M., 1975, The function of the ectodermal apical ridge and distinctive characteristics of adjacent distal mesoderm in the avian wing-bud, *J. Embryol. Exp. Morphol.* **34**:155.

Caplan, A. I., and Koutroupas, S., 1973, The control of muscle and cartilage development in the chick limb: The role of differential vascularization, *J. Embryol. Exp. Morphol.* **29**:571.

Caplan, A. I., and Rosenberg, M. J., 1975, Interrelationship between poly (ADP-Rib) synthesis, intracellular NAD levels, and muscle and cartilage differentiation from mesodermal cells of embryonic chick limb, *Proc. Natl. Acad. Sci. U.S.A.* **72**:1852.

Caplan, A. I., and Stoolmiller, A. C., 1973, Control of chondrogenic expression in mesodermal cells of embryonic chick limb, *Proc. Natl. Acad. Sci. U.S.A.* **70**:1713.

Chaube, S., Kreis, W., Uchida, K., and Murphy, M. L., 1968, The teratogenic effect of 1-β-D-arabino-furanosylcytosine in the rat, *Biochem. Pharmacol.* **17**:1213.

Dienstman, S. R., Biehl, J., Holtzer, S., and Holtzer, H., 1974, Myogenic and chondrogenic lineages in developing limb buds grown *in vitro*, *Dev. Biol.* **39**:83.

Ede, D. A., 1972, Cell behavior and embryonic development, *Int. J. Neurosci.* **3**:165.

Ede, D. A., and Flint, O. P., 1975a, Intercellular adhesion and formation of aggregates in normal and *talpid³* mutant chick limb mesenchyme, *J. Cell Sci.* **18**:97.

Ede, D. A., and Flint, O. P., 1975b, Cell movement and adhesion in the developing chick wing bud: Studies on cultured mesenchyme cells from normal and *talpid³* mutant embryos, *J. Cell. Sci.* **18**:301.

Frantz, C. H., and O'Rahilly, R., 1961, Congenital skeletal limb deficiencies, *J. Bone Joint Surg.* **43A**:1202.

Gasseling, M. T., and Saunders, J. W., 1961, Effects of the apical ectodermal ridge on growth of the versene-stripped chick limb bud, *Dev. Biol.* **3**:1.

Gebhardt, D. O. E., 1970, The embryolethal and teratogenic effects of cyclophosphamide on mouse embryos, *Teratology* **3**:273.

Goetinck, P. F., 1966, Genetic aspects of skin and limb development, *Curr. Top. Dev. Biol.* **1**:253.

Goetinck, P. F., Pennypacker, J. P., and Royal, P. D., 1974, Proteochondroitin sulfate synthesis and chondrogenic expression, *Exp. Cell Res.* **87**:241.

Goff, R. A., 1962, The relation of developmental status of limb formation to X-radiation sensitivity in chick embryos. I. Gross Study, *J. Exp. Zool.* **151**:177.

Grüneberg, H., 1963, *The Pathology of Development: A Study of Inherited Skeletal Disorders in Animals*, Blackwell Scientific Publications, Oxford, U.K.

Grüneberg, H., and Lee, A. J., 1973, The anatomy and development of brachypodism in the mouse, *J. Embryol. Exp. Morphol.* **30**:119.

Hamburger, V., and Hamilton, H., 1951, A series of normal stages in the development of the chick embryo, *J. Morphol.* **88**:49.

Hampé, A., 1960, La compétition entre les elements osseux du zeugopode de Poulet, *J. Exbryol. Exp. Morphol.* **8**:241.

Hinchliffe, J. R., and Ede, D. A., 1967, Limb development in the polydactylous *talpid*³ mutant of the fowl, *J. Embryol. Exp. Morphol.* **17**:385.

Hinchliffe, J. R., and Ede, D. A., 1973, Cell death and the development of limb form and skeletal pattern in normal and wingless (ws) chick embryos, *J. Embryol. Exp. Morphol.* **30**:753.

Hinchliffe, J. R., and Thorogood, P. V., 1974, Genetic inhibition of mesenchymal cell death and the development of form and skeletal pattern in the limbs of *talpid*³ (*ta*³) mutant chick embryos, *J. Embryol. Exp. Morphol.* **31**:747.

Holtzer, H., Weintraub, H., Mayne, R., and Mochan, B., 1972, The cell cycle, cell lineages, and cell differentiation, *Curr. Top. Dev. Biol.* **7**:229.

Hornbruch, A., and Wolpert, L., 1970, Cell division in the early growth and morphogenesis of the chick limb, *Nature (London)* **226**:764.

Janners, M. Y., and Searls, R. L., 1970, Changes in rate of cellular proliferation during the differentiation of cartilage and muscle in the mesenchyme of the embryonic chick wing, *Dev. Biol.* **23**:136.

Kochhar, D. M., 1973, Limb development in mouse embryos. I. Analysis of teratogenic effects of retinoic acid, *Teratology* **7**:289.

Kochhar, D. M., 1975a, Assessment of teratogenic response in cultured postimplantation mouse embryos: Effects of hydroxyurea, *in: New Approaches to the Evaluation of Abnormal Embryonic Development. Second Symposium on Prenatal Development* (D. Neubert and H. J. Merker, eds.), pp. 250–277, Georg Thieme, Stuttgart.

Kochhar, D. M., 1975b, The use of *in vitro* procedures in teratology, *Teratology* **11**:273.

Kochhar, D. M., 1976a, Cellular basis of congenital limb deformity induced in mice by vitamin A, *in: Morphogenesis and Malformation of the Limb* (D. Bergsma and W. Lenz, eds.), The National Foundation-March of Dimes, *Birth Defects Orig. Art. Ser.* (in press).

Kochhar, D. M., 1976b, Transplacental passage of label after administration of [³H]retinoic acid (vitamin A acid) to pregnant mice, *Teratology* **14**:53.

Kochhar, D. M., and Aydelotte, M. B., 1974, Susceptible stages and abnormal morphogenesis in the developing mouse limb, analyzed in organ culture after transplacental exposure to vitamin A (retinoic acid), *J. Embryol. Exp. Morphol.* **31**:721.

Kwasigroch, T. E., and Kochhar, D. M., 1975, Locomotory behavior of limb bud cells: Effect of excess vitamin A *in vivo* and *in vitro*, *Exp. Cell Res.* **95**:269.

Landauer, W., 1969, Dynamic aspects of hereditary and induced limb malformations, *in: Limb Development and Deformity: Problems of Evaluation and Rehabilitation* (C. A. Swinyard, ed.), pp. 120–135, Charles C Thomas, Springfield, Ill.

Lane, P. W., and Dickie, M. M., 1968, Three recessive mutations producing disproportionate dwarfing in mice, *J. Hered.* **59**:300.

Lenz, W., 1964, Chemicals and malformations in man, *in: Second International Conference on Congenital Malformations*, pp. 263–276, International Medical Congress, New York.

Levitt, D., and Dorfman, A., 1973, Control of chondrogenesis in limb-bud cell cultures by bromodeoxyuridine, *Proc. Natl. Acad. Sci. U.S.A.* **70**:2201.

Lewis, J. H., 1975, Fate maps and the pattern of cell division: A calculation for the chick wing-bud, *J. Embryol. Exp. Morphol.* **33**:419.

Linsenmayer, T. F., Toole, B. P., and Trelstad, R. L., 1973, Temporal and spatial transitions in collagen types during embryonic chick limb development, *Dev. Biol.* **35**:232.

Milaire, J., 1965, Aspects of limb morphogenesis in mammals, *in: Organogenesis* (R. DeHaan and H. Ursprung, eds.), pp. 283–300, Holt, Rinehart and Winston, New York.

Murakami, U., and Kameyama, Y., 1965, Malformations of the mouse fetus caused by hypervitaminosis-A of the mother during pregnancy, *Arch. Environ. Health* **10**:732.

Murphy, D. P., and DeRenyi, M., 1930, Postconception pelvic irradiation of the albino rat (*Mus norwegicus*); its effect upon the offspring, *Surg. Gynecol. Obstet.* **50**:861.

O'Rahilly, R., 1969, The origins, development, and implications of the nomenclature of congenital skeletal limb deficiencies, *in: Limb Development and Deformity: Problems of Evaluation and Rehabilitation* (C. A. Swinyard, ed.), pp. 500–503, Charles C Thomas, Springfield, Ill.

O'Rahilly, R., and Gardner, E., 1975, The timing and sequence of events in the development of the limbs in the human embryo, *Anat. Embryol.* **148:**1.

Orkin, R. W., Pratt, R. M., and Martin, G. R., 1976, Undersulfated chondroitin sulfate in the cartilage matrix of brachymorphic mice, *Dev. Biol.* **50:**82.

Pawelek, J. M., 1969, Effects of thyroxine and low oxygen tension on chondrogenic expression in cell culture, *Dev. Biol.* **19:**52.

Pennypacker, J. P., and Goetinck, P. F., 1976, Biochemical and ultrastructural studies of collagen and proteochondroitin sulfate in normal and nanomelic cartilage, *Dev. Biol.* **50:**35.

Rhodes, R. K., and Elmer, W. A., 1975, Aberrant metabolism of matrix components in neonatal fibular cartilage of brachypod (bp^H) mice, *Dev. Biol.* **46:**14.

Russell, L. B., and Russell, W. L., 1954, An analysis of the changing radiation response of the developing mouse embryo, *J. Cell. Comp. Physiol.* **43**(51):103.

Sadler, T. W., and Kochhar, D. M., 1976, Chlorambucil-induced cell death in embryonic mouse limb-buds, *Toxicol. Appl. Pharmacol.* **37:**237.

Sadler, T. W., Kochhar, D. M., and Cardell, R. R., 1976, Effects of hydroxyurea on ultrastructure of neuroepithelial cells in mouse embryos, *Teratology* **13:**35A.

Saunders, J. W., Jr., 1948, The proximo-distal sequence of origin of the parts of the chick wing and the role of the ectoderm, *J. Exp. Zool.* **108:**363.

Scott, W. J., Ritter, E. J., and Wilson, J. G., 1975, Studies on induction of polydactyly in rats with cytosine arabinoside, *Dev. Biol.* **45:**103.

Searls, R. L., 1973, Newer knowledge of chondrogenesis, *Clin. Orthop. Relat. Res.* **96:**327.

Searls, R. L., and Janners, M. Y., 1969, The stabilization of cartilage properties in the cartilage-forming mesenchyme of the embryonic chick limb, *J. Exp. Zool.* **170:**365.

Searls, R. L., Hilfer, S. R., and Mirow, S. M., 1972, An ultrastructural study of early chondrogenesis in the chick wing bud, *Dev. Biol.* **28:**123.

Seegmiller, R. E., Fraser, F. C., and Sheldon, H., 1971, A new chondrodystrophic mutant in mice electron microscopy of normal and abnormal chondrogenesis, *J. Cell Biol.* **48:**580.

Seegmiller, R. E., Overman, D. O., and Runner, M. N., 1972, Histological and fine structural changes during chondrogenesis in micromelia induced by 6-aminonicotinamide, *Dev. Biol.* **28:**555.

Shepard, T. H., 1971, Organ-culture studies of achondroplastic rabbit cartilage: Evidence for a metabolic defect in glucose utilization, *J. Embryol. Exp. Morphol.* **25:**347.

Sledge, C. B., 1966, Some morphologic and experimental aspects of limb development, *Clin. Orthop. Relat. Res.* **44:**241.

Stark, R. J., and Searls, R. L., 1973, A description of chick wing bud development and a model of limb morphogenesis, *Dev. Biol.* **33:**138.

Stark, R. J., and Searls, R. L., 1974, The establishment of the cartilage pattern in the embryonic chick wing and evidence for a role of the dorsal and ventral ectoderm in normal wing development, *Dev. Biol.* **38:**51.

Stocum, D. L., 1975, Outgrowth and pattern formation during limb ontogeny and regeneration, *Differentiation* **3:**167.

Summerbell, D., 1973, Growth and Regulation in the development of the chick limb, Ph.D. thesis, University of London.

Summerbell, D., 1976, A descriptive study of the rate of elongation and differentiation of the skeleton of the developing chick wing, *J. Embryol. Exp. Morphol.* **35:**241.

Summerbell, D., Lewis, J. H., and Wolpert, L., 1973, Positional information in chick limb morphogenesis, *Nature (London)* **244:**492.

Swinyard, C. A., 1969, *Limb Development and Deformity: Problems of Evaluation and Rehabilitation,* Charles C Thomas, Springfield, Ill.

Warkany, J., 1969, Experimental production of mammalian limb malformations, *in: Limb Development and Deformity: Problems of Evaluation and Rehabilitation* (C. A. Swinyard, ed.), pp. 140–160, Charles C Thomas, Springfield, Ill.

Warkany, J., 1971, *Congenital Malformations: Notes and Comments,* Year Book Medical Publishers, Chicago.

Warkany, J., and Schraffenberger, E., 1947, Congenital malformations induced in rats by roentgen rays, *Am. J. Roentgenol.* **57:**455.

Weicker, H., 1965, 100 children with thalidomide embryopathy, XIth International Congress of Peciatrics, Tokyo.

Wilson, J. G., 1954, Differentiation and the reaction of rat embryos to radiation, *J. Cell Comp. Physiol.* **43**(S1):11.

Wolpert, L., 1976. Mechanisms of limb development and malformation, *Br. Med. Bull.* **32:**65.

Zwilling, E., 1956, Interaction between limb bud ectoderm in the chick embryo. IV. Experiments which a wingless mutant, *J. Exp. Zool* **132:**241.

Zwilling, E., 1961, Limb morphogenesis, *Adv. Morphog.* **1:**301.

Zwilling, E., 1968, Morphogenetic phases in development, *Dev. Biol. Suppl.* **2:**184.

Zwilling, E., 1969, Abnormal morphogenesis in limb development, *in: Limb Development and Deformity: Problems of Evaluation and Rehabilitation* (C. A. Swinyard, ed.), pp. 100–115, Charles C Thomas, Springfield, Ill.

Index